Youth Culture and Net Culture:
Online Social Practices

Elza Dunkels
Umeå University, Sweden

Gun-Marie Frånberg
Umeå University, Sweden

Camilla Hällgren
Umeå University, Sweden

INFORMATION SCIENCE REFERENCE
Hershey · New York

Senior Editorial Director: Kristin Klinger
Director of Book Publications: Julia Mosemann
Editorial Director: Lindsay Johnston
Acquisitions Editor: Erika Carter
Development Editor: Myla Harty
Production Coordinator: Jamie Snavely
Typesetters: Casey Conapitski & Deanna Zombro
Cover Design: Nick Newcomer

Published in the United States of America by
Information Science Reference (an imprint of IGI Global)
701 E. Chocolate Avenue
Hershey PA 17033
Tel: 717-533-8845
Fax: 717-533-8661
E-mail: cust@igi-global.com
Web site: http://www.igi-global.com

Library of Congress Cataloging-in-Publication Data

Youth culture and net culture : online social practices / Elza Dunkels, Gun-Marie Frånberg and Camilla Hällgren, editors.
p. cm.
Includes bibliographical references and index.
Summary: "This book engages the complex relationship between technology and youth culture, while outlining the details of various online social activities"--Provided by publisher.
ISBN 978-1-60960-209-3 (hardcover) -- ISBN 978-1-60960-211-6 (ebook) 1. Internet and teenagers. 2. Teenagers--Social networks. 3. Youth--Social aspects. 4. Technology and youth--Social aspects. 5. Online social networks. I. Dunkels, Elza, 1960- II. Frånberg, Gun-Marie. III. Hällgren, Camilla, 1973-
HQ799.2.I5.Y667 2011
004.67'80835--dc22
2010050364

British Cataloguing in Publication Data
A Cataloguing in Publication record for this book is available from the British Library.

Editorial Advisory Board

Table of Contents

Detailed Table of Contents

Section 3
Identity and Sexuality

This chapter describes the process, explains the aspects, analyses the experiences and considers the social policy implications, of cybersex among young people from the sexually conservative Mauritian society. It is argued that internet-based technologies are further breaking down 'the traditional and moral values', which some politicians, religious leaders and parents want to preserve through social policy related to sexuality education in Mauritius. The author suggests that an appropriate formal sex education for young people should take into account the Net Culture context.

In many western cultures and societies sexuality is surrounded by shame and guilt and often restricted to the private areas of life. Prior research on the use of the internet for sexual purposes has primarily focused on its negative and problematic aspects, such as compulsivity and addiction. Thus, little is known about any possible benefits. The purpose of the chapter is to focus on how young people aged 12-24 use the internet as a source of knowledge about sexuality.

This chapter considers the attitudes of adolescents with African background in relation to online dating. Important findings are that there is no significant difference in the attitudes of male and female adolescents in relation to online dating. Also, the perception of parents in relation to adolescents getting involved in online dating activities is in tandem with the African belief that adolescents are too young to get tangled in any aspect of sexual activities, online dating inclusive. It is also revealed that both male and female adolescents have coping strategies for dealing with online dating activities.

Analyzing newsgroup's messages, the chapter focuses on an Israeli newsgroup that appeals to GLBT (gay, lesbian, bisexual, transgender) youth and operates within the most popular user-generated con-

Foreword

The issues raised in this book are relatively new, but important. Children and young people's use of media and their presence on the Internet are also issues I have prioritised as Minister for Culture. We have provided additional resources to the Swedish Media Council and we have also organised an 'experience and understanding' conference which we have called Digital Tourist. The conference has travelled round the country the way tourists do and has attracted a great deal of attention. Researchers and experts have shared their knowledge. But more importantly, the natural-born experts on the subject – groups of upper secondary school pupils from each location – have acted as digital guides and used a workshop to walk the adult conference participants through aspects of their everyday use of media. At a number of stations they have generously responded to questions from interested visitors and shared their expertise and knowledge. This model of knowledge transfer is simple, but it has worked extremely well. We should realise that it takes a certain amount of courage on the part of us adults to admit that we are novices and young people are the experts.

In organising the Digital Tourist conference, we wanted to inspire people and get more people involved in the issue of media skills. The background is that many adults have far too little knowledge of children and young people's digital activities. This has become clear to the Swedish Media Council, which has noticed increased interest in the past few years. Many people are turning to them for information about children and young people's media reality. Therefore, this book is a welcome contribution.

We are currently facing a situation where young people's knowledge of digital media and the Internet – in terms of both opportunities and risks – is much greater than parents' and other adults' knowledge. I believe we need a greater level of knowledge transfer between the generations. A knowledge transfer that goes in both directions. Adults need to understand children's reality, and children would benefit greatly from the life experience and judgement of an adult. The Internet is like life in general – it harbours both opportunities and dangers.

There is a tangible difference in how different generations use and relate to new technical opportunities. The difference can be expressed by saying that adults use new technology – whereas young people live it, it is an integrated part of their lives. Adults are, and will remain, tourists – thirsty for knowledge, but tourists nonetheless. This difference is often called a digital generation gap – one that can cause alienation and create obstacles.

In order to be supportive and beneficial to our children and young people – in this area, too – we adults need more knowledge about their digital lives and media reality. At the same time, we must continue to draw attention to their vulnerability and exposure in this new media landscape.

I believe that this development carries with it something that is fundamentally positive. These media offer unprecedented opportunities for the children and young people of today to be creative and express themselves, and to participate in society. This development is beneficial, both for creativity and ingenuity, and for the democratic process. As Minister for Culture, I am pleased to note that the children and young people of today have many more opportunities to be creative and to participate than young people had just ten years ago. Digital media are becoming new tools for information, for the free expression of opinions and for participating in society on one's own terms. They have created a culture that provides greater scope for creativity, social contact across national borders and new ways of creating or seeking out an identity.

If the adult world is able to meet young people in the arena where they spend their time and, to a certain extent, speak their language, with similar knowledge and concepts, we lay the foundation for better decisions concerning children and young people, and their situation.

This is why I welcome this book, which provides greater understanding and enables the transfer of knowledge between experts. I hope that the conclusions can also benefit a broader public.

Lena Adelsohn Liljeroth
Minister for Culture

Preface

The subject area of youth and online social practices deals with research that concerns children and youth and their use of interactive media. This research consists of a broad range of different perspectives such as quantitative and descriptive studies of what goes on in the contemporary media landscape, psychological and sociological research on online social practices, pedagogical research on formal and non-formal learning strategies, to mention a few. The aim of the book is to outline this emerging research area, evolving around young people and contemporary digital arenas. The field is growing in size, shape and complexity and the need for study is urgent. This book is a valuable contributions by providing critical perspectives and a broad overview.

The book, *Youth Culture and Net Culture: Online Social Practices* , covers current areas of research on young people's use of interactive media. The different chapters represent cutting edge research with a critical perspective as a common denominator. As editors we were pleased that our call for chapters had such good results. All authors were interested in the critical perspective that we emphasized in our call. So you will find chapters presenting novel ideas, different aspects of young people's net cultures and ground-breaking research that will be of great value to the academic society as well as to policy makers. The book aims at providing relevant theoretical frameworks and the latest empirical research findings in the area.

The target audience of the book will be composed of students, professionals and researchers working in the field of young people and the internet in various disciplines (e.g., education, library and information science, psychology, sociology, computer science, linguistics, informatics, media and communication science). The book may serve as literature at an undergraduate level and provide an overview of the area for researchers, teachers, students and policy makers. It is written for professionals as well as students who want to improve their understanding of online social practices from a young people's perspective. *Youth Culture and Net Culture: Online Social Practices* focuses on online social practices from a youth culture viewpoint. During our planning process we tried to find a way to involve young people in the making of the book. The aim was to make young people not only the objects of studies but also subjects. The idea was to find students who were willing to give their side of the story. This is not as easy as it may seem; we needed to find students who were sufficiently good at English, who were interested in reading a substantial amount of academic text and who had the time to go through a writing process together with us. And since the chapters from the adult authors were to be submitted in spring 2010, we had to ask the students to do this work during the busiest period of school. Luckily enough, there is an International Baccalaureate Diploma Programme in Umeå, Sweden, quite close to the university. Their English teacher Neil Duncan presented our idea to one of his classes and some of the students wanted to participate. The contributors are students in the IB1 class of the International Baccalaureate Diploma

Programme at Östra Gymnasiet in Umeå. All IB students have CAS – Creativity, Action, Service as part of their diploma, and these students volunteered their contributions here as Service – helping other people who need or ask for help – in this case by giving their time and sharing their ideas, opinions and experiences in print with us when editing this book. The students were first presented with the abstracts of every adult chapter. After reading through the abstracts some of the students wanted to read entire chapters, which at this stage were in the form of drafts. Our instructions for the writing were deliberately vague, because we wanted to avoid steering these young authors into writing in a traditional academic style. Even though there may be educational winnings from training young people in academic writing, we wanted their tone to be authentic since they were to write about their own thoughts and feelings. After the first reading of the adult chapters, we presented the students with questions like "What are your thoughts about this? Did anything particular catch your attention, either in a positive or a negative way? Do you have any ideas what to write about this?" We also pointed out that the students did not have to comment on the adults' chapters, their chapters could just as well be separate texts. Today, we are very proud to present this way of getting young people's views on research concerning the internet and we hope that the readers will appreciate this vantage point. We also encourage other editors and publishers to follow our example, since it has been a valuable experience for us and these students and hopefully beneficial to the book.

The contributors of, *Youth Culture and Net Culture: Online Social Practices* come from practically all over the world. In fact, the globalization that serves as a backdrop to research on contemporary media became almost too clear to us during the editing process. Two of the current international tragedies in 2010 affected authors in the book. First, there was the military coup in Nigeria in February 2010. Some of the authors in the book are researchers at Nigerian universities and their writing process was delayed for obvious reasons. Also, it became clear to us and the authors how much we have come to depend on the internet and that its services are always on. The Nigerian crisis affected infrastructure and there were a few weeks during which we could not communicate as planned. The second tragedy was the airplane crash in Russia in April 2010 when the Polish President died along with 100 other people, including the First Lady and many prominent Polish officials. The passengers were on their way to Smolensk in Western Russia to commemorate the massacre in Katyn during the Second World War. This accident had severe repercussions on Polish everyday life and our authors were naturally affected. We are grateful that the authors from Poland and Nigeria still had the strength to finalize their work.

The present volume is dedicated to research on young people's social activities online. There is a focus on generational aspects of online social practices, as the reader will notice, but there are also other facets, such as gender and social class. As mentioned above, globalization is an inevitable theme in many of the chapters. When we say inevitable we mean it both because the subject of the internet is global in itself but also because this book has attracted authors from many parts of our world. This richness in approaches provides us with valuable knowledge in a time when the educational system, policy makers and non-governmental organizations call for this knowledge. Thus we let *Youth Culture and Net Culture: Online Social Practices* start out with young people's own views of their online social practices.

In the first section of this book the young authors from a Swedish upper secondary school describe and discuss their own and their peers' online activities. They write about differences between generations, their own and others' blogging and creativity.

In the second section the context is in focus; how can we describe the setting in which the online social practices take place? What does this world look like and what approaches can we develop? The opening chapter of this section is the editors' own, *Young People and Online Risk*. The chapter problematizes online risk and questions safety measures that often follow discussions of online risk. The chapter also aims at initiating a discussion on how we approach risk in a general sense. In the next chapter, *Youth and Online Social Networking: From Local Experiences to Public Discourses* Malene Charlotte Larsen and Thomas Ryberg present the results of a study of Danish young people's internet use. The authors discuss how young people construct themselves as users of social network sites both in relation to concrete and local experiences and in relation to mediated discourses. To sum up the first section, Håkan Selg's chapter *Swedish Students Online: An Inquiry into Differing Cultures on the Internet*, presents the results from a major survey among Swedish university students regarding their internet use. Selg concludes that the study does not support earlier claims of a net generation or digital natives. Instead, Selg points out that the adaption of internet use is far more complex than a generations approach suggests.

The third section focuses on identity. From various vantage points the different chapters address questions pertaining to identity construction in a contemporary media landscape. The opening chapter in this section is Ann-Charlotte Palmgren's chapter *Fat Talk, Constructing the Body Through Eating Disorders Online among Swedish Girls*. The chapter explores how young women construct their body by writing about eating disorders in blogs. Palmgren uses blog entries from a Swedish net community to study the construction of girlhood through the construction of the ideal body online. In the next chapter, *To be Continued ... Fan fiction and the Constructing of Identity*, Patrik Wikström and Christina Olin-Scheller report the findings from a study on fan fiction in a Swedish context. The authors contextualize the fan fiction phenomenon as a part of a larger change where media consumers' role as collaborative cultural producers grows ever stronger. In *Digital Neighbourhoods – A Sociological Perspective on the Forming of Self-feeling Online*, Ulrik Lögdlund and Marcin de Kaminski discuss how young people in Sweden relate to the internet. The authors report findings from a study revolving around the notion of self-feeling in relation to how young people develop and maintain relations online. *The Use of Interactive Media in Identity Construction by Female Undergraduates in a Nigerian University* by Oyewole Jaiyeola Aramide examines how interactive media is used in identity construction processes by female undergraduates. This chapter concludes that the use of the internet for identity construction cannot be ignored and that it is unavoidable for youths to employ the internet for self gratification, considering its advantages. The next chapter, *The Representation of Female Friendships on Young Women's MySpace Profiles: the All-female World and the Feminine 'Other'* by Amy Shields Dobson presents the results of a study of MySpace profiles owned by young Australian women. With this data as a starting point Dobson discusses constructions of female friendship and points out the complexity of female self-representation in a contemporary media landscape. S. Faye Hendrick's and Simon Lindgren's chapter is called *YouTube as a Performative Arena – How Swedish Youth are Negotiating Space, Community Membership, and Gender Identities Through the art of Parkour*. The authors explore how Swedish youth involved in urban sport activities, parkour, use the video sharing site Youtube as a performative arena. They present the results of a case study of communities on YouTube, using a variety of on- and offline ethnographic data.

The fourth section is a variation on the thirs; it details sexuality as one aspect if identity. The first chapter in this section is *Young People and Cybersex in a Sexually Conservative Society: A Case Study from Mauritius* by Komalsingh Rambaree. The author discusses the process, aspects and experiences of cybersex in Mauritius, a country which is transforming into a cyber-island and at the same time still wants to preserve its views on young people's sexuality. The second chapter is Kristian Daneback's and Cecilia Löfberg's *Youth, Sexuality and the Internet - Young People's Use of the Internet to Learn about Sexuality*. The chapter discusses how young people use the internet to learn about and understand more about sex and sexuality. The authors conclude that the internet is an important arena for discussing sexual matters and that this arena in many ways differs from offline arenas. Olugbenga David Ojo's chapter *Adolescents and Online Dating Attitudes* deals with the attitudes of adolescents with an African background in relation to online dating. Ojo contextualizes young people's online dating in relation to African culture and tradition and describes that parents in general see adolescents as being too young for sexual activities. Further, Ojo accounts for some coping strategies that the adolescents undertake. Avi Marciano's chapter *The Role of Internet Newsgroups in the Coming-Out Process of Gay Male Youth: an Israeli Case Study* examines the role of the internet in the complex coming-out process of gay male youth. The findings indicate that the internet can serve as a social arena where the informants for the first time in their lives are free of moral judgment on their sexuality.

In the fifth and closing section of *Youth Culture and Net Culture: Online Social Practices* we want to focus on contemporary challenges; chapters that draw trajectories into the future. These chapters present challenges that come from the fact that our world is a changing place and that we are in the middle of a great shift or a minor turn, depending on how much importance you place in the changing media landscape. Opening this section, *The Competent Youth's Exposure of Teachers at YouTube.se* by Marcus Samuelsson takes its starting point in movies of exposed Swedish teachers on YouTube. The chapter investigates how adults react when youth use their knowledge of the internet to expose what they have experienced in school. In their chapter *Moving from Cyber-bullying to Cyber-kindness: What do Students, Educators and Parents say*? Wanda Cassidy, Karen Brown and Margaret Jackson explore cyber-bullying and its opposite cyber-kindness from the views of students, teachers and parents. The authors present the results of two studies carried out in British Columbia, Canada and conclude that much of the cyber-bullying is, as the authors put it, "happening under the radar of school staff and parents". On the same subject, cyber-bullying, Jacek Pyżalski problematizes aggression on the internet in his chapter *Electronic Aggression among Adolescents: an Old House with a New Facade (Or Even a Number of Houses)*. As the title suggests, he looks into differences between electronic and traditional aggression and describes the diversity of electronic aggression. Furthermore, he introduces a model for comparing electronic and traditional aggression, the ABACUS model. Yet another contemporary challenge is presented by the following chapter, *Ways of ICT Usage Among Mildly Intellectually Disabled Adolescents – Potential Risks and Advantages* by Piotr Plichta. This chapter explores patterns of ICT usage among mildly intellectually disabled adolescents. Plichta stresses the importance of leisure activities to, and rehabilitation of the disabled, as well as the risk of digital exclusion. Violent content in video games is the theme of Eva-Maria Schiller's, Marie-Thérèse Schultes', Dagmar Strohmeier's and Christiane Spiel's chapter *Gaming and Aggression: The Importance of Age-Appropriateness in Violent Video Games*. The authors present a study on pre-adolescents in which violent content of games was categorized based on Pan European Game Information (PEGI) descriptors. Furthermore, practical implications for adolescents and for parents are discussed. The next chapter is Kareena McAloney's and Joanne Wilson's *Young People, Sexual Content and Solicitation Online*. The chapter provides an up to date overview of knowledge regarding young people's exposure to and experiences of sexual material and sexual predators online. The authors also address current mechanisms designed to protect children and young people as they engage in online activities. Closing this section and the entire book, we find *Spirituality in Cybercrime (YahooYahoo) Activities among Youths in South West Nigeria* by Agunbiade Ojo Melvin and Titilayo Ayotunde. The chapter provides a socio-cultural analysis of the relevance of spirituality among Nigerian youths in relation to cybercrime.

All these chapters pose questions that we, the editors and the authors, hope that the reader will be eager to further explore after reading this book. So please follow us to the engaging, entertaining, worrying, sad and joyful youth culture of today.

Elza Dunkels
Umeå University, Sweden

Gun-Marie Frånberg
Umeå University, Sweden

Camilla Hällgren
Umeå University, Sweden

Acknowledgment

During the editing of this book there have of course been a number of obstacles to overcome and we would like to thank those people who have helped us with this. First of all, we would like to thank the authors who heeded our call for chapters. They represent the finest, most important research at this turbulent period of time and we are happy that they chose to come together in our book. Among the authors we would like to extend a special thank you to the student authors. They showed such commitment to their task and also proved to have the courage to write about their own thoughts in an academic, perhaps sometimes intimidating, context. Furthermore, they had to finalize their chapters at the end of the semester when all the tests and assignments were due as well. If it had not been for their teacher Neil Duncan, who freely gave us and the students of his time, this task would have been so much harder if not impossible. We would also like to extend our gratitude to the Editorial Advisory Board who patiently and diligently assisted us with the important questions of publishing. A problem when recruiting a board is that we are not the only ones who appreciate the work of these professors and consequently they have been very busy. Nevertheless they have helped us selflessly and we are deeply grateful for this. The Swedish Minister for Culture, Lena Adelsohn Liljeroth took time during a busy period to read the chapters and write an insightful foreword concerning an area she has taken under her wings as being responsible for culture and media in the Swedish government.

As in any book project there are also a number of people who may think that they have done nothing to help but they have; by providing time, by excusing us from meetings, by cooking and cleaning at home and by encouraging us. So we want to thank our colleagues at the Department of Applied Educational Science, Umeå University, Sweden and, not least, our husbands and children who patiently wait for us to send this manuscript.

Elza Dunkels
Umeå University, Sweden

Gun-Marie Frånberg
Umeå University, Sweden

Camilla Hällgren
Umeå University, Sweden

Young Authors' Thoughts on Net Culture

THE NEOMILLENNIALS VS. PREVIOUS GENERATIONS

Susanna Junyent

The natural process of a family is that the parents are responsible for the kids until they are 18 years old and able to be responsible for themselves.

Traditionally when children are growing up, the parents show the kids day by day what is wrong and what is right, and how to do things. But with the neomillenial generation things have changed.

Parents do not know how to turn on the DVD for their children so the 4 year-old child turns on the DVD for his/her parents. This is the case with my brother who is four years old: he shows my dad how to manage the DVD player.

Another example is when a mother gets stuck with her cellphone and doesn't know how to quit an application and asks her child, about 5 or 6 years old, how to do it. I saw that just the other day on the bus and it shocked me. It felt unnatural.

This can be seen as both positive and negative. The positive aspect is that Neomillienial kids learn technology really fast, things like DVD, game consoles, cellphones etc…

One negative point is that the natural way of parents teaching and showing their kids is lost in this perspective. So the fact that they are the parent/adult no longer means that they know more than the kid as the child is already familiar with all the new technology and the parent/adult may not be so.

On the other hand, neomillennial children grow up with all this technology which means that they feel familiar with it. For them it is something normal in everyday life, so it becomes a need too. This is in contrast to the adult, who did not grow up with technology and for whom it could be like a luxury, and s/he sees it as an extra.

In conclusion, the advance of technology makes lives easier in many respects and is an opportunity for neomillennial kids who grow up with it, but parents should be updated to be able to help the neomillennial kids, because otherwise it could lead to an imbalance or disequilibrium.

MY BLOG: EXPRESSIONS & EMOTIONS

Hibak M. Dubad

It is safe to say that today's youth use the internet a lot! But most importantly they use it as a tool to express themselves and the ways for doing that are many. Also, the youth today, have access to all kinds of different resources that can help us explain how we feel or even what we are doing at a certain point in time. I am no exception in this matter.

Although there is a huge variety in how and where we can express ourselves, websites like Youtube, Twitter, and Facebook are all tools with an aim for doing just that. However, I only use my blog! I do this by posts that lets the reader know how I'm doing and perhaps what I'm up to.

I've been blogging on a regular basis for nearly four years now and the sole reason for this is because I feel like it is the only way out there for me to express myself – without any rules or regulations attached to it. That is also why I think a lot of young people do this today, especially girls, as we get to formulate our own opinion in our own words. And just the freedom of being able to post almost everything you want appeals to the blogger as well.

I would say that the majority of teenage girls out there have their own blog and those that don't, they at least read them. I think the reason for this is because of how easy it is to stay in connection to your own thoughts and opinions. I say this because the reader often tends to read blogs that immediately speak to them and so therefore, they feel like there is some type of understanding in what the writer is saying.

The "blog-world" is huge due to the fact that there are no specific age-groups needed to read or write a blog. Therefore almost anyone can take part in the many different topics or issues being addressed within them. I personally see this as a good thing as it widens and strengthens the target group of the readers or even writers.

In my blog I tend to express myself in a way that justifies my mood for one particular moment. I also include emotions and thoughts in my posts, as they explain how I feel the best. This can be done by just writing it down in words, or by posting a song that you feel is in relation to what you are going through. Basically anything that reflects the reality of your situation.

In contrast to this, I personally do not feel comfortable in giving away too much and that is why I make use of poetry and music in my posts for example, as they do not clearly state nor specify exactly how I feel and to what extent. This is a comfort in itself, because the blogger is aware of the fact that the only person that truly knows how he/she is feeling is the person themselves – leaving space for the reader to speculate and form ideas. Also, this is an underlying fact that most bloggers don't speak about, as it will lead to the reader coming back to read your posts out of curiosity.

BLOGGING

Bekhta Allal

Blogging has become very popular in recent years. I don't really remember the first time I was in contact with a blog, but I started my first blog at the beginning of 2007 and I've been blogging since.

Why do I do it? Well, there are so many different kinds of blogs out there. Fashion blogs, personal style blogs, private blogs about music, sexuality or pretty much anything you are interested in or feel you need or want to write about.

I started blogging mostly for my own benefit. Like a diary, but because it's on the internet I don't put my deepest thoughts in the blog. Blogging is for people to read! And that's pretty much one reason – I want people to know what I'm up to. They may not be interested, but like I said before, blogs are for reading. And bloggers read other blogs.

How do I write my blog? It's fun to write, it feels good to write down your thoughts though it may be very cryptic. I don't want people to know everything about how I feel and I absolutely never put negative thoughts on my blog if I've for example have had a fight with a friend or just someone that I know, and this person reads my blog from time to time. Then this person would probably know what/who I'm talking about. So maybe I'll put a picture (without text) from Google that expresses what I feel, but readers will just take this picture as any random picture. It sounds pointless but at least I get it out from my head.

I usually write my blog every day. Not always when I have a lot to do, but blogging is a way for me to put my thoughts down and that way I can go on with my day. So often I start by writing what day it is, what will happen or what has happened. I put pictures in and in the end I write in colours when I want to point out something, like something that I'm excited about or just a goodbye.

I have a diary where I put my absolute deepest thoughts, but blogging is a way for me to get my words out about what I feel and how I'm doing. I post my blog-address on Facebook and then I know for sure that people will read it. I don't put stuff like "I have a problem with this person" or "I don't know what I will do about this, because he's so annoying" when it's something that is going on right now. Close friends will ask what I'm talking about, and that person I may be writing about will probably read my blog because I posted the address on Face Book. So no, I don't do it like that. I'm more secretive, I just put a small text in the beginning and then start writing about my day or whatever's on my mind.

There's also this thing that I discovered not so long ago – that I can put YouTube video clips on my blog. That way I can express myself even more – through music. I absolutely love music and lyrics and all that goes with it. So music for me is a good way to express. If I'm sad, happy or just like the song, it's a good way to get your feelings through and show emotions. If I really want to point something out, but not make myself too obvious, I write a small piece from the lyrics of the song under the clip, and then that it's a very good song, for example. I don't like to be too obvious with what I feel. But by writing, being secretive and putting out video clips with songs, I get my thoughts out in a good way – people read even if they may not understand everything, but I get my word out and I won't have to be carrying thoughts inside. But of course, I think that if something really bothers you or you have a problem, talk to your friends or just someone you feel you can talk to. I talk to my friends about anything really.

But writing in my blog is really a way for me to get my thoughts out, whatever I'm feeling for the moment, I just post it and then I'm done.

HOW DO HEALTH BLOGS AFFECT YOUNG GIRLS AND THEIR WAY OF LOOKING AT THEMSELVES AND THEIR BODIES?

Melika Medimagh-Svedjedal

The blogs being focused on are blogs which came up when the search words "Health" and "Anorexia" were typed in on the blog portal bloglovin.com. The blogs are being written by three different girls who represent different ways of looking at their bodies, health and losing weight. How do these three bloggers affect young girls with their way of writing and the pictures they post on their blogs? Which blogger could have the most influence on girls wanting to lose weight - the anorexic blogger, the blogger who wants to lose weight, or the healthy blogger? In order to protect the bloggers, I have not written out their blog addresses.

Here are some quotes so that you can get a picture of their language and what they stand for.

The anorexic girl:

"I'm sitting in school and I feel so fat. I gained one kilo today, exactly one kilo even thought I only ate one grapefruit and some prawns with lettuce leaves. Okay, I managed a few spoonfuls meat sauce straight from the pot on the stove as well, but it came up again straight away, in a plastic bag in my room. Fresh, as always. "

The girl who wants to lose weight:

"Thank you for your comments on different approaches but I don't want to go on a diet. I don't want to start eating different low calorie soups and start with different diets etc. I know that it would probably lead to even more weight loss but then I would also lose my purpose. I want to learn to eat healthy, just enough and right, not just lose weight as fast as possible."

The healthy girl:

Today's lunch was the classic spaghetti bolognese! I put a lot of grated carrot and tomato cubes to freshen it up a little but also to have some extra vitamins and nutrition.

As is quite evident, the first blogger focuses a lot on her weight and what she is eating; she also gives us evidence that she has an eating disorder, with her last sentence "it came up again straight away". Young girls who are thinking a lot about their weight, and what their body looks like, and who might already be on their way into an eating disorder, could get a "push" in the wrong direction from reading this blog. She tells us about her disorder in detail and it is quite interesting to see how she, through her blog, has all this control over herself and over every single calorie she eats in a day. When I started to look at the comments I saw that the girls (from what I can see, there are only girls commenting) reading her blog have the same thoughts as she does. They give her encouragement and agree with her point of view, some also ask her questions like "How do you manage to throw up? I've tried so many times but I've only managed a few times". The girls commenting on her blog are about 13-16 years old.

The second blogger I looked at is a chubby girl who is trying to lose weight in a healthy way. She gives the impression that she really wants to live healthily but she is struggling with it. She explains that when she is feeling down, she eats, and that that is one of her biggest problems. This girl actually gives me the feeling that she is just trying to be healthy and that she is not interested in diets, or just eating nothing, which is something she says when people are giving her advice about diets and how to lose weight fast. On one of her posts, where she is comparing pictures of her body from March 8 and May 3, where she has lost approximately 2 kg, 27 girls commented. They are all happy for her and congratulate her and tell her to keep up the good work. I think this girl is a good role model for girls who want to lose weight in a healthy way. Unfortunately, I think that girls with body issues in the 10-15 age range would rather look at the first blogger than the second because young girls are usually impatient and do not want to be healthy. Young girls, or young people on the whole, don't like vegetables and "healthy" food, they would probably rather eat nothing.

The third blogger, the healthy girl is not, as far as I can see, talking about weight in her blog. She is more interested in being healthy and she is probably happy with her weight. She does not give any advice on how to lose weight, but more on how to give your "normal" food extra vitamins and nutrients. She has a lot of recipes on her blog for healthy dishes. Some of her recipes include "Creamy minced meat and spelt mix", "Spelt-pizza" and Risotto. When I looked at the comments on this girl's blog, I got the impression that the people reading her blog are also really into the healthy life, and many thank her for a good blog and tell her that she gives them good inspiration both on living healthily but also that they often use her recipes and get inspired by them.

After looking through the three blogs and the comments on them, I would say that the anorexic blogger could have most influence on young girls. The language she uses in her blog when talking about weight probably starts making young girls think about if their bodies are "normal", if they should lose some weight, if they are fat, what other people think of them and things like that. This blogger really stands out from the other two, which makes her blog the blog you read. People want to read something that stands out, and since she is so shocking in what she says, she is probably the one with the most influence on young girls than the other two, who I think influence older girls.

ONLINE ART AND MUSIC

Victoria Kawashima Altice

The internet provides a blank canvas for anyone to showcase practically anything they want. One thing many people online post is their music and art. This allows people to not only express themselves, but also to show the world what they enjoy and receive comments and criticism from all different points of view.

I feel that posting music or art that you create online is less intimidating than showing it in real life. The internet seems like a place where people won't judge as much because they don't look at you face-to-face. Sites that are specialized for music and art give everybody a fair start. When I'm on those websites, I can imagine somebody on the other side of the net working on their latest creation; spending long hours getting ready to present it for all to see. I can feel the hard work that's put into the finished result, and I become motivated to try writing one myself. You can feel the emotion put into the song through the screen if that person has put in the effort. I go online to watch people sing sometimes, to brighten up my day. Or if I'm feeling blue, I can find the words I can't find in myself, in somebody's song. The internet is more than a talent show. It lets people be more honest and the artist more relaxed. Because, really, what's there to lose?

It can be a place where people care purely about the music as compared to the person that sings it and how famous they are. Because people worldwide have such a variety of tastes and styles, the internet is also a place to be inspired. And if you get inspired, it's possible that you are simultaneously inspiring someone else, and that's always a good thought. People upload songs about anything and everything. When they're sad, happy, frustrated, inspired, or even hungry. Whether it's the melody to a song, or the way they sang it so passionately, inspiration from it can help you over the block and on with your endeavours.

An artist could also get recognized online. Singers recording their songs and uploading them for their own pleasure can be seen by recording labels everywhere. And it doesn't matter where you start;

a good voice is a good voice. Many artists online develop a fan base large enough to fill the sea and some even tour the country.

With all the support from people online, the internet is a place to inspire and be inspired; to listen, and to be listened to.

Section 1
Context of Net Cultures

Chapter 1
Young People and Online Risk

Elza Dunkels
Umeå University, Sweden

Gun-Marie Frånberg
Umeå University, Sweden

Camilla Hällgren
Umeå University, Sweden

ABSTRACT

The authors suspect that the young perspective has been left out when online risk and safety are discussed in contemporary research. The aim of this chapter is to give a critical approach to this matter and question fear as a driving force for protecting young people online. Interviews with children about their views of internet use (Dunkels, 2007) and a study of safe use guides from European countries conducted in 2008 (Lüders et al, 2009) form the empirical base. The discussion in the chapter is underpinned by ideas of childhood as a social construction, emerging ideas of power relations pertaining to age and theories of technology reception. The authors also introduce a metaphor, the layer cake, to better understand how the same action can be viewed from different vantage points.

INTRODUCTION

but I dont want you to tell anyone, not even my mum and dad!

The above quote comes from a 12-year old girl (Dunkels, 2007) who had just told the researcher how she handles a situation that was part of her everyday life – unknown men making dirty comments about pictures she had posted online. She explained that she had sufficient control over the situation but that she was worried that her parents would get troubled and perhaps restrict her internet use as a consequence.

This is a crucial instance that urges us to propose a critical approach to online risk and safety and question fear as a driving force for protecting young people online. The interviewed girl makes a valuable point when she assumes that her parents will not be able to handle her online reality; this might be interpreted as fear paralyzing her parents. In the following we will present alternative views

DOI: 10.4018/978-1-60960-209-3.ch001

of what constitutes risk and how online safety might be shaped.

Internet use has gone through substantial change since the mid 1990's when the internet mass use began (Findahl, 2009). One change is how the social network sites differ from the early internet days' virtual communities (Larsen, 2009). One main difference is that the early meeting places aimed at helping users find new relations while today's social networking sites mainly help users organize existing relationships. Virtual communities as described by among others Rheingold (1993) build upon their users' common interests while social network sites are constructed around the users' personalities. Another difference is the use of open chat rooms that peaked around the millennial and a phenomenon that to a great extent is marginalized today. Applications that require identification are more common today than when the first safe use guides were written. The level of openness, i.e. how much of the interaction is public and how much is private differs greatly among these different applications as shown in the Table 1.

Also, the research approach has changed. Early research on computer mediated communication was predominantly focused on the differences between online and offline. Words like *virtual* and *cyber* were used to mark this distinction. There were studies showing that online interaction lacks the ability to convey social information (Yao & Flanagin, 2006). Furthermore, studies were often exoticizing youth and their internet communication, describing the young and their activities as greatly different from earlier generations (Herring, 2008).

The trilogy of the prevailing discourse of fear, what we know about contemporary online interaction and young peoples' experiences seem asymmetrical. There is a growing suspicion that we might have left out the young perspective when we discuss online risk and safety.

EMPIRICAL AND THEORETICAL CONSIDERATIONS

Empirically this chapter is based upon interviews with children about their views of internet use (Dunkels, 2007) and a study of safe use guides from European countries conducted in 2008 (Lüders et al, 2009).

Theoretically the chapter is underpinned by ideas of childhood as a social construction (Eriksson, 2008), emerging ideas of power relations pertaining to age (Alderson, 2005) and theories of technology reception (Dunkels, 2007). Childhood as a social construction has been developed mainly

Table 1. The table shows different levels of openness, i.e. how much of the interaction is public and how much is private, in some of the most used interactive applications among young people (Dunkels, 2007)

Application	**Open chat rooms**	**Net communities**	**Instant messaging applications**
Degree of openness	Public (Open)	Public and private (Both open and closed)	Private (Closed)
Description	Authentication is not compulsory. You enter a chat room to see who else is there and start communicating. The conversation is public.	Membership is required. Many net communities offer different levels of openness within the community. Some interaction is public - often friends' comments and blogs - and some is closed - often e-mail and instant messages.	Relationships are approved by both parties. Creating a buddy list is one of the central functions. Contact attempts can be accepted or declined. If both parties agree they can monitor each other when on line. All interaction is private.
Examples	http://www.teenchat.com	Facebook Myspace.com	MSN Messenger ICQ (E-mail)

in a sociological context, describing patterns where a child is constructed not only in relation to her age but also in relation to parameters connected to age. Hence, a child can be described as dependent on adults, in need of care and guidance, not fully developed physically and mentally, etc. Some of these traits can be ascribed to age while some can be seen as mere construction – something we have agreed upon in our common definition of the concept *child* (Eriksson, 2008).

During the first phases of mass use of the internet there was a strong focus on generational aspects (e.g. Buckingham, 2002; Oblinger & Oblinger, 2005; Prensky, 2001, 2006; Tapscott, 1998). This approach has later been criticized by Bennett et al. (2008), Bayne & Ross (2007) and Herring (2008), among others, for its one-dimensioned view on generations; not everyone born in the same era has equal experiences or living conditions. However vital it is to make the matter more complex, giving weight to gender, ethnicity, social class among other intersecting perspectives, we find age interesting as a viewpoint when addressing online safety issues. Age can be seen as one of many parameters constructing and upholding power structures in modern society. In the same way that we have learned to recognize power structures built on gender and ethnicity, we can also identify structures that build on age. This structure has been called *childism* (Alderson, 2005). To develop knowledge about this structure in order to unveil and explain the stereotyping and discrimination of young people is urgently needed (Dunkels, 2007). Childism is not as straightforward as sexism or racism, since there does exist a natural, and probably desired, power relation between adults and children. Adults can thus both use and abuse their power advantage.

The political weakness of childhood (Qvortrup, 2003) is that as soon as their members reach the age when they potentially may be strong enough to establish conscious actions, they are no longer children and the incitement to act against existing power structures disappears. Another complication for children's self organization is according to Qvortrup (2003) the fact that children are divided into small groups, usually families, and thus not organized generationally. This complicates the solidarity bonds. Qvortrup (2003) also points out that the introduction of a generational perspective is not an attempt to reduce other relations, such as class, gender and ethnicity.

Dunkels (2007) attempts to explain the discrepancies between adults' and young people's views of contemporary technology use. The explanation model uses historical technology shifts, as described by Drotner (1999) among others, to describe how technology itself stands in the way of adults understanding problems connected with young people's media use. In short, adults focus on the technology and connect its use to the development of unwanted phenomena. Instead of focusing on the use of technology, which is more common among young people, adults blame technology and look for technological solutions to the problems that arise. This model helps our understanding of society's negative reception of media-related new phenomena among the young. Emotional responses to new technologies are what Drotner (1999) calls *media panics* and Marwick (2008) calls *technopanics*. These responses have some common traits. They typically pathologize young people's use of the medium in question and this concern leads to attempts to change, or control, young people's use of the medium. Herring (2008, p. 74) claims that media often describe young internet users as being "vulnerable and in need of societal protection and direction".

In the next section we will discuss safe use guides, i.e. tips for parents and children on how to keep safe on the internet.

SAFE USE GUIDES

Safe use guides aimed at adults and children are strikingly similar even though they may be published in different countries (Lüders et al,

2009). The first sightings of these guidelines are from the 1990's and they have changed very little since then, even though internet use has changed significantly. In fact, safe use guides can be traced as far back as 1995 (Safe Surf, 2007) and many of the safety tips are the same today (Lüders et al, 2009). Thus it seems that today's tips have been reproduced from guides as old as 15 years. If this is the case then it is crucial that we make sure that original guides were based upon relevant information regarding what constitutes risk.

If the safe use guides circulated today were once written in a media landscape that has changed, but the guides themselves in most part have been reproduced, this calls for some discussion. Which of these recycled tips are meaningful in a contemporary media setting?

A study of European safe use guides (Lüders et al, 2009) shows that these guides typically recommend that children be anonymous online. Most safe use guides contain advice not to give out any personal information online, arguing that anonymity can serve as protection against online dangers. But what assumptions underpin this advice? And what does current research say about online anonymity?

This advice has to be contextualized in relation to the fact that most young people socialize online with people they already know (cf. Larsen, 2009). In a context where you know the other parties, anonymity loses its possible protective power. In many of the current online environments, such as social networking sites it is not an option to leave out your real name or your physical location. Boyd & Ellison (2007) define social network sites as "web-based services that allow individuals to (1) construct a public or semi-public profile within a bounded system, (2) articulate a list of other users with whom they share a connection, and (3) view and traverse their list of connections and those made by others within the system". This would then mean that identifying oneself is in fact a vital part of online social networking. If, then, the users are not allowed to identify themselves then the interaction can be seen as partly meaningless. A 12-year old girl in Dunkels' study (2008) replied to the question if she ever visited open chat rooms:

a yeh, one time but only 4 bout 10 min.
elza why only for 10 minutes?
a there was nuthin 2 talk bout

Also Moinian (2007) found that most of her informants were not interested in being anonymous online, since this would inhibit the interaction.

Most of Dunkels' (2008) informants said that they would tell their parents if anything unsettling happened online, but some of the children said that they would never tell their parents. Revealing what they themselves considered to be quite ordinary situations such as getting dirty comments about posted pictures, they all had counter strategies. These strategies were for instance blocking people, avoiding certain web sites and net communities, only adding friends who could prove some kind of connection – a friend of a friend or having met in real life at some point (Dunkels, 2008).

As the opening quote of this chapter reveals the informant was worried that if her parents found out about her online experiences they would cut her off the internet. She knew that her parents would do this out of concern for her but she felt that she could handle the situation herself. When discussing this further it turned out that the informant herself was not particularly worried about these contact attempts because she knew how to block these people out. In view of this informant's reasoning, it can be argued that safety tips impossible to follow might create a more dangerous situation than if she had not been given that particular piece of advice.

In Sweden Shannon (2007) conducted a study of 315 reports to the police concerning sex crimes where the victim was under the age of 18 and where some of the contact had been online. The study shows that most victims had a troubled real life situation. They were members of a dysfunctional family or were being bullied at school and had no adult they could rely on. These children have proved more vulnerable to the kind of manipulative relations that online perpetrators often try to build with possible victims. This is an example of a substantiated risk that is not covered by the standard safety tips that are circulated. In fact, if the most common risk factor is having a troubled life, one solution is to make sure that every child feels good about herself and that she has adults to rely on. This would then be a task for the surrounding world and not for the child herself to arrange. Furthermore, this piece of information does not fit into a safe use guide. In fact, omitting this very important information in communication with children can be seen as yet another way of letting children down. It could, however, be argued that the individual also needs to take responsibility for her actions and that the safety tips may serve a purpose guiding children in this respect. However important it is to develop personal safety strategies, it is crucial that the main strategy on a societal level is not aimed at simply placing responsibility on the young users themselves. One argument against focusing on personal safety strategies is that these risk placing the blame on the individual should something unwanted occur. This could be interpreted as an escape route for adults, and we risk transferring blame from adults to children instead of the other way around.

Rules that are not possible to follow because they do not take into consideration all the contextual components that surround them can thus create a situation where the child chooses to hide her actions from adults. If that happens we have created more dangerous conditions for young people's online activities, rather than, as intended, reducing the dangers.

In the following section we will discuss how the views on risk and danger can be seen as culturally determined.

CULTURE AND RISK

Some researchers have pointed out that there can be a significant difference between adults' and children's views of internet use and that this difference may be the cause of some current problems (Larsen, 2009; Dunkels, 2008).

Larsen (2008) uses Scollon's Geographies of Discourses to describe different levels of actions: local discourses describe our daily lives and every-day activities, and global discourses are the mediated or public descriptions of our activities. Larsen claims that most young people act on a local discourse level, i.e. when communicating with friends, be it offline or online. Larsen (2008) exemplifies by describing a young girl who publishes a self-portrait on a social web site. This activity is mostly aimed at her friends and contemporaries, although adults in general and media in particular on their hand interpret her action in a global discourse. In this discourse the girl's action ends up in a context where pedophiles and others constitute potential threats. Larsen presents this as an explanation to the differing views on risk between generations when it comes to the internet, claiming that negative experiences on the internet are not normally part of young people's daily lives. This matches Dunkels' (2007) conclusions that young people are aware of the dangers of the internet but that the negative sides are not very prominent when children talk about their internet use.

In contradistinction to what children express, adults tend talk about online interactions and children in terms of uncertainty and anxiety. Dunkels (2007) outlines how some openness is promoted by the internet and how the older generation sometimes expresses concern regarding this. Raynes-Goldie (2010, p. 4) describes how attitudes towards social privacy has changed since the dawn of social media; there is a big increase in privacy pragmatists - "people who are concerned about their privacy but are willing to trade some of it for something beneficial". Madden et al (2007) also reach the conclusion that most people are not very concerned about their information being public online.

Altheide & Michalowski (1999) point out that it is important to separate danger from fear. Danger is the actual threat; "we can deal with danger, we can be educated about it, take steps to avoid it or minimize its impact." (Altheide & Michalowski, 1999, p. 501) Fear, however, does not come as a necessary response to danger but as a part of "the risky society" (Altheide & Michalowski, 1999, p. 476). In fact, Altheide & Michalowski show that fear has increased even in periods of decreasing crime rate, so there seems to be no correlation between increased fear and more crime. They also point out that the discourse of fear can be used to justify more formal control. In fact, many politicians use children in their rhetoric to rationalize control; when the aim is to protect children the means may be justified by this indisputably good goal. The European Data Retention Directive (Directive 2006/24/EC) and its associated debates, may serve as a recent example of how the protection of children's vulnerability is used politically, in this case as an argument to enforce access to and storage of individual communication data. Qvortrup (2003) also encourages us to discuss status offence with reference to age. He uses the curfew rules to illustrate status offences; how children can be put in house arrest merely on account of their young age, with their good as objective.

So it seems that views on risk can be seen as varying with culture. The discourse of risk and safety may also be employed in political contexts. If we widen the concept of culture so that it also embraces generations we might find tools to think of risk from the vantage points of both adults and children.

With this as a starting point, what does current research tell us about online risk? In the following sections we will discuss sex crimes against young people and bullying among and against young people.

SEX CRIMES

Wolak et al (2008) point out that sex crimes initiated over the internet in reality are more complex than media reports imply. There is no simple link between a child's actions and her vulnerability to offenders and therefore practitioners need to have accurate information since they may encounter these crimes in many different contexts. In Wolak et al's (2006) study there was no evidence that personal information posted online led to victims being stalked or abducted. Still, one of the most persistent online safety tips is not leaving out personal information online (Lüders et al, 2009).

Internet has indeed changed the arena for child molesters considerably. According to Wolak et al (2008) online communication allows for private conversations that can take place very often and sometimes round the clock. Furthermore, Suler (2005) describes psychological differences between online and offline communication. One difference is that our thoughts have a tendency to merge with others when we communicate in writing. Suler (2005) calls this *solipsistic introjection* and this phenomenon occurs because we need to read the words to ourselves and in that process we use our own voice if the other person's voice is not known to us. The fact that our own voice generally does not pose a threat to us makes us a little less on our guard and the words we read make a little more sense than they might do if we were talking face to face to a stranger. This and other factors deriving from traits in computer mediated communication affect conditions online.

The Online Safety and Technology Working Group's report from 2010 shows that sexual predation is not as widespread as once believed (OSTWG, 2010). Another American study shows that both offline and online stranger crimes against girls have declined since the mid 1990's (Cassell & Cramer, 2008). However, monitoring media reports during the same period of time gives a completely different picture of young people's lives; bottomless danger and a constant threat to every child who goes online. And perhaps most of all, an increasing risk. One possible explanation to the inconsistency between adults' views and statistical facts is lack of information. Adults in general have very little information about young people's activities online and the information they do have often comes from media. Media on their side have an interest in presenting this information in a spectacular way, as they need to capture the readers' interest. Media in general does not make a distinction between likely risk and worst-case risk, something Sharples et al (2009) point out we need to do. Wolak et al (2006, 2008) remark that internet initiated sex offences are a small portion of all sex crimes against children. Cassell & Cramer (2008) also point out that when it comes to assaults on girls aged 0 to 17 years old in 98 per cent of the cases the offender is a family member or an acquaintance. This means that the walled gardens that are created as a safety measure can in fact be a trap for some children.

According to Wolak et al. (2008) only five % of the offenders actually pretended to be teenagers. The rest were truthful about their age. Also, many offenders accurately confess their intention to have sex with the victim and as much as 73% of the victims in one study (Wolak et al., 2008) met their offender more than once. In a report from a three year project with young people who have been abused online Jonsson et al (2009) show that it is unusual that perpetrators lie about their age or their agenda to have sex with the child. Furthermore, Jonsson et al (2009) found that sexual curiosity had been a driving force among many of the victims in their study. Many of the children were insecure about their sexual identity and used the internet to find answers to questions about normality, etc. Seeing children as sexual beings is a theme that is rarely found in the literature concerning online safety in an educational setting. Neither is it visible in media reporting. This fact is important and raises questions about how we construct children and how we construct victims. Does a victim need to behave in a certain way to retain her position as a victim? Do we generally have problems with adding sexuality as a parameter in our construction of the child? Monitoring the media debate about young people as victims of sexual offences we sense that there might be a terminology problem, or rather a question of what is associated with terms like victim and offender. Defining a child as a sexual being is often done from an adult perspective and used by people who want to sexualize the child to justify their abuse of children or teenagers. This strong association to abusive practices makes it hard for people who condemn adults' criminal actions to recognize the youth perspective of sexuality. Also, this terminology problem could have other long withstanding roots such as the image of the child as being innocent, as a pure being for the adults to mould, etc. Jonsson et al. (2009) claim that it is an exception that perpetrators lie about their age or their intention to have sex with the child. Only recently, research has taken interest in looking at children as active in sexual abuse situations. The absence of early research on this theme may be explained by the subject bordering on taboo subjects; if the child is active, is the adult still to be seen as a criminal? A consequence of our tendency not to recognize children's sexual development and not discussing children's own actions in sexual abuse is that children might blame themselves for what happened.

BULLYING

Most teachers and students are aware of bullying as a problematic phenomenon difficult to handle and this has severe consequences. Schools are also taking actions against bullying in different more or less successful ways (Frånberg, 2009). The fact that many students are being harassed through electronic communication is an increasing problem that adults feel even more uncertain about (Li, 2006, Kiriakidis & Kavoura, 2010). First; adults know less about young people's internet use and second; they are also uncertain about how serious cyber bullying is and how grave cyber threats actually are. To be able to deal with this problem, we require more knowledge about the issue.

Willard (2007, p. 265) defines cyber bullying as "sending or posting harmful or cruel material or engaging in other forms of social aggression using the internet or other digital technologies". The division into traditional bullying and cyber bullying can be seen as an expression of our lingering idea that the online and the offline are two different worlds. However, not all young people embrace this thought (Dunkels, 2009). Instead, the online and the offline can be seen as different rooms or different dimensions of the same. Policy makers and researchers have made some effort trying to identify certain characteristics that are specific to bullying taking place on the internet and we would like to critically examine some of these characteristics. One distinguishing characteristics of cyber bullying, according to Shariff (2008) and Kiriakidis & Kavoura (2010), is anonymity; the internet allows targeting of victims without being easily detected. The perpetrator's identity can for instance be protected by screen names. Another feature is the infinite audience, referring to the fact that the internet supports great numbers of perpetrators and bystanders to get involved in the abuse. Thirdly, Shariff (2008) lists permanence of expression as a distinguishing factor; once posted online abusive material about a person is difficult to remove.

It is however possible to problematize these characteristics. The concept of anonymity is not as straightforward as it might seem. Often when we talk of online anonymity we actually refer to pseudonymity which means that the person has taken on a screen name. These two forms can in fact be present at the same time but occasionally the parties in a computer mediated communication are known to each other even though they may use screen names (Zhao et al, 2008). So this anonymity trait may not be of such great importance as it seems to be on a superficial level. In fact, as pointed out earlier most online connections occur between people who also know each other from real life. However, according to psychologist Suler (2005) the possibility to be invisible to the other party amplifies the *online disinhibition effect*, the fact that we act in a less inhibited fashion online. Zhao et al (2008) use the expression *nonymity* to describe online relationships that are anchored offline; family members, co-workers, class mates, etc.

The idea of an infinite audience is also possible to problematize. There have in fact been media reports of grave bullying incidents where nearly the whole world has been involved or silent bystanders. One such incident is known as the Star Wars Kid (Globe & Mail, 2007) where a young boy was severely harassed subsequent to a film of him was widely circulated on the internet. Still, we cannot conclude from these unusual, albeit horrifying, events that this is a typical characteristics of online bullying. The fact that the audience is possibly infinite does not automatically mean that it is in effect infinite. In many cases there has to be an incitement involved in order to catch the interest of the audience. There also needs to be a connection between the victim and the perpetrator to make the bullying interesting for the bully.

The third attribute of cyber bullying according to Shariff (2008), the permanence of expression, is perhaps not as easy to problematize. The information we have today indicates that we can expect most of the information posted online to remain public under the foreseeable future. However, we might think about how we conceive of this fact. Today's adults were raised in a world where it was not only possible but also desirable to separate the public from the private and this is likely to affect how we look upon the permanence of expression online. Children, who were raised with the greater agreement between private and public that the internet allows, are likely to view this in another way. The permanence of expression can also be seen as abuse in itself because the abusive incident is repeated every time someone watches it again. Again, this can be interpreted as one of the internet's affordances that can be used and abused. The fact that most bullying incidents that occur online constitute their own documentation can be exploited by the educational and law systems in order to follow up the incident or prosecute the perpetrator.

It is also important to nuance a fourth attribute used to separate cyber bullying from traditional bullying, the idea of the constant presence; that cyber bullying has the potential to go on the whole day and follow the child in school as well as in their homes (Kiriakidis & Kavoura, 2010). This theory argues that victims of online bullying have no place to hide and can be targeted anytime and anywhere (Tippet et al, 2009). This is in fact what some victims of cyber bullying claim but we need to question if this really is a distinguishing factor. Even traditional bullying follows the victim home although it is perhaps not as visible to the bystander. In fact, research on bullying shows that it has serious and lifelong consequences for the victim. By adding the constant presence as a distinguishing parameter, we risk devaluating bullying as a phenomenon.

It might even be interesting to discuss who might benefit from these distinguishing parameters. What is hidden in the will to focus on the dissimilarities rather than considering bullying as a whole? There is an apparent risk that this division into cyber bullying and traditional bullying is yet another escape route from having to take informed responsibility as adults.

TECHNOLOGICAL SOLUTIONS

For technological solutions to work there needs to be a classification of the content in order to give the technological tool parameters to work with. Traditional classification of content, such as video ratings, for example, has been undertaken by humans (Oswell, 1999). Today's technological solutions are in fact the first attempts at carrying out this task with the help of machines. To date we have not constructed machines as complex as the human mind and this fact alone can be seen as a reason not to attempt to use technological solutions to this kind of social problems. But there are other problems as well. The monitoring and regulating of a child's internet use is underpinned by the construction of the relationship between children and adults. Consequently this construction needs to be discussed. In Oswell's (1999, p. 57) words, "the panopticon of the living room" is created if the relation between child and parents is seen as a relation of supervision.

All technological solutions have some basic problems according to Price & Verhulst (2005). The user will always have to set parameters from a list that the constructor provides. This list will need to be very long if the software is to fit all kinds of family structures, cultures, ideologies, etc. If the list is not long, the user will have to trust the constructor's choice of parameters. In fact, according to Price & Verhulst (2005) this is the second problem, the fact that the choice of what content is filtered out is subject to ideological biases. This bias is however not visible to the user. Oswell (1999) argues that video classifications "are derived from particular normative (and normalizing) discourses of the child and family" which is important to consider if these classifications are to be used as recommendations for parents.

A study of 35 schools by the British Office for Standards in Education, Children's Services and Skills (Ofsted, 2010) shows that 13 schools chose to lock down their system, making certain websites inaccessible. The study found two main problems with this procedure in these schools. Firstly, it took up valuable learning time and secondly, it did not encourage pupils to feel accountable for their actions.

There is also the risk that the practical managing of the technological solution, such as setting the parameters, takes up valuable face-to-face time in families or schools. Tynes (2007) suggests that young people be helped to develop an "exit strategy", using blocking, reporting to trusted persons or to a cyber tip line. The development of such exit strategies can be done in collaboration with the young internet users and consequently helps us in creating partnership rather than top-down rules. Furthermore, developing strategies does not require any expensive technological tool.

LAYER CAKE METAPHOR

The apparent discrepancy between media coverage and young people's own narratives of the risk and safety phenomena can be interpreted as a clash between different vantage points. Speaking with Scollon (2007) the discrepancy could also depend on adults and young people thinking about the world from different geographical locations of discourses. To expand this line of reasoning we have developed a metaphor that aims at understanding and explaining the above described clash.

We are inspired by Larsens (2008) way to use Scollon's geographies of discourse when we try to understand the discrepancy between different perspectives on the same phenomena and associated consequences of this gap. We argue that it is possible to simultaneously theorize and act on different levels; we claim that these levels can be symbolized by a layer cake (Figure 1). Further, we suggest that acting and theorizing from separate perspectives might be pictured as two different pieces of this cake of discourses.

When talking about young people's internet use there are a number of levels to consider. There is the personal level where for instance many young people act when publishing images on the internet, communicating online, etc. There is also the societal level where Government authorities act and reason about young people's internet use; laws are passed, strategic decisions are made, etc. Between these layers there might also be several layers such as the municipal level, the school level, the family level and different group levels.

At the societal level there is an expectation to abide by rules and values that are common to society as a whole. It will not be considered appropriate to pass a law that is in clear conflict with agreed upon strives for equity, for instance. At the personal level, however, it might be relevant to act in a way that contradicts social values in order to

Figure 1. Layer cake metaphor

achieve personal goals. This way of thinking may provide us with tools to interpret young people's online actions not as acting against social values but as the ability to move between layers and adjust their practice accordingly. If we look at a young girl publishing sexually explicit self-portraits on a social networking site, we may view them from the societal and adult level and interpret her actions as undesirable. However, her actions do not in any way prove that she fails to embrace the values she contradicts, rather her actions reveal her ability to move between the layers as she sees fit. The ability to move between the layers can be seen as a skill equivalent to the way we adjust our language depending on who we address (Trousdale, 2010). While she is acting at the personal or local layer of the cake, her actions are interpreted and judged from a societal layer. Instead of viewing her as naive we use the layer cake metaphor and suggest that she is a competent navigator of several layers of discourses, maybe even that she is discourse sensitive. Furthermore, in the same way as it is possible to act at different levels it is likely that we may theorize on different levels.

We believe that actions and thoughts at different layers of the discourse cake illustrate tensions and asymmetries in human relations. The conflicts that this chapter has outlined might be explained by the fact that we lack terminology to describe the conflict that arises around phenomena such as online risk and safety. We might use the layer cake metaphor to understand this particular conflict. We also think the metaphor is useful to us in order to find new questions to ask when outlining research and developing policy. Furthermore, we believe that the layer cake metaphor is useful in everyday interaction with young people. It may prove fruitful to critically question whether a description or an explanation makes sense on all levels before moving on to decisions based upon the description.

POLICY MAKING

This chapter has outlined some problematizations regarding young people and online risk. We argue that there is too little research on these important and widely discussed issues. Furthermore, we have found that some of the present research may in fact have started out with biased assumptions and therefore the results may not be relevant. Is there, however, some knowledge that research today agrees upon? Is there a lowest common denominator that can be found in online victims or perpetrators? Jonsson et al. (2009) found that most children were well informed about online dangers and yet they displayed risky conduct online. Should this be a general situation, then we must ask ourselves if information about online risks really is the solution. Jonsson et al.'s observation implies that something else is at play and that the problem is not that young people are naïve or ill informed. Also, the claims that one common denominator among victims is that many have a troubled social situation, i.e. their offline situation is problematic; imply that the solution is to be found elsewhere. Jonsson et al.conclude that we need to discuss how to protect children without judging their actions. Norwegian Social Research (NOVA, 2006) issued guidelines for counselors handling youth who have been selling or trading sex. They point out that it is vital not to show disgust or dismay when young people inform adults about sexual abuse. Furthermore, they stress the importance of not placing guilt upon the child and of avoiding moralizing. The non-moralizing approach can be seen as one way of keeping the information channels open between adults and children. If, in fact, the only thing we know for certain is that children who are troubled offline also are in the danger zone online, then this information exchange is vital. We need to create an atmosphere of trust, in which young people feel comfortable enough to tell us if anything bad happens. This may actually be our only way to protect children, given the fact that technological solutions seem to be crude instruments.

Oswell (1999) argues that policy rarely places the child in the position of an agent, but mostly as the object of regulation. Dunkels (2007) agrees and calls attention to the need to contextualize policy, i.e. taking on the perspective of the young internet user as one way of ensuring the policy's validity.

As Sharples et al (2009, p. 73) point out, policy makers have two different approaches to choose from when developing an internet safety strategy. One is to set out to protect against the worst case scenario and the other is to base policy on "continually assessed levels of acceptable risk". This chapter has argued that basing strategies on a worst case scenario may not protect as much as we might hope. Instead it may create an unsafe environment for children in the sense that they become afraid to tell adults about unsettling experiences.

Ybarra et al (2008) conclude from their extensive research into the internet safety field that the most important is not to restrict access to different interaction tools. Instead, we need "mental health interventions for vulnerable youth" and provide internet safety education that takes into consideration all types of online communications (Ybarra et al, 2008, p. 356). Jonsson et al. (2009) request more research into the area as well as laws that are adjusted to the contemporary media situation.

O'Connell et al. (2004, p.29) point out that it may not be such a good idea to advise young people to report incidents to their parents since "teenagers are hardly renowned at sharing personal experiences with their parents". Instead, they urge us to acknowledge the learners' existing knowledge and their capacity to develop and use strategies for safe use.

The Ofsted (2010) report concludes that the schools that were judged "outstanding for e-safety" all hade adjusted their e-safety strategies to their own situation. This strengthens the thought that there can be no one-size-fit-all solution to the online safety issue. Instead, the work to develop solutions should be seen as a continuous process involving everyone; students, parents, school staff at all levels and policy makers.

Sharples et al (2008) put risk in a historical perspective pointing out the fact that we do not forbid our children to walk home from school even though they can in fact be abducted or injured. Neither do we prevent access to the school playground even though it is a renowned area for bullying.

In these areas, policy has evolved over time to balance the likely risks against the benefits to children of exercise and creative play, also taking account of pragmatic issues such as difficulty of prevention and the value of getting children out of the school buildings over break time. (Sharples et al, 2008, p. 33)

As Hinduja & Patchin (2008) point out, having a private profile is not a guarantee for the child's safety. Since most child abuse and sexual assault is committed by family members or other people the child should be able to trust and only a few percent of these crimes actually are committed by strangers, a private profile may instead be a trap. Just like the walled garden is not really a protection against burglars, since the wall also protects the criminal from being spotted from the outside, the private profile can be in the way of someone discovering the child's situation.

Ybarra et al (2008), who are heavily cited because of their extensive research on online risk, conclude that technologies will continue to evolve into new applications. Therefore we should focus on children's online conduct and their psychosocial situation rather than on a certain technological tool. O'Connell et al. (2004) claim that online safety programs must draw upon evidence-based practice and research that take into consideration the different ways in which young people use the internet for their recreation, and, just as important, the ways in which perpetrators use the internet as an instrument of abuse. Furthermore, they argue that because technology constantly evolves, it is vital that any advice on internet safety be updated regularly, perhaps as often as every six months.

Marwick (2008) describes how researchers tend to deny problems connected to moral panics regarding new media, in attempts to level the emotional outbursts. This is probably not helping the matter. Academia needs to find ways of expounding facts and interpretations of situations that are likely to be subject to emotional reactions.

CONCLUDING REMARKS

This chapter has attempted to critically examine safety measures aimed at protecting children and young people online. In this process we have looked at the concepts of risk, danger and safety as well as the rhetoric concerning children and adults. We conclude that even though an entire generation has passed since the early days of internet massification much remains to be done. We need more research into the area – research that contextualizes internet use and systematically seeks the vantage points of the young users. We also need to increase the philosophical and pedagogical debate about children, danger, risk and safety. Even if this chapter deals with online risk the debate does have general implications that in fact go beyond internet use.

Young people and online risk is indeed a complex area. A complex problem calls for complex solutions and it is our responsibility as researchers to seek these complex solutions and not fall into the pitfall of simple solutions of remedies to cure all ails. A recurring theme among the cited research in this chapter has been communication; communication between the young internet users and the adults that surround them. However, we have not in any detail touched upon what must be done meanwhile, while we are busy creating a world where young people communicate their troubles to adults. Young people are in fact being abused as a result of their being online, even though these cases clearly are not as many as once believed. Now, the simple fact is that every victim is one too many and therefore it is important that no efforts are spared in order to ensure that every child is safe, offline as well as online. It is, however, not within our professional knowledge to give detailed advice on this. Instead, we end this chapter repeating, and agreeing with, what the Norwegian Social Research (NOVA, 2006) points out in their guidelines for counselors. They stress the importance of adults not panicking when young people choose to let them know that they are being abused and that it is the adults' responsibility not to moralize.

REFERENCES

Alderson, P. (2005). Generation Inequalities. *UK Health Watch, 2005*, 47–52.

Altheide, D., & Michalowski, S. (1999). Fear in the News: A Discourse of Control. *The Sociological Quarterly*, *40*(3), 475–503. doi:10.1111/j.1533-8525.1999.tb01730.x

Bayne, S., & Ross, J. (2007). *The 'digital native' and 'digital immigrant': a dangerous opposition*. Paper presented at the Annual Conference of the Society for Research into Higher Education (SRHE) December 2007.

Bennett, S., Maton, K., & Kervin, L. (2008). The 'digital natives' debate: A critical review of the evidence. *British Journal of Educational Technology, 39*(5), 775–786. doi:10.1111/j.1467-8535.2007.00793.x

boyd, d. m., & Ellison, N. B. (2007). Social network sites: Definition, history, and scholarship. *Journal of Computer-Mediated Communication, 13*(1), article 11.

Buckingham, D. (2002). The Electronic Generation? Children and New Media. In Lievrouw, L., & Livingstone, S. (Eds.), *The Handbook of New Media* (pp. 77–89). London: Sage.

Cassell, J., & Cramer, M. (2008). High Tech or High Risk: Moral Panics about Girls Online. In McPherson, T. (Ed.), *Digital Young, Innovation, and the Unexpected. The John D. and Catherine T. MacArthur Foundation Series on Digital Media and Learning*. Cambridge, MA: MIT Press.

Directive 2006/24/EC. Directive 2006/24/EC of the European Parliament and of the Council on the retention of data generated or processed in connection with the provision of publicly available electronic communications services or of public communications networks and amending Directive 2002/58/EC.

Drotner, K. (1999). Dangerous Media? Panic Discourses and Dilemmas of Modernity. *Paedagogica Historica, 35*(3), 593–619. doi:10.1080/0030923990350303

Dunkels, E. (2007). *Bridging the Distance – Children's Strategies on the Internet*. Umea, Sweden: Umeå university.

Dunkels, E. (2008). Children's Strategies on the Internet. *Critical Studies in Education, 49*(2). doi:10.1080/17508480802123914

Dunkels, E. (2009). När kan vi tala om nätmobbning? [When can we speak of cyber bullying?]. *Locus (Denton, Tex.), 2009*(2).

Eriksson, M. (2008). *Barns röster om våld: att tolka och förstå* [*Children's vioces about violence: to interpret and understand*]. Malmö, Sweden: Gleerups.

Findahl, O. (2009*). Internet 15 år [Internet 15 years]*. Stockholm, Sweden: Stiftelsen för internetinfrastruktur.

Frånberg, G., & Gill, P. (2009). Vad är mobbning? [What is bullying?] In *På tal om mobbning – och det som görs*. Stockholm, Sweden: Skolverket.

Globe & Mail. (2007-04-07). *'Star Wars Kid' cuts a deal with his tormentors*. www.theglobeandmail.com/servlet/story/RTGAM.20060407.wxstarwars07/BNStory/National/home

Herring, S. (2008). Questioning the Generational Divide: Technological Exoticism and Adult Construction of Youth Identity. In Buckingham, D. (Ed.) *Youth, Identity and Digital Media.* (pp. 71-92).John D. and Catherine T. MacArthur Foundation Series on Digital Media and Learning. Cambridge, MA: MIT Press.

Hinduja, S., & Patchin, J. (2008). Personal information of adolescents on the Internet: A quantitative content analysis of MySpace. *Journal of Adolescence, 31*(1), 125–146. doi:10.1016/j.adolescence.2007.05.004

Jonsson, L., Warfvinge, C., & Banck, L. (2009). *Children and Sexual Abuse via IT*. Linköping, Sweden: BUP-Elefanten.

Kiriakidis, S. P., & Kavoura, A. (2010). Cyberbullying: a review of the literature on harassment through the internet and other electronic means. *Family & Community Health, 33*(2), 82–93.

Larsen, M. (2008). *Online Social Networking: From Local Experiences to Global Discourses*. Paper presented at Internet Research 9.0: Rethinking Community, Rethinking Place, The IT University, Copenhagen, Denmark 20081015–18.

Larsen, M. (2009). Sociale netværkssider og digital ungdomskultur: Når unge praktiserer venskab på nettet [Social networking sites and digital youth culture: When young people practice friendship online]. *MedieKultur, 47*, 45–65.

Li, Q. (2006). Cyberbullying in Schools: A research of Gender Differences. *School Psychology International, 27*(2), 157–170. doi:10.1177/0143034306064547

Lüders, M., Bae Brandtzæg, P., & Dunkels, E. (2009). Risky Contacts. In Livingstone, S., & Haddon, L. (Eds.), *Kids Online*. London: Policy Press.

Madden, M., Fox, S., Smith, A., & Vitak, J. (2007). *Digital Footprints – Online identity management and search in the age of transparency*. Washington: PEW/Internet.

Marwick, A. (2008). To catch a predator? *The MySpace moral panic*. *First Monday*, *13*(6).

Moinian, F. (2007). *Negotiating Identities*. Stockholm, Sweden: Stockholm Institute of Education Press.

NOVA. (2006). *Ungdom som selger eller bytter sex– en faglig veileder til hjelpeapparatet* [Youth Selling or Trading Sex]. Oslo, Norway: Akademika AS.

O'Connell, R., Price, J., & Barrow, C. (2004). *Cyber Stalking, Abusive Cyber Sex and Online Grooming: A Programme of Education for Teenagers*. University of Central Lancashire.

Oblinger, D., & Oblinger, J. (2005). *Educating the Net Generation*. Boulder, CO: Educause.

Ofsted (2010). The safe use of new technologies. Report no. 090231. *The Office for Standards in Education, Children's Services and Skills*. Manchester, UK: Ofsted.

OSTWG Online Safety and Technology Working Group. (2010). *Youth Safety on a Living Internet*. Washington, DC: National Telecommunications and Information Administration.

Oswell, D. (1999). The Dark Side of Cyberspace: Internet Content Regulation and Child Protection. *Convergence*, *5*(4), 42–62.

Prensky, M. (2001). *Digital Natives, Digital Immigrants*. On the Horizon. NCB University Press, 5.

Price, M., & Verhulst, S. (2005). *Self-regulation and the Internet*. Hague, the Netherlands: Kluwer Law International.

Qvortrup, J. (2003). *Barndom i et sociologiskt generationsperspektiv [Childhood from a sociological generations perspective]*. (Working paper no. 123-03, Centre for Cultural Research, University of Aarhus). http://www.hum.au.dk/ckulturf/pages/publications/jq/barndom.htm

Raynes-Goldie, K. (2010). Aliases, creeping, and wall cleaning: Understanding privacy in the age of Facebook. *First Monday*, *15*, 1–4.

Rheingold, H. (1993). *The virtual community: Homesteading on the electronic frontier*. Cambridge, MA: MIT Press.

Scollon, R. (2007). *Geographies of Discourse*. Draft available at www.aptalaska.net/~ron/ron/downloads/Geographies%20of%20Discourse%20download%20draft.doc

Shannon, D. (2007). *Vuxnas sexuella kontakter med barn via Internet [Adults' sexual contacts with children online]*. Stockholm, Sweden: Brottsförebyggande rådet.

Shariff, S. (2008). *Cyberbullying. Issues and solutions for the school, the classroom and the home*. New York: Routledge.

Sharples, M., Graber, R., Harrison, C & Logan K. (2008) E-safety and Web 2.0. *Web 2.0 technologies for learning at Key Stages 3 and 4*. Becta report.

Sharples, M., Graber, R., Harrison, C., & Logan, K. (2009). E-safety and Web 2.0 for children aged 11–16. *Journal of Computer Assisted Learning*, *25*, 70–84. doi:10.1111/j.1365-2729.2008.00304.x

Suler, J. (2005)... *International Journal of Applied Psychoanalytic Studies*, *2*(2), 184–188. doi:10.1002/aps.42

Surf, S. (2007). Retrieved from http://www.safesurf.com/time.htm

Tapscott, D. (1998). *Growing up digital: The rise of the Net Generation*. New York: McGraw-Hill.

Tippet, N., Thompson, F., & Smith, P. K. (2009). Research on Cyberbullying: Key findings and practical suggestions. *Education.com*. http://www.education.com/reference/article/cyberbullying-research

Trousdale, G. (2010). *An Introduction to English Sociolinguistics*. Edinburgh, UK: Edinburgh University Press.

Tynes, B. (2006). Internet Safety Gone Wild? Sacrificing the Educational and Psychosocial Benefits of Online Social Environments. *Journal of Adolescent Research*, *22*(6), 575–584. doi:10.1177/0743558407303979

Willard, N. (2007). *Cyberbullying and cyberthreats responding to the challenge of online social aggression, threats, and distress*. Champaign, IL: Research Press.

Wolak, J., Finkelhor, D., & Mitchell, K. (2006). *Trends in arrests of "online predators"*. Durham, NC: Crimes Against Children Research Center.

Wolak, J., Finkelhor, D., Mitchell, K. J., & Ybarra, M. L. (2008). Online "Predators" and their victims. Myths, realities, and Implications for prevention and Treatment. *The American Psychologist*, *63*, 111–128. doi:10.1037/0003-066X.63.2.111

Yao, M., & Flanagin, A. (2006). A self-awareness approach to computer-mediated communication. *Computers in Human Behavior*, *22*, 518–544. doi:10.1016/j.chb.2004.10.008

Ybarra, M., & Mitchell, K. (2008). How Risky Are Social Networking Sites? A Comparison of Places Online Where Youth Sexual Solicitation and Harassment Occurs. *Pediatrics*, *121*(2), 350–358. doi:10.1542/peds.2007-0693

Zhao, S., Grasmuck, S., & Martin, J. (2008). Identity construction on Facebook: Digital empowerment in anchored relationships. *Computers in Human Behavior*, *24*(5), 1816–1836. doi:10.1016/j.chb.2008.02.012

Chapter 2
Youth and Online Social Networking:
From Local Experiences to Public Discourses

Malene Charlotte Larsen
Aalborg University, Denmark

Thomas Ryberg
Aalborg University, Denmark

ABSTRACT

Often, young people do not have a voice in the public debate on internet safety and online social networking, but as this chapter will demonstrate that does not mean they do not have an opinion. Based on responses from 2400 Danish adolescents to an open-ended questionnaire, the authors discuss their accounts of good and bad experiences with social network sites. Furthermore, they analyse how youth (aged 12 to 18) position themselves as users of social network sites both in relation to very concrete and local experiences from their everyday life, and in relation to public media discourses. They discuss how they portray themselves as 'responsible young people' by distancing themselves from the public or "grown up" discourses represented by e.g. their parents or the news media.

INTRODUCTION

"There are many who believe that young people act without thinking online and communicate with one perverted man after the other. But some of us just use the net to communicate with the people we already know, and we are careful."[1]

DOI: 10.4018/978-1-60960-209-3.ch002

The quotation was written by a 14-year-old girl as a comment in an online questionnaire which provides part of the empirical basis of this chapter. The quote captures quite well the essence of what the chapter is about, since it highlights the tension between what young people experience online on a local and day-to-day basis (communicating and hanging out with friends), and then wider public discourses which surround their use of new media (which tend to highlight the pitfalls and dangers for young people). As suggested by Herring (2008) public discourses on young people's use of technology often seem to be expressions of adults' "hopes and dreams" or "fears and anxieties", rather than reflecting adequately young people's own experiences. Herring (2008) identifies three forms of public discourse (Media Production and Advertising, Media Commentary and new media research) which have the common traits that they: produce or re-produce overly optimistic (utopian) or negative (dystopian) accounts of young people's use of media, and represent an outsider view (adult) on youth's online practices. She explores the ways in which young people contest or condone of these public discourses governed and produced largely by adults. She suggests that youth are characterised by a dual consciousness or awareness where: "young people are aware of adult representations of their generation and orient to them, while simultaneously orienting to their own experiences" (Herring, 2008, p. 78).

In this chapter we analyse and discuss these tensions or dual awareness from the point of view of young people themselves, and with basis in different types of empirical data. The aim of our analysis is to emphasise and report from an "insider perspective" by drawing on ethnographic research, and by giving voice to youth's own statements and reflections on online practices, as they were expressed in an online questionnaire.

We find the notion of 'dual consciousness' interesting, and while Herring (2008) seems to ground the concept primarily in a broader, more general reading with only some concrete examples, we will discuss the concept in relation to a number of empirical examples. Thorough empirically based knowledge within this area is an important research contribution these years. Particularly because the topic of young people's online social practices is subject of much discussion and concern in the general public debate. As Sonia Livingstone has pointed out there has been "a notable discrepancy between the high levels of public concern over children and young people's use of new media and the paucity of empirical research conducted" (Livingstone, 2002, p. 3). With that in mind, this chapter is based on five years of ethnographic research, a comprehensive open-ended questionnaire about Danish adolescents' use of social network sites and a media content survey and analysis.

The chapter represents a particular aspect of a research project concerned with Danish adolescents' use of social network sites (an investigation carried out by the first author). The overall aim of the research project is to understand how social network sites are integrated into the everyday lives of 12-18-year-old Danes, and how everyday teenage life is represented, mediated and acted out in the digital space. The research process started back in late 2004 where focus group interviews with young people about their use of technology and media revealed that a particular social network site (Arto.dk) was hugely popular among teenagers. Simultaneously, media stories portraying this site as crawling with sexual predators and bullies began to surface. The main author became interested in why the site was so popular and loved among adolescents; but also because of a growing concern with what seemed to be a huge gap between youths' experiences, and then the stories portrayed in the news media.

In order to explore this, a child-centred approach has been adopted[2] (Hake, 1999; Kampmann, 1998) and several types of data have been collected, including five years of participant observation, interviews, informal conversations, an open-ended online questionnaire and a media content analysis[3]. The project took its departure in intensive participant observation and engagement with online youth practices (which also included monitoring news media, as we shall return to). In particular observations have been carried out on the site Arto.dk, but has been expanded to include other sites as well (e.g. as participants have moved). To critically examine and extend findings and observations from the ethnographic engagement a large-scale open-ended online questionnaire was designed and launched in 2007. As the data from the survey further highlighted the gap between youth's own experiences and public discourses the main author conducted a media content survey and analysis of Danish print media's coverage of young people's use of social network sites from 2005 to 2006. We use this to ground and demonstrate our claims about the most predominant public discourses present in the period before the online survey was launched.

With basis in these types of data this chapter explores and analyses the experiences young people have online (positive as well as negative). Secondly, we discuss how the young users articulate and account for these experiences and orient and position themselves in relation to the public (adult) discourses. With particular reference to the latter we focus on how young people display 'dual consciousness', and position themselves in relation to what we call 'local experiences' and 'public discourses' when accounting for their use of social network sites. In the analysis we emphasise data from the open-ended questionnaire and we demonstrate how the respondents used the open-ended questions as an opportunity and strategy to be heard. We analyse how they in their own accounts of negative experiences simultaneously tried to distance themselves from the negative experiences, and the public concerns represented by e.g. their parents or the news media. They did so by stressing their positive experiences and by explicitly positioning themselves as responsible young people. Furthermore, we argue that a very pronounced finding in relation to the positive experiences is a very social and supportive 'love' discourse, which stand in stark contrast to the 'harassment and bullying' discourses often emphasised in public news media discourses. Based on the analysis, we argue that young people have a need for their online social practices to be understood and taken seriously; particularly at a point in time where those practices have been surrounded by a high level of concern or moral panic.

We start the chapter by situating the topic of youth and online social networking within a Danish context, and by describing (some) of the major transitions within this area, which have also affected the research project. Subsequently, we describe the methodological and theoretical approach followed in the research project, and how data have been collected, coded and analysed before moving onto the analysis and discussion.

YOUTH AND ONLINE SOCIAL NETWORKING IN A DANISH CONTEXT

Social network sites (SNSs) are web-based spaces, where users via personalized profiles can: link to each other; list each other as friends or contacts; communicate and socialise across physical and temporal boundaries. Social network sites are often described as 'egocentric', in so far as they operate "with the individual at the center of their own community" (boyd & Ellison, 2007). In this way (some) SNSs distinguish themselves from earlier online communities, as they are friendship-driven rather than interest-driven[4]. Furthermore, with basis in the long-term observations we believe it can be useful to include a focus on the kind of activities that take place on SNSs, and we suggest four overarching categories that cover different features, aspects or activities on social network sites:

- **The personal and branding related features**

(such as a profile, a picture gallery, a blog, a notice board etc.).

- **The social and contact enabling features**

(such as a guest book/wall, a debate forum, clubs/groups etc.).

- **Entertainment**

(such as games, videos, jokes etc.).

- **Support and practical information**

(such as rules, safety guidelines and support sections)

(Larsen, 2005, 2007b, 2009)

The boundaries between the four aspects or types of activities are malleable and dynamic, and with direct reference to observational data from Arto.dk it has been interesting to see how features that were originally intended (and perceived) as personal or branding related (e.g. a profile text or a user name) were also used for friendship maintenance, or how social or contact enabling features, such as the guest book, were used by users to profile or draw attention to themselves. For instance, it became a common practice among users to delete the less interesting messages from their guest book, and leave only the ones that contain flattering remarks from their best friends. We return briefly to the categories in our analysis, and merely want to stress that the kinds of social network sites we are focusing on are 'friendship-driven' sites, such as Facebook or Arto, where in particular the first two types of activities are prominent.

As is often the case with new media, teenagers have been the first to take these friendship driven social network sites to their hearts. As several studies have shown, their use of these sites is related to everyday offline life, as they mostly use their profiles to communicate and stay in touch with existing friends or someone they "already know" (e.g. boyd, 2008; Dunkels, 2005; Ellison, Steinfield, & Lampe, 2007; Ito et al., 2010; Lampe, Ellison, & Steinfield, 2006; Larsen, 2005). A point strongly emphasised by the quote from the 14-year-old girl in the introduction to this chapter. These findings also question and to some degree dissolve the historical dichotomy between 'online' and 'offline' in relation to youth and sociability online. Online and offline practices merge and for adolescents the former can be as valuable as the latter. Thus, we must understand social network sites as "embodiments, stabilizations, and concretizations of existing social structure and cultural meanings" (Ito, 2008, p. 402).

Online social networking has existed in a Danish context for quite some time. The first Danish social network site (Arto.dk) emerged around the late 1990ties. Today, recent data show that approximately half of the Danish population use social network sites, and more specifically within the age-group of 16-19 it is 86%. The most popular site is Facebook, where 95% of Danish SNS users between the age of 16 and 74 have a profile (Danmarks Statistik, 2009).[5] Recently, Facebook has also become increasingly popular among the youngest part of the population (including children under the age of 13 which is the minimum age for users of the site). However, a few years ago the Danish site called Arto.dk (now Arto.com) was without comparison Danish teenagers' preferred venue on the Internet with a huge group of core users between the age of 12 and 17. At one point it was estimated that 85% of all Danish teenager had a profile on the site, although these numbers were (at that time) subject to uncertainty (Larsen, 2005).

Arto.dk was the first social network site to launch in Denmark in 1998 and was created by an 18-year old high school student as a spare time activity. In the beginning the site was not really, by definition, a social network site but merely a repository for jokes (also it should be noted that the very term social network site (Danish: social netværksside) did not really surface until around 2006 where MySpace entered public awareness in Denmark). After a while the creator thought it would be fun to introduce a guest book where the users of the site could comment on the jokes, submit their own, and send messages back and forth. Instantly, the guest book function became more popular than the jokes themselves, and from that point Arto developed into a social network site, and grew bigger and bigger just by word of mouth between Danish youngsters. In this way, Arto is a typical first generation social network site and it has a similar history as other national SNSs such as Swedish LunarStorm, Chinese QQ, Korean Cyworld or French Skyblog. These started out as rather specific tools or services, but slowly turned into SNSs thus gaining a huge popularity among teens in their country (boyd & Ellison, 2007).

Because of its huge popularity Arto was initially the main field site for studying Danish teenagers' online social networking practices during the period of 2005 – 2009. The main author has had her own profile on the site, and has conducted an extensive virtual ethnography on the site (Hine, 2000, 1998; Lindlof & Shatzer, 1998). The objective of having the profile was openly communicated to the other users and observations and experiences have been documented in field notes and screen dumps. Today, Arto is threatened by Facebook's major popularity in Denmark, and during 2009 Arto shrunk to a third of its size with a decrease in particularly the group of 15-19 year old users (Lund, 2009). Even though Arto has lost many users to Facebook, the site is still one of the most popular sites among teenagers in Denmark, who primarily use the site to communicate with their existing friends from school, or within their local environment. Secondly, teenagers use the site to find new friends with whom they, however, seldom meet face to face (Larsen, 2005, 2007b). In particular, the 'personal and branding' and 'social and contact enabling' related features are popular on Arto (and Facebook).

Because Arto has been so popular during much of the research period, the site features prominently in terms of data collection, and consequently also in this chapter. However, ethnographic observations have included other sites as well, although not as intense and focused as with Arto. Furthermore, the online survey, which the analysis in this chapter draws most directly on, includes responses and data concerning other sites (even though many answers deal with user experiences from Arto). In any case, a specific focus on Arto is necessary to include in this chapter when focusing on public discourses, since the site has been heavily (almost exclusively) represented in the Danish news media in stories about sexual predators, paedophiles, online bullying or harassment (resembling very much the MySpace moral panic in the United States, as reported in (Marwick, 2008)). We return to this when discussing the media content survey in more detail where we demonstrate how the printed press during 2005 and 2006 has painted a predominantly negative and one-sided picture of young Danes' activities on Arto. First, however, we present the theoretical and methodological concept underpinning this research project, as to explain the more specific framework shaping the ethnographic research, but also the online survey and media content analysis.

THEORETICAL FRAMEWORKS: MEDIATED DISCOURSE ANALYSIS, NEXUS ANALYSIS AND GEOGRAPHIES OF DISCOURSE

The research project draws on different frameworks within discourse studies and ethnographic research, all initially developed by Ron Scollon: Mediated Discourse Analysis (Norris & Jones, 2005; R. Scollon, 2001a, 2001b), Nexus Analysis (R. Scollon & S. W. Scollon, 2004, 2007) and Geographies of Discourse (R. Scollon, 2007). These frameworks distinguish themselves from other discourse studies (like Critical Discourse Analysis) by focusing on social actions, rather than focusing mainly on written text or language. As a consequence, the unit of analysis is moved towards a focus on the crucial social actions carried out by the central social actors within the field of study:

By not privileging discourse or social action but, rather, seeing discourse as one of many available tools with which people take action, either along with discourse or separate from it, MDA strives to preserve the complexity of the social situation. It provides a way of understanding how all of the objects and all of the language and all of the actions taken with these various meditational means intersect at a nexus of multiple social practices and the trajectories of multiple histories and storylines that reproduce social identities and social groups.

(Jones & Norris, 2005, p. 4)

Nexus Analysis (Scollon and Scollon, 2004), which is closely connected to both Mediated Discourse Analysis (MDA) (R. Scollon, 2001a, 2001b) and ethnography, serves as the underlying theoretical and methodological framework in the research project. In a nexus analysis one analyses a Nexus of Practice which is defined as "a recognizable grouping of a set of mediated actions. [...] ... the concept of the nexus of practice simultaneously signifies a genre of activity and the group of people who engage in that activity." (R. Scollon, 2001a, p. 150). A nexus of practice is not necessarily a 'place', but every linkage of a set of repeatable actions, which are recognized by a social group, can be understood and analysed as a nexus of practice. In this sense a social network site can be regarded as a nexus of practice consisting of multiple overlapping, repeated mediated actions. The idea of a nexus of practice can be seen as a scale-free, fractal-like analytic concept where a nexus of practice might include multiple smaller nexuses of practice e.g. one might want to focus more exclusively on a particular group within a SNS, the use of a specific game on a SNS, or the use of a 'wall-feature' among the users. In this project the concept is used as an entrance into identifying broader patterns of practice (mediated actions).

The purpose of Nexus Analysis (and Mediated Discourse Analysis for that matter) is to map out the most central – but not nessecearily discursive – practices within any given nexus of practice. This is achieved from an understanding of social actions as consisting of three elements: 1) the historical bodies (or habitus) of the participants in that action (Bourdieu, 1977, 1998; Nishida, 1958), 2) the interaction order which they mutually produce among themselves and (Goffman, 1983) 3) the discourses in place which enable that action or are used by the participants as mediational means in their action (R. Scollon & S. W. Scollon, 2004, p. 153). In a nexus analysis concrete social actions are always analysed within a broader sociocultural frame, which is also why Scollon and Scollon recommend that the researcher conducts a media content survey as part of the investigation.

Methodologically the process of carrying out a Nexus Analysis consists of three phases: engaging, navigating and changing the nexus of practice. In the first phase one is concerned with collecting a diverse set of data material, obtaining a 'zone of identification' and having one's research activities merge with one's participation activities:

"In order to do a nexus analysis you must establish a zone of identification with a nexus of practice. That is, you must find a nexus in which you have or can take a place as an accepted legitimate participant. Within this zone of identification you can begin to analyse the social practices of the nexus not in a distant or objective fashion but in order to change the nexus of practice."

(R. Scollon & S. W. Scollon, 2004, p. 11)

This first phase of the nexus analysis is about finding the central social actors, observing the interaction order and establishing the most prominent cycles of discourse within the nexus of practice one is studying (Scollon and Scollon 2004). For example the topics presented in this chapter are some of the predominant cycles of discourse that have been identified through the engagement with nexus(es) of practice (Arto in particular, but also other sites), and through observing the central social actions on these sites. In the second phase of a nexus analysis one navigates and maps the data material applying for instance discourse analysis (e.g. MDA), motive analysis and content analysis. In the third phase the researcher looks at how her analysis has changed the field of study. The latter is, however, beyond the scope of this chapter.

In the chapter we also make reference to other concepts developed by Scollon. Our argument is that the moral panics and the news stories about Arto and other social network sites can be considered "global discourses" (public discourses), whereas the everyday life on these sites are "local experiences". Here we are inspired by the notion of 'Geographies of Discourse' (R. Scollon, 2007)[6], where, the term 'discourse' is divided into "the local" and "the global". "Local discourses" are seen as our daily lives and moment-by-moment real-time activities or practices. "Global discourses" on the other hand represent mediated or public discourses which give us news about actions, activities or things which have happened, or which we are worried about, but are distant from our daily lives (R. Scollon, 2007).

The analytic goal of the concepts is to analyse how concrete local discourses are related to global discourses. One example that Scollon uses is hearing an explosion from your office window and then reading about it in the newspaper the next day (R. Scollon, 2007). Equally, we can think about how young people's actions on social network sites are inscribed in more overarching global discourses represented by the news media. While local experiences and public discourses might live largely separate lives and only rarely flow concretely together they do sometimes coincide. As observed in the course of this research project (and by Herring (2008)), young people might discuss news stories about a SNS in the forums of a SNS or other online venue. However, an even more concrete example is from a national Danish newspaper article where a 14-year old Arto user with the profile name "Wildcat", was portrayed. The article was entitled "Young girls strip naked" and it reads:

"She calls herself Wildcat' and she looks like one. Pouting mouth and tight-fitting sweater that barely hides her big breasts. 20 years old? 25 years old? One of the innumerable porn sites on the internet? No, Wildcat' is 14 years old, she is Danish – and she entices on Arto, the most popular social network site for thousands of young people. More than half a million have a 'profile' on Arto."

(Avisen.dk, 2007, my translation)

The online version of the article on the newspaper's website was illustrated with screen-shots taken from the girl's profile on Arto. Here, we see an example of how local actions become part of a global discourse. The profile and the pictures that this girl had uploaded in a local context for her friends to see were suddenly (mis)used as part of a news story on how young girls strip naked and entice older men online. When reading the article the main author immediately looked up Wildcat's profile on Arto, and she looked like a normal 14 year old girl communicating with her friends. However, in the news story Wildcat' became a representative of a whole group of young girls apparently behaving badly online. This was a relatively isolated (but quite horrifying) example of a journalist's/news paper's complete disregard of copyright to pictures posted online, as wells as any concerns on the girl's privacy. However, it is not the only time that specific user names have been published in the Danish news media in relation to stories about young people's online behaviour (and since profiles on Arto are public it was quite easy locating the individual mentioned). Interestingly in this case, a user named Wildcat' commented on the article on the newspaper's website stating that she did not behave as it was portrayed in the article. Also, friends of the girl with the profile name Wildcat' commented on the article backing up the fact that she did not strip naked or tried to entice older men. It is obviously difficult to verify whether these persons were in fact Wildcat' and her friends, but in any case it is quite interesting that the localised voices of (seemingly) young people from Arto seized the opportunity to interfere with the public discourse.

The aim of referring to Geographies of Discourse is to capture how, in cases like this, a direct link between local experiences and public discourses is established, when it comes to young people's use of social network sites. In this chapter, the "local discourses" cover the many different and concrete experiences of Danish teenagers on social network sites. The "global discourses or public discourses" are the ongoing public debates and news stories (claiming to be) portraying young people's general use of social network sites. By using these concepts we can, however, demonstrate some of the routes and links between "local experiences" and global, public discourses.

A MEDIA CONTENT SURVEY OF DANISH PRINT MEDIA'S COVERAGE OF YOUTH AND ONLINE SOCIAL NETWORKING

While studying online social networking practices during a period of time where it became increasingly popular and attracted particularly young people, the main author also conducted a media content survey, as to better ground and highlight how the news media dealt with this topic. As part of the participant observation news stories concerning SNSs have continuously been collected and stored. As pointed out by several authors, gathering knowledge on how the news media treat your topic can be rewarding for any research project, and add new or forgotten thoughts to your existing knowledge (e.g. Rapley, 2007; R. Scollon & S. W. Scollon, 2004). This is particularly relevant in this case, as youth and new media is a subject that historically gets a lot of attention in the news media (Drotner, 1999). Furthermore, texts produced by the press will often affect other actors (Critcher, 2003, pp. 137-138; Golding & Middleton, 1979, p. 19), and when studying how young people account for their online experiences this perspective seems inevitable. As Scollon writes:

"...whatever the importance an issue might have on a broad social scale, it remains to be made clear how this issue is being taken up by some identified members of society. [...]... we need not only an analysis of the texts of public discourse (though we do need that), and not only an analysis of the social actions [...], but also an analysis of the indirect and complex linkages..."

(R. Scollon, 2001a, pp. 159-161)

The news stories have later been more thoroughly re-counted, re-collected and analysed (coded and categorised) to verify and expand the immediate understandings and readings of the public discourses. This was done, as it became evident from analysing responses to the online survey that the gap between local and public discourses was an even more important issue than anticipated. The main author decided to limit the detailed media content survey and analysis to the period of 2005-2006 for several reasons. First of all, the online survey was released early 2007, and the media coverage up to this point would therefore be necessary to contextualise, analyse and interpret the responses. Secondly, the period was chosen because it was around this time the news media started reporting on young people's uses of social network sites. Originally, the intention was to include press coverage of different SNSs used by young people in the media content survey and analysis. However, it turned out that Arto was more or less the only site that the Danish news media covered (understandably, as it was the first popular and most used SNS in the country among youth at the time).[7] Therefore, the results from the media content and analysis only concern news stories featuring Arto.

The Danish database Infomedia (Infomedia, 2010) was used to conduct the search for articles as it contains data from all national newspapers. A decision was made to focus on the printed press because Infomedia is more comprehensive in relation to news from printed media, whereas the inclusion of web sources, TV and Radio is partial and fragmented. Furthermore, a list of criteria was designed; for instance the article should be specifically about the use of Arto among youth (for which reason articles where Arto was just mentioned as a foot note were filtered out). The media content survey resulted in a list of 155 articles that covered Arto as part of their main story, and these were spread across nine national newspapers. Based on the articles' main content and angle they were coded in the data coding program Nvivo 8 by using an open coding strategy (Gibbs, 2002). The coding was carried out as an iterative process where the material was re-assessed and re-coded several times in order to ensure the most accurate picture. In the end, the result was 14 different categories or themes cutting across the newspaper articles:

1. Paedophilia or sexual offences against teens (29 coding references)
2. Bullying and harassment among users (27 coding references)
3. Sexual behaviour and sexual expressions by users (24 coding references)
4. "The Chat Consultant Case"[8] (20 coding references)
5. Safety, web ethics and good advice (15 coding references)
6. Threats and violence among users (14 coding references)
7. Arto's economy and ownership (8 coding references)
8. Youth expressing grief online (7 coding references)
9. Calls for parental involvement (7 coding references)
10. Possibilities on Arto (7 coding references)
11. Schools' blocking of Arto (6 coding references)
12. Nazis recruitments' of members (6 coding references)
13. Young people are careful (2 coding references)
14. Time consumption (1 coding reference)

In the coding work it was not assessed whether the articles were positive or negative, as it has been done in similar media content surveys.[9] However, as it can be seen from the themes, most of the articles were negative towards young people's use of Arto. This pattern is characteristic for how the topic of youth and new media is treated in the news during a period with moral panic (Critcher, 2003; Drotner, 1999, 1992; McRobbie & Thornton, 1995; Springhall, 1998).[10]

As Drotner (1999) has demonstrated, basically all new types of media have given rise to media specific moral panics, but "every new panic develops as if it were the first time such issues were debated in public, and yet the debates are strikingly similar" (Drotner, 1992, p. 52). The focus is often on children and young people because their development is the most decisive and vulnerable, but also because childhood is a 'paradise lost' to adults who simply have a hard time relating to the experiences of children and young people (Drotner, 1992). During a time of moral panic certain discourses become dominant (as can be seen in the list above), and the news media often take up single events or cases which are argued to represent the general state of affairs. Furthermore, such stories tend to disregard the fact the media usage of most children and young people is part of everyday practices characterised by diversity, not one-sidedness (Drotner, 1999). Following from this the media coverage of Arto, as presented above is a classic example of a moral panic. In many of the articles isolated incidents of e.g. bullying or sexual harassment were presented, as if they reflected young people's general use of and experiences with the site.

In Denmark, at that time, the media coverage of Arto resulted in an intense public debate and different political[11] initiatives or processes of moral regulation[12] (Critcher, 2008). The focal point was often that online social networking is dangerous, and it was discussed what technical precautions could be taken, or how to legally regulate or prevent bad things from happening. As Marwick (2008) has noted in relation to the MySpace moral panic in the US, "breathless negative coverage of technology frightens parents, prevents teenagers from learning responsible use, and fuels panics, resulting in misguided or unconstitutional legislation." (Marwick, 2008). More or less the same seemed to be exactly the case with Arto in Denmark at that time. The rather one-sided media coverage let some parents, teachers or politicians to demand that the site should be closed down. Many Danish schools (tried to) blocked the site, and a lot of parents banned their children from using the site forcing them to delete their profiles (instead of teaching them responsible online behaviour). As a researcher it was interesting to see how young people reacted to this online. On Arto, the media coverage was heavily debated among the users and many of them opposed the negative coverage. As a 15-year-old boy wrote in the debate section: "Why don't they write about all of us who are happy with using the site?".

THE ONLINE QUESTIONNAIRE

The large-scale online questionnaire was conducted in 2007 in collaboration with The Danish Media Council for Children and Young People[13], and consisted of primarily open-ended questions. Apart from collecting some basic statistical data, the main purpose of the questionnaire was to gain an insight into, and provide an overview of the different experiences that young people between 12 and 18 years old have with social network sites. These types of data are what that Scollon calls "individual members experiences" (2001a). The age group was chosen, as studies have shown that Danish youth start using the internet for online social networking, chatting and communication with their friends from the age of 12 (Eurobarometer, 2007; Rattleff & Tønnesen, 2007). Also, many of the popular social network sites in Denmark, like Arto, have an age limit of 12 years for their users.

The questionnaire contained 18 questions divided into three sections: "About you", "What do you think?" and "What do your parents think?" The first section consisted of a few factual questions and quantitative categories about media habits and use. The second section, "What do you think?" contained a number of open-ended questions to which the respondents could answer qualitatively (essay style). These questions focused in particular on the respondents' experiences and thoughts on their use of social network sites. In many of the questions the respondents were asked to provide examples from their profile, e.g. messages from their guest book or comments from their picture gallery, and afterwards they were asked to reflect upon them. The questions were formulated on the basis of the previous studies and ethnography on Arto. Respondents were therefore asked to reflect upon what had proven to be the most central social actions and typical uses of social network sites among Danish teenagers (in particular in relation to the personal and branding related features and social and contact enabling features). The third section consisted in a few predominantly quantitative questions about their parents' views on, and knowledge about, their online habits. On the final page of the questionnaire there was an option where the respondents could write any further comments they might have.

2400 youngsters between the age of 12 and 18 years old answered the questionnaire with personal views on and examples from their online experiences.[14] An equal amount of boys and girls participated in the survey, which was launched at

'Safer Internet day 2007' (Ins@fe, 2007). It was available on a number of popular Danish social network sites for children and young people, but also websites for libraries and public institutions. In this way, we (deliberately) did not design the survey to be representative (e.g. by also distributing paper copies). The respondents were not chosen to participate, but did so on their own initiative by following the links that were placed on the different SNSs or websites. As a consequence, the participants were not selected on the basis of demographic or socio-economic variables, as one can do when conducting a more classic survey. Therefore, this questionnaire may not be entirely representative for all Danish young people between the age of 12 and 18. Rather, it represents users of SNSs and other internet sites in the age group (the internet penetration and use in Denmark is however quite widespread). We decided to focus on the desirability of having many relevant respondents even if it may not match entirely the general population of teens in Denmark (Schrøder, 2000, p. 83).

In the survey 88,3% of the respondents stated that they use social network sites on the internet.[15] The term 'social network site' was not explained or defined in detail in the questionnaire so the respondents answered the question from their own perspective and interpretation of what constitutes a SNS (we did, however use three different words in Danish to explain the term: 'chat portals', 'communities' and 'online meeting place'). Among the sites they stated to use most often were first and foremost Arto and subsequently other sites such as NationX, MySpace, Jubii and Habbo. In addition some of them considered MSN Messenger to be a social network site as well. When asked about why they use the sites the majority of the respondents stated that they did so in order to communicate and keep in contact with their friends. Generally, their answers point to the fact that they use the digital space as a continuation of their everyday physical settings. This is in line with what other studies on the subject have also shown; namely that online social networking is most often used by youth to sustain existing, offline relationships, as also mentioned in the beginning of the chapter (e.g. boyd, 2008; Dunkels, 2005; Ellison et al., 2007; Lampe et al., 2006; Larsen, 2005). The respondents generally stated that they spend a lot of time on their favourite social network sites. More than half of them reported spending more than 2 hours a day on social network sites:

In the category "Other" some wrote that they use social network sites "All the time" or stated that they spent up to four, five or even ten hours a day on social network sites.

The following analysis will take its point of departure in the open-ended questions from the second section, "What do you think?", where respondents were asked to write about and give examples of their online experiences. These answers have been coded via Nvivo 8 in a similar manner as described in the media content survey. Because of the huge body of material, only answers from the question categories that seem most relevant to examine more closely have (so far) been coded (relevance has been established on basis of the participant observation and nexus analysis). The coding of the answers has been conducted through several iterations, as to identify categories and patterns related to the most relevant cycles of discourse. In the first part of the analysis we highlight some of the most significant results related to what the respondents reported as posi-

Table 1. (Respondents' answers to the question, "How often do you use communities or social network sites?")

Seldom	2,0%
Once a week	2,7%
2-3 times a week	9,7%
Less than an hour a day	16,8%
1-2 hours a day	29,5%
More than two hours a day	30,9%
Other	8,5%

tive and negative experiences on social network sites. When relevant, we supplement with some quantitative results. In addition, we include comments from the "further comments" box of the questionnaire in the analysis because many of the respondents used this option to express their general views on internet safety, press coverage or adults' opinions towards their use of social network sites. Also, they commented on the questionnaire itself, and how they thought their answers should be analysed or understood. These comments are particularly interesting to include when analysing and reflecting on the relations and gaps between local experiences and public discourses.

ANALYSIS: GOOD AND BAD EXPERIENCES

In the questionnaire we asked the respondents about both positive and negative experiences they have had online. We did not define what "a good" or "a bad" experience might be, as we wanted the respondents' view of it. Based on the coded data there seemed to be a strong consensus among the respondents about what constitutes a good experience. When asked to write about such an experience, their stories particularly related to maintaining contact with existing friends or finding new friends or boyfriends/girlfriends. The following are examples of some of the most common answers to the question "Tell us about a good experience you have had online":

1. *"I got a net-friend from Copenhagen and we write together, but we have no plans for meeting up." (Girl, 12 years old)*
2. *"I have had many, thanks to the internet I have gotten a number of new friends. I have also gotten a better relationship with the friends I had before." (Boy, 15 years old)*
3. *"When I was about to start boarding school, I wrote in the microphone at NX and asked if anyone else were about to start at this school, and in this way I started to write with some of the people I was supposed to go to boarding school with. I have also written with many people from my local area, who I have then met at for examples parties and then I have started to know them better .." (Girl, 15 years old)*
4. *"It was when I met my present ex-girlfriend. We both played handball, then I saw her at a tournament in Italy, but I did not dare to go and talk to her, so I wrote to her on Arto when we returned back home." (Boy, 14 years old)*
5. *"It is always a good experience when I write with my friends. I cannot find exactly one, as there are many. I always get happy when I write with my friends." (Girl, 13 years old)*
6. *"The day I wrote 'Hi' to my now beloved Liv:-) (Author: Liv = name of girl) That turned out to be the best thing I have ever done in my life.:)" (Boy, 17 years old)*
7. *"A lot of the people we surround ourselves with in everyday life, we don't really get to talk to. So if you find them at Arto, you can leave a message saying "Hi sweetie, what's up?" And in this way you can strengthen the contact with other people." (Girl, 16 years old)*

(Examples of answers to the question "Tell us about a good experience you have had online.")

As can be seen from the extract above, the positive experiences revolve around local experiences and moment-by-moment activities from everyday life. Some of the respondents found it difficult to name only *one* good experience, as they generally experience many good things online, as one girl points out (example 5). Overall, when looking at the different statements from the respondents about what they consider to be positive online experiences, it becomes clear that their experiences can be divided into four overarching categories:

- Staying in touch with friends
- Meeting old friends or acquaintance
- Finding new friends or boyfriends/ girlfriends
- Receiving sweet messages and comments from friends

The fact that it is very important for young people to receive sweet messages and comments from their friends also became clear in the answers to one of the other questions in the questionnaire. Here, we asked to respondents to give an example of a message they had received in their guestbook/ on their wall which they were particularly happy with. The message "I love you" was by far the most used example. This supports the observation that there is a distinct love discourse between teenagers; especially on Arto (Larsen, 2005, 2007b, 2007a). Also, messages stating that the user is missed, or what the user means to someone else, are among the most common messages which the respondents are happy to receive in their profiles:

8. *"I love you, and I miss you so much, you are my best friend" (From 16 year old girl's guest book)*
9. *"HI SWEETY I LOVE YOU CAN'T LIVE WITHOUT YOU" (From 12 year old girl's guestbook)*
10. *"You are simply one of the happiest and nicest people I know. (From 17 year old boy's guest book)*
11. *"I Love you my best friend!" (From 12 year old boy's guest book)*
12. *"I Love my Brow, I Miss My Brow, I Respect My Brow - I Support My Brow 112% - I Will Always Be There, For My Brow" (From 15 year old boy's guest book))*

(Examples of answers to the question "Give an example of a message you have received that you were pleased with".)

Even though writing 'I love you' messages seems more pronounced among girls and younger users (12-15 years old), an interesting development witnessed during the years of ethnographic observations is that the boys also write messages with the wording "I love you" to each other. In the beginning of the research period the boys seemed more reluctant to write this kind of affectionate messages and would even make fun of the girls who did so (Larsen, 2005), but today this seems to be a more common practice also between teenage boys. Often, however, the 'I love you messages' are accompanied by masculine (street) slang expressions, such as 'Brow', 'Homie' or simultaneous use of irony to downgrade the emotional content. In the questionnaire, when explaining why those specific messages are nice to receive the respondents (both male and female) emphasise that it means a lot to them to receive positive messages, and that it makes them feel valued by their friends.

A high number of the respondents agreed that messages as those mentioned above are among the best messages they have received. As a follow up question we asked the respondents if the message in question came from 1) someone unknown, 2) their best friend, 3) their boyfriend or girlfriend or 4) someone else who the user knows. In 47,8% of the cases the message was sent from their best friend, in 23,8% it was sent from a girl- or boyfriend or, as in 23,1% of the cases, by someone else familiar to the user. Only in 5,2% of the cases the nice message was from someone unknown to the user. In this way, the positive messages are primarily related to the users' IRL-friends (In Real Life), their physical environment and everyday setting (whereas negative experiences, as we shall see, are mostly attributed to strangers and unknowns).

When using different social networking sites young people do have unpleasant experiences as well. As Livingstone (2008) has pointed out there is a fine line between opportunities and risks on social network sites which young people must balance. In the survey, 31,5% of the respondents stated that they had had unpleasant experiences online, while 57,1% expressed that they had never experienced anything unpleasant. 11,4% answered "Do not know" to this question. Among the 31,5%, who stated that they had had unpleasant experiences, it varied what they considered to be unpleasant. The following are some of the most common examples given by respondents who had had unpleasant experiences:

13. *"Harassment, but I usually report it and/ or block the user so that he will not have to possibility to write to me again." (Boy, 12 years old)*
14. *"Received threats from my x-boyfriend ., Showed it all to my mom and my big brother." (Girl, 17 years old)*
15. *"Hmm.. That was when a guy who was 18 wanted to meet up with me. I kept on saying no, as I did not know the person." (Girl, 12 years old)*
16. *"A girlfriend had gotten hold of my code to Arto.dk, and then she changed my profile into a demeaning site (opposite to what I had written myself), and broke up with me. We had a hard time getting the profile blocked." (Boy, 14 years old)*
17. *"I have been bullied and smeared. I didn't do that much, I just ignored it." (Girl, 16 years old)*
18. *"There are many people on www.arto.dk who ask if you want to "cam" dirty. This bothers me a lot, as I am online in order to meet people I can have a nice time with, by writing together about interest etc." (Girl, 17 years old)*
19. *"Threats are always unpleasant, but mostly I have dealt with it calmly and as a joke." (Boy, 18 years old)*
20. *"Some guys who write and ask you private questions, about sex and stuff like that. I block them and report them." (Girl, 15 years old)*

(Examples of answers to the question"If yes, what was it about (Author: the bad experience) and what did you do about it?")

When looking at the examples provided by the respondents there did not seem to be a distinct shared understanding of what is considered to be unpleasant, to the same degree as was the case with the positive experiences. Being threatened or harassed are among some of the most common examples, but some of the respondents considered viruses or spam in their guest books to be unpleasant. In general, many of the examples were described as isolated events, rather than something the respondents experience often. However, one distinct pattern in their responses was that the sources of unpleasant experiences were strangers and peripheral persons, rather than being associated with their regular network. Furthermore, the positive experiences were described as something they related to on a personal level (being about them and as important parts of their life). In contrast, they seem to distance themselves from, the (occasional) negative experiences, which were described and perceived as exceptions and located at a more general, impersonal and distant level. Thus, the negative experiences are not necessarily perceived as "local discourses" by the respondents, in the sense that they are not described as being part of their moment-by-moment activities and daily lives. Bad experiences are attributed to the fact that there are "idiots *out there*" (our emphasis), that comments are sent by *'some* guys' and that it is impossible to avoid coming into contact with 'them' from time to time. In the few cases where the negative experiences are related to people they know IRL (in real life) it is often connected to prior offline conflicts (e.g. example 14 and 16). When explaining about their negative experiences, whether caused by strangers or someone they know, many of the respondents highlight that they know how to handle the (occasional) bad experiences, e.g. by blocking or reporting the users who harass them (example 13), by showing the messages to grownups (example 14) or by simply ignoring it (example 17).

The respondents were also asked to provide examples of specific messages they had received which they did not like. Even though their examples are fairly different, their answers mostly concern sexual comments or invitations from strangers such as "Are you dirty?" or "Shall we chat/cam dirty?". Among other unpleasant messages the respondents mention threats and degrading comments about the profile owner e.g. being ugly, disgusting, a whore, gay etc. The following are messages or comments which the respondents have copied from their profiles:

21. *"can I suck your dick" (From 12 year old boy's guest book)*
22. *"Hey beautiful! What are you doing? I'm just sitting here masturbating." (From 15 year old girl's guest book)*
23. *"fuck you I'll knock the daylights out of you if you look at my profile!" (From 14 year old boy's guest book)*
24. *"Can I have your phone number"* (*From 17 year old girl's guest book*)
25. *"Hey good-looking do you want to chat dirty?" (From 15 year old girl's guest book)*

(Examples of answers to the question "Give an example of a message you have received that you did not like".)

As it was also the case with reports of their bad experiences, the examples of negative messages display a higher degree of variation than do the positive messages, and there is not the same amount of agreement among the respondents as to what constitutes an unpleasant message. For example some state that they do not like messages from random users saying "Just popped by" or "What are you doing?" or so-called 'spam-messages', where a user has sent the same message to a lot of people at the same time.

In 70,1% of the cases the unpleasant message came from an unknown person. Only in 2,8% of the cases the message was written by a best friend and in 1,3% of the cases by a girl- or boyfriend. In 25,9% by someone else familiar to the user. In this way, the messages which the users found most unpleasant were more often sent by someone they did not know. This is an interesting point, particularly in relation to previous surveys that have shown that online bullying among schoolchildren is a big problem on social network sites (e.g. Rasmussen & Hansen, 2005). However, the answers from this questionnaire seem to suggest that bullying from class- or schoolmates is not considered a big problem among the users of social network sites. Rather, negative messages are more often connected to URL-people (people they do not know In Real Life). This could be explained by the fact that young people do not necessarily see bullying online as separated from bullying offline, but rather as part of a local conflict in school or other physical environments (Rasmussen & Hansen, 2005).

Generally, the results from the questionnaire show that there is a marked difference in where, and from whom, good and bad experiences originate. The good experiences are more often associated with best friends, boyfriends or girlfriends, whereas bad experiences often arise from being contacted by strangers. Furthermore, in cases of harassment and bullying (whether from strangers or friends) it is important to remember that (various) social network sites are part of a complex social life, where online and offline activities and practices merge. Therefore, the experiences of young people on social network sites are connected to local activities, friendships, conflicts and problems – just like teenage life in general – and they should not be analysed in isolation from that e.g. as dangers and risks only looming on the internet (as terms such as cyber-bullying could suggest).

FROM LOCAL EXPERIENCES TO PUBLIC DISCOURSES

When filling out the questionnaire many of the respondents commented on their own responses and, in doing so, answered a number of questions we did not ask them. Throughout the questionnaire some respondents directly addressed us (the main author and the Danish Media Council for Children and Young People) as senders of the questionnaire and as adults. They commented on our questions or on their own answers, as to make sure that we understood their statements in the right way. For instance this 15 year old boy addressed us when answering the question dealing with unpleasant online experiences:

26. *"(I do not understand unpleasant experiences the same way you do. We sit in front a screen – there is no need to turn it off, if someone sits on the other side (Author: of the screen) acting up or trying to threaten you?) – So no, I have not (Author: had unpleasant experiences)."*

Equally, this 16 year old girl felt the need to "set the record straight" when answering the question "Give an example of a message you have received that you did not like". She wrote:

27. *"Well, the fact is that if you are not old enough to know that there are unpleasant people around, then you are not even old enough to use cyberspace. Now I want to say something to you that I hope you read – Don't you think it is better to receive an unpleasant message OVER THE INTERNET rather than meeting an unpleasant man, when you are waiting for the bus?"*

Similar responses can be found throughout the survey. Here, we see how the respondents seem to have strong ideas on how grownups will interpret their statements; and some feel obliged to point out, what they seem to consider a truism; namely that being online is considerably less dangerous than real life (and they are protected behind the screens). This probably has to do with their historical body (R. Scollon & S. W. Scollon, 2004) and how they have previously experienced adults having (over) interpreted stories they have heard about young people's online encounters. When reading their different accounts, it becomes clear that many of the respondents want to explain how social network sites are just part of their everyday life in which bad experiences are inevitable. From their point of view occasional unpleasant experiences or messages are part of being online, and many of the respondents emphasise that they do not care about the unpleasant messages, that they delete them immediately or simply block the sender. Their answers indicate that their experiences on social network sites are mostly positive, and that they are sick of hearing otherwise.

The comments from the respondents also indicate that they are using the questionnaire as a strategy and an opportunity to be heard. A total of 336 respondents used the "further comments" option to write comments. A few were frivolous remarks such as "Big tits" or "This sucks", but most of them were serious comments dealing with the subject of the questionnaire or the questionnaire itself. For example, some were trying to guide us in our analysis, asking us not to focus on "the bad stuff" and in doing so they also commented (implicit or explicit) on the public discourses:

28. *"And please do not just take all the bad examples and make a big deal out of them…I'll bet you there are more positive than negative…" (Boy, 15 years old)*
29. *"Do not punish arto.dk, it is the girls' own fault that they meet up with "old men", I would never do that." (Girl, 12 years old)*
30. *"[…] Besides that, I think that the media often have a tendency to focus way too much on the negative stuff about the internet. It seems as if the internet and especially arto.dk just HAVE to be turned into something dangerous. They totally forget to count in all the positive things, and compare it with all the bad stuff that normally also occur outside the internet, and which have always occurred. In my opinion they have a totally*

wrong approach, and they cover it waaay too subjective."." (Boy, 16 years old)*

31. *"People have too many prejudices concerning the internet: How hard can it be to just turn off the monitor?" (Girl, 14 years old)*
32. *"Even though the advice you give young people and especially young people's parents are somewhat sensible, a lot of it also scares parents and children more than it guides them. The internet is not just a playground with evil children. If you use your head and KNOW something about how computers work, it is very difficult to get bad experiences. [...]." (Boy, 17 years old)*

(Examples of answers to the question "If you have any further comments, you can write them here".)

Many of the comments seem to be directly referring to the public, adult discourses mentioned in the media content survey and analysis. For one thing the many news stories on Arto, but also grownups' general conceptions of 'the internet' and even the advice provided to parents and young people by various institutions and persons (e.g. from the media council or the main author)). Some of the respondents, however also expressed their appreciation with the questionnaire, and it is interesting to see how they seemed really grateful that someone was (finally) listening to *them*:

33. *"Nice that someone is asking us what's going on:)" (Girl, 16 years old)*
34. *"Thank you, it was fun to "participate":)" (Boy, 13 years old)*
35. *"I think it is nice that questionnaires concerning use of internet are being sent out, because I think that adults/parents often 'label' young people as irresponsible etc...." (Girl, 17 years old)*
36. *"Thank you... it is nice that there is someone who bothers to listen to us..(:" (Girl, 14 years old)*

(Examples of answers to the question:"If you have any further comments, you can write them here".)

Furthermore, some of the respondents appreciated the open-ended questions where they were able to report on their own experiences and views on online behaviour:

37. *"I just want to say that I think this survey has been good compared to many others, as you are given the opportunity to formulate your own answers and elaborate on them, instead of just selecting between predefined categories. [...]" (Boy, 16 years old)*
38. *"It was some good questions that made me remember things I thought I had forgotten all about:) Thank you very much:)" (Boy, 13 years old)*

(Examples of answers to the question:"If you have any further comments, you can write them here".)

This kind of appreciation is very similar to how the users of Arto reacted to the researcher during the online ethnography, where they seemed genuinely interested in the research and the results.

As mentioned, the survey confirms what previous studies have also shown; that online social networking is primarily used by youth to sustain existing, offline relationships. What is interesting is that this point is stressed by some of the respondents in order to distance themselves from the public discourses or cultural hegemony saying that young people are irresponsible, and that social network sites are (solely) dangerous:

39. *"Understand that the Internet is the future. The Internet is full of possibilities for young people - I did not learn to spell and read in school, I learned that when I started using the Internet. I had a lot of trouble in fourth grade and then I started using a chat portal named Netstationen (www.n.dk), I got some friends and in this way I learnt to write and*

spell, and today I get a least an A in Spelling." *(Boy, 16 years old)*

40. "*To all parents .. There is so much bad stuff about Arto in the media, but they forget to tell all the good stuff. Arto has many good qualities. For example you can learn about HTML codes, find games, talk to your friends and sometimes write with famous people in the live chat." (Girl, 14 years old)* (Examples of answers to the question:"If you have any further comments, you can write them here".)

The questionnaire did not include any open-ended questions on what their parents or other adults might think of their use of social network sites, or how they perceive the media coverage. However, as their comments show they clearly felt the need to express their views on this. The respondents seem to be aware of the public debates and the moral panics concerning their use of social network sites. Their statements reflect that they are sick of hearing about this, and being portrayed in the media without being asked their opinion.

This paints a picture of a group of young people who want to be taken seriously. Even though they are not adults, they do not want to be perceived as small children who act "*without thinking*", and do not know how to behave properly online. Rather, they want to be perceived as experienced internet users and responsible young people. In the questionnaire we see how the young people are struggling to position themselves and appear in this way, addressing both us (the main author and The Media Council), the news media and their parents. Some even underline the learning potential of social network sites, in order to get their message across (example 39 and 40). Throughout the survey and particularly in these final comments we see a group of young people positioning themselves as 'responsible young people' who want to be taken seriously.

CONCLUDING REMARKS

This chapter has demonstrated how young people position themselves as users of social network sites, both in relation to very concrete and local experiences from their everyday life and in relation to public discourses. It is quite evident that (particularly around the time of the launch of the questionnaire) there is a pronounced gap between youth's experiences with social networking sites and then how these sites were presented in the media (and thereby disseminated to the wider public). On the one hand, the public discourses and media stories emphasise, rather one-sidedly, the risks and dangers (bullying, harassment, sexual predators, risky sexual content and actions etc.). On the other hand, what the young people experience on a day-to-day basis is an environment where they can communicate with their real-life friends, find new friends and receive loads of social and emotional affirmation. Although some of this (e.g. finding new friends) might seem risky to adults, this can be among the opportunities children and young people seek online, as pointed out by Livingstone and Helsper (2007).

In the survey the respondents seemed to be using the open-ended questions as an opportunity to be heard and to position themselves as 'responsible young people' and 'skilled users of social network sites'. In particular, by distancing themselves from the public and "grown up" discourses represented by e.g. their parents, teachers or the media, thereby displaying and explicitly discussing their own 'dual awareness'. When accounting for their negative experiences, we can also see how risks and dangers, such as threats or sexual comments are treated as 'distant' or manageable problems. In a sense they appear as 'global discourses' not directly related to their own, personal online life, but are perceived as remote problems or 'something that goes on in cyberspace'. Some acknowledge that risky or sexually explicit behaviour does occur, but attribute it to other users, and distance themselves from this (example 29). Others pragmatically state that one will eventually come across creeps and idiots and that this is safer than meeting them in real life (and if you are not ready to handle this you should not be online (example 27)). Even when they come into contact with bullies, sexual comments etc. the source or origin of bad experiences are depicted as coming from abstract, generalised 'others', strangers or outsiders, and considered more as annoyances than actual dangers (which they state they are capable of managing). This account and localisation of the source of bad experiences seem also to be supported by their report of from whom unpleasant messages originate, as these most often were received from either strangers or more peripheral persons in their network (not recognised as friend). In stark contrast to the negative accounts the positive, and local, experiences are perceived as something very tightly connected to their real, everyday life, and as something, which is meaningful and important to them as individuals.

The majority of the experiences young people have on social network sites do not add up with the picture painted in the public debate and the news media. Often, the glimpses of 'youth online' that we as adults get from reading, viewing or listening to news stories do not reflect the general experiences that young people have online. Rather, young people view such dangers and risks as remote, abstract problems far removed from their day-to-day experiences of being online. Therefore, as a society (politicians, parents, researchers, media etc.) we should be careful about focusing one-sidedly on the dangers and risks of online social networking. We do not mean to argue that these should be ignored, but it seems quite evident from the questionnaire that an exaggerated, hyped (sensationalist) portrayal of the looming dangers of 'cyberspace' alienate and repulse young people, rather than helping them to manage the risks or unpleasantries they are at times confronted with. This one-sided focus seems to leave young people somewhat alienated, as not having a voice and not being taken seriously as members of society. Some of the potential dangers of this could be that they retract and distance themselves from parental guidance or adult advice or play down unpleasant experiences they may actually struggle with. For example, studies have shown that young people are reluctant to seek help from their parents or other adults when they encounter something unpleasant online (Eurobarometer, 2007). Likewise, they may overstate or be too confident with their own capacity to deal with unpleasant or even potentially dangerous situations and need adult guidance and help. However, this also seem to require that adults acknowledge and recognise the existing experiences of children and young people and meet them at eye-level.

REFERENCES

Avisen.dk. (2007, January 15). Unge piger smider tøjet. *Nyhedsavisen*. Retrieved from http:/www.avisen.dk

Baym, N. K. (2007). The new shape of online community: The example of Swedish independent music fandom. *First Monday, 12*(8). Retrieved from URL: http://firstmonday.org/issues/issue12_8/baym/index.html

Bourdieu, P. (1977). *Outline of a Theory of Practice*. Cambridge, UK: Cambridge University Press.

Bourdieu, P. (1998). *Pracrical reason: On the theory of action*. Stanford, CA: Stanford University Press.

boyd, D., & Ellison, N. B. (2007). Social Network Sites: Definition, History, and Scholarship. *Journal of Computer-Mediated Communication, 13*(1), 210-230.

boyd, D. (2008). Why Youth (Heart) Social Network Sites: The Role of Networked Publics in Teenage Social Life. In *Youth, Identity, and Digital Media*, The John D. and Catherine T. MacArthur Foundation Series on Digital Media and Learning (pp. 119-142). Cambridge, MA: The MIT Press.

Cohen, S. (1972). *Folk Devils and Moral Panics: the Creation of the Mods and Rockers*. London: MacGibbon and Kee.

Critcher, C. (2003). *Moral panics and the media. Issues in cultural and media studies*. Buckingham, UK: Open University Press.

Critcher, C. (2008). Making Waves: Historical Aspects of Public Debates about Children and Mass Media. In *The International Handbook of Children, Media and Culture* (pp. 91-104). Los Angeles, CA: SAGE.

Danmarks Statistik. (2009). *Befolkningens brug af internet 2009*. Danmarks Statistik. Retrieved from http://www.dst.dk/publikation.aspx?cid=14039

Drotner, K. (1992). Modernity and Media Pancis. In Skovmand, M., & Schrøder, K. C. (Eds.), *Media Cultures. Reappraising Transnational Media* (pp. 42–62). London: Routledge.

Drotner, K. (1999). *Unge, medier og modernitet - pejlinger i et foranderligt landskab*. Valby, Denmark: Borgen.

Dunkels, E. (2005). Nätkulturer - Vad gör barn och unga på Internet? [Net cultures - What do Children and Young People do on the Internet?]. *Tidskrift för lärarutbildning och forskning, 1-2*, 41-49.

Ellison, N. B., Steinfield, C., & Lampe, C. (2007). The benefits of Facebook "friends:" Social capital and college students' use of online social network sites. *Journal of Computer-Mediated Communication, 12*(4), article 1.

Eurobarometer. (2007). *Safer Internet for Children, Qualitative study in 29 European countries, National Analysis: Denmark*. Hellerup: European Commission, Directorate-General Information Society and Media. Retrieved from http://andk.medieraadet.dk/upload/denmark_report.pdf

FDIM. (2009). Charmærket. *Foreningen af Danske Interaktive Medier*. Retrieved January 15, 2010, from http://www.fdim.dk/?pageid=48

Gibbs, G. R. (2002). *Qualitative data analysis: explorations with NVivo. Understanding social research*. Buckingham, UK: Open University Press.

Goffman, E. (1983). The interaction ritual. *Americal Sociological Review*, (48), 1-19.

Golding, P., & Middleton, S. (1979). Making claims: news media and the welfare state. *Media Culture & Society*, *1*(1), 5–21. doi:10.1177/016344377900100102

Haddon, L., & Stald, G. (2009). *A cross-national European analysis of press coverage of children and the internet. LSE*. London: EU Kids Online.

Hake, K. (1999). Barneperspektivet - en forskningsstrategi. In *Børn, unge og medier. Nordiske forskningsperspektiver* (pp. 193–208). Göteborg, Sweden: Nordicom.

Herring, D. (2008). Questioning the Generational Divide: Technological Exoticism and Adult Constructions of Online Youth Identity. In Buckingham, D. (Ed.), *Youth, Identity, and Digital Media, The John D. and Catherine T. MacArthur Foundation Series on Digital Media and Learning* (pp. 71–92). Cambridge, MA: The MIT Press.

Hine, C. (1998). Virtual Ethnography. In *IRISS '98: Conference Papers*. Retrieved from http://www.intute.ac.uk/socialsciences/archive/iriss/papers/paper16.htm

Hine, C. (2000). *Virtual Ethnography*. London: Sage Publications.

Hunt, A. (1999). *Governing Morals: A Social History of Moral Regulation. Cambridge studies in law and society*. Cambridge, UK: Cambridge University Press.

Infomedia. (2010). *Infomedia*. Retrieved January 15, 2010, from http://www.infomedia.dk

Ins@fe. (2007). Safer Internet Day 2007. *Safer Internet Day 2007*. Retrieved January 15, 2010, from http://old.saferinternet.org/ww/en/pub/insafe/mediaroom/sid2007.htm

Ito, M. (2008). Mobilizing the Imagination in Everyday Play: The Case of Japanese Media Mixes. In *The International Handbook of Children, Media and Culture* (pp. 397-412). Los Angeles, CA: SAGE.

Ito, M., Baumer, S., & Bittanti, M. boyd, D., Cody, R., Herr, B., Horst, H. A., et al. (2010). *Hanging Out, Messing Around, Geeking Out: Living and Learning with New Media*. The John D. and Catherine T. MacArthur Foundation Series on Digital Media and Learning. Cambridge, MA: MIT Press.

Jones, R. H., & Norris, S. (2005). Discourse as action/discourse in action. In Norris, S., & Jones, R. H. (Eds.), *Discourse in Action. Introducing mediated discourse analysis* (pp. 3–14). London: Routledge.

Kampmann, J. (1998). *Børneperspektiv og børn som informanter*. København, Denmark: Børnerådet.

Lampe, C., Ellison, N., & Steinfield, C. (2006). A face(book) in the crowd: social Searching vs. social browsing. In *Proceedings of the 2006 20th Anniversary Conference on Computer Supported Cooperative Work*. Association for Computing Machinery.

Larsen, M. C. (2005). *Ungdom, venskab og identitet - en etnografisk undersøgelse af unges brug af hjemmesiden Arto* (Upubliceret specialeafhandling). Institut for Kommunikation, Aalborg Universitet. Retrieved from http://www.ell.aau.dk/fileadmin/user_upload/documents/staff/Malene_Larsen_-_Documents/Ungdom__venskab_og_identitet_Malene_Charlotte_Larsen.pdf

Larsen, M. C. (2007a). Kærlighed og venskab på Arto.dk (Love and friendship on Arto.dk). *ungdomsforskning, 6*(1), 11.

Larsen, M. C. (2007b). Understanding Social Networking: On Young People's Construction and Co-construction of Identity Online. In *Proceedings from the conference Internet Research 8.0: Let's Play, Association of Internet Researchers, Vancouver.* Association of Internet Researchers.

Larsen, M. C. (2009). Sociale netværkssider og digital ungdomskultur: Når unge praktiserer venskab på nettet. *MedieKultur, 47*, 45–65.

Lindlof, T. R., & Shatzer, M. (1998). Media ethnography in virtual space: Strategies, limits, and possibilities. *Journal of Broadcasting & Electronic Media*.

Livingstone, S. (2002). *Young People and New Media*. London: SAGE Publications.

Livingstone, S. (2008). Taking risky opportunities in youthful content creation: teenagers' use of social networking sites for intimacy, privacy and self-expression. *New Media & Society, 10*(3), 393–411. doi:10.1177/1461444808089415

Livingstone, S., & Bober, M. (2005). *UK Children Go Online. Final report of key project findings*. Department of Media and Communications, The London School of Economics and Political Science. Retrieved from www.children-go-online.net

Livingstone, S., & Helsper, E. J. (2007). Taking risks when communicating on the Internet: the role of offline social-psychological factors in young people's vulnerability to online risks. *Information Communication and Society, 10*(5), 619. doi:10.1080/13691180701657998

Lull, J., & Hinerman, S. (1997). The Search for the Scandal. In Lull, J., & Hinerman, S. (Eds.), *Media Scandals: Morality and Desire in the Popular Culture Marketplace*. Cambridge, UK: Polity Press.

Lund, J. (2009). *Digital view: Life on the Danish Internet, August 17-23 2009 – arto.com vs facebook*. København, Denmark: Jon Lund. Retrieved from http://jon-lund.com/main/digital-view-life-on-the-danish-internet-august-17-23-2009-artocom-vs-facebook/

Marwick, A. E. (2008). To Catch a Predator: The MySpace Moral Panic. *First Monday, 13*, 6–2.

McRobbie, A., & Thornton, S. L. (1995). Rethinking 'Moral Panic' for Multi-Mediated Social Worlds. *The British Journal of Sociology*, *46*(4), 559–574. doi:10.2307/591571

Nishida, K. (1958). *Intelligibility and the Philosophy of Nothingness*. Tokyo: Maruzen Co. Ltd.

Norris, S., & Jones, R. H. (Eds.). (2005). *Discourse in Action: Introducing Mediated Discourse Analysis*. London: Routledge.

Rapley, T. (2007). *Doing Conversation, Discourse and Document Analysis. The SAGE Qualitative Research Kit*. London: Sage Publications.

Rasmussen, D., & Hansen, H. R. (2005). *Chat, chikane og mobning blandt børn og unge. Afsluttende rapport om skoleelevers brug af chat i deres interaktive sociale liv*. Konsulentgruppen AMOK. Retrieved from http://www.livsmodlab.dk/txt/chat_chikane_og_mobning_blandt_boern_og_unge.pdf

Rattleff, P., & Tønnesen, P. H. (2007). *Børn og unges brug af internettet i fritiden*. København, Denmark: The Danish School of Education - University of Aarhus, The Danish Media Council for Children and Young People. Retrieved from http://andk.medieraadet.dk/upload/brugafinternettetrapport.pdf

Safer Internet Programme. (2009). *Safer Internet Programme: Homepage - Europa - Information Society*. Retrieved January 15, 2010, from http://ec.europa.eu/information_society/activities/sip/index_en.htm

Schrøder, K. (2000). Pionérdagene er forbi! Hvor går receptionsforskningen hen? *MedieKultur*, Medieforskning til tiden: Rapport fra Statens Humanistiske Forskningsråds konference om dansk film- og medieforskning, (Vol. 31).

Scollon, R. (2001a). Action and Text: Towards an integrated understanding of the place of text in social (inter)action, mediated discourse analysis and the problem of social action. In Wodak, R., & Meyer, M. (Eds.), *Methods of Critical Discourse Analysis* (pp. 139–183). London: Sage Publications.

Scollon, R. (2001b). *Mediated discourse: The Nexus of Practice*. New York: Routledge.

Scollon, R. (2007, July). *Geographies of Discourse*. Website reading draft.

Scollon, R., & Scollon, S. W. (2004). *Nexus Analysis: Discourse and the Emerging Internet*. New York: Routledge.

Scollon, R., & Scollon, S. W. (2007). Nexus analysis: Refocusing ethnography on action. *Journal of Sociolinguistics*, *11*, 608–625. doi:10.1111/j.1467-9841.2007.00342.x

Springhall, J. (1998). *Youth, popular culture and moral panics: penny gaffs to gangsta rap, 1830-1997*. Basingstoke, UK: Macmillan.

ENDNOTES

1. All quotes from respondents are translated from Danish into English by the authors.
2. Within media studies Livingstone and Bober describe a child-centred approach as one that "invites children's own understandings of their daily lives. It regards children as active, motivated and imaginative (though not necessarily sophisticated) agents who shape the meanings and consequences of the 'new' through the lens of their established social practices." (2005, p. 6).
3. All the empirical work and collection of data has been carried out by the main author, whereas the second author has primarily been involved in discussions of and reflections on the data, as to bring new perspectives

into the analytic work! The inclusive 'we' is therefore used for style and consistency, as to avoid shifts between I and we e.g. 'we discuss', 'I collected'.

[4] However, as pointed out by Baym, social network sites have not replaced online communities, but the latter have taken a new form resembling both the site-based online group and the egocentric network (2007).

[5] Within this age group 6% use MySpace, 9% use LinkedIn and 7% has profiles on other sites or social networking services (Danmarks Statistik, 2009).

[6] Sadly, there is only a website reading draft of this work available, as Ron Scollon died in 2008, before he had the chance to publish it.

[7] However, with few exceptions of stories about MySpace, but those stories did not deal with how Danish youth used the site, but covered MySpace as a music distribution site.

[8] "The Chat Consultant Case" refers to a special case in Denmark, where a man who was employed by the municipality to function as a "safe chat consultant" was caught on hidden camera by a national TV channel exposing his intentions to meet young girls on Arto and lure them into having sex with him in homemade porn videos. The case created quite a stir, also in the Danish news media, which had promoted him (extensively) as an expert in stories about paedophiles or young girls' sexual behaviour on Arto. In the programme that caught the man red-handed, he was filmed trying to blackmail the people behind Arto to stop his criticism of the site. In this media content survey, all stories about Arto and the chat consultant are listed under this category. The Chat Consultant case could be labelled a media scandal. According to Lull and Hinerman, media scandals occur when "private acts that disgrace or offend the idealized, dominant morality of a social community are made public and narrativized by the media, producing a range of effects from ideological and cultural retrenchment to disruption and change." (1997, p. 3). Media scandals differ discursively from moral panics, in so far that they can be traced back to real persons who are held responsible for their acts (Lull & Hinerman, 1997, p. 4). We have not included direct references to the case, as to refrain from exposing his name.

[9] For instance, a group of researchers under the research project "EU Kids Go Online" have conducted a "cross-national European analysis of the press coverage of children and the internet" (Haddon & Stald, 2009), in which the criteria of positive/negative/mixed/descriptive were used in the coding work of media coverage from printed newspapers across 14 countries (including Denmark). Among other things, this survey showed that the newspaper coverage of the topic overall has been negative because of a focus on risks in the articles. Even though this survey is interesting, it can be criticised for having a very small sample of articles and covering only two months. Since the purpose of the survey was to assess patterns in media coverage concerning the topic 'children and the internet' across 14 countries, one can argue that the empirical foundation is too thin. The researchers reflect on this point themselves: "The limitation of this analysis, relevant for interpreting the national differences in particular, is that this study was a snapshot, and therefore there is a question of how normal, or robust over time, these patterns will be." (Haddon & Stald, 2009, p. 35). Furthermore, we would argue that the media coverage of children and internet usage often seems to cluster and having a short collection period can therefore create a misguiding picture of the patterns in the coverage.

[10] Stanley Cohen was the first to coin the term moral panic (1972), but later on the term has been used to describe especially the news media's coverage of young people and new media.

[11] For instance, The Ministry of Science, Technology and Innovation along with other interest groups, service providers and organisations, created a codex called "ChatCheck", which requires providers of social network sites for youth to fulfil a list of technical demands and provide automated surveillance of communication and activities on their sites (FDIM, 2009).

[12] Critcher refers to Hunt's definition of 'moral regulation' as "practices whereby some social agents problematize some aspects of the conduct, values or culture of others on moral grounds and seek to impose regulation upon them." (Hunt, 1999).

[13] The Danish Media Council for Children and Young People is appointed national Awareness Node under the European network Insafe. Insafe is a network of national nodes that coordinate internet safety awareness in Europe. The network is set up and co-funded within the framework of the European Commission's Safer Internet plus Programme (Safer Internet Programme, 2009).

[14] Only 1960 responded fully to all questions, whereas some did not answer all questions or abandoned the questionnaire before completion. Partially completed questionnaires have been included in the qualitative analyses, but left out in statistical, quantitative measurements.

[15] This figure should not be confused with the average number of Danish children and young people between 12 and 18 who use social network sites, as the questionnaire was linked to from different Danish social network sites and therefore might have an overrepresentation of actual users. The figure of 88,3% could therefore be higher than the average amount of users in Denmark within this age group. However, recent data from Danmark statistik (2009) show that 86% percent of adolescents between 16-19 use a social network site.

Chapter 3
Swedish Students Online:
An Inquiry into Differing Cultures on the Internet

Håkan Selg
Uppsala University, Sweden

ABSTRACT

Results from a major survey among Internet users at Swedish universities indicate fundamental differences in patterns of usage. The "Web 2.0 culture" is socially driven and characterised by interactivity and participation. In the "Web 1.0 culture", the Internet is considered more of a tool for the rationalising of duties and tasks in everyday life. A strong age element can be observed in the sense that a majority of the Web 2.0 culture adherents have grown up in a digital environment with broadband access while those belonging to the Web 1.0 culture generally adopted Internet as adults. However, the findings do not support the claims made by early commentators of a "Net Generation", or "Digital Natives", with a set of common characteristics. The considerable variations within the age groups indicate that the process of appropriation of the Internet by the individual is far more complex than what is embedded in the generations approach.

INTRODUCTION

Open-Ended Technologies

A main characteristic of information and communication technologies (ICT) is their open-ended nature. In contrast to many other technologies ICT is neither determined in design and development nor in consumption and use (Silverstone, 2005).

DOI: 10.4018/978-1-60960-209-3.ch003

The particular complexity of ICT lies in the double articulation as both objects of consumption and as media of consumption. As key technologies in social and individual actions of everyday life ICT is socially constructed in the sense of being continuously defined and redefined through the human capacity to create meaning and order in the world (*Ibid*).

To claim open-endedness is to consider present expressions of ICT as something provisional, something that will undergo continuous developments by the actions of the users. The subjects of technological change will switch from individual innovators to the user communities, resulting in unintended directions and consequences (*Ibid*). We are left with a genuine feeling of uncertainty about future user patterns of ICT and what kind of effects on the already established ways of organising our societies – if any – that will be brought about. The traditional way of governments, major companies and journalists to turn to technical experts for visionary outlooks and forecasts is not a solution this time. Then, who will provide an understanding of the network society of tomorrow? The answer is as always: those with most experience. Who are they?

The Generation Approach

In early 2000, Marc Prensky, a specialist in digital game-based learning coined the terms Digital Natives/Digital Immigrants (2001). He used a language metaphor to illustrate the idea that the (younger) generations that grew up with computers, video games and the Internet, "the Digital Natives", master the digital language in a way that older generations never will be able to do. Those of us who were not born into the digital world but at some later point in our lives adopted the new technology are, and always will be, "Digital Immigrants". The importance of the distinction is this: As Digital Immigrants learn – like all immigrants, some better than others – to adapt to their environment, they always retain their "accent", that is, their foot in the past.

Prensky is neither the first nor the only one to identify those having grown up with digital technologies at home as a key user group. A few years earlier, Don Tapscott published *Growing up Digital* (1998) where he concludes that with the Net-Generation – or N-Gen –, the children have become the authorities over their parents for the first time. The idea of a web-generation creating its own culture or lifestyle through, and by means of, ICTs has also been suggested by several other authors (for an overview see Hartmann, 2005).

Among the new qualities of those having grown up with access to ICT is the parallel – or hypertextual – processing in learning activities, compared to the step-by-step linear approach of their parents. Closely related to these abilities is the assumed advanced multitasking capacity (Prensky, 2001; Tapscott, 1998). Another feature is their superior networking skills. Additionally the N-Gen members are characterised as open-minded, tolerant and inclusive, and after years of intense practicing, with full control of the "netiquette" and social codes. At the same time they are considered impatient with respect to slow responses and what they conceive as lengthy procedures (*Ibid*).

Both authors predict revolutionary consequences for the traditional educational systems. This is partly due to fundamental changes in the abilities and attitudes among the youth, partly due to failure of adaption among traditional educators. In addition, when the N-Gen culture is extended into society according to Tapscott, every institution will have to change. As N-Gens prefer networks, conventional business companies will have to transform into less hierarchical and bureaucratic forms, the same holds for governments as well. A new openness will infuse the marketplace, as power and authority will shift to web knowledgeable consumers with new online tools for comparison of product and services offerings.

The Swedish Situation

The Digital Natives metaphor also has expanded into the Swedish arena of public debate, although with some delay. A search on "Digital Natives" translated into Swedish provided 17 900 hits, the earliest in 2007 (Google, 2010). However, young Swedes online activities are of an earlier date and can be traced back to the early 1990s. At that time though, very little attention was attracted to the phenomenon from the world outside the group of enthusiasts. One of the first scholarly documents was about boys at the computer (Nissen, 1993) and the first doctoral dissertations appeared in the early 2000. Issues studied comprised the managing of virtual communities (Pargman, 2000), web chats (Sveningsson Elm, 2001), economy on the "warez" scene (Rehn, 2001), gender specific embodiments in virtual worlds (Sundén, 2002) and children's' strategies on the Internet (Dunkels, 2007).

In parallel to the scholarly activities, mostly carried out though ethnographic approaches, quantitative studies have been produced as well. National surveys on a yearly basis were made by institutions such as NORDICOM and Statistics Sweden. In the late 1990s, it was observed that younger age groups distinguished themselves in terms of frequency of use, particularly with regards to communicative tools such as chat and e-mail (Bergström, 2001). Similar results were reported in the yearly longitudinal survey starting in 2000, "Internet in Sweden" (Findahl, 2001). Therefore our preliminary impression is that a similar generation characteristic of Internet use as launched by Prensky and Tapscott could be observed in Sweden.

Effects of New User Patterns

Irrespective of a possible revolutionary nature of abilities and attitudes among those who have grown up with access to ICT, the web has gone through an important transformation during its short time of existence. Initially it was composed of non-interactive websites where users were confined to the passive viewing of information that was provided to them. Since then the Internet has been reshaped into a platform with sites, where users are allowed to interact with other users and to change website content. This transformation is summarised into the term Web 2.0 that was popularized by the O'Reilly Media conference in 2004 (O'Reilly, 2005). Notwithstanding that the Web 1.0/Web 2.0 concepts most often are associated to user interactivity and user participation, the subheading of O'Reilly's article is *Design patterns and business models for the next generation of software*. This indicates that in O'Reilly's perspective the key distinction lies in the different business models of web-based software.

In a Web 1.0 environment the challenge to generate revenues goes via the creation of market dominance. Such dominance could be achieved by getting control over standards for high-priced server products. In that respect the web software is sharing the same commercial ideas as Microsoft and Encyclopaedia Britannica Online, to give a few examples (O'Reilly, 2005). As a contrast the Web 2.0 models are relying on user participation and user-created content. The role of the software company is to provide a service – the platform – for the users to interact with each other. Network effects from user contributions pave the way to market dominance in the Web 2.0 era. The value of the software is proportional to the scale and dynamism of the data it helps to manage (*Ibid*). Examples of Web 2.0 are Google, social-networking sites such as Facebook, video-sharing sites, wikis, blogs, mashups and folksonomies. Web 2.0 is criticized as an unclear concept, but warrants attention due to its increasing presence across professional and popular practice (see e.g. Hughes, 2009).

In a broader perspective of what sociologists call the network society we are witnessing a change in the role of established companies and organisations as their customers become increasingly active users in networks (Castells, 1996; Wellman, 1988). Examples of its manifestations are found in open source software (Lakhani & von Hippel, 2003; Weber, 2004), product co-design and testing (Jeppesen & Molin, 2003) and the reshaping of popular culture (Jenkins, 2006). User participative activities have brought discontinuity and disruptive changes in the structure of social processes (Shirky, 2009), with impact on the organization of production (Benkler, 2006), and with repercussions on management requirements (Hamel, 2007).

As with all change, this situation offers both opportunities and risks. Virtual communities can be an inexhaustible source of information that can be used to improve products and services. At the same time, many business companies watch with mixed feelings as their control over important resources slowly erodes. In such a business environment early information about new patterns of use can be crucial to success. Most importantly, companies need to be one step ahead in the development and adaptation of products, services and business models. Knowledge about early users is therefore a matter of strategic importance.

The Research Questions

There are reasons to believe that the future directions of ICT applications will be designed by the combined activities of user communities rather than by individual innovators. Several U.S. based commentators argue that practitioners of interactive and participatory ICT applications are found among those who have grown up with access to digital games, media and communicative means rather than among those who have acquired ICT skills at an adult stage. Our research questions could then be formulated:

1. Can a distinct user pattern, characterised by interactive and participatory elements corresponding to the Web2.0 concept, be observed among Swedish users?
2. Can an age component associated to different user patterns be observed?

METHOD

Points of Departure

"User pattern" is an amorphous term, it has various meanings. It can be used to describe the sum of the activities by an individual user on the Internet. "User profile" would then be a synonymous expression. Furthermore there is an implicit understanding of the term as relating to groups of users, and to the observed differences between individual users, or groups of users, that appear to be systematic. As a practical matter, we want to look at the user profile of each individual and then allocate individuals with similar user profiles into distinct groups. In statistical terms it is expressed in the form of correlations between different net activities performed by an individual user so that his or her user pattern – or user profile – is defined by a set of higher and lower correlation coefficients. The users with congruent sets of correlation coefficients are then grouped together. This is actually a description in ordinary language of the combined factor and cluster analyses that has been implemented in this study.

With the aim to investigate possible age-related divergences in the user patterns, it follows that our study has to include all age groups, not only the younger generations. With age as a parameter indicative of degrees of differentiation of a population, a quantitative structural analysis will be applied (Blau, 1977).

Quantitative data serve as indicator of the scale of the Web 2.0 usage. This is important when we want to discuss its possible impact on for example established structures of education, business models and media consumption. However, quantitative data usually fall short when trying to explain why patterns of usage may differ between individuals or groups. Then qualitative data are more apt if the goal is to create an understanding why users and non-users behave as they do. Focus groups interviews are suggested as means to shed light on quantitative data already collected (Krueger & Casey, 2009). Therefore, quantitative and qualitative methods should be considered complementary. "One is not sufficient without the other" (Anderson & Tracey, 2001, p. 472).

Considering that our study focuses on actions and attitudes related to new technologies, we must also take into account that individuals in a social system do not all adopt an innovation at the same time. According to Rogers (2003), the adoption process, when plotted over time on a frequency basis, usually follows a normal, bell-shaped curve (p. 272). The implication for our study is that if the entire Swedish population should be considered, our questions may be addressed at a too early stage in the diffusion process to get consistent replies. It leaves us with the challenge to identify the relatively early adopters and to direct our inquiry to them.

The Population

In 2003 a survey of the use of Internet in Swedish universities and university colleges was carried out (Selg, 2003). The questionnaire design was co-ordinated with the national survey "Internet in Sweden" (Findahl, 2001). A comparison of the results indicated that universities are in the forefront regarding the use of the Internet. Quantitative and qualitative studies in the United States have yielded similar conclusions, namely that Internet penetration among students is much higher than in the general population (Jones & Madden, 2002; McMillan & Morrison, 2006; Robinson, Neustadtl et al., 2001). Based on these experiences, it seems to be an appropriate strategy to direct our inquiry to the academic world in order to comply with the condition of addressing early adopters.

A survey among students and staff at 32 Swedish universities and university colleges was carried out in 2007-2008. The survey was conducted by the Swedish IT-User Centre, NITA, at Uppsala University with funding from VINNOVA, the Swedish Governmental Agency for Innovation Systems. The subjects of study were found in three different sub-populations; 1) staff at universities, 2) post-graduate students and 3) students in undergraduate education. The aggregated population amounted to 285 000.

The Sample

The respondents were selected by means of stratified sample. The two student categories were distributed in eight strata according to academic discipline (Swedish University Nomenclature). A ninth stratum comprised university staff. The sample size of each stratum was calculated from the requirement of an error margin of at most ±6% with a 95% confidence interval for an estimated share of 50% within the stratum. A sample size of approximately 400 respondents in each stratum was derived. A higher precision requirement was made for two of the strata (Computer and Engineering sciences) ending up in sample sizes of 600 respondents. The weighting of these samples was motivated by the experience from survey in 2003 mentioned above, where Internet users from these two disciplines presented the most advanced user patterns. Weights for each respondent were calculated in order to allow for an estimation of the population.

A postal questionnaire was sent to the 4,000 respondents. Data were collected between November 2007 and January 2008. A response rate of 50% was achieved. The distribution of respondents is displayed in Figure 1. The age span is 18 to 70 years and with 64% in the categories 18 to 30 years. The mode is 22 years.

The Questionnaire

A number of activities on Internet were examined and with a particular focus on communicative and interactive usage. The questionnaire comprised 107 numbered question, many of them with subquestions.

The introductory parts related to the access to and use of computers and the Internet. One section dealt with the use of mobile phone including connecting to the Internet. Interactive and participatory activities were examined under a number of headers: File sharing, communities, open content databases, open source software, open educational resources and blogs/websites. Additional sets of questions related to educational situation such as academic discipline and credits obtained, to plans or experience from starting a

Table 1. Population and samples with distribution on strata

	Population		Sample	
		%		%
TOTAL	284 524	100,0	4 000	100
Humanities	22 503	7,9	400	10
Social science	80 897	28,4	400	10
Natural science excl. computer science	12 390	4,4	400	10
Computer science	6 494	2,3	600	15
Engineering science	42 008	14,8	600	15
Teacher training	38 462	13,5	400	10
Medical/paramedical education	43 668	15,3	400	10
Other	21 844	7,7%	400	10
Staffs	16 258	5,7%	400	10

company, and to social background factors such as family situation and living arrangements.

All together approximately 350 variables were generated. The design of the questionnaire was co-ordinated with the national survey "Internet in Sweden" in order to compare the results. A draft version of the questionnaire was tested by a group of specialists at Statistics Sweden.

Data Analysis

The data was analysed by the using of SPSS software. Cross tabulations with respect to gender, age and academic disciplines were made. The selection of these attributes was based on observations from the 2003 survey (Selg, 2003). According to that survey, men appear in the early adoption stages to a greater extent than women and so do younger users in comparison to middle-aged. The same was observed with students from technically-oriented disciplines compared to students from other academic fields. In addition, it was discovered that file sharing practices serve as a marker for other Internet activities; individuals with advanced Internet user patterns in general are found to be frequent files sharers and vice versa.

All together, 86 variables concerned the use of Internet – type of activity, frequency of use,

Figure 1. Age distribution of respondents (n=1913)

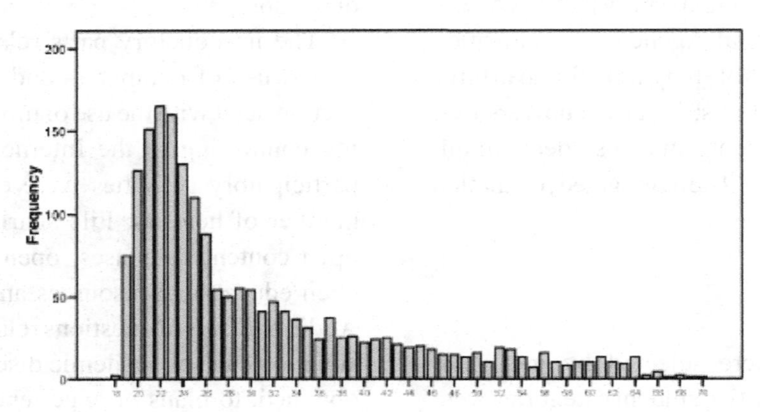

and purpose. In order to exploit the potential correlations between these variables a factor analysis (Principal Component Analysis) was made. However, due to few users, a number of the original 86 variables/Internet activities did not satisfy the conditions of a normal distribution. Those variables were omitted from the factor analysis. The remaining 48 variables were concentrated into 11 factors.

The 11 factors were then processed in a TwoStep Clustering procedure with automatic determination of the number of clusters. Two clusters were obtained as a result. A test of significance with a 95% confidence interval was made for each of the factors on the two clusters respectively.

Additionally, the results for 18 selected variables were compared with the corresponding results from the national study "Internet in Sweden 2008". The variables included the use of communicative tools such as e-mail and instant messaging, search and retrieval activities, e-commerce and social networks participation (Selg & Findahl, 2008).

Focus Groups

The purpose of focus groups interviews is to uncover factors that influence opinions, behaviour or motivation. Focus groups can provide insight into complicated topics when opinions or attitudes are conditional or when the area of concern relates to multifaceted behaviour or motivation (Krueger & Casey, 2009). A merit of the interview in comparison with the questionnaire is that relevant responses may be uncovered whether or not these have been anticipated by the inquirer (Merton et al., 1990). Through his/her familiarity with the topics of discussion, the interviewer is better prepared "to recognize symbolic or functional silences, distortions, avoidances, or blockings and is, consequently, better prepared to explore their implications" (*Ibid*, p. 4).

Three focus groups were interviewed with in all 16 panellists and 2 moderators involved. The groups were recruited among students on different levels and employees at Uppsala University and at the Royal Institute of Technology (KTH), Stockholm:

1) Master students at the Department of Information Technology (Uppsala),
2) Staff (lecturer, researcher, systems administrator) and PhD students at the Department of Informatics and Media (Uppsala)
3) Under-graduate students at the Department of Media Technology and Graphics Art (KTH).

Data were collected in September and October 2008. The discussions were documented by notes and with a recording device as a back-up. The notes from each group were edited in a draft document that was distributed by e-mail to the participants for comments and corrections. I have translated all quotes from the original Swedish material.

RESULTS FROM QUESTIONNAIRE

Interactive and Participatory User Pattern

Results from the combined factor and cluster analysis indicate two distinct patterns of usage among Internet users. As presented in Table 2, a number of Internet activities are common for the two groups, i.e. statistically significant variations were not observed ($p \leq 0{,}050$, n=1485).

The net activities, common to the two clusters, are characterised by one-way communication, such as to search for information or to make orders and payments. The communication flows are "vertical" as they connect business companies with customers and public agencies with the citizens. The e-mail protocol serves as the predominant tool for communication. We will further refer to this category of users as the "Web 1.0 cluster".

The differentiating group of net activities, which are statistically significant, contains a num-

Table 2. Common and differentiating activities on the Internet among the two clusters

Common activities	Differentiating activities
Study and work	Communities
Travels and events	Entertainment
e-Commerce	Blogs
e-Government	News
e-Literature	On-line encyclopaedias
Entrepreneurial activities	

ber of actions with the characterising feature of a two-way communication pattern between users. This is the world of blogs and on-line communities, the file sharing of popular culture and other forms of social computing. Here we can observe "horizontal" communication flows between the actors that at the same time appear as consumers and producers of digital content. Different forms of instant messaging are favourite protocols. This group of users is named the "Web 2.0 cluster".

We thus conclude that a distinct user pattern, characterised by interactive and participatory elements corresponding to the Web 2.0 concept, can be observed among Swedish users. Therefore the answer to our first research question is yes.

The Age Component

As indicated in Table 3, age makes up a strong distinguishing feature between the two clusters. A majority of Internet users from 31 years and above belong to the Web 1.0 cluster, while an equally predominant share of users up to 30 years are found in the Web 2.0 cluster.

On our second research question about an age component associated to different user patterns, again the answer is yes and the findings presented in Table 3 are very clear on this matter. However it is necessary to underline that there are important variations within the younger age groups as well as among the adult age groups. 25% of the respondents in the group up to 25 years present a traditional "Web 1.0" user pattern while a participatory and interactive "Web 2.0" patterns is found among 19% in the group 36-45 years.

Gender, Academic Fields and File Sharing Practices

Some gender tendencies could be observed as well (Table 4): Among the men, 63 percent belong to

Table 3. Age distribution of cluster participants (p=0.000, n=1485)

Age category	Web 1.0 cluster	Web 2.0 cluster	Sum
– 25 years	25%	75%	100%
26 – 30 years	42%	58%	100%
31 – 35 years	72%	28%	100%
36 – 45 years	81%	19%	100%
46 – 55 years	95%	5%	100%
> 55 years	100%	0%	100%
All (est. number = 208 000)	44%	56%	100%

Table 4. Gender distribution of cluster participants (p=0.000, n=1485)

Gender	Web 1.0 cluster	Web 2.0 cluster	Sum
Females	48%	52%	100%
Males	37%	63%	100%
All (est. number = 208 000)	44%	56%	100%

the Web 2.0 cluster compared with 52 percent of the women.

Table 5 presents the distribution on academic disciplines. The Web 2.0 cluster is predominant in most of the fields, with the exception of Teacher training and Medical educations with a 50/50 distribution. Although Computer and Engineering sciences still count for a higher Web 2.0 share than for example Humanities, the difference is less than expected given the tendencies of the survey from 2003. This may be an indication that the advanced usage patterns now have been extended over the student population independent of technical orientations.

In conformity with the tendencies in the 2003 survey, the use of file sharing practices remain an indicator of advanced Internet activities (Table 6) and with a strong majority in the Web 2.0 cluster. Also illustrative are the distributions of the two non-user categories; former users of file sharing services are to a much higher extent found in the Web 2.0 cluster than those who never have practised file sharing.

Table 5. Distribution of cluster participants on academic disciplines (p=0.000, n=1287)

Academic discipline	Web 1.0 cluster	Web 2.0 cluster	Sum
Humanities	38%	62%	100%
Social science	38%	62%	100%
Natural science	35%	65%	100%
Computer science	27%	73%	100%
Engineering science	32%	68%	100%
Teacher training	50%	50%	100%
Medical/paramedical education	51%	49%	100%
All (est. number = 193 000)	44%	56%	100%

Table 6. Distribution of cluster participants on file sharing categories (p=0.000, n=1485)

File sharing category	Web 1.0 cluster	Web 2.0 cluster	Sum
Heavy user	10%	90%	100%
Moderate user	20%	80%	100%
Former user, not anymore	47%	53%	100%
Never been user	69%	31%	100%
All (est. number = 208 000)	44%	56%	100%

Synchronous/Asynchronous Communications

Communications on the Internet can be synchronous or asynchronous. Synchronous discussions use instant messaging – or chat – protocols. E-mail is one example of asynchronous communication where people can port notes that others can read at any time. SMS is a communicative tool, very popular among young people. Essentially asynchronous, SMS in the way it is used by the youth has similar synchronous properties as instant messaging.

From the survey data the use of different communicative tools (e-mail, instant messaging and SMS) with distribution on the two clusters, is reported in Table 7. What they have in common is the use of e-mail, the asynchronous tool. No significant difference between the clusters with regards to e-mail use is observed.

Instead the main deviations relate to the usage of synchronous communicative devices, the IM protocol and the SMS. As earlier stated, the IM protocol serves as a marker of the interactive user patterns of the Web 2.0 cluster. Within the Web 1.0 cluster, a substantial use of SMS is observed as well, however still on a more moderate level than among members of the Web 2.0 cluster.

An interesting aspect is the versatile communication pattern of the Web 2.0 cluster, with a very frequent use of e-mail, IM protocols and SMS respectively. Rather than being competing alternatives, the three communication tools should be considered complementary.

Representative Results?

Now a crucial question remains: Are the results obtained representative for the Swedish public in general? From surveys in 2003 a distinct divergence in the degrees of Internet adoption between the university population and the Swedes in general were observed (Selg & Findahl, 2006). In order to check how representative these findings are, we have compared our data with the results from the national survey "Internet in Sweden" from the same year (2008). As a general impression, the observed differences from 2003 have been substantially reduced, and for some activities have completely disappeared. The findings indicate that the use of Internet for most people has become an ordinary element in everyday life (Selg & Findahl, 2008).

In the case of participation in social networks sites (SNS), the diffusion patterns reported in the two surveys are the same, however with obvious differences in communicative intensity. Students visit the SNSs much more frequently than what other users do (*Ibid)*. Therefore the results mirror the observations from other scholars about the university studies as a period in life with large

Table 7. Use of different tools for communication among clusters (n=1485)

Tools	Cluster	Several times a day	Every-day	A few times a week	A few times a month	Less often or never	Sum
E-mail	Web 1.0						
(p=0.120)	Web 2.0						
	All	40%	34%	22%	4%	0%	100%
IM	Web 1.0	9%	14%	16%	11%	50%	100%
(p=0.000)	Web 2.0	43%	23%	21%	6%	7%	100%
SMS	Web 1.0	23%	29%	33%	10%	5%	100%
(p=0.000)	Web 2.0	44%	36%	17%	2%	1%	100%

amounts of unstructured time (Arnett, 2006; Ellison, 2008; Quan-Haase & Collins, 2008).

RESULTS FROM FOCUS GROUPS

The Origins of Interactive and Participatory Usage

Discussions with students in focus groups provided an overview of the origins social computing. Most of the participating students had grown up with access to computers and the Internet since the age of 10-12 years. Their first activity of importance was the chat function – IRC, ICQ, MSN or Hotmail – later on "fun software" such as games.

"At the start you chat with friends and class mates about secrets that you don't tell at school" (student).

The next stage in the communicative process was to chat with young people outside the group of friends that they met on a daily basis; first in the home town, then in the same region, later on *"with people in Stockholm or in the rest of Sweden" (student).*

Being curious to see what the others look like, they used to post pictures of themselves, often followed by the exchanging of favourite music recordings. With a telephone-based 56k modem, the transmission of a single tune could take several hours. If there were restrictions to go on-line imposed by the parents, the computer(s) at school could be used for chat and downloading.

"Nor in school was it allowed but the control was less strict" (student).

Especially among the boys, the computer became the main tool in a social process where the exchange of various forms of popular culture – music, films, games and software – played a predominant role. The main driving force was the pursuit of the most recent recordings in for example pop music, with a certain prestige to be among the first. These findings are in line with results from interviews with experts made in connection with the 2003 survey where the importance of social drivers and popular culture were stressed (Selg & Findahl, 2006).

The findings clearly indicate the critical importance of online communication as a motive for young people to start using the Internet. The same observation was made by Tapscott (1998): "With the advent of the Web, millions of children around the world are routinely gathering online to chat…." (p.56).

Online Communities and Social Network Sites

From the focus groups we learned that on-line communities and SNSs providing games and forums are very popular among teen-agers. Among the younger ages it is commonplace to be a member in various such sites, while gradually reducing the number with increasing age.

"It is a matter of personal development; for a while you enjoy being part of a community, with time you lose interest" (master student).

Online communities and SNS often contain embedded ranking systems that attract some users. In Lunarstorm scores are obtained in relation to the number of posted comments and in Facebook the number of friends may serve a similar purpose. The perceived importance of such rankings seems to vary with age.

"Before I had more than 300 friends on my Facebook account but after a while I got tired of keeping track of all old pals from high school. Now I have reduced the number of friends to some 80" (PhD student).

64% of Swedish students visit social sites on a recurrent basis, with Facebook without comparison as the most popular. Lunarstorm, that at the turn of the century attracted a majority of the younger Swedes, has lost considerable popularity during recent years (Selg, 2008). U.S. studies report that 85% of the students in join at least one SNS, with Facebook and MySpace as the most popular (Ellison, 2008).

All the panel participants from the university staff – generally in their thirties and upwards –were aware of new social software sites such as online communities and SNS. Their own children, students and mass media served as sources of information. Many of them had also received invitations to sites such as Facebook or LinkedIn. However they shared an attitude of whereby there was a lack of understanding of the benefits from the social networks.

"I'm not a member of Facebook, cannot grasp the purpose with membership and participation" (administrator).

"I accepted to enter LinkedIn, but I'm still not clear about the point" (lecturer).

Among the experienced students various motives for participation in social networks online were suggested.

"Sites such as Facebook and LinkedIn serve as tools for keeping track of people rather than for communicate; a kind of online address book" (master student).

Everybody did not agree.

"The mail function in Facebook competes with ordinary e-mail clients, the addressing is smarter, which makes it easier to find each other.....and the majority of those you want to communicate with are all in Facebook" (student).

"Social sites function according to the rule that the more you contribute, the more you will get out from it. You start posting something that may attract the attention of someone else who in turn may provide interesting links" (PhD student).

One major impression from the discussions is that both the way online communities and SNS are used, and the motives behind the use, may vary considerably among the users. With the words of Silverstone (2005) these software tools are "socially constructed in the sense of being continuously and individually defined and redefined in order to create meaning and order in everyday life" (p. 5).

Time Conflicts

The lack of available time was suggested as being one of the reasons for abstaining from being active in online social sites by various participants.

"I'm reading the e-mails in the morning in my office and in the evening back home again as well. For newsletters and things like that there is just no time considering that also work has to be done" (lecturer).

"To have time available is obviously a condition for participating in social networks sites. This is not the case when you have a job and children" (administrator).

Is it true that having a job and a family implies such severe time restrictions on online social activities? Or could it be explained in terms of different multitasking capacities as suggested by some commentators?

Beginning with the first job/family criteria, our survey data allows a comparison between respondents with or without children. Therefore it is possible to study the importance of the child factor on the distribution between the Web 1.0 and the Web 2.0 clusters. Testing for the job factor makes less sense considering that the majority of the respondents are students. Common-sense makes us expect that having children is something age dependent, at least moving from the twenties to the thirties and forties. The same counts for the cluster distributions according to Table 3. Therefore we shall limit the test to those younger age groups where a majority is found in the Web 2.0 cluster, that is, up to 30 years.

Assuming the Web 2.0 cluster as a proxy for participation in online communities and SNS, Table 8 indicates a negative relation between having children and social software activities. However, this tendency is not very strong, and almost half of the respondents with children manifest the interactive user pattern of the Web 2.0 cluster.

Table 8. Cluster distribution of respondents 18-30 years with our without children (p=0.002, n=985)

Household status	Web 1.0 cluster	Web 2.0 cluster	Sum
No children	32%	68%	100%
Children	54%	46%	100%
All (est. number = 152 000)	33%	67%	100%

Let us now turn to the opinions of the panel participants who actually are frequent users of social sites.

"After having logged in on Facebook you keep it in the background while studying. Each time a friend posts something there is a reminder which allows for a nice little break in the preps" (student).

"Micro time-outs are necessary when you are working in a concentrated manner in front of the screen. Then Facebook serves a good purpose" (student).

These two students refer to the breaks brought about by SNS as a positive element in their struggles. Other students express a more ambiguous attitude.

"The general problem is to concentrate, to resist the reflex to check if there are incoming messages or mails" (master student).

"I have great problems to refrain from connecting to the net, it almost an addiction" (PhD student).

Several panel participants present explicit strategies to manage the situation.

"When you are accessible all the time, it is necessary to master the technology so you can focus on one issue if needed" (master student).

"The telephone can be put on silent mood, and storing incoming messages. With the laptop in a full screen mood, msn and e-mails are not discovered" (student).

"Sometimes, control could be maintained by carrying out tasks immediately instead of doing them in a certain order, but then there is always the risk of split-attention" (PhD student)

"When working with an academic document to be published, msn and e-mail clients have to be closed down. The same with your mobile: when too many incoming calls, switch it off!" (master student).

The impression provided by the discussions is that of a communication technology that offers its users distraction by social contact. In many cases this is perceived as rewarding, sometimes as disturbing. The stronger the intellectual challenge of a task to be carried out, the greater is the need of control of the technology. There is nothing in the discussions that supports the claims of a superior multitasking capacity among the student compared to their lecturers and other university employees. Nor do the job and children arguments seem convincing. What differs between students and staff in the focus groups is that the students after years of training and practice of different applications of social software have acquired skills and strategies, i.e. how and when to use such communicative tools.

Privacy Conflicts

A matter of concern expressed by several members of the staff was the blurring of borders between the professional, or public, life on the one hand, and the private life on the other. Two dimensions of this conflict were discussed; 1) working hours versus leisure time, and 2) the publishing of information that traditionally has been considered personal.

Regarding the intrusion of job tasks on staff members' leisure time it was recognised that this is something inherent in academic life, particularly in the research process. However, new communicative tools have accentuated the tendency. The attitudes and claims of immediate reactions from students formed by synchronous communicative means have already been discussed. Furthermore the asynchronous e-mail tool is contributing as well.

"The e-mailing is part of the job and I find it quite rewarding. But as a result it tends to spread out at the expense of my private life" (lecturer).

A general attitude among the staff was the hesitation to publish details of private nature that might be accessed by the general public. In several cases this was one major argument against social network sites.

"I am sceptical about such social sites, where does all the information go?" (administrator).

"If you have grown up with the web and are used to socialize and to be seen there, you consider it probably as something normal. Being somewhere in the borderland myself I am wary about publish private matters" (researcher).

The common opinion among all the panel participants was that young people are less reluctant than for example their parents to publish private matters on the web. Why is it so?

"A motive is the desire to be seen by other people. The thrill of exposing one's opinions, particularly if they are a little extreme" (student).

In psychological terms university students are characterised as "emerging adults". This is a period from roughly age 18 to their mid-twenties, where most identity exploration takes place, that is, they learn more about who they are and what they want out of life (Arnett, 2006). During this time, individuals are relatively free from the social expectations they experienced as teenagers or will experience as adult. They can explore different orientations toward work, interpersonal relationships, and the world in general. Encountering new ideas, engaging in new experiences, and meeting diverse kinds of people are especially important during this time because they expose individuals to different life experiences and paths (Ellison, 2008).

With this in mind we should perhaps not be surprised by the publishing activities online of teen-agers and young students. Then the interesting question stands: Will they maintain these extrovert tendencies as they get older, or will they assume the more prudent attitudes of their parents?

"The attitude to the publishing of personal information use to vary a lot, depending on the personality" (student).

"There is a tendency that people are more reluctant to publish details on themselves today than before. Instead such information is often restricted to friends. But there is an age dimension involved so that younger people care less about going public; the older you get, the wiser" (master student).

The attitudes and activities are influenced by a personality component with some individuals being more timid than others, and an age element. Other circumstances also contribute to make the picture more complex.

"When participating in discussions or other net activities, nick names are generally used. If everybody had to appear with their correct names, the Internet forums would look different" (master student).

"An alias or nickname is something that you acquire in an early stage, then it is a normal thing to keep it" (student).

Nicknames seem to play a critical role in social activities online.

"Quite often people appear on Facebook under different identities, one under an alias with lots of personal material, and another under the correct name – for the employer – with only little personal information" (student).

As already mentioned, the personality seems to influence the attitudes.

"It does not matter very much if you try to conceal things about yourself, people will find out what they want to know about you anyway" (student).

"When everybody is posting personal information there will be no dramatic effects left" (student).

To summarise, there are indications from our discussions, and supported by survey data, that new communicative user patterns, both synchronous and asynchronous, may contribute to the blurring of borders between working hours and leisure time.

Regarding the publishing online of personal information, thus making private matters public, the discussion leaves us with the impression that attitudes may have changed less than what we previously imagined. Young people seem to reflect on what publicly is known about them in the same way as those in their middle ages. What differs – again – is that the youth have developed strategies how to manage online publishing tools, while many older users have not. Lacking this competence, many Internet users choose to stay out. We may then conclude that behaviours and attitudes vary a lot among individuals, even within the same age categories.

Reactions on Response Delays

Among the properties attributed to the N-Gen or Digital Natives, by Tapscott (1998) and Prensky (2001) there is a relative impatience with respect to slow responses and what they conceive as lengthy procedures. Gratifications should be immediate. One is inclined to assume a relation between such attitudes and the use of synchronous communication. In one of the focus groups the issue was spontaneously addressed by a member of the staff.

"The students seem to believe that we only are giving one course and that is all. Since some years it is necessary to explain to them that we actually have other things to do as well" (lecturer).

A colleague reported similar experiences.

"If I haven't answered within one hour they often phone me. And trying to restrict the contacts to certain hours is just to forget" (lecturer).

Several of the participating students confirmed that from contacts with friends they were used to rapid reactions. The attitudes that had been formed in that way during their spare time have then been transferred into the studies.

The Information Overload

The situation of an information overload was brought up by several participants. On the one hand, individuals experience an increased exposure to information of all kinds; on the other hand the time available for information consumption has not changed, there are still only 24 hours in a day. The real value produced by an information provider comes in locating, filtering, and communicating what is useful to the consumer (Shapiro & Varian, 1999). As Herbert Simon (1997) puts it, "What information consumes is rather obvious: it consumes the attention of its recipients. Hence a wealth of information creates a poverty of attention".

This conflict was commented upon by many participants.

"The richness in access to information creates a sensation of powerlessness; how can you be able to take in everything?" (lecturer).

"I experience the same in my research activities. Before, the university library put distinct limits to the reference literature, but on the net there are no limits" (researcher).

While the problems were formulated by members of the staff, the suggested solutions were presented by the students. Some of their comments refer to filtering techniques.

"To actively select your sources of information is absolutely necessary. For the filtering process there are online-based methods and packaging tools" (PhD student).

"The linking to favourite blogs and other preferred sources of information is a key function" (student).

Other suggestions relate to the potential of the social network.

"An easy way is to take advantage of the aggregated knowledge of your Facebook friends" (student).

"Before I buy anything, particularly in the case of digital products, I use to enter online communities to get their opinions" (PhD student).

"For people living in foreign countries the blogs offer a tool for useful information about living abroad. You share your experience and you learn a lot about everyday situations. From the blogs you get more information than from books" (master student).

"For most kinds of information, a good alternative is to post a question on a net forum, wait two days, and then pick up the recommendations" (student).

"From community experience you have learnt how to communicate with people from other countries and cultures, to know how to ask people you don't know for solutions. Faster than Google and faster than reading articles" (master student).

Once again the panel discussions indicate a line of demarcation between the students and the members of the staff, or between the Web 2.0 and the Web 1.0 clusters. Regarding the information overload due to Internet access, the students have learnt to manage the ICT-related technical and social tools to orient themselves, while their teachers are frustrated when their traditional filtering methods are perceived as insufficient.

Discussion

Our first research question was to examine if a distinct user pattern, characterised by interactive and participatory elements corresponding to the Web 2.0 concept, can be observed among Swedish users. The answer is yes, we have found that the participants of the survey are divided into two clusters. In the first cluster, we find those users who chiefly use the Internet for one-way communication activities. In the second cluster, besides being as frequent users of one-way actions, they also engage in two-way interactive and participatory activities in a substantial way. Therefore, the statements about the Internet as a platform for user interaction seem to be relevant also for the Swedish user environment.

Our second research question was to investigate an age component associated to different user patterns can be observed. Again, the answer is yes and the findings presented are very clear on this matter. However it is necessary to underline that there are important variations within the younger age groups as well as among the adult age groups. We may therefore conclude that despite a clear age component, it is less appropriate to speak of the differences in user patterns in terms of generations. The individual variations are simply too big.

Other scholars, critical to the generations approach, underline the danger with the generalizations about common characteristics of a whole generation of young people. One argument is that cognitive differences in young people of different ages and variations within age groups will not be taken into account. Also the potential impact of socio-economic and cultural factors tends to be overlooked. In reality there may be as much variation *within* the Digital Native generation as *between* the generations (Bennett et al., 2008).

Turning to our qualitative data it must first of all be emphasized that the word "impression" is the recurrent expression to characterize the outcomes from the discussions in the focus groups. The participants were relatively few and the interpretative element in the selection of statements is considerable.

Evident from the focus groups discussions are the developments of skills by young users from years of practicing of various kinds of social software. These social learning processes have their origins in the online communicative tools. The training coincides with a period in life with large amounts of unstructured time and when most of identity exploration takes place (Arnett, 2006; Ellison, 2008; Quan-Haase, 2008). These experiences, acquired in an intense social environment, are of crucial importance as they constitute the basis for a set of strategies that are necessary in order to face and to get by with some of the challenges of the information society; time conflicts, privacy matters and information overload.

Social processes may be explained and understood by networks analysis and the concepts of strong and weak ties (Granovetter, 1973). Strong ties are maintained between family members and close friends while weak ties are associated with acquaintances. Granovetter argues, perhaps contrary to our intuitive understanding, that the weak ties are very important in our social lives. Phenomena as diffusion of ideas and information, social mobility such as getting a job, political organisation and social cohesion in general, can be explained by the use of network analysis of weak ties (*Ibid*). Weak ties have a decisive role for community building, and the seeming paradox of the possible meaning in having several hundred Facebook friends gets its explanation (Ellison, 2008). We may therefore conclude that the results of our findings are in line with the statements of Prensky and Tapscott regarding superior networking skills and knowledge of social codes.

The claimed impatient attitudes from younger people with respect to slow responses and what they conceive as lengthy procedures make sense out of our findings. From the survey we observed that the Web 2.0 cluster makes use of all the available tools in the communicative repertoire in a complementary way. That means they will use the tool that appears to be the most appropriate in the current situation – chat, e-mail or SMS. It also signals a tendency away from the traditional telephone culture with the origins in a situation where the telephone was considered a collective resource of the office or the household. In the work place the use was restricted by office hours. At home, privacy considerations put a limit to calls during certain hours. For example, to phone somebody after nine o'clock in the evening is considerate an intrusion by many Swedes.

When both the mobile phone and the laptop are perfectly individualised such social rules do not seem to make any sense. To refrain from communication due to unsuitable hours is not longer necessary; if a (synchronous) voice call is not suitable, than (synchronous) chat or SMS, alternatively an (asynchronous) e-mail could be chosen. With the practice to always carry the mobile phone – to an increasing extent also the laptop with a mobile Internet access – then the response from the addressed person often will come shortly. Therefore from having been a device for discrete use, communicative tools are now used in a continuous way. Attitudes are adapting accordingly.

Turning to the multitasking qualifications, our findings do not support the idea of a generation with mental qualities alien to us. Rather our discussions mirror acquired skills and strategies regarding how and when to use communicative tools. The idea of unique multitasking capacities among Digital Natives has also been criticized by other scholars both for lacking empirical evidence, or to be supported by anecdotes and appeals to common-sense beliefs. The often used example of a young person doing homework while engaged in other activities has been compared with the pre-Internet complaints about teen-agers doing their homework in front of the television, while talking on the telephone and having a friend in the room (Bennett et al., 2008; Jones & Madden, 2002).

CONCLUSIONS: TWO CULTURES ON THE INTERNET

Our research questions related to user patterns on the Internet, i.e. what could be observed and displayed by our survey statistics. By including our qualitative data, we have been able to gather some understanding about the individual and social processes behind the statistical numbers. These processes are influenced by motives and attitudes, skills and strategies, at the same time as they are producing such motives and attitudes, skills and strategies in a reflexive manner.

Behind the user patterns we find such individual and social properties that are comprised in the concept of culture. Culture hence is considered as a set of skills or habits rather than values, norms and wants (Swidler, 1986). The causal explanation of cultural influence on action is then that an individual chooses such a line of action for which she or he already has the cultural resources, such as skills and competence. Thus a culture is not a unified system that pushes the actions in a consistent direction. Rather is more like a "tool kit" or repertoire from which actors select different pieces (Hannerz, 1969; Swidler, 1986).

This notion of culture enables us to explain the differences observed among the two clusters of Internet users. The reason why the users in the Web 1.0 cluster do not take advantage of the opportunities of social communication is not that they cling to cultural values distinctive of those of the Web 2.0 cluster. It is because they are reluctant to abandon familiar strategies of action for which they have the cultural equipment (Swidler, 1986). From this perspective our study demonstrates a "cultural lag" between groups of Internet users.

Rather than to discuss in terms of generations, the concept of different cultures on the Internet is to prefer, and for two reasons: Firstly, young people exhibit a full scheme of individualities and must not be considered a homogenous mass. Different kinds a people make differing use of the range of applications and service – "tool kits" – that the Internet supports, probably also for differing reasons. "The average Internet user simply does not exist" (Anderson & Tracey, 2001, p. 462). Secondly, because an individual Internet user may adhere to both cultures at the same time. Actually this is what the data tell us: those belonging to the Web 2.0 culture are Web 1.0 users as well.

FUTURE RESEARCH DIRECTIONS

Motives and attitudes behind the decision to participate in online social medias, or to refrain from such participation, constitute a major field of research which our panel discussions only have touched upon. This is a topic for recurrent investigation both through quantitative and qualitative means, not least due to, at least so far, its rapid development and change.

Furthermore, up to now ICT has been a family of different and distinct communicative tools and hardware; e-mail, instant messaging and SNS forums on computers, and SMS and telephony on mobile phones. From both hardware camps, attempts have been taken to intrude on the other's domain, for example with IP-telephony on the computers and with e-mail on the mobile phones. We are now witnessing a convergence of hardware in the sense that mobile phone and handheld computers tend to merge into the same device. What will then happen with those previously well-defined tools for communication? Will some of them merge as well? What will be the effects on how we communicate between us?

From the perspective of synchronous/asynchronous communication, the development poses very interesting questions. So far we have seen how (synchronous) IM protocols are more popular among younger users than among adults. Wellman et al. (2001) suggest that many professionals during their labour day cannot be stuck to their computers which makes the (asynchronous) e-mail more attractive. Will the new digital devices imply a change in the balance between two communicative rationales in favour of synchronous communication? Will there be consequences for which groups in society that will act as leaders in the exploration of new digital applications? Will the adults recapture the initiative?

REFERENCES

Anderson, B., & Tracey, K. (2001). Digital Living: The Impact (or Otherwise) of the Internet on Everyday Life. *The American Behavioral Scientist*, *45*(3), 456–475. doi:10.1177/00027640121957295

Arnett, J. J. (2006). *Emerging Adulthood The Winding Road from the Late Teens Through the Twenties*. New York: Oxford University Press.

Bennett, S., Maton, K., & Kervin, L. (2008). The 'digital natives' debate: A critical review of the evidence. *British Journal of Educational Technology, 39*(5), 775–786. doi:10.1111/j.1467-8535.2007.00793.x

Bergström, A. (2001). Internet - från revolution till vardagsanvändning. In S. Holmberg & L. Weibull (Eds.). *Land, Du Välsignade?: SOM-Undersökningen 2000* (pp. 283-296). Göteborg: SOM-institutet, Univ.

Blau, P. M. (1977). A Macrosociological Theory of Social Structure. *American Journal of Sociology, 83*(1), 26–54. doi:10.1086/226505

Castells, M. (1996). *The Information Age: Economy, Society and Culture*. Malden, MA: Blackwell.

Digitala infödingar (n.d.). Google search for "Digitala infödingar", March 3, 2010.

Dunkels, E. (2007). *Bridging the Distance: Children's Strategies on the Internet*. Doctoral dissertation. Department of Interactive Media and Learning, Umeå University.

Ellison, N. B. (2008). Introduction: Reshaping Campus Communication and Community through Social Network Sites. In *The ECAR Study of Undergraduate Students and Information Technology, 2008*, ECAR Research Study Vol. 8, pp. 19-32. EDUCAUSE Center for Applied Research. Retrieved from http://www.educause.edu/ecar.

Findahl, O. (2001). *Svenskarna och Internet 2000*. World Internet Institute. Retrieved from http://www.worldinternetinstitute.org.

Findahl, O. (2008). *Svenskarna och Internet 2008*. World Internet Institute. Retrieved from http://www.worldinternetinstitute.org.

Granovetter, M. S. (1973). The Strength of Weak Ties. *American Journal of Sociology, 78*(6), 1360–1380. doi:10.1086/225469

Hartmann, M. (2005). The Discourse of the Perfect Future - Young People and New Technologies. In R. Silverstone (Ed.), *Media, Technology, and Everyday Life in Europe: From Information to Communication* (pp. 141-158). Aldershot, Hants, England: Ashgate.

Haythornthwaite, C. (2001). Introduction: The Internet in Everyday Life. *The American Behavioral Scientist, 45*(3), 363–382. doi:10.1177/00027640121957240

Hughes, B. (2009). What is Web 2.0, and what's not: A Roadmap for Research Relevance. Presented at *2009 EURAM (European Academy of Management) Conference on Renaissance and Renewal in Management Studies*, Liverpool.

Jones, S., & Madden, M. (2002). The Internet Goes to College: how students are living in the future with today's technology\. *Pew Internet and American Life Project*. Retrieved December 31, 2009, from http://www.pewinternet.org/~/media//Files/Reports/2002/PIP_College_Report.pdf.pdf.

Krueger, R. A., & Casey, M. A. (2009). *Focus Groups* (4th ed.). Thousand Oaks, CA: SAGE Inc.

McMillan, S. J., & Morrison, M. (2006). Coming of age with the internet: A qualitative exploration of how the internet has become an integral part of young people's lives. *New Media & Society, 8*(1), 73–95. doi:10.1177/1461444806059871

Merton, R. K., Fiske, M., & Kendall, P. L. (1990). *The Focused Interview: A Manual of Problems and Procedures* (2nd ed.). New York: The Free Press.

O'Reilly, T. (2005). *What Is Web 2.0: Design patterns and business models for the next generation of software*. Retrieved Januari 6, 2010, from http://oreilly.com/pub/a/web2/archive/what-is-web-20.html?page=1.

Prensky, M. (2001a). Digital Natives, Digital Immigrants Part 1. *Horizon, 9*(5), 1–6. doi:10.1108/10748120110424816

Prensky, M. (2001b). Digital Natives, Digital Immigrants Part 2: Do They Really Think Differently? *Horizon*, *9*(6), 1–6. doi:10.1108/10748120110424843

Quan-Haase, A., & Collins, J. L. (2008). I'm there, but I might not want to talk you. *Information Communication and Society*, *11*(4), 526–543. doi:10.1080/13691180801999043

Robinson, J. P., Neustadtl, A., & Kestnbaum, M. (2001). An Online Data Web Site for Internet Research: Some Features and an Example. *The American Behavioral Scientist*, *45*(3), 565–568. doi:10.1177/00027640121957222

Rogers, E. M. (2003). *Diffusion of innovations* (5th ed.). New York: Free Press.

Selg, H. (2003). *Internet i den svenska högskolan våren 2003 (Eng. Internet at Swedish universities spring 2003)*. Användarstudier inom SUNET. Retrieved from http://basun.sunet.se/anvandare.pdf.

Selg, H., & Findahl, O. (2006). *File sharing in peer-to-peer networks - actors, motives and effects.* Broadband technologies transforming busienss models and challenging regulatory framworks - lessons from the music industry, Deliverable 4. Retrieved from http://w1.nada.kth.se/media/Research/MusicLessons/Reports/MusicLessons-DL4.pdf.

Selg, H., & Findahl, O. (2008). *Nya användarmönster. Jämförande analys av två användarstudier*. InternetExplorers, Delrapport 6. Nationellt IT-användarcentrum NITA, Uppsala universitet. Retrieved from http://www.internetexplorers.se/.

Shapiro, C., & Varian, H. R. (1999). *Information Rules: A Strategic Guide to the Network Economy*. Boston: Harvard Business School.

Silverstone, R. (2005). Introduction. In R. Silverstone (Ed.), *Media, Technology, and Everyday Life in Europe: From Information to Communication*. (pp. 1-18). Aldershot, Hants, England: Ashgate.

Silverstone, R. (Ed.). (2005). *Media, Technology, and Everyday Life in Europe: From Information to Communication. Aldershot*. Hants, England: Ashgate.

Swidler, A. (1986). Culture in Action: Symbols and Strategies. *American Sociological Review*, *51*(2), 273–286. doi:10.2307/2095521

Tapscott, D. (1998). *Growing up Digital: The Rise of the Net Generation*. London: McGraw-Hill.

Wellman, B. (1988). Structural analysis: from method and metaphor to theory and substance. I B. Wellman & S. D. Berkowitz (Eds.), *Social Structures: A Network Approach*, Structural analysis in the social sciences. Cambridge, UK: Cambridge Univ. Press.

Wellman, B., Haase, A. Q., Witte, J., & Hampton, K. (2001). Does the Internet Increase, Decrease, or Supplement Social Capital?: Social Networks, Participation, and Community Commitment. *The American Behavioral Scientist*, *45*(3), 436–455. doi:10.1177/00027640121957286

ADDITIONAL READING

Benkler, Y. (2006). *The Wealth of Networks: How Social Production Transforms Markets and Freedom*. New Haven, CT: Yale University Press.

boyd, d. m., & Ellison, N. B. (2007). Social Network Sites: Definition, History, and Scholarship. *Journal of Computer-Mediated Communication, 13(1)*, 210-230.

Cohen, S. (2002). *Folk devils and moral panics* (3rd ed.). New York: Routledge.

Deuze, M. (2006). Participation, Remediation, Bricolage: Considering Principal Components of a Digital Culture. *The Information Society*, *22*(2), 63–75. doi:10.1080/01972240600567170

Gresham, J. L. Jr. (1994). From invisible college to cyberspace college: Computer conferencing and the transformation of informal scholarly communication networks. *Interpersonal Computing and Technology*, *2*(4), 37–52.

Haddon, R. (1992). Explaining ICT consumption: The case of the home computer. In Silverstone, R., & Hirsch, E. (Eds.), *Consuming technologies: media and information in domestic spaces* (pp. 82–96). London: Routledge. doi:10.4324/9780203401491_chapter_5

Hamel, G. (2007). *The Future of Management*. Boston: Harvard Business School Press.

Hannerz, U. (1969). *Soulside: Inquiries into Getto Culture and Community*. New York: Columbia University Press.

Howard, P. E. N., Rainie, L., & Jones, S. (2001). Days and Nights on the Internet: The Impact of a Diffusing Technology. *The American Behavioral Scientist*, *45*(3), 383–404.

Ito, M. (2007). Education vs. Entertainment: A Cultural History of Children's Software. *The John D. and Catherine T. MacArthur Foundation Series on Digital Media and Learning*, 89-116.

Jenkins, H. (2006). *Convergence Culture: Where Old and New Media Collide*. New York: New York University Press.

Jeppesen, L. B., & Molin, M. J. (2003). Consumers as Co-developers: Learning and Innovation Outside the Firm. *Technology Analysis and Strategic Management*, *15*(3), 363–384. doi:10.1080/0953732031000160153 1

Kaplan, A. M., & Haenlein, M. (2010). Users of the world, unite! The challenges and opportunities of Social Media. *Business Horizons*, *53*(1), 59–68. doi:10.1016/j.bushor.2009.09.003

Koku, E., Nazer, N., & Wellman, B. (2001). Netting Scholars: Online and Offline. *The American Behavioral Scientist*, *44*(10), 1752–1774. doi:10.1177/00027640121958023

Lakhani, K. R., & von Hippel, E. (2003). How open source software works: "free" user-to-user assistance. *Research Policy*, *32*(6), 923–943. doi:10.1016/S0048-7333(02)00095-1

Nie, N. H., & Erbring, L. (2002). Internet and Society: A Preliminary Report. *IT & Society, 1(1)*, 275-283. Retrieved January 13, 2010 from http://www.stanford.edu/group/siqss/cgi-bin/typ_page.php?page_number=3

Nissen, J. (1993). *Pojkarna vid datorn: Unga entusiaster i datateknikens värld.* Linköping studies in arts and science. Stockholm, Sweden: Symposion graduale.

Pargman, D. (2000). *Code Begets Community: On Social and Technical Aspects of Managing a Virtual Community. Linköping studies in arts and science* (1st ed.). Linköping, Sweden: Tema, Univ.

Punie, Y., Bogdanowics, M., Berg, A., Pauwels, C., & Burgelman, J. (2005). Consumption and Quality of Life in a Digital World. In R. Silverstone (Ed.), *Media, Technology, and Everyday Life in Europe: From Information to Communication* (pp. 93-106). Aldershot, Hants, England: Ashgate.

Rehn, A. (2001). *Electronic Potlatch: A Study of New Technologies and Primitive Economic Behavior. Trita-IEO*. Stockholm, Sweden: KTH.

Rheingold, H. (1991). *The Virtual Community: Homesteading on the electronic frontier (Electronic edition)*. Retrieved from http://www.rheingold.com/vc/book/.

Shirky, C. (2009). *Here Comes Everybody: The Power of Organizing Without Organizations*. New York: Penguin.

Silverstone, R., & Hirsch, E. (Eds.). (1992). *Consuming technologies: media and information in domestic spaces*. London: Routledge. doi:10.4324/9780203401491

Silverstone, R., Hirsch, E., & Morley, D. (1992). Information and communcation technologies and the moral economy of the household. In Silverstone, R., & Hirsch, E. (Eds.), *Consuming technologies: media and information in domestic spaces* (pp. 15–31). London: Routledge.

Simon, H. (1997). Designing Organizations for an Information-Rich World. In Lamberton, D. M. (Ed.), *The Economics of Communication and Information*. Cheltenham, England: Edward Elgar.

Sundén, J. (2002). *Material Virtualities: Approaching Online Textual Embodiment. Linköping studies in arts and science* (1st ed.). Linköping, Sweden: Tema, Univ.

Sveningsson Elm, M. (2001). *Creating a Sense of Community: Experiences from a Swedish Web Chat. Linköping studies in arts and science* (1st ed.). Linköping, Sweden: Tema, Univ.

Weber, S. (2004). *The Success of Open Source*. Cambridge, MA: Harvard University Press.

KEY TERMS AND DEFINITIONS

ICT: Information and communication technologies.

Asynchronous/Synchronous Communication: Direct communication, where all parties involved in the communication are present at the same time is a form of synchronous communication. Examples include a telephone conversation and instant messaging. Asynchronous communication does not require that all parties involved in the communication need to be present and available at the same time. Examples of this include e-mail, discussion boards and SMS.

SMS (Short Message Service): Exchange of short text messages between mobile phones.

Computer Mediated Communication (CMC): Any communicative transaction that occurs through the use of two or more networked computers.

Web 2.0: Web applications that facilitate interactive information sharing, interoperability, user-centered design, and collaboration.

Virtual Communities: A social network of individuals who interact through specific media, potentially crossing geographical and political boundaries in order to pursue mutual interests or goals.

Online Community: A virtual community that exists online whose members enable its existence through taking part in membership rituals

Social Network Sites: Web-based services that allow individuals to (1) construct a public or semi-public profile within a bounded system, (2) articulate a list of other users with whom they share a connection, and (3) view and traverse their list of connections and those made by others within the system (boyd & Ellison, 2007).

Social Media: A group of Internet-based applications that build on the ideological and technological foundations of Web 2.0, and that allow the creation and exchange of user-generated content (Kaplan & Haenlein, 2010)

Mashups: A web page or application that uses or combines data or functionality from two or many more external sources to create a new service.

Folksonomy: A system of classification derived from the practice and method of collaboratively creating and managing tags to annotate and categorize content; this practice is also known as collaborative tagging, social classification, social indexing, and social tagging

Lunarstorm: A Swedish social networking website used mostly by teenagers.

Section 2
Identity

Chapter 4
Fat Talk:
Constructing the Body through Eating Disorders Online among Swedish Girls

Ann-Charlotte Palmgren
Åbo Akademi University, Finland

ABSTRACT

The purpose of this chapter is to study how young women in a Swedish context construct their body by writing about eating disorders in blogs. Connected to the body and eating disorders a construction of girlhood can be seen. The blogs studied are all part of the online community ungdomar.se. The chapter begins with a background to eating disorders, blogs and girlhood and youth in a cultural context. The main focus is on examples from thirteen blogs. The content and typographical emphasis in the blogs are analysed and discussed. The study shows that the process of becoming or constructing a certain body and blogging is both social and collective because of the interaction between the blogger, the community and blog commentators. The body is not only constructed by teenaged girls by striving to a certain type of a female body, but also by mastering the talk about one's own body, dissatisfaction with it and by typographical emphasises in the blogs.

INTRODUCTION

In this chapter the focus is on the body, eating disorders and blogs. Since the material consist of blogs written by girls, the chapter is connected to the construction of girlhood among the girls studied. How is the body staged in these textual performances online? What typographical emphasises are used? How do the blogger discuss eating, food and feelings? How do the bloggers write about normality and abnormality?

I will begin with a brief background to eating disorders in a cultural context and a short conceptual and theoretical background to girlhood and youth as identity constructions. This is followed by a general introduction to blogs, the online community, the blogs analyzed and methodological concerns. The main focus in the chapter is on examples from blogs and the examination of how the body and girlhood is constructed through writ-

DOI: 10.4018/978-1-60960-209-3.ch004

ing about eating disorders. I will end with some concluding reflections on the specific questions stated earlier.

EATING DISORDERS OFFLINE AND ONLINE

Eating disorders and young girls have been studied in psychology, medicine, health studies, cultural studies and social sciences. Eating disorders are not a new phenomenon, but have been dated back to the middle ages (Bell 1985) and the Victorian era (Brumberg 1988 and Schwartz 1985). The American general public knew virtually nothing about them until the 1970s, when the popular press began to feature stories about young women who refused to eat despite available and plentiful food (Brumberg 1988, 8). Much of the research conducted in the 1980s had a medical or psychological perspective. Bordo's (1993) work on eating disorders was seen as ground breaking because of the feminist perspective that most of the earlier research was lacking. In the 1990s researchers became aware of how widespread women's problems with food, eating and body images were (Bordo 1993, 64). During the 2000s several studies appeared about eating disorders focusing on pro-anorexia, in connection to web pages, web forums and blogs.[1] Pro-anorexia, also called pro-ana, is an internet based movement where eating disorders are seen as a choice of life style instead of an illness (Gavin, Rodham & Poyer 2008). In July 2001 an American eating disorder advocacy group, ANAD (Anorexia Nervosa and Associated Disorders), made pleas to servers such as Yahoo to take down sites about pro-anorexia. Four days later 115 sites were shut down (Reaves 2001).

In a Swedish context the first treatment in institutional care was developed during the 1970s and during 1993 the Swedish Association for Anorexia and Bulimia (Svenska-Anorexi-Bulimi sällskapet) was founded, an association for people who worked with or studied eating disorders (Clinton & Norring, 2002). During the 1990s and 2000s several fictional and autobiographical books about eating disorders, written by young Swedish women, were published.[2] Young women wrote in youth magazines about the fixation with appearance and hysteria with thinness. Feminist literature by young women was published[3] and the advertisement for women's underwear by the company H&M was sabotaged (Ambjörnsson 2004, 176). Scholars argued that the phenomena that earlier had been described as young women's superficial engagement with the body and the appearance, should not been seen as an individual problem or as a sign of vanity, but as a phenomena that reflected the social system as a whole (Solheim 2001, 112). From the middle of the 1990s one could talk about a public discourse about young women, the appearance and the societies influence (Ambjörnsson 2004, 177).

The blogs I study in this chapter deal with eating disorders to some extent. I am interested in how they write about eating disorders and how eating disorders and dissatisfaction with the body are expressed in an online environment. My aim is not to study reasons for eating disorders and I will therefore not speculate about possible causes. There is little consensus on what causes eating disorders. Explanations generally fit within one of three models: the medical, the psychological or the cultural.

GIRLHOOD AND YOUTH AS A SOCIAL AND CULTURAL CATEGORY

This chapter is written from a constructivist perspective, which means that I see girlhood and youth as socially and culturally constructed. They are created by being performed: they are not a cause of, but rather an effect of different sorts of actions (Butler 1990, 140). According to Butler (1990, 151) the constructed divide between men and women has been developed from what she calls the heterosexual matrix. This matrix can be described as a cultural filter that makes bodies, gender, sexuality and desire comprehendible, separates masculinity from femininity and later ties them together through the acts of the heterosexual desire (Ambjörnsson 2004, 15). People constantly make performances that display characteristics that are thought of as masculine or feminine, and in doing so, they construct themselves to fit into one of the categories men or women (Sveningsson Elm 2007, 106). However, it is not enough to do one, or a few performances, but gender needs to be worked upon, re-created and staged through constantly repeated performances (Butler 1990). This applies to girlhood and the body as well, which I will show different examples of and discuss later in this chapter. In other words, there are many ways to be a girl, and these forms depend not only on the material bodies that perform girlhood, but also on the specific social and historical contexts in which those bodies are located.

Girlhood is not only a period determined by psychological development or biological processes. Girlhood is something that is constructed through negotiated processes (Aapola 2005, 1). Even if age is a noticeable factor for determination of a girl's social space, girlhood cannot be determined exclusively by age (Ojanen 2008).

In an international context girlhood studies were developed during the 1970s and 1980s. Despite the publication of several groundbreaking girl-centered studies during the 1970s and 1980s, as Kearney (2009) writes, much research at that time which involved female youth was focused on understanding women more than girls and only a handful of scholars demonstrated a consisted commitment to researching girlhood and girls' culture as unique social formations.[4] During the 1990s a dramatic increase in research about girlhood and girls' culture can be seen and in 2008 the first girl-centered academic journal, Girlhood Studies: An Interdisciplinary Journal was launched. In a Scandinavian context, according to Ambjörnsson (2004, 17), researchers in youth studies started to publish books and articles specifically about girls' culture and girlhood during the 1980s and 1990s.[5] Later, during the 2000s Scandinavian girlhood studies has grown and the body and/or sexuality have frequently been connected to this research.[6]

Regarding youth and online identities in general, Stern (2008, 98) has shown that online publications can provide important opportunities for managing the complex situations and shifting self-expectations that characterize adolescence. In descriptions of their decisions about what to reveal, exaggerate and omit in their online communication, youth authors reveal a highly conscious process of self-inquiry. In boyd's (2008, 128) study about teen's profiles created in social network sites the process of writing oneself into being, forces teens to work through identity in new ways. Teens have to work out how they invision themselves and how they want to be seen. Liu (2007) states that through the use of text, images and design, teens create profiles that signal information about their identities. The mediated self-representations that they create reveal both what they have in common and how they distinguish themselves from those around them.

STUDYING BLOGS

A blog can be categorized as a frequent, chronological publication of personal thoughts and web links. In practice, blogs have become experiments in self-expression, with people reading and cross-linking to other blogs, thus creating blog communities (Dijck 2005). As David Chaney (2002:52) observes, everyday life is a creative project because all though, it has the predictability of mundane expectations, it is simultaneously being worked at both in the doing and in retrospective reconsideration. The blogs can also be seen as creative projects of this type. Quotidian acts such as diary writing should not only be regarded as stilled reflections of life, but as ways of constructing life. Blogging, besides being an act of self-disclosure, is also a ritual of exchange: bloggers expect to be signaled and perhaps to be responded to (Dijck 2005), which is studied in this chapter.

One of the most early studies of blogs (Kawaura, Kawakami and Yamashita 1998) departs from the notion that blogs can be classified according to their content as records of fact, or expression of sentiment, according to their directionality as written for oneself or written for others. Later studies continued to classify blogs along axes of self and others and of personal and public (Blood 2000, Herring 2004). Earlier research about blogs seems to have been concentrated on characterization and the structure of a blog. Other aspects have been the relationship between the blogger and the reader and language (for example McNeill 2003, Sorapue 2003, Serfaty 2004).

The discussion of the ethical problems of online research has been discussed to a great extent, but researchers have not come to a conclusion. First, there are scholars who argue that archived material on the Internet is publicly available and therefore participant consent is not necessary for a study of them (see for example Walther, 2002). Second, some researchers claim that online postings, though publicly accessible, are written with an expectation of privacy and should be treated as such (see for example King, 1996, Scharf, 1999). Lastly, there are those who argue that online interaction defies clearcut prescription as either public or private. Waskul and Douglas (1996: 131) argue that cyberspace is simultaneously 'publicly-private and privately- public'. I decided to ask the bloggers for permission to quote their blogs.

The blogs examined in this chapter have all been selected from the Swedish online community ungdomar.se (in English youth).[7] The blogs were found by searching the word kropp (body in English) and choosing the 38 first bloggers that used the word in a blog entry. Even if the members of the community consist of both boys and girls, all of the bloggers using the word body stated that they were girls. I contacted the 38 bloggers asking for permission to study their blogs. Of these 18 answered that they were willing to be quoted. One answered that she did not allow me to quote her since she did not want people to think that it was normal to have eating disorders. Of the 18 bloggers 13 wrote to a great extent about eating disorders or dissatisfaction with the body. The 13 bloggers had written between 8 and 468 blog entries. One of the bloggers who gave her permission to be cited deleted her blog a couple of days after our contact and before I had the time to copy her entries. The majority of the bloggers asked to be anonymous and did not want me to publish the name of their blog or the address to their blog, which I respected. The 13 blogs I found and corresponded with are my empirical material in this chapter.

I chose to look for blogs belonging to a youth community since I was interested in contemporary youth culture online. In general I wanted to study how the body is presented in blogs, since early Internet research had come to the conclusion that the body is not present in the body free medium, blogs. Because of this I simply searched for the word body in blogs, but I was not specifically searching for the eating disordered body. I had expected a more heterogeneous body in the blogs and I had not restricted my search for blogs written by girls. In the end the material was restricted to girl's blogs since they were the only ones I found consisting of the word body.

FATFATFAT IS FATTER THAN FAT

Danet (2001) suggest that digital culture provokes and sustains an interest not only in the content of the text but also in the aesthetics and form of the text. In this subchapter I will concentrate on the form of the text in the blogs that I have studied. How are they typographed? Which of the words are emphasised? With what intentions are the blog entries typographed?

In several of the blog entries certain words where emphasized by writing the word several times without space. The word bigger becomes even bigger by writing it several times, the dread is worse when the word is written four times in a row without space between the words. Sometimes a letter in the word can be written several times so that 'haaaaate' is more 'hate' than hatefull. Another example is given by the blogger who do not want to 'die', but to 'diiiie'.

Exaggerating some words or a whole sentence can be seen as screaming. This is applied by using capital letters. For example "I will never be able to get thin! I CAN'T stop eating", "I haaaate being SO awfully d i s g u s t i n g l y u g l y and f a t" or "My weight is just under 60 kilograms, I'm 170 cm tall and it is NORMAL". In some examples where part of the entries, or part of the comments were written in capital letters it is a question of emphasising a certain word more than shouting, which can be seen in the following two examples:

THEY DON'T LIE! And this makes me sad. Feels like you are fading away from me <3

CAN PEOPLE JUST STOP LYING?

Using capital letters could be seen as a form of expressing emotions by using typography, or text formatting. Typography essentially includes font (type, size, colour, background colour, etc.) and typesetting (such as bold, italics, bold-italics). Making certain words bold is very common in blogs. Here is one example:

shit, I have to loose more.

I want to live like a normal person.

no food.

I want to feel good.

loose more, loose more, loose more.

By writing some parts in bold the blogger emphasizes them. She wants to live like a normal human being who feels good, but less than someone who looses more weight. Emphasizing is a way of calling attention and conveying importance. In several of the blog entries that I have studied the blogger expresses a feeling of being two persons or personalities in the same body. The quote above can be analyzed as an example of this, where the two persons with different desires are in the same body. The three sentences in bold can be categorized as belonging to a category of eating disorderly longing to loose more weight, while the other two sentences are connected to longing for a normal life, which in this quote equals with feeling good.

Characteristic for the blogs that belong to the community ungdomar.se is that they use punctuations, spaces and commas in the middle of a word. This phenomenon which is specific for blogs on ungdomar.se, can be said to be a consequence of the rules of the community. They state for example the following:

§ 10. So called thinspiration is not allowed on the web site. This includes pictures, avatars, calculators for calories, tips on how to hide your state of ill-health, links to pro-ana sites, daily food routines etc. All such material will consistently be erased. We refer those who suffer from eating disorders to our friends Anorexia/Bulimia-contact.

The bloggers of the community write about their eating disorders, weight, eating habits and food, but try to hide their posts by different means in the blog, usually by misspelling certain words or having spaces between the letters. In this way the blog entries are not found by typing the words in the community's search engine. Quite often you may read in a blog that the moderators have erased blog entries that they feel that violate the rules. We have seen several examples of this in this chapter already but here is one where one can see that words related to the anorectic body are misspelled:

Why do you say those things?

Why do you tell me things that I can't bare to hear?

If there is one unwritten rule among a'n'o'r'e'c't'i'c's it is this.

Not to trigger.

To *not* tell how good it feels when your legs hurt and how much you exercise during a week. To not throw away food when I see and then pretend to really make an effort.

Like it would be okay.

Like I didn't spare you from such a thing.

Like I wouldn't v'o'm'i't and cheat and exercise too much when it is difficult for me.

A strategy for hiding forbidden text in the blogs is used when the text is written in white on a white background with the statement 'hidden text' in the beginning of the text. To read the text the reader has to mark the text so it is visible. In the following example only the first sentence is readable at a first glance. You can see the text "hidden text", but nothing after it, if you don't mark the space. In this chapter I have chosen to highlight the text that is hidden to illustrate it.

I know that I promised myself and mum to eat more, but lunch is still scaring me a lot. Hidden text: I don't like the feeling I get from food in my stomach, food that makes my body miserable and transforms me into the world's biggest monster.

The strategy does not make the text disappear in the search engine like punctuation does, but by hiding the text the blogger may feel that she is not promoting eating disorders, thinness or not eating, which is prohibited according to the rule.

EMOTICONS IN BLOGS

An increasingly larger part of online communication is social (Sproull & Kiesler, 1991). However, in early research scholars argued that the modalities of online social interaction were much more limited than those of face-to-face social interaction and that they provided no good way to convey nonverbal cues (see for example Kiesler, Siegel & McGuire, 1984). Later, researchers have found that online communities have appropriated the use of emoticons and other textual representations to help address this issue, but the lack of a standard set of expressive and versatile representations introduces ambiguity, which limits their usefulness (Sarbaugh-Thompson & Feldman, 1998).

Emoticons are used in the studied blogs, but not to a great extent, which could be understood as Drenks, Bos and von Grubbkow (2008) suggest for the use of emoticons; emoticons are used more consciously than actual nonverbal behaviour, which implies that there is more control over the message a person wants to convey and, as a consequence, it might be easier to regulate emotions. Following are some examples both taken from blog entries and the comments to blog entries:

This makes me want to go on a d i e t or something. I want to be healthy, but I don't want to be assaulted:(

Don't destroy it by starving:(

You have a great body. Be satisfied with it:)

I don't think before I eat, I just get to damn hungry and ooh >< I don't know, I'm thinking of surrendering to the crap..

The three first examples above express happiness and unhappiness, The meaning of the sentences are clearly connected to the emoticons that are used. The emoticon in the fourth example expresses anger. Anger can also be seen as frustration, when it appears after 'ooh' and before 'I don't know'. It is fairly easy to understand the emoticon in connection to the words. These examples of the use of emoticons can all be seen as attempts to strengthen the message and to show the mood of the blogger. The following three examples are different:

You who thinks that ”bones are beautiful”, do you think that my body is ugly/disgusting/whatever? I'm just wondering:P

hey, you are perfect:):P

I do think that my body is okay:D (at least today ^^)

In the first and second examples the emoticon ':P' is used, but in somewhat different ways. In the first example a question is followed by the emoticon that represents a tongue sticking out, a bit teasing. The blogger is critical in her blog entry and has a serious question but makes the question and criticism more harmless by using the emoticon. The emoticon can also stand for sarcasm or a joke. A sarcasm might be communicated using positive words but a negative tone or the other way around (Planalp & Knie, 2002). In the second example the blogger might express mixed feelings or a joke. Walther et al. (2001) argue that, in mixed messages, the valences of verbal and nonverbal messages may cancel each other out, resulting in a more neutral interpretation overall. The last example can also be seen as a mixed message because of the emoticons. The first express laughter and the second amusement by symbolizing heaved eyebrows. Connected to these more traditional emoticons is the combination of a '<' and '3', so that it forms a heart. Bellow we find some examples of this:

I don't think that they are lying <3

THEY DON'T LIE! <3<3<3<3<3

THEY DON'T LIE! And this makes me sad. Feels like you are fading away from me <3

Sweetheart, they are honest. I wish that you could see how thin you are. Falafel is good, so I hope that you ate, ;) Did it feel better after seeing you sister? <3<3<3<3

The heart expresses love in these examples. The emoticon is fairly new and is almost always used in the blogs studied in positive comments after a negative comment. The heart is not directly connected to the words, but connects the blogger and the blog reader.

THE BODY AS NORMAL AND ABNORMAL

I'm not thin but I couldn't care less. (the picture was taken today). I'm not thin, but I'm not fat either. My weight is just under 60 kilograms, I'm 170 cm tall and it is NORMAL. I don't want to be thinner. Maybe have a bit more well-trained body but definitely not thinner. It is disgustingly ugly and unhealthy. I wish everyone would think so and stop dieting! What the hell is good about dieting anyway? Fdjigjfdig, can't explain myself. You who think that ”bones are beautiful”, do you think that my body is ugly/disgusting/whatever? I'm just wondering:P

The themes of feeling normal or abnormal in relation to the body and others are prominent experiences in the blogs. In the blog entry above the blogger has published a picture of herself in a bathing suite standing in front of a mirror. It is very uncommon for the blogs in this study to consist of photographs of the blogger. In a period of one and a half year she has three similar entries, writing about her measurements, having a picture of herself in a bathing suite standing in front of a mirror and writing about what is seen as normal. The emphasis in the entry cited above is on the word normal, written in capital letters. When writing that she is normal or feel normal she assumes that there is something that is abnormal. To be abnormal is to deviate from the norm, which raises the questions: what is considered normal and whose norm. One way of seeing normal is to rely on measurement and statistical deviations. When calculation the bloggers body mass index, which is a number, calculated from a person's height and weight, one can come to the conclusion that her weight status is normal, but close to underweight. The blogger states repeatedly that she is not thin and do not want to become thinner. She sees thinness as something ugly and unhealthy. However, in blog entries earlier she has written about behaviour that would be considered as belonging to someone with eating disorders. In the blog entry the thin body could be seen as the abnormal body. The three similar entries can be analysed as her recovering from eating disorders. Aligned with a changing body image, eating and dieting is an attitudinal change about the ideal body. She wishes that everyone would stop dieting. She receives four anonymous comments for the entry:

I think that you should be satisfied with your body! It is really fine! Don't destroy it by starving:(I know from experience. I hope that you learn to accept your body as well.

Your body is really beautiful. Don't diet if you can. It would make me happy.

Good if you don't want to loose more weight

You have a great body. Be satisfied with it:)

The commentators are all answering the question she ends her blog entry with: "do you think that my body is ugly/disgusting/whatever?" The commentators all say that she should be satisfied with her body, that it is fine, beautiful and great. The first comment is interesting since it seem to be a contradiction to the blog entry. It seems like the commentator does not read that the blogger does not want to diet and want others as well to stop dieting. The commentator even writes that she hopes that the blogger would learn to accept her body. It seems like the commentator actually refer the all the other blog entries, instead of this entry. The second commentator also addresses dieting in an interesting way since the comment seems to assume that there are circumstances where dieting is necessary.

According to Harjunen (2009) social acceptability of the body is often evaluated in regard to its appearance. But the only thing being judged is not the appearance, as the physical body is habitually seen as a reflection of one's personal characteristics, life-style, morals, values and behaviour (e.g. Featherstone1982). With this idea in mind the blogger does not only ask if her body is ugly or disgusting, but also if she is a nice person. According to popular and stereotypical understanding, the thin body relates to a number of positive qualities such as intelligence, being in control, effectiveness, healthiness and beauty. In contrast the fat body is associated with very negative values such as laziness, being out of control, and stupidity (Ogden 1992 cited in Harjunen 2009, 16). Here, the normal are the things that are considered as good qualities and the abnormal as the qualities that one should avoid. The blog entry was uncommon since it consisted of a photograph of the blogger, but also because it expressed a kind of satisfaction with the body, even if the blogger was not totally satisfied and wanted to be more well-trained than she was. In the blog entry she wanted the blog readers to comment on her body, while she is in one of the later entries comparing her to other girls. She starts her entry by saying that she is not fat and ends it with wondering if she is overweight and wanting to go on a diet:

My weight is 60 k i l o g r a m s and I'm 170 cm. Even so, I am not f a t, I'm okay with my body even if I have some extra f a t. But my self-esteem isn't THAT good, I can't watch a lot of super t h i n girls saying that they are f a t, because if they are f a t, then I'm like totally o v e r w e i g h t? This makes me want to go on a d i e t or something. I want to be healthy, but I don't want to be harassed:(

Ambjörnsson (2004, 180) found that the girls that she studied were all irritated with one girl. This girl did not express dissatisfaction with her body. Ambjörnsson could only remember once during her field work when one of the girls expressed satisfaction with her body. One of the girls in Ambjörnsson's study stated that it is not easy for a girl to be satisfied with her body, and even less easy to be accepted by other girls if you are satisfied. As Ambjörnsson states the unwillingness to express that they were proud of their bodies do not mean that some of the girls thought that their body was okay. Rather it shows the importance to know where and how you talk about your body, regardless of how you feel about it. Considering this it could on the one hand be seen as strange that one of the bloggers I analyze says that she is normal. On the other hand the blogger does not express satisfaction with her body, which the comments suggest that she should feel.

As stated earlier, writing about the normal body is assuming that there are an abnormal body. Normal could be seen as being not too thin or too fat. When talking about the normal, abnormal and norm one always have to see in relationship to power structures. Someone decide what is normal. In the blogs the commentators appear to be the ones with the power to judge what a normal body is. As Harjunen (2009) writes, the construction of the norm also constructs the other. Furthermore, various social actors and institutions are involved in building the boundaries between the normative and non-normative bodies and categorizing them as such. As a consequence, healthy bodies become separated from the unhealthy, able-bodies from the disabled, and bodies that exceed or fall behind the body norm, from the normative-sized bodies among others. (Harjunen 2009, 14) Figueroa Sarriera (2006, 116) states that contemporary deconstructionist theory reminds us of the ways in which language simultaneously includes presence and absence. The conflicting relations between the self and the other have an important position in contemporary sociopolitical discourses. According to Michel Foucault (1980) the asymmetrical relationships between the self and the other produce technologies of power. The main purpose of these technologies is the objectification of the subject in order to subjugate it to specific assumptions of what is considered as normal. (Figueroa Sarriera 2006, 116) In the third blog entry with the photograph of the blogger in her bathing suite she writes:

I took a picture of myself today and came to the conclusion that I'm not as f,at as I thought! I don't like how my stomach looks like from the front, it looks... well, big on the width; it is true, but well. I do think that my body is okay:D (at least today ^^)

She then receives the following comments:

wow, really nice body! =) be proud and take care of it <3

you are too thin

hey, you are perfect:):P

This shows how not being fat means that the blogger is okay. Only one of the commentator's states that the blogger is too thin, something that is fairly uncommon in the blogs. Rich, Harjunen and Evans (2006) found that "too big" and "too small" bodies were targeted by intense normalizing discourses and practices in school. Peer- and social pressure to become "normal" was felt by both very thin and fat girls. When commenting that the blogger is too thin it can be means to "put the blogger on the ground". It is not okay for the blogger to say that her body is okay. Another blogger writes about becoming normal:

God, let me become normal, normal and good, normal and nice. Hidden text: Normal and thin.

I did a medical examination today; I had eaten breakfast, some lunch and an apple and water. Still the scales showed a lower weight than earlier, it may never happen again!

After measuring we talked about the result of the blood test, I'm dehydrated, have a mineral deficiency and a lot of other stuff. My heart beats too slow, I think that my ECG was 47 if I remember correctly.

It looks like I have to start with nutrition drinks after all, but it feels okay. At least I don't have to eat my self to death then.

At Åhléns [a department store], in one of the fitting rooms, I tried a nice green striped sweater. When I saw my body I realized how ugly I am. It isn't pretty to look like an... I don't want to have to put it on, hm.

Anorectic, *so to speak.*

Here the things that are normal are the things that are good. Normal is good, normal is nice and normal is thin. The later statement she hides, probably because it is forbidden to promote thinness and eating disorders. Related to the discussion about what is normal, and often seen as synonymous, is the discussion about what is natural in regards of the body. One of the commentators to the blog entry above writes: "No, nobody dresses in anorexia. It is completely different if one is "naturally thin" =/ But enough is as good as a feast" Here the person who writes a comment uses the word 'naturally thin', which implies that unnaturally thin exists. The natural is at the same time connected to the 'normal'. The use of 'naturally thin' may refer to the blogger who has been to a doctor and found out that her results from the medical examination are not normal, even if she doesn't use the word 'normal' herself. The blog entry is very interesting since the blogger for the first time use the word 'anorectic'. She even emphasises the word by writing it in bold. Even if most of the blog entries that I have read are about eating disorders the words eating disorder or anorexia is seldom written. In one bloggers first blog entry she writes:

[...] One of the latest things I've done is to loose 20 kg in 6 months. And the more I loose, the more faults I find on my self. My goal is 5 more kilograms. But it is quite obvious that it will not end there. But well, I'm not exaggeratedly thin and I don't have anorexia. If I judge.

A couple of weeks later the same blogger writes that she has read her own blog entries and come to the conclusion that it sounds like she has a problem and she states that her weight is about 47 kg which she earlier have said that isn't that thin.

Research, about eating disorder, show that an often critical step in recovery from eating disorders can be exposing oneself to others. This is something that the bloggers constantly go through. This statement is in ways also a contradiction since it can lead to recovery, but it can also lead to exposure to pro-anorexia and comments from blog readers that the blogger is not thin enough, but should go on a diet. What is normal and what feels normal, changes, as read in the blog entries. Research about teen girls (for example Ambjörnsson 2004) show that when a thin girl says that she is fat, she receives response from other girls and friends who state that she is not fat at all. If a girl that is considered overweight says that she is fat the response is not the same. If she receives response the response is negative. It seems that the girl in blogs is always told by the readers that she is not fat at all, regardless of if the blog reader has seen her or not. The Internet provides, according to Duffy (2003), the context in which persuasive messages can reach a large group and where one can be part of a community while maintaining anonymity. There is a tension between the desire to go public with the stories about eating disorders and the fear of how these stories may be distorted and misrepresented as deviant. In my data one of the blogger did not want me to quote her because of this fear.

FOOD, FEELINGS AND FEEDING THE BODY

Walstrom (2000) found that the label 'having eating disorders' is an accountable identity that requires credibility. In Dias (2003) research about pro-anorexia and narratives, he found that some themes occurred repeatedly on pro-anorexia websites. These themes were: not feeling understood by those around them, feeling out of control (even though they were trying to control food and their weight), feeling isolated and in pain, using the eating disorder as a form of coping and a security blanket, recognizing that they still need that security blanket even though they were aware of the potential dangers of anorexia, needing support and connection, feeling ambivalent towards both pro-anorexia and recovery, and resisting dominant interpretations of their experiences of eating disorder. As mentioned in the earlier subchapter the bloggers write about their body as normal and abnormal, but other prominent themes are different and particular feelings, as Dias shows. The blogs studied write about feeling sick, disgusted, guilty, ashamed and sad, to name a few. The feelings are always connected to the body and usually related to food and eating. One of the blogger writes:

I feel sick looking at all those candy wrappers.

I wasn't going to,

Then I ate one, then two,

And now I don't want to count.

I'm an unsuccessful anorectic.

It is even written on a paper that arrived today.

Diagnosis: Atypical anorexia nervosa.

In the quote the blogger expresses feeling sick and unsuccessful at the same time as feeling out of control. The blogger had earlier received a diagnosis stating that she suffered from atypical anorexia nervosa. In the blog entry the sick feeling is not a result of the candy that she has eaten, but the candy wrappers that are left as a reminder of her loosing control and eating. She understands atypical anorexia nervosa as not being a good eating disordered, even if it probably means something else for the doctor who has written it on the paper. The things that you put in the mouth is said to constitute your identity. By eating more candy than she wants to count she becomes an unsuccessful anorectic. One's relationship to food is supposed to reveal much about one as an individual. In the blog entry below the blogger constructs both her body and her identity with the food she eats:

I disgust myself. Today I've eaten

Breakfast: one portion of cornflakes with milk

Lunch: a glass of soured milk and one toast.

Dinner: (three dishes!) one toast Skagen sandwich, one potato and meat and one fairly big brownie! + and one whole lot of candy.

In other words I'm so disgusted by with myself and my body, that I can't even describe it.

The blogger is repulsed with herself and her own body. She becomes disgusted by eating and by the act of putting certain things in her mouth. This kind of list is prohibited in the community ungdomar.se. This would suggest that a list of what one eats is always (according to the maintainers of ungdomar.se) something that is wrong. They seem to assume that lists of food are always connected to a problematic and unhealthy relationship to food. Mentions of disgust are always connected to food and eating in the blogs that I have studied. Another example of this is:

I will never be able to get thin! I CAN'T stop eating. I may as well just start to learn to live with my disgusting and ugly body, which I get filled with anguish from while looking at it! I get seized with dread from eating, but I still can't stop eating. I don't think before I eat, I just get to damn hungry and ooh >< I don't know, I'm thinking of surrendering to the crap.. just live a normal life instead.. if I don't starve I will start to cut my self and if I don't hurt my self in some way, I will have the fattest dread every single fucking night.. I can't take it. I just want to diiiie.

Disgust can be seen as an emotion of avoidance, or more as an emotion after not being able to avoid something. Lack of avoidance is connected to lack of control in the blogs. A very prominent way of controlling one's body is controlling the practice of eating. Everyone has to eat. But the bloggers often mention a wish to stop eating. Some well known slogans for pro-anorexia are "Hunger hurts, but starving works", "Nothing tastes as good as thin feels" and "Stay Strong. Starve On", which are a contradiction to the fact that everyone has to eat to live. Eating can be seen as something individual, but at the same time we are constantly instructed how to eat and what to eat. There are social, political, and economic pressures on girls. Society, friends and families perpetuate notions and norms that equate thinness with attractiveness (Hesse-Biber, 1996) and what is beautiful with what is good (Feingold, 1992). Historically and culturally specific codes are attached to foods and the rituals that surround them. While hunger is supposedly a natural impulse, it is a highly regulated behaviour structured by moral dictates about moderation and control. One of the bloggers writes about her relationship to her mother and food:

I know that I promised myself and mum to eat more, but lunch is still scaring me a lot. Hidden text: I don't like the feeling I get from food in my stomach, food that makes my body miserable and transforms me into the world's biggest monster.

By hiding some of the text it could be seen as both hiding the feeling and the food. The feeling makes food an enemy to the body. At the same time as bloggers often write about how little they eat (but not as little as they would want to) and that they occasionally binge, they seldom list food amounts that would be considered binge eating by medical experts. One could say that the blogs about eating disorders and the goal of thinness is mostly consisting of talk about the fat body and fatness. In general one could say that the blogs about thinness and eating disorders mention fat more often than thin. Here is one example were fat and fatness prominent even if the blogger earlier showed a photograph of herself and showing a very thin body:

It feels like everyone is looking at my disgusting stomach and fat tights. I hate being so damn fat and not lithe all the time. I really want to be thin! You that can eat how much sugar and fat as you like, without getting fat. YOU DON'T KNOW HOW JEALOUS I AM!

Research shows that the over eater is constructed as someone not very bright who just need to learn how to control herself, to be sensible, and avoid no-no foods. She lacks the feminine appeal of the thin. (Keane 2002) In research about eating disorders control is also a frequent theme, as we have seen. The person with eating disorders aims to be in control, to control their body. While others maybe see them as thin and as in control, they see themselves as fat and out of control. One of the bloggers writes:

Yesterday was perfect. I didn't eat breakfast or lunch and for dinner I only ate about 300 kcal.

Today didn't start so good. No breakfast and a medium lunch, for dinner I ate one of those meat soups in a can (I have recently discovered that they only consist of about 15 kcal/100g if you attenuate it), but then a couple of hours passed and BANG I ate non-stop. First an apple, then corn flakes and milk, then a sandwich and BANG I ate about 4-5 cupcakes!

It feels so damn... I don't know! I want to go out and run. But it isn't possible. I have to take a walk tomorrow and try to eat in small portions along the day perhaps.

I had been thinking about talking to someone about having difficulties with eating. But no, if I can eat that much in two hours it can't be that I have difficulties eating.

In the blog entry control over the body and eating is a good thing, while the feeling of being out of control is something bad, something that has to be taken care of. In the entry the blogger write about eating non-stop for two hours, starting with an apple and ending with cupcakes. Regardless of what the blogger weight is, she seems to see herself as fat and over weight. According to Elspeth Probyn (2000) the shame-filled performance of eating for the fat woman displays an identity constituted by food and her relationship to it. One of the ways in which the fat woman attempts to circumvent negative responses to her body and its practices is by eating in private. Yet a paradox arises out of hiding one's feeding performance from public eyes, since this is then interpreted as addictive behaviour, an isolated and perverted indulgence: the private fat body that eats is constituted as engaging in addictive behaviours and permitting the fulfilment of excessive desires, and yet the public "fat" body is revieled, and the feeding performance is seen as a display of wanton desires, lack of control and disregard of social codes of feminine restraint. This is interesting in a blog context. The eating may be done in private, but when writing in your blog what you eat it becomes public.

Another feeling connected to the body, food and eating, that often is expressed in the blogs studied is the feeling of not being understood. The feeling of not being understood in the blogs that I have studied is portrayed in the following blog entry:

Hidden text: "you have a body of a model"

"my friends think that you are cute and have a great body"

"you have a great body J, really great"

"you have to e at more, you are starting to be so thin"

CAN PEOPLE JUST STOP LYING?

I know that the things that you are saying aren't true, so why do you even say it?

you just make me sad.

I can't take it anymore.

stop lying to me.

[...]

I have a falafel here beside me. I don't want to. But I have to, otherwise I will become so drunk tonight again.

The blog entry received five comments, none of them anonymous and all but one of them are saying more or less that the blogger's friends are not lying:

Be careful with what you are. When one is conscious of how thin one is. Hell, it isn't nice to hear it all the time. You prefer them calling you fat? Because that makes you more motivated to eat less and vomit more? But to call you thin confirms that you have reached your goal. Only you are keeping yourself from being happy.

The comment is interesting since it is different from the others and can be seen as criticising the blogger. It does not only criticise the blogger but it also questions the bloggers intention. Amjörnsson's (2004) study showed that the interaction that leads to the unsuccessful body followed certain patterns. One of the patterns was that expressions of dissatisfaction were expected to be greeted with compliments which in turn were repaid with dismissals. The first comment above does not follow this pattern, while the other four do. This is one of the few instances in the blogs studied when complaint is followed by something else than compliments. However, it is very rare that the blogger replies to compliments with any kind of answer.

CONCLUSION: THE PROCESS OF CONSTRUCTING THE BODY IN BLOGS

The aim of this study has been to analyse how the eating disordered body is staged in textual performances online and specifically in blogs among Swedish girls in the community ungdomar.

se. Searching for the word body in blogs in the youth community ungdomar.se resulted in finding that about 80% of the blog entries were about a problematic relationship to the body, most often eating disorders, and always unsatisfaction of some sort.

I have found that the texts about the body and eating disorder consist of typographical emphasises and emoticons. The bloggers emphasis certain words related to the body, for example fat becomes even fatter by writing it in bold, big becomes bigger by writing it as biiiiig. The community have several rules and one of them is directly connected to eating disorders and pro-anorexia, an online community where eating disorders are seen as a life style instead of an illness. According to the rules of the community is forbidden to write about pro-anorexia. I found that the bloggers use different techniques to hide their posts about the body, thinness and fatness. One example is that they use punctuations in words so that it will be difficult to find blog entries consisting of lists of food, measurements and thoughts about dieting.

The themes of feeling normal or abnormal in relation to the body and others are expressed in the blogs. The bloggers write about being normal, about being naturally thin, but also about eating as an abnormal practise. Food is often mentioned in blogs about eating disorders. The bloggers mention that they eat in the quiet. Eating is something private but becomes public by writing about it in a blog. It seems like you have to blog in a certain way and about a certain body image or relationship to your body in order to get comments.

The purpose is not to argue that all girls are occupied with these kind of thoughts about their bodies, but the self is very much constructed through the body, weight, body image and eating disorders for the girls studied. A girl should mention dissatisfaction with her body to receive comments from other girls. Because of this I argue that I study the construction of girlhood through the construction of the ideal body online. At the same time blogging for these girls is a ritual of exchange. The construction of the body and girlhood becomes social and collective because of the interaction that is possible in the blogs.

REFERENCES

Aapola, S., Gonick, M., & Harris, A. (2005). *Young Feminity. Girlhood, Power and Social Change*. New York: Palgrave Macmillan.

Ambjörnsson, F. (2004). *I en klass för sig. Genus, klass och sexualitet bland gymnasietjejer*. Stockholm, Sweden: Ordfront.

Arkhem, H. (2005) *Jätten i spegeln*. Richters förlag.

Baron, N. (n.d.). *The Myth of Impoverished Signal: Dispelling the Spoken-Language Fallacy for Emoticons in Online Communication*. Retrieved December 5, 2009, from: http://www1.american.edu/tesol/Scholarly%20Documents,%20Articles,%20Research/Baron_Emoticons%5B1%5D.pdf.

Bell, R. (1985). *Holy Anorexia*. Chicago: University of Chicago Press.

Blood, R. (2000). *Weblogs: a History and Perspective*. Retrieved 13 September 2007, found at http://www.rebeccablood.net/essays/weblog_history.html.

Bordo, S. (1993). *Unbearable Weight. Feminism, Western Culture and the Body*. Berkeley, CA: California University Press.

boyd, d. (2008). *Taken Out of Context American Teen Sociality in Networked Publics*. Unpublished doctoral dissertation, University of California, Berkley.

Brumberg, J. (1988). *Fasting Girls. The Emergence of Anorexia Nervosa as a Modern Disease*. Cambridge, MA: Harvard University Press.

Burton, G. (2004). *Media and Society: Critical Perspectives*. Berkshire/GBR: McGraw-Hill Education.

Butler, J. (1990). *Gender trouble. Feminism and the subversion of identity*. New York: Routledge.

Chaney, D. (2002). *Cultural Change and Everyday Life*. Basingstoke, UK: Palgrave.

Daher-Larsson, A. (1990). *Ormen och jag. Om en kamp mot anorexia*. Södra Sandby: Atlantis.

Dahlberg, C. (2000). *I himlen får jag äta*. St. Petersburg: Ord & Visor förlag.

Danet, B. (2001). *Cyberpl@y: Communicating Online*. Oxford, UK: Berg.

Denks, D., Bos, A. & von Grumbkow (2008). Emoticons and Online Message Interpretation.

Dias, K. (2003). The Ana Sanctuary: Women's Pro-Anorexia Narratives in Cyberspace. *Journal of International Women's Studies*, *4*(2).

Dijck van, J. (2005). Composing the Self: Of Diaries and Lifelogs. *Fibreculture*, (3).

Drotner, K., & Rudberg, M. (Eds.). (1993). *Dobbeltblikk på det moderne: Unge kvinners hverdagsliv og kultur i Norden*. Oslo, Sweden: Universitetforlaget.

Duffy, M. E. (2003). Web of Hate: A Fantasy Theme Analysis of the Rhetorical Vision of Hate Groups Online. *The Journal of Communication Inquiry*, *27*, 291–312.

Ehn, A. (1995). *Vårfrost*. Stockholm, Sweden: Norstedt.

Featherstone, M. (1982). The Body in Consumer Culture'. *Theory, Culture & Society*, *1*(2), 18–33. doi:10.1177/026327648200100203

Feingold, A. (1992). Good-looking people are not what we think. *Psychological Bulletin*, *111*, 304–341. doi:10.1037/0033-2909.111.2.304

Figueroa Sarriera, H. (2006). Connecting the Selves Computer-Mediated Identification Processes. In Silver, D. (Ed.), *Critical Cyberculture Studies*. New York: NYU Press.

Foucault, M., & Gordon, C. (Eds.). (1980). *Power/Knowledge. Selected Interviews and other Writings 1972-1977*. Brighton, UK: Harvester Press.

Franck, M. (2009). *Frigjord oskuld. Heterosexuellt mognadsimperativ i svensk ungdomsroman*. Åbo, Finland: Åbo Akademi University Press.

Ganetz, H., & Lövgren, K. (Eds.). (1991). *Om unga kvinnor*. Lund, Sweden: Studentlitteratur.

Gavin, J., Rodham, K., & Poyer, H. (2008, March). The Presentation of "Pro-Anorexia" in Online Group Interactions. *Qualitative Health Research*, *18*(3), 325–333. doi:10.1177/1049732307311640

Giles, D. (2006). Constructing identities in cyberspace: The case of eating disorders. *The British Journal of Social Psychology*, *45*, 463–477. doi:10.1348/014466605X53596

Göthlund, A. (1997). *Bilder av tonårsflickor. Om estetik och identitetsarbete*. Tema Kommunikation: Linköpings universitet.

Harjunen, H. (2009). *Women and Fat. Approaches to the Social Study of Fatness*. Jyväskylä, Finland: Jyväskylä University Press.

Herring, S. C. (2004). Slouching toward the ordinary: Current trends in computer-mediated communication. *New Media & Society*, *6*(1). doi:10.1177/1461444804039906

Hesse-Biber, S. (1996). *Am I Thin Enough Yet? The Cult of Thinness and the Commercialization of Identity*. New York: Oxford University Press.

Kawaura, Kawakami & Yamashita. (1998). Keeping A Diary in Cyberspace. *The Japanese Psychological Research*, *40*(4), 234–245. doi:10.1111/1468-5884.00097

Keane, H. (2002). *What's Wrong with Addiction?*Victoria, Australia: Melbourne University Press.

Kearney, M. (2009). Coalescing: The Development of Girls' Studies. *NWSA Journal, 21*(1), 1–28.

Kiesler, S., Siegel, J., & McGuire, T. W. (1984). Social psychological aspects of computermediated communication. *The American Psychologist, 39*, 1123–1134. doi:10.1037/0003-066X.39.10.1123

King, S. A. (1996). Researching Internet communities: Proposed ethical guidelines for the reporting of results. *The Information Society, 12*(2), 119–128. doi:10.1080/713856145

Kleven, K. (1992). *Jentekultur som kyskhetsbelte. Om kuturelle, samfunnsmässige og psykologiske endringer i unge jenters verden*. Oslo, Sweden: Universitetsforlaget.

Langellier, K. (1998). Voiceless Bodies, Bodiless Voices: The Future of Personal Narrative Performance. In Dailey, S. J. (Ed.), *The Future of Performance Studies: Visions and Revisions* (pp. 207–213). Annandale, VA: National Communication Association.

Langellier, K., & Peterson, E. (2004). *Storytelling in Daily Life: Performing Narrative*. Philadelphia, PA: Temple University Press.

Liu, H. (2007). Social Network Profiles as Taste Performances. *Journal of Computer-Mediated Communication, 13*(1), 13.

McCabe, J. (2009). Resisting Alienation: The Social Construction of Internet Communities Supporting Eating Disorders. *Communication Studies, 60*(1), 1–16. doi:10.1080/10510970802623542

McNeill, L. (2003). Teaching an Old Genre New Tricks: The Diary on the Internet. *Biography, 26*(1). doi:10.1353/bio.2003.0028

McRobbie, A. (1978). 'Working Class Girls and the Culture of Femininity'. In Women'sStudies Group, Centre for Contemporary Cultural Studies (eds). *Women Take Issue: Aspects of Women's Subordination*, pp. 96–108. London: Hutchinson.

McRobbie, A. (1991). *Feminism and Youth Culture: From "Jackie" to "Just Seventeen."*. Boston: Unwin Hyman.

Näre, S., & Lähteenmaa, J. (1992). *Letit liehumaan! Tyttökulttuuri*. Helsinki: Suomalaisen Kirjallisuuden Seura.

Norring, D., & Engström, I. (2002). Ätstörningarnas förekomst. In Clinton, C. (Ed.) *Ätstörningar. Bakgrund och aktuella behandlingsmetoder*. Stockholm: Natur och Kultur.

Ogden, J. (1992). *Fat Chance! The Myth of Dieting Explained*. London: Routledge.

Oinas, E. (2001). *Making Sense of the Teenage Body–Sociological Perspectives on Girls, Changing Bodies, and Knowledge*. Åbo, Finland: Abo Akademi University Press.

Ojanen, K. (2005). Tallitytöt. Harrastus tyttöyksien tekemisenä. In Saarikoski, H. (Ed.), *Leikkikentiltä. Lastenperinteen tutkimuksia 2000-luvulta* (pp. 202–138). Helsinki, Finland: Suomalaisen Kirjallisuuden Seura.

Ojanen, K. (2008). Tyttötutkimuksen tytöt: keskusteluja moninaisuudesta ja tyttöjen vallasta. *Elore, 15*, 1/08.

Probyn, E. (2000). *Carnal Appetites: FoodSexIdentities*. London, New York: Routledge.

Reaves, J. (2001). *Anorexia Goes High Tech*. Time Magazine 31 July 2001. Retreived October 13, 2009, from http://www.time.com.

Rich, E., Harjunen, H., & Evans, J. (2006) Normal gone bad' – Health Discourses, Schools and the Female Body. In Peter Twohig & Vera Kalitzkus (Eds) *Bordering Biomedicine Interdisciplinary Perspectives on Health, Illness and Disease*. New York: Rodopi.

Saarikoski, H. (2001). *Mistä on huonot tytöt tehty?* Helsinki, Finland: Tammi.

Sarbaugh-Thompson, J. S., & Feldman, M. S. (1998). Electronic mail and organizational communication: Does saying "hi" really matter? *Organization Science*, *9*(6), 685–698. doi:10.1287/orsc.9.6.685

Schwartz, H. (1986). *Never Satisfied. A Cultural History of Diets, Fantasies and Fat*. New York: Free Press.

Serfaty, V. (2004). Online Diaries: Towards a Structural Approach. *Journal of American Studies*, *38*(3). doi:10.1017/S0021875804008746

Sharf, B. F. (1997). Communicating breast cancer online: Support and empowerment on the Internet. *Women & Health*, *26*(1), 65–84. doi:10.1300/J013v26n01_05

Social Science Computer Review, *26*(3), 379–388. doi:10.1177/0894439307311611

Solheim, J. (2001). *Den öppna kroppen. Om könssymbolik i modern kultur*. Göteborg, Sweden: Daidalos.

Sorapure, M. (2003). Screening Moments, Scrolling Lives: Diary Writing on the Web. *Biography*, *26*(1). doi:10.1353/bio.2003.0034

Sproull, L., & Kiesler, S. (1991). *Connections*. Cambridge, MA: MIT Press.

Stern, S. (2008). Producing Sites, Exploring Identities: Youth Online Authorship. In Buckingham, D. (Ed.), *Youth, Identity, and Digital Media* (pp. 95–118). Cambridge, MA: The MIT Press.

Strandberg, L. (2000) *När mörkret kom – en bok om anorexia*. Falun: Författarhuset.

Sveninsson, M. (Ed.). (2007). *Cyberfeminism in Northern lights: digital media and gender in a Nordic context*. Newcastle-upon-Tyne: Cambridge Scholars.

Thulin, K., & Östergren, J. (1997). *X-märkt: flickornas guide till verkligheten*. Stockholm, Sweden: Rabén & Sjögren.

Tierney, S. (2006). The Dangers and Draw of Online Communication: Pro- Anorexia Websites and their Implications for Users, Practitioners, and Researchers. *Eating Disorders*, *14*(3), 181–190. doi:10.1080/10640260600638865

Walstrom, M. K. (2000). "You know, Who's the Thinnest?": Combating surveillance and creating safety in coping with eating disorders online. *CyberPsychology & Behaviour*, *3*, 761–783. doi:10.1089/10949310050191755

Walther, J. B. (2002). Research ethics in Internet-enabled research: Human subjects issues and methodological myopia. *Ethics and Information Technology*, *4*, 205–216. doi:10.1023/A:1021368426115

Ward, K. (2007). 'I Love You to the Bones': Constructing the Anorexic Body in 'Pro-Ana' Message Boards. *Sociological Research Online*, *12*(2). doi:10.5153/sro.1220

Waskul, D., & Douglas, M. (1996). Considering the electronic participant: Some polemical observations on the ethics of online research. *The Information Society*, *12*(2), 129–139. doi:10.1080/713856142

ADDITIONAL READING

Adams, N., & Bettis, P. (2005). *Geographies of girlhood. Identities in-between*. New York: Taylor & Francis.

Jiwani, Y., Steenbergen, C., & Mitchell, C. (2006). *Girlhood: redefining the limits*. Montrael, Canada: Black Rose Books.

Kennedy, H. (2003). Technobiography: Researching Lives, Online and Off. *Biography*, *26*(1), 120–139. doi:10.1353/bio.2003.0024

Nugent, C. (2010). *Conceptualizing the sexualization of girlhood through online public discourse*. Unpublished M.A. dissertation, Boston College, United States.

Thiel Stern, S. (2007). *Instant Identity: Adolescent Girls and the World of Instant Messaging*. New York: Peter Lang.

Willis, J. L. (2007). Girls Constructing Identity and Transforming "Femininity": Intersections between Empirical and Theoretical Understandings of 21st Century Girlhood" *Paper presented at the annual meeting of the National Women's Studies Association, TBA, St. Charles, IL, Pheasant Run.*

KEY TERMS AND DEFINITIONS

Blog: A frequent, chronological publication of personal thoughts and web links.

Girlhood: A culturally and socially constructed category.

Online Community: a group of people with similar interests who connect on a web site.

Eating Disorder: a psychological disorder that involve insufficient or excessive food intake

Emoticon: keyboard letters and symbols, usually expressing a facial expression

ENDNOTES

1. For example Dias (2003), Tierney (2006), Ward (2007) and Gavin, Rodham & Poyer (2008),
2. For example Daher-Larsson (1990), Ehn (1995), Strandberg (2000), Dahlberg (2000) and Arkhem (2005).
3. For example Thulin et al (1997) and Elf Karlén et al (2003).
4. For example for example McRobbie 1978 and 1991.
5. For example Ganetz & Lövgren 1991, Kleven 1992, Näre & Lähteenmaa 1992, Drotner & Rudberg 1993 and Göthlund 1997
6. Oinas (2001) studied girlhood and menstration, Saarikoski (2001) studied girlhood and verbal bullying among girls in schools, Ambjörnsson (2004) studied how desire and relationships are related to the becoming of a girl, Ojanen (2005) studied the construction of girlhood in horse stables and Franck (2009) studied girlhood and heterosexual developmental imperative in Swedish Young Adult Novels.
7. According to the community, ungdomar.se started as an online version of a youth-friendly clinic where young people had the opportunity to ask questions. In the autumn of 1999, an initiative to start a home page, where doctors, psychologists, midwives, study and vocational counselors could give advice, was done. The home page was first named ungdomsmottagningen.com (in English youthfriendlyclinic) and was opened one year later. In 2009 ungdomar.se was described as one of the biggest communities in Sweden for young people. Ungdomar.se consist of forums, articles, "friends on duty", photograph albums, blogs, chat and web television. The members are between 13 and 30 years old and 59% of the members are young women. Ungdomar.se is today privately owned and free-standing initiative. The community is developed and administered by Youmeet Technology AB.

Chapter 5
To Be Continued... Fan Fiction and the Constructing of Identity

Patrik Wikström
Jönköping International Business School, Sweden

Christina Olin-Scheller
Karlstad University, Sweden

ABSTRACT

This chapter contributes to the existing body of knowledge on fan fiction by reporting the findings from a quantitative and qualitative study on fan fiction in a Swedish context. The authors contextualize the fan fiction phenomenon as a part of a larger transformation of the media sphere and the society in general where media consumers' role as collaborative cultural producers grows ever stronger. They explore what kind of stories inspire the writers and conclude that as in many other parts of the entertainment industry, fan fiction is dominated by a small number of international media brands. The authors show how fan fiction can play an important role in the development of adolescents' literacies and identities and how their pastime works as a vehicle for personal growth.

INTRODUCTION

Storytelling is a fundamental part of human culture. Throughout the history of human civilization storytelling have allowed communities and societies to convey knowledge, norms and values between generations, and it has served as an important tool for people – young and old – in the crafting their identities. The act of creating and sharing stories is by necessity based on some kind of communication technology and the character of these technologies influences *how* stories are created and shared and to some extent *what kind* of stories are created and shared. Clay tablets, codices, eReaders, and the remaining plethora of more or less influential communication technologies all contribute to the development of storytelling.

During the last centuries, traditional tools for writing and distributing literature have established a romantic image of the male author who creates and shares his stories with the world, and an admiring audience which listens and appreciates the works of the genius. It seems as the latest major communication technology – the Internet

DOI: 10.4018/978-1-60960-209-3.ch005

– questions this basic structure and enables the establishment of a somewhat less unidirectional structure for storytelling. The Internet allows the formerly passive audience to actively contribute to the stories which are shared among the members of global and local communities. This chapter explores how one specific aspect of the Internet transforms an equally specific aspect of storytelling. At the centre of our attention lies a phenomenon known as *fan fiction* – stories based on well-established[1] characters and structures, but written by the fans of these well-established characters rather than by the original author. While fan fiction is far from a new phenomenon, we will look at how the Internet has released this form of expression from the shackles of the offline world and enabled it to develop into an integral and vital part of Internet culture.

After introducing concepts, theoretical approaches, and methods, the chapter will present the findings from a quantitative and qualitative study on fan fiction and fan fiction writers. Questions regarding who is writing fan fiction and what kind of stories they are inspired by are followed by an exploration of how the development of literacies and identities are stimulated by reading and writing fan fiction. Lastly, the chapter makes general conclusions and suggests opportunities for future research.

BACKGROUND

After the industrial revolution and before the advent of digital communication technologies, the capability to create and disseminate information throughout society was concentrated to a limited number of fairly powerful organizations. These organizations were strong enough to be able to acquire the scarce resources and the expensive equipment required in order to operate a newspaper, a television station, or a book publisher to be reckoned with. Common people were more less shut out from cultural production and were sentenced to the role of the consumer – passively watching, reading and listening to the works of others.

However, during the second half of the last century, technological development contributed to a radical increase in the accessibility of these scarce communication resources. More and more people were able to create their own stories and share their creations with the world. New practices for using media content emerged and a culture sometimes referred to as a 'remix culture' (e.g. Lessig, 2008) or a 'participatory culture' (e.g. Jenkins, 2006a) became part of the normal way of life in the digital world. Today, the world's six most visited web sites (excluding search related sites) can be categorized as *peer* media i.e. media services which are (1) interactive; (2) the most valuable content is generated by amateurs rather than by professionals; (3) and the emphasis is placed on contact and community elements rather than on information per se (Küng, 2008: 86; http://www.alexa.com).

The transformation of consumers of media into *producers* of media has been noted by many scholars and can be considered a part of a general societal transformation (e.g. Firat, 1987). Several scholars have explored this transformation from a media perspective and have used terms such as 'prosumption' (Toffler, 1980) or 'produsage' (Burgess & Green, 2009) to label this kind of media consumer. Others have suggested the concept of 'contribution' as a way to de-emphasize the otherwise somewhat troubling production-consumption dichotomy (e.g. Olin-Scheller & Wikström, 2009).

Fan fiction is a phenomenon which has a longer history but which nevertheless can be seen as a specific manifestation of these trends. Fan fiction is normally written without a commercial purpose and without the approval of the author of the original text. Throughout history there are numerous examples of fans who have been inspired to write their own stories based on the characters which they treasure. One early example is the group of Jane Austin followers who during the end of the

19th century began to refer to themselves as Janeites and engaged in the writing and reading of fan fiction based on Austin's stories and characters. The origin of fan fiction as it can be seen today can usually be traced back to the 1960s and Anglo-Saxon *Star Trek* fan communities and Japanese manga[2] fan communities (e.g. Hellekson & Busse, 2006; Jenkins, 1992; Pugh, 2005). Almost half a century after its modern origin, Anglo-Saxon and Japanese popular culture continue to dominate the field and manga and science fiction are still able to inspire the fan fiction communities.

While the history of fan fiction is long, it is the emergence of digital communication technologies which have allowed the phenomenon to develop into a fictional genre which requires to be taken seriously. Before the Internet fan fiction was a fictional genre which was produced more or less as any other kind of fiction literature – by a single author in solitude. After the advent of the Internet the genre has transformed into an communal activity where texts are shaped and created by several writers which are collaborating and communicating online.

Several scholars have examined fan culture and fan fiction in particular since the end of the last century. Primarily there has been an interest from scholars in the field of in the field of cultural studies and sociology, such as Baker & Brooks, (1998), Hills (2002), Jenkins (1992, 2006a, 2006b), Lewis (1992), and Sandvoss (2005). The phenomenon has also been explored by scholars in other disciplines. For instance, Coombe (1998) has explored fan fiction from a copyright perspective and Kozinets (1999) has linked fan culture to copyright related issues.

IDENTITIES AND LITERACIES

Drotner (1991), Goffman (1959) and others argue that identity work shall not be considered as something static and unchangeable. On the contrary, identity is something that is constructed through different actions and in the interaction with others. By comparing your own actions and your own identities with others', learning and personal development can be enabled and stimulated (Ashcroft, Griffiths, & Tiffin, 2007). Online communities strengthen this process since such environments make it possible for users to perform and play with different identities in much more unconventional and liberated ways than what would have been possible in the offline world (e.g. Hellekson & Busse, 2006; Hällgren, 2006; Robinson, 2007; Turkle, 1997).

In many online communities, the most important part of the users' identity work takes place on the users' 'profile pages' (e.g. Turkle, 1997). On these pages they display different artifacts and objects which are supposed to represent some kind of identity and communicate a meaningful message to other community members. However, even though profile pages also exist within many fan fiction sites these profiles play a relatively minor role in the users' identity work. Rather, the users' identities are primarily shaped through the fiction they write, and the comments, the chats, and the discussions they engage in with their peers.

As a consequence, the act of reading and writing becomes a vital part of the users' identity work, and the users' ability to perform these acts – their literacy – has a strong impact on their ability to shape and experiment with their identities. Traditionally literacy have often been associated with a person's technical skills of reading, writing and calculating, but today the meaning of the term literacy has been expanded to refer to "the ability to identify, understand, interpret, create, communicate and compute, using printed and written materials associated with varying contexts" (UNESCO, 2004: 13). It is necessary to talk about a range of different communicative abilities, and to refer to these as litera*cies* rather than literacy.

It is possible to think of fan fiction writers' development of their literacies and their identities as a complex but nevertheless unified process.

As the writers' literacies develop, they acquire a wider toolbox which allows them to write more sophisticated stories and to engage in more advanced interaction with their community peers. Through this process they are also able to meet and explore alternative behaviours, values, and roles which allow them to grow and develop their identities (Erikson, 1959). It is important to recognize that this process is deeply communal in its character. The development of a fan fiction writer's literacies and identities is embedded within corresponding processes on community level. Norms, values and practices for how fiction texts should be interpreted, used and discussed are continuously evolving through a negotiation between community members. Such a community is often referred to as an *interpretive* community, based on a common understanding of how texts ought to be written and constructed for different purposes (Fish, 1976; Olin-Scheller, 2008).

Based on this reasoning we explore the intertwined development of identities and literacies. Similar studies have been done in other settings, for instance by Jimenez (2000) who examined how Latina/o students in Midwestern United States developed their identities and their literacy in English. One way of studying identity and literacy development is to examine the texts which the writers produce during an extended time period and to analyze these texts in order to establish the character of the writers' development. Such textual analyses are indeed useful and valuable, but in this chapter we present findings based on another kind of empirical data. Rather than using texts per se we use a phenomenologically inspired approach and conduct interviews with the writers in order to understand how they think of their fan fiction practices and their own personal development. When using this approach we follow in the footsteps of several other scholars who have studied identity formation using introspective interview data as empirical material, such as Adams & Montemayor (1983), Allison & Schultz (2001), Flum (1994), Grotevant, Thorbecke, & Meyer (1982), Harter (1990), Josselson, Greenberger, & McConochie (1977a & 1977b), Meilman (1979), and Paikoff & Brooks-Gunn (1991).

A NOTE ON RESEARCH METHODS

The empirical data used in this study consist of the aforementioned interviews in combination with data from an online survey and observations of a fan fiction website. All these three data sources will be briefly discussed in this section.

The purpose of the quantitative online survey was to map the extent and basic characteristics of the field fan fiction. The survey was answered by 932 Internet users in the age group 15 to 24 years old. The questions in the survey concerned *fan works* rather than to fan fiction alone. Fan works is a term which refers to all kinds of artifacts created by fans: literature, videos, music, games, images, etc. By using this approach we were able to compare fiction writing with other forms of expression and were able to provide a richer context to our study. We used Sweden as the geographical context, but due to the very international character of the fan fiction phenomenon we argue that the findings are relevant for fan fiction in most parts of the world. It is important to note that we primarily use the results from the survey study for descriptive purposes and only to guide the most important part of our project, i.e. the qualitative interview study. We do not use the findings from the survey to explain our observations or to establish any causal relationships.

Based on the findings from the survey we concluded that the web site *fanfiction.net* was the largest and most important archive and community for fan fiction. *Fanfiction.net* is a US based website where most of the published texts and the conversations take place in English. Nevertheless, there are fan fiction writers from all over the world publishing stories in their own language on *fanfiction.net*. From *fanfiction.net* we were able to gather descriptive data showing the number of published stories, which characters were more popular than others, etc. Some of this data is presented later in this chapter. However, the most important use of the website was to locate the informants for our interview study.

For practical reasons we wanted to find Swedish writers publishing on *fanfiction.net*. Using the website search engine we identified a number of writers who had published texts in Swedish. Since as in most communities, the members are able to present themselves with a 'profile page' we were able to find out some more basic information about these writers. Two important parts of the profiles are the 'Friends' and 'Favorites' section. In these sections, writers link to other fan fiction writers to which they feel some kind of connection. Using these links we used a cascading technique to find additional Swedish fan fiction writers which we considered to be potential informants.

In total we were able to identify 81 fan fiction writers which were contacted via email or via internal messages within the website. We soon realized that it was difficult to find and get in touch with male fan fiction writers. Out of the 81 writers on our original list, we were able to identify 5 writers as boys, and the 31 writers who eventually participated in our interview study were all girls. There are several ways to explain the female domination within our sample. For instance, one reason might be that we by coincidence happened to find a group of girls who only collaborate with other girls. If we had looked a bit closer we might have been able to find several male fan fiction writers. However, we have indeed searched from several different entry points starting in several different fandoms[3], without finding a single boy willing to participate in our study. Hence we claim that if there had been a large group of male fan fiction writers active on the website – we would have found them.

A second explanation might be that male fan fiction writers publish their works on other websites. In order to reduce this particular risk we have also tried to find informants on other websites, and we have nevertheless ended up with the similar result.

A third explanation might be that fan fiction writing boys think their pursuit is too private and they are not willing to engage in a discussion with curious researchers. This explanation might very well be true. None of the boys who we were able to locate were willing to participate in our study.

A fourth reason could be that boys perhaps for some reason do not engage in fan fiction writing in the same extent as girls. Based on our findings we conclude that this probably is the most likely explanation to the lack of boys in our study. Previous studies (e.g. Carlsson, 2010) have pointed at the female domination of the field of literary consumption and a similar gender imbalance have been observed within the field of fan fiction (e.g. Gray, 2008). Hence, the result of our search for informants may simply be yet another manifestation of this situation.

Out of the 31 interviews, 9 interviews was answered via email, 20 were conducted via telephone, and 2 at real-life meetings. There are primarily two reasons behind the different interview modes. One is geography and logistics and the other reason is personal integrity – some of the writers did not want to talk in real-time but preferred to answer in writing.

The semi-structured interviews were conducted during fall 2008. For obvious reasons, the email interviews were more structured and less interactive than the real-time interviews, but otherwise we were unable to observe any systematic differences between the interview results based on the interview mode.

The first real-time interview was jointly made by the two authors in order to synchronize the authors and ensure that both would conduct the remaining interviews in a similar style. The remaining informants were arbitrarily split into two groups which were interviewed by the authors on a one-on-one basis. The interviews, which in average lasted for 30 minutes, were recorded and transcribed independently by the two authors. The data was then merged and both authors analyzed the entire material from all 31 interviews in parallel. The authors' findings were then discussed and any dissonances between the two analyses were sorted out. This process ensured that the data that was created is of high quality and reliability.

We use quotes from the interviews to illustrate and to strengthen a certain reasoning. It should be noted that the interviews were conducted in Swedish and translated into English which is why expressions and grammar in the quotes might be somewhat unorthodox.

THE FAN FICTION COMMUNITY

There are different ways to be involved in fan fiction such as reading the works of other fans; giving feedback and discussing these texts; or writing and publishing you own fan fiction texts. In order to give feedback to a text it is obviously necessary to read it, but otherwise there are no dependencies between the three activities. During our study we found fan fiction users of all kinds, including those who were exclusively reading or exclusively writing fan fiction and those who were engaged in all three activities.

Our online survey showed that approximately six percent of all Internet users within our age group (15 to 24 years old) regularly read some kind of fan work, two percent regularly commented on these texts and one percent regularly created and published their own texts. These findings correspond well with those presented by other studies on user behavior, content contribution, and online communities (e.g. Horowitz, 2006).

The survey, which also covered other forms of expression besides literature, showed that there are no significant differences between boys' and girls' involvement in the production and consumption of fan created content. However, as already mentioned in the methods section, we were able to find significant differences in *what kind* of media the boys and girls prefer. The survey showed that girls are more interested in written text while the boys preferred visual forms of expressions such as pictures and videos. These findings are confirmed by the findings from other studies on media habits and gender, e.g. Findahl (2009) and Carlsson (2010).

THE STORIES THAT INSPIRE

Even though more or less any story, character, or fictional universe can inspire to fan fiction writing, some stories are significantly more popular than others. There are more than three million fan fiction stories on *fanfiction.net* inspired by a plethora of different books, movies, TV series, actual persons, etc. However, even though there seem to be a considerable diversity, the field of fan fiction is dominated by a few very influential stories. The pie chart in figure 1 illustrates that the ten most popular stories constitute almost forty percent of all stories published on *fanfiction.net* and the fifty most popular stories constitute two thirds of the published fan fiction stories (Figure 1).

A basic examination of the fifty most influential stories show that they share some important characteristics. For instance, all of them are high-profile international media franchises, supported by major multinational media companies. We can also note that besides the two British media franchises (*Harry Potter* and *Lord of the Rings*) on the list, all the other ones are either of Japanese or American origin.

The examination of the stories published on *fanfiction.net* also showed that forty of the fifty most popular stories either belong to the manga, the science fiction or the fantasy genre. Apparently, stories based on imaginative worlds and characters are more popular than those which are at least visually closer to reality[4]. One such imaginative world which has dominated the fan fiction genre for many years and inspired to approximately fifteen percent of all fan fiction stories published on *fanfiction.net* is J.K. Rowl-

Figure 1. Two thirds of all fan fiction stories published on fanfiction.net are based on 50 different stories, while the remaining third is based on thousands of stories, TV series, movies, video games, etc.

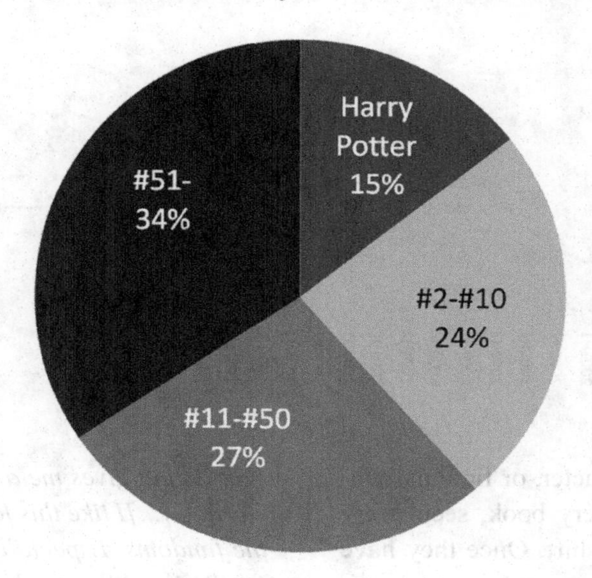

ing's *Harry Potter*. As illustrated by the graph below, the rise of *Harry Potter* as the leading fandom within the field began in the beginning of the 2000s, about the same time as the first novels in the series premiered as film adaptations. It seems as in order for a fandom to really take off, it is not sufficient for it to be a popular book, it also has to be available as a film, a TV series, or a video game. This generalization is true for all the fandoms on the top ten list and for most of the fandoms on the top fifty.

Table 1. The ten most popular media franchises on fanfiction.net

1	*Harry Potter*
2	*Naruto*
3	*Twilight*
4	*Inuyasha*
5	*Final Fantasy*
6	*Yu-Gi-Oh*
7	*Kingdom Hearts*
8	*Lord of the Rings*
9	*Gundam Wing*
10	*CSI*

Another media franchise which also has been packaged as several different media products and adaptations is Stephanie Meyer's immensely popular *Twilight* series. As illustrated by the graph in Figure 2, the *Twilight* series has within only a few years been able to pass *Harry Potter* in terms of stories published during a six-month-period. During 2009, the fandom actually was the most productive fandom of all on *fanfiction.net*.

FAN FICTION AS A SOCIAL PRACTICE

The fan fiction writers who participated in our interview study spend several hours every week reading and writing fan fiction. Fan fiction is one of their most important pastimes and comparable to other entertainment activities such as watching television and reading books. One way of explaining their fascination for fan fiction is to recognize that the fans more than anything are passionate

Figure 2. Harry Potter and Twilight are two of the most popular fandoms

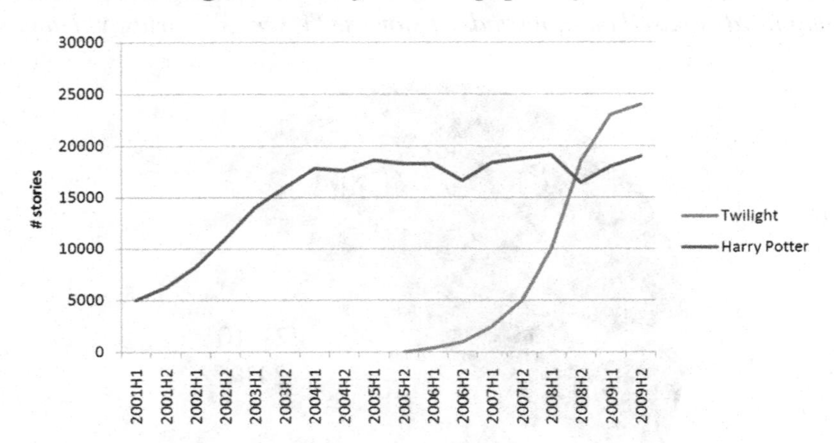

about a certain story, character, or fictional universe. They have read every book, seen every movie, and bought the T-shirt. Once they have realized that there are no more products or experiences within the fandom which they can immerse themselves into, they search in ever wider circles and eventually they discover fan fiction. Reading and writing fan fiction becomes an additional leisurely and entertaining way for them to stay within the fandom, to read others' stories about their favorite characters and to invent entirely new stories themselves. Informant #1 explains:

The thing is that in fan fiction you get what you really want. If I like Faramir in The Lord of the Rings best, I can choose to read stories where he is the main character. The demands are also low, this is liberating. It's like day-dreaming or junk-food, it doesn't have to be high quality – it's just fan fiction!

Another perhaps even more fundamental aspect of fan fiction is its communal features. As most manifestations of fan culture, fan fiction is a social practice where likeminded people meet to discuss and deliberate the intricacies of their shared passion. Informant #2 reflects on the fan fiction community:

Fan fiction gives me a lot of friends all over the world... [...]I like this feeling of solidarity within the fandoms. Especially the small genres within fan fiction, like people interested in slash[5] come together or people preferring certain types of pairings and things like that. This creates a whole culture just between those parts..."

Besides the community and entertainment aspects of fan fiction, the informants in our study often reflected on how their fan fiction interest has given them new skills and abilities, and allowed them to grow as a person. Some of the skills which they learn are fairly technical, such as spelling, grammar, and narratology. Informant #3 reflects on her development: "When I started six years ago I was really only good at writing dialogues. Now I'm better at describing characters, emotions, thoughts, settings etc." Informant #4 follows a similar reasoning and explains that her texts have changed a lot since she started writing and reading fan fiction.

In the beginning the chapters were often very short, and the text was basically based on dialogues. Now I have more telling texts and the chapters are longer and longer. I also try to write more parallel stories, and describe the characters more deeply, i.e. I do not only focus on the main character.

Another specific skills which is part of the fan fiction writers' literacies concern the relationship between the original story and the fan fiction story. There are very explicit norms and rules within the fan fiction community which guides this relationship. A character in a fan fiction story which does not adhere to its original traits and features is considered to be "Out-of-Character" – a serious breach against fan fiction convention. On the other hand the fan fiction story should not be too close to the original story for it to be interesting and creative. One of our informants (#5) explained during the interviews how she in her early days as a fan fiction writer, would be happy for feedback that told her that her writing was similar to J.K. Rowling's. She continued by explaining that her understanding of her own texts and the original texts had changed and that "now my ambition is to develop my own style and I want everyone to be sure that it isn't Rowling".

Another informant (#6) explains the link between literacies and identities by referring to another behavior which is not very highly regarded in the fan fiction community. The issue concerns the use of *Mary Sue*-characters in the story – i.e. to include an often overly romantic character that is an ideal image of the writer herself (Gregson, 2005). "After a while" the informant says, "you develop as a writer and also as a person and then you can create more genuine characters which are far from being Mary Sueish."

By reading, discussing and writing fan fiction, the informants are confronted with attitudes and values which may be quite different to the world they are normally used to. One of our informants told us that "previously, I wanted everyone to write the way I wanted and do what I wanted. Now I respect others' opinions and the fact that they not always want to change what they think and write." Informant #7 also reflects on how she has matured and how that influences the way she understands and interprets the texts which she reads:

In the beginning you thought like 'what's the matter? Ok, she was hit by her boyfriend, but it just a fan fic!' But you get more mature. Then it was a bigger shock if you read about domestic violence because you realize that the one writing the story might have experienced the same thing herself.

We have also seen how fan fiction, can have "therapeutical" impact on the readers and writers. Several writers mentioned how the feedback from peers – which according to fan fiction convention always should be positive and constructive – improves their self-esteem and encourage them to continue to express their thoughts and feelings in writing. An informant explains: "Mostly people say a lot of positive things which makes you so happy you go through the roof!". Another informant reflects on how fan fiction writing helps her deal with her everyday problems and worries:

You can get rid of a lot of impressions ... and hard feelings... you can let the character be like very depressed and something drastic can happen! You can sort of write away everything that sucks – that is very comforting...

One fan fiction subgenre which has received much attention during the last decades is a genre known as *slash* or *slash fiction*. A slash story gravitates around a relationship – often erotic – between two boys. It seems like this subgenre has had a big impact in the identity constructing process for many of our informants. Informant #1 talks about slash fiction: "I think I like slash because then I don't have to identify with a girl". Informant #8 also likes slash, primarily within a specific manga fandom. "I don't quite know what makes me read slash", she says and continues: "Maybe it's because girls who read about men are interested in and attracted to men. I think men are interesting and sort of exciting and I want to write about them." She points out that she is interested in slash since the female characters in the fan fiction genre often are poorly elaborated.

She explains that "the girls in the stories are either silly or very girlish so they are not good to use for pairings with the guys. So I take a girlish man and make a pairing with an ordinary man."

There also seems to be a common understanding among the informants that slash give them possibilities to experiment and stretch the envelope a bit and that the genre makes them more open-minded towards minorities. Informant #9 contemplates:

There are no books about two gay 15-year-olds having a relation where everyone thinks it's normal! I think I have become very tolerant to all kinds of sexual relationships and relations just by reading a lot of fan fics about the subject.

Informant #10 who is a very experienced fan fiction contributor, explains what fan fiction can mean during adolescence.

I've used fan fics differently while I was growing up and during adolescence. Boys and girls just entering adolescence write quite erotic stories. If you've just started discovering your sexuality you have to examine it! Fan fiction generally, and slash specifically, are good and safe ways of expressing and exploring sexuality. As a teenager you are curious and slash is different. Here you dare to go beyond the limits a bit more...

DISCUSSION AND CONCLUSION

The purpose of this chapter has been to present fan fiction as a social Internet based practice which grows in importance and deserves the attention from scholarly research. We started out by contextualizing the fan fiction phenomenon as a part of a larger transformation of the media sphere which strengthens the media consumers' role as collaborative cultural producers. We revealed the contours of the field of fan fiction by illustrating the characteristics of the fan fiction writers and the stories which inspire them to engage in the practice. The study also reported on the character of the stories which are able to inspire these young writers to spend so much of their leisure time online. Among other things we concluded that as in many other parts of the entertainment industry a limited number of international media brands dominate the field. Another interesting observation was that even though we learned from our informants that they perceived the books as more important than the film adaptations as a source of inspiration we could also see that the number of published fan fiction stories does not really take off until the books are available as films. This finding opens up for different interpretations. For instance, it could be that it actually is the visual media which truly inspire while the books provide the details required to produce fan fiction which is well-grounded in the original story. Another explanation may be that even though the books are preferred by fan fiction writers, the enormous marketing pressure created during the launch of a multimillion dollar film production promotes not only the film but the entire story as such and thereby also has implications for exotic media practices such as fan fiction.

We continued by reporting findings from our interview study which showed how fan fiction can play an important role in the development of young peoples' literacies and identities. The informants often spontaneously mentioned the learning and development they had experienced – both in terms of learning new skills and in terms of developing as a person. However, it should be noted that many of the informants in our study are in their adolescence which is a very turbulent stage of their lives and it is difficult to identify what change is driven by their fan fiction practices. But nevertheless, it is an important finding that according to the informants' own statements, fan fiction has at least played a very important role in the development they experience.

OPPORTUNITIES FOR FUTURE RESEARCH

Fan fiction has been studied from numerous perspectives during the last decades, but our study nevertheless raises questions which need to be addressed. One example of such a question relates to the gender imbalance. Where are the boys? Indications from the survey shows that boys are as interested in fan *works* as girls but that they choose to exhibit this interest in other ways than the girls. Much online research has been devoted to girls, but the research on boys social online behavior is still inadequate and deserves additional scholarly attention.

Another interesting opportunity concerns the role of fan fiction in relation to the publishers. Is fan fiction a threat, an opportunity or only a peripheral phenomenon which does not need to be considered? We believe that fan fiction is an opportunity for publishers, but further research is required in order to examine how publishers and other major media producers relate to the phenomenon.

Another relevant opportunity for further research relates to the complex interaction between literacies, identities, and the norms which are negotiated within the fan fiction community. It is interesting to note that many norms actually are fairly restrictive and conservative. Even though there are several unconventional subgenres, often the narratological structure and the characters used in the stories are very conventional and can often reiterate traditional societal roles and hierarchies. It is an interesting fact that the conservatism and traditionalism is so strong in an environment which one might expect could constitute a very liberal arena, free from the constraints and traditions reigning the real world. Our study is only exploratory in its nature and a larger study would be needed in order to further illuminate the important area in the intersection of youth culture and net culture.

REFERENCES

Adams, G. R., & Montemayor, R. (1983). Identify Formation During Early Adolescence. *The Journal of Early Adolescence*, *3*(3), 193–202. doi:10.1177/0272431683033002

Allison, B. N., & Schultz, J. B. (2001). Interpersonal Identity Formation During Early Adolescence. *Adolescence*, *36*(143), 509–523.

Ashcroft, B., Griffiths, G., & Tiffin, H. (2007). *Post-Colonial Studies: The Key Concepts* (2 ed.). Oxon, UK: Routledge.

Baker, M., & Brooks, K. (1998). *Knowing Audiences: Judge Dredd, its friends, fans and foes*. Luton, UK: University of Luton Press.

Burgess, J. E., & Green, J. B. (2009). *YouTube: Online Video and Participatory Culture*. Cambridge, UK: Polity Press.

Carlsson, U. (2010). *Mediesverige 2010*. Göteborg, UK: Nordicom.

Coombe, R. (1998). *The cultural life of intellectual properties – Authorship, Appropriation, and the Law*. Durham, NC: Duke University Press.

Drotner, K. (1991). *At skabe sig - selv: ungdom, æstetik, pædagogik* (1 ed.). Copenhagen, Denmark: Gyldendal.

Findahl, O. (2009). *Svenskarna och Internet 2009*. Stockholm.

Firat, A. F. (1987). Towards a deeper understanding of consumption experiences: The underlying dimensions. *Advances in Consumer Research. Association for Consumer Research (U. S.)*, *14*(1), 342–346.

Fish, S. M. (1976). Interpreting the "Variorum.". *Critical Inquiry*, *2*(3), 465–485. Retrieved from http://www.jstor.org/stable/1342862. doi:10.1086/447852

Flum, H. (1994). Styles of Identity Formation in Early and Middle Adolescence. *Genetic, Social, and General Psychology Monographs, 120*(4), 435–468.

Goffman, E. (1959). *The Presentation of Self in Everyday Life* (1 ed.). New York: Anchor Books.

Gray, L. (2008). A Fangirl's Crush. *Houston Chronicle*. Retrieved from http://www.chron.com/disp/story.mpl/moms/5611220.html.

Gregson, K. (2005). What if the Lead Character Looks Like Me? Girl Fans of Shoujo Anime and Their Web Sites. In S. Mazzarella, *Girl Wide Web. Girls, the Internet, Negotioation of Identity*. New York: Peter Lang.

Grotevant, H. D., Thorbecke, W., & Meyer, M. L. (1982). An extension of Marcia's Identity Status Interview into the interpersonal domain. *Journal of Youth and Adolescence, 11*(1), 33–47. .doi:10.1007/BF01537815

Hällgren, C. (2006). *Researching and developing Swedkid: a Swedish case study at the intersection of the web, racism and education /Diss./*. Fakulteten för lärarutbildning vid Umeå universitet.

Harter, S. (1990). Self and identity development. In S. S. Feldman & G. R. Elliott, *At the threshold: The developing adolescent* (1 ed., pp. 352-387). Cambridge, MA: Harvard University Press.

Hellekson, K., & Busse, K. (2006). *Fan fiction and fan communities in the age of the Internet: new essays*. Jefferson, NC: McFarland & Co.

Hills, M. (2002). *Fan Cultures*. London: Routledge.

Horowitz, B. (2006). Creators, Synthezisers and Consumers. *Elatable*. Retrieved from http://blog.elatable.com/2006/02/creators-synthesizers-and-consumers.html.

Jenkins, H. (1992). *Textual poachers: television fans & participatory culture*. New York: Routledge.

Jenkins, H. (2006). *Convergence Culture: Where old and new media collide*. New York: New York University Press.

Jenkins, H. (2006). *Fans, Bloggers, and Gamers: Exploring Participatory Culture*. New York: New York University Press.

Jimenez, R. T. (2000). Literacy and the Identity Development of Latina/o Students. *American Educational Research Journal, 37*(4), 971–1000. doi:.doi:10.3102/00028312037004971

Josselson, R., Greenberger, E., & McConochie, D. (1977). Phenomenological aspects of psychosocial maturity in adolescence. Part I. Boys. *Journal of Youth and Adolescence, 6*(1), 25–55. doi:10.1007/BF02138922

Josselson, R., Greenberger, E., & McConochie, D. (1977). Phenomenological aspects of psychosocial maturity in adolescence. Part II. Girls. *Journal of Youth and Adolescence, 6*(2), 145–167. doi:10.1007/BF02139081

Kozinets, R. (1999). E-Tribalized Marketing?: The Strategic Implications of Virtual Communities of Consumption. *European Management Journal, 17*(3), 252–264. doi:10.1016/S0263-2373(99)00004-3

Küng, L. (2008). *Strategic Management in the Media: Theory to Practice*. London: SAGE Publications.

Lessig, L. (2008). *Remix: Making Art and Commerce Thrive in the Hybrid Economy*. New York: Penguin Press.

Lewis, L. A. (1992). *The Adoring Audience: Fan Culture and Popular Media*. London: Routledge. doi:10.4324/9780203181539

Meilman, P. W. (1979). Cross-sectional age changes in ego identity status during adolescence. *Developmental Psychology, 15*(2), 230–231. doi:10.1037/0012-1649.15.2.230

Olin-Scheller, C. (2008). Trollkarl eller mugglare? Tolkningsgemenskaper i ett nytt medielandskap. *Didaktikens forum, 3*, 44-58.

Olin-Scheller, C., & Wikström, P. (2009). *Beyond the Boundaries of the Book - Young peoples' encounters with web based fiction*. Karlstad: NordMedia09, the 19th Nordic Conference for Media and Communication Research.

Olin-Scheller, C., & Wikström, P. (2010). *Författande fans*. Lund, Sweden: Studentlitteratur.

Paikoff, R. L., & Brooks-Gunn, J. (1991). Do parent-child relationships change during puberty? *Psychological Bulletin, 110*(1), 47–66. doi:10.1037/0033-2909.110.1.47

Pugh, S. (2005). *The Democratic Genre: Fan fiction in a Literary Context*. Glawgow, UK: Seren.

Robinson, L. (2007). The cyberself: the self-ing project goes online, symbolic interaction in the digital age. *New Media & Society, 9*(1), 93–110. .doi:10.1177/1461444807072216

Sandvoss, C. (2005). *Fans*. London: Polity Press.

Schodt, F. L. (1996). *Dreamland Japan: writings on modern manga*. Berkeley, CA: Stone Bridge Press.

Toffler, A. (1980). *The Third Wave*. London: Collins.

Turkle, S. (1997). *Life on the screen: identity in the age of the Internet*. New York: Touchstone.

UNESCO. (2004). *The Plurality of Literacy and its Implications for Policies and Programmes*. Paris: United Nations Educational Scientific and Cultural Organization.

ADDITIONAL READING

Appleyard, J. A. (1991). *Becoming a Reader. The Experience of Fiction from Childhood to Adulthood*. Cambridge, UK: Cambridge University Press. doi:10.1017/CBO9780511527609

Buckingham, D. (2000). *After the Death of Childhood. Growing up in the Age of Electronic Media*. Cambridge, UK: Polity Press.

Fornäs, J. (2006). *Digital borderlands: cultural studies of identity and interactivity on the Internet*. New York: Peter Lang.

Kustritz, A. (2003). Slashing the Romance Narrative. *Journal of American Culture, 26*(3), 371–384. doi:10.1111/1542-734X.00098

Lancaster, K. (2001). *Interacting with Babylon 5*. Austin, TX: University of Texas Press.

Mazzarella, S. R. (2005). *Girl Wide Web. Girls, the Internet, Negotiation of Identity*. New York: Peter Lang.

McRobbie, A. (1991). *Feminism and Youth Culture: from "Jackie" to "just seventeen"*. Basingstoke, UK: Macmillan Education.

Millard, E. (1997). *Differently Literate. Boys, Girls and the Schooling of Literacy*. London: Falmer Press.

Parrish, J. J. (2007). *Inventing a Universe: Reading and Writing Internet Fan Fiction*. Pittsburgh, PA: University of Pittsburgh.

Street, B. V. (2001). *Literacy and development: ethnographic perspectives*. London: Routledge. doi:10.4324/9780203468418

Thorburn, D., & Jenkins, H. (Eds.). (2003). *Rethinking Media Change: The Aesthetics of Transition*. Cambridge, MA: The MIT Press.

Verba, J. M. (1996). *Boldly Writing: A Trekker fan and zine history, 1967-1987*. (2 ed.). Minnesota: FLT Publications.

ENDNOTES

1. By "well-established" we mean that the stories are well-known within their target audiences. We do not mean that the stories are recognized by literary scholars as important parts of a national or international literary treasure.
2. Manga is a term for a certain kind of comics originating from Japan (e.g. Schodt, 1996).
3. Actually the term 'fandom' does not refer to the story or 'canon' as such but to the community of fans which admire the story. However, the term is often used as a reference to the entire ecosystem swirling around a certain media brand such as 'the Harry Potter fandom' or 'the Naruto fandom'.
4. The question why some fandoms are more popular than others is for instance examined by Olin-Scheller & Wikström (2010) but lies outside the scope of this treatise.
5. *Slash* is a fan fiction sub genre where the main plot involves the romantic or erotic relationship between two characters of the same sex, usually male characters.

Chapter 6
Digital Neighbourhoods:
A Sociological Perspective on the Forming of Self-Feeling Online

Ulrik Lögdlund
Linköping University, Sweden

Marcin de Kaminski
Lund University, Sweden

ABSTRACT

The aim of this study was to discuss how young people in Sweden relate to the Internet trying to picture how they developed and maintained relations online. The study revolves around the notion of self-feeling holding a few basic questions at the stake; how much time do young people spend online and in what different contexts? How does interaction take place in online communities and how is self-feeling constructed? The study takes on a sociological perspective on online interaction and leans on an interpretive approach represented by Charles Cooley. The study is based on eleven qualitative semi-structured interviews with eleven Swedish young people made on the Internet in 2009. The results of the study point out that an excessive amount of time was spent by the interviewees interacting in minor online communities and services. The character of communication was mainly personal and intimate. Analyses make visible three aspects of why spending time online; to prevent loneliness, to create opportunities of inclusiveness and to avoid exclusion. Finally, the study discusses the construction of self-feeling as part of a socialisation process taking place in contemporary digital neighbourhoods.

INTRODUCTION

The notion of information society refers to major changes within communication and labour as well as the contents of work (Masuda, 1983; Webster, 1995). The 'third wave' of societal change has dramatically changed the society stressing the importance to posses, handling and analyse information (Toffler, 1980). Much sociological research has been made from a macro perspective, taking the society as the point of departure, analysing the information society and information technologies. For instance one of the major sources in Sociology that focuses net cultures is Castells (1996) work from the mid 90ths on the organising of the Network Society.

DOI: 10.4018/978-1-60960-209-3.ch006

It can be argued that the field of net cultures lacks studies taking on a micro-sociological perspective focusing interactions between individuals and groups. This is a plausible standpoint concerning studies made in a Swedish context. However, the past five years the interest in the Internet has evolved as a multidisciplinary field seeing that social sciences put much effort into explaining and describing the Internet as an environment in which people lives with and in (Sveningsson *et al.,* 2003). The Internet is an arena that contains more than merely a technical infrastructure and in recent years the scope has been widen to involve the emerging field of digital arenas of young people occupied in identity forming activities (Buckingham, 2008), learning (Tapscott, 1999) and abusive practices (Dunkels, 2008).

In this study we will outline a sociological perspective on online interaction and net cultures in a Swedish context. The aim of the study is to discuss how young people are engaged in social interaction in the context of online communities. How much time do young people spend online and in what contexts? How does interaction of young people take place in online communities? How do young people construct self-feeling in online interaction? The study is based on sociological theories and in this matter we lean towards an interpretive approach seeing Cooley as central to how we can understand societal and cultural movements in our contemporaries. The sociological perspective of research may be useful to study human communication, whether online or not, and our study may contribute to a micro sociological perspective that focuses on the construction of self-feeling in online communities on the Internet.

NET CULTURES AND DIGITAL NATIVES

The Internet and digital media have become increasingly embedded in our daily lives. One central notion within the framework of net based activities is 'net cultures' (Dunkels, 2007, 2009). It can be argued that 'net cultures' is a notion difficult to define due to its volatility. It can also be said that the inherent complexity of the concept depicts a distinction between net cultures and everyday cultures. Buckingham (1999) argues net cultures concern different fields or domains which are not limited to the area of Internet exclusively. The term net cultures have also been used as a tool (Tapscott, 1998) abandoning the demarcation between the net and the everyday life. Tapscott argues net cultures as a kind of cultural interaction which is displayed on the Internet. The approach of cultural interaction also makes the Internet a wider field of exploration embracing the contexts of learning, development and socialisation (ibid.).

The concepts of the 'digital natives' and 'digital immigrants' were first introduced by Prensky (2001). Digital natives can be described as the generation that has grown up in the digital society being considered natives since they are fluent speakers of the digital language. In contrast 'digital immigrants' is a concept that refers to users who do not have a native understanding of the Internet and the digital world (ibid.). Digital immigrants are therefore forced to grasp knowledge about information and communication technologies to keep up with the developments of the technological field.

The discussion concerning digital natives and immigrants has also been addressed as an educational topic focusing the new challenges of the digital generation (Danielsson & Axelsson, 2007). The authors argue in the digital age many diverse opportunities have aroused for communication and learning. Also new questions have been stated, such as how identities are formed and how we may understand socialisation. It can be argued that research in this field has been trying to understand the conditions under which the new digital generation live, learn and work in the shadow of an old 'analogue' society. In this Tapscott (1999) stresses a change between the analogue society and the digital as a shift from a traditional linear to a modern interactive and non-sequential model. The Internet can be seen as a multi dimensional environment influencing the conditions of learning. This viewpoint has in turn caused segmentation between the digital and the analogue generations. Tapscott (1999) explains that the shift between generations is not merely a question of technology as much as it is an issue of how we look upon the world in a broader perspective.

Online Communities

Online communities are sometimes mentioned in a sense that makes us think of an escape from everyday life. Donath (1998) argues the ways in which self-feeling are presented are different when comparing the world apart from Internet with the Internet itself. In the physical world identities are easy to relate to since one can attach one's identity to one's body. On the Internet this argument is no longer self-contained and net based self-presentations are volatile and difficult to grip. Zhao (2007) states the understanding of the Internet has to involve the idea of a real, but incredibly dynamic, digital world. Zhao argues what is easily taken for 'identity play' is more of a tacit agreement among the digital natives. Further more, Zhao compares interactive online processes to a 'collective aging' seeing the digital reality has been made up of common experiences that we share with strangers. The dynamic processes of the online world make digital phenomena crucial to how we understand the formation and utilisation of online communities. In this matter there are cultural differences between the ways digital natives and immigrants relate to the Internet. For those born into a digital world the Internet is considered as real with real social connections and interactions. The Internet can be described as a more mobile and interactive reality than the world apart from the Internet (Tapscott, 1997). However, there are also common features between the two worlds, for instance the striving for collectiveness and the forming of self-feeling.

In discussions concerning online communities the conflict between digital natives and immigrants are likely to be exposed. The digital immigrants tend to see online communities as something abstract hidden in an inexplicable cloud and only perceivable through a computer monitor. The digital natives on the other hand experience messages, ongoing conversations and relations between people as things actually taking place (Tapscott, 1997). It can be argued that digital natives transfer the means of social interactions to a digital context. Enochsson (2005) argues that the computer is not a tool digital natives need to shape everyday life. The computer is rather a 'cognitive prosthesis' or an extension of consciousness (ibid.) never to be disregarded as a necessity to master social life.

Wellman & Guli (1999) discuss how sociologists have tried to analyse how changes in technology has affected social communities. The notion of social online communities is now part of the field of digital technology. We can for instance observe communication in online communities as new fields to explore and investigate by sociologists. The question is from what perspective communication in online communities can be scrutinised. In the following section we will depict different sociological perspectives on communication.

Sociological Perspectives on Communication

Communication is a diversified theoretical field in Sociology and we may discern three major perspectives; the functionalist, the critical and the interpretivist. The functionalist approach sees communication as a chain, linking the sender and the receiver. Communication is action, intended or unintended, and messages are regarded as concrete things with properties that can be measured. The functionalist perspective approach communication as a "metaphorical pipeline" through which information is transmitted (Bratton 2007, p. 324) using verbal (face-to-face or electronically by means of videoconferencing), non-verbal (gestures and facial expressions) or written methods.

The second perspective, the critical approach, seeks to expose the pervasive power influences that post-industrial society have over individuals. Critical theorists are concerned with examining communication as a source of power analysing the questions of who can speak, when and in what way. Communication becomes the expression, or the suppression of the member's voices, using the metaphors of culture and political systems (Bratton 2007). Significant to the critical approach is discourse analyses made by Foucault. Foucault theorises culture seeing culture as a social process rather than something being static and common to all members of a nation or an ethnic group. Communication and language becomes more or less hegemonic in interpretations of the world (Winther Jorgensen and Phillips 1999).

Finally, the interpretive perspective sees communication as the transference of understanding and meaning. The construction of meaning is not only affected by the skill, attitude and knowledge of participants, but also by the socio-cultural context in which the communication takes place (Bratton, 2001). We may say that the interpretive approach sees culture as constructed and that meaning shared by members serves to create and shape social reality (Berger & Luckmann, 1966).

In the interpretive approach the exchange of information as well as the transmission of meaning is essential to identity forming activities. According to the symbolic interactionism, a major influence of the interpretive perspective, individuals are made through interaction with others (Cooley, 1922). Language can be seen as a system of signification and a source of symbolic meaning in human social life (Mead, 1967). Language enables individuals to become self-conscious and aware of their own individuality (Blumer, 1969). The key element in this process of becoming self-conscious is the symbol. One example of symbols is words people use to refer to different objects. The use of symbols in social encounters necessarily involves other people in order to receive and interpret the meaning of them. The 'dualism' of communication is significant both to communication between individuals as well as on a macro sociological societal layer. However, language is ambiguous and changes over time signifying a multiplicity of meanings and ideas at any point of time (Charon, 2007).

In the following section we will present one sociological theorists that represent the interpretive approach introduced above. First we will briefly illustrate a general view of Charles Cooley's theory on the society and second describe two main concepts useful to our study.

Society, Primary Groups and Self-Feeling

Cooley (1922) portrays the society as a living totality in which the individual is attached to the collective inseparably. The individual becomes a member of a social whole made up of differentiated persons seeing that society is the sum of the separate members. The two-way relation between the individual and the society as well as the processes in which the individual becomes the society is the leading principle of the theories of Cooley.

Self and society are twin-born, we know one immediately as we know the other, and the notion of a separate and independent ego is an illusion (Cooley, 1909, p 5).

Cooley (1922) criticises contemporary ideas of individualism that depict the individual as separated from the society. Individualism is rather based on different levels of sociality. Cooley deprecates the idea of a single individual as an independent actor whose actions are the origin of social phenomena. Individuality does not invalidate the social since we can attribute a social heritage to individualism that derives from everyday life. The experience of what is individual can be rather obvious to one person, but the apprehension of the larger group or the society make more abstract conceptions (1922, p. 42). Cooley explains there are no emotions that are completely individual since we are always part of groupings and social contexts. The human being is profoundly attached to the collective through biological and social heritage (Cooley, 1922, p. 3).

A separate individual is an abstraction unknown to experience, and so likewise is society when regarded as something apart from individuals (Cooley, 1902, p. 36)

Cooley states individuality is rather a question of functional distinction. Individuals are attached and dependent to one another by functionality which gives every person in the society a particular meaning (1922, p 36). For instance, a member of a digital community needs other members to define the existence of the group. The community will also recognise the collaborative effort of the members as a collective feedback received on an individual level. When talking about the online community as a single experience we do apart from the members. On the other hand if we talk about a single member we deliberately set the whole picture aside. To temporarily neglect the idea of the community is to limit the importance of the complete image. Cooley (1922) argues the full consciousness of the relation between members and community creates a better understanding what it means to be a member in a group or society. The apprehension of the individual functionality in groups emerges as a consequence of direct and sincere interplay between people in every day life. The context of interaction is elaborated in two fundamental concepts by Cooley; primary groups and the theory of looking-glass self.

Cooley (1909) argues primary groups are fundamental in forming the social nature and the ideals of the individual in times of economic and social confusion. Primary groups can be characterised by intimate face-to-face association and cooperation. The primary groups are also a differentiated and competitive unit socialised by sympathy and disciplined of a common spirit. Cooley (1909) exemplifies the primary groups as the family, the play-group of children and the neighbourhood and argues the primary groups belonging to all times and all stages of development universal in human nature and human ideals (p. 24). The family and the neighbourhood are more influential than other groups and give the individual his earliest and most complete experience of social unity. Cooley (1909) argues within the primary groups the individual minds are merged and the capacities of the members find adequate expression. Primary groups do not change in the same degree as other relations, but form a comparatively permanent moral unit in order to fostering the principles of loyalty, truth, service, kindness, lawfulness and freedom.

(...) human nature is not something existing separately in the individual, but a group-nature or primary phase of society, a relatively simple and general condition of the social mind (Cooley, 1909, p. 29).

Cooley argues the self-feeling being an aspect of the social self. In the theory of the looking-glass self others are important to how we come to realise ourselves. Bringing the self at focus we have to distinguish between three aspects; a) the physical body, b) the inner consciousness and c) the social self. In general we see our body as a reference of the self. Cooley (1922) states the localisation of the self-feeling to our body as natural since we have to localise such feelings somewhere. The apprehension of the self is only an imaginary conception attached to the physical body in the same way as clothing and jewellery.

Cooley (1902) argues the self-feeling gathers the thoughts and the activities of the individual. This function is the most important to the self-feeling. Idleness creates a weaker self-feeling while activities create a stronger. Activity is important to Cooley since the self-feeling is an emotion discovered by ourselves but confirmed as proof by others. The self is an 'empirical observation' of the self confirmed by activities in every day life. The self-feeling is related to different emotions which cannot be meaningful unless attached to you, her, him or them. It is by thinking of others we experience ourselves (Cooley, 1902).

The social self may be apprehended by an individual but also in form of a collective activity of a large number of members within a community. The individual self can hardly be separated from the self-feeling since our consciousness refers so strongly to the group. For instance in team game it is difficult to distinguish the performance of an individual from the group. The self-feeling is based on the viewpoint of the group and the individual's performance and function in the group will determine the social self. Cooley (1902) makes social self distant from the psychic idea of the self as primarily an intrinsic emotion and argues the idea of the self can never do apart from other people or all the relations that exists between people (Cooley, 1909, p 5).

(...) the imagination of one's self (...) appears in a particular mind, and the kind of self-feeling on has is determined by the attitude toward this attributed that other mind (Cooley, p. 183).

The idea of the social self can be described as a process in which the self-feeling emerge by means of interaction with others. The apprehension of oneself is dependent on how we apprehend and interpret what others think of us. The reflections lead to the creation of the self-feeling that, not only embrace the view of others, but in fact is completely depended on the view of others. Thus, the process of develop a self-feeling can be described in four parts; a) we sense the ideas of others towards ourselves, b) we apprehend the concepts of others, c) we create a sense of ourselves, d) we feel pride or shame. This is what seems to be fundamental to the theory of the looking-glass self by Cooley.

The Design of the Study

Focus of the study is the common users of the Internet. The selection of interviewees is based on two additional requirements; they were supposed to spend more than one hour in average online and that they should be eighteen years of age. According to the Swedish Media Council (2008) approximately seventy percent of the Swedish youth are online more than one hour per day. Our sample becomes important to exemplify the construction of self-feeling among a group of young people, rather than trying to generalise the results to the larger population. The sample we used in this study is an experienced group of young people using the Internet several hours per day. The sample consists of eleven individuals (mixed men and women) who use the Internet five hours per day in average during the interview study in 2009. The time spent online spans from one hour to eight hours per day and some of interviewees are likely to use the Internet extensively if they lack time restrictions. The second requirement, the age variable, was launched in order to avoid excessive preparations concerning ethical restrictions common to social science research. In general, to involve minors or children should have required a written permission executed by the children's parents. As a consequence we address our sample young people rather than youth.

In the study one individual fall off due to problems of communication taking on a technical character. The remaining eleven individuals are all part of the study, even though we do not exemplify all of their statements in our excerpts.

The interviews were all of a semi-structured type (Patton, 1990) putting the narratives of the interviewees in centre. In the initial phase of the interviews we made clear that the interviewees did not have to follow the stated questions. We believe we have been 'open' and kept a responsive ear to personal topics, interest and matters important to the interviewees.

The incentives of using net base interviews instead of traditional interviews conducted face-to-face were many; it has been possible to keep several discussions going on simultaneously (Sveningsson *et al.,* 2003), to follow up and resume discussion that have been started earlier and to keep record of what have been said continuously (Dunkels & Enochsson, 2008). The set backs of the net based interview is mainly the risk of brief replies. To some people it is easier to set forth answers and unfold statements verbally than in written text. The written message may also be valued differently from verbal communication and it seems like people in general express themselves shorter trying to be more precise in written communication. Additional set backs of the interviews on the Internet is the lack of body language (Dunkels & Enochsson, 2008), which may help the researcher to interpret the statements of the interviewees.

In the following section we will present the main features of the narratives of the eleven interviewees. The statements of the interviewees have been analysed and categorised in three different topics. Each subject refers to the main questions of the study; the time spent in different contexts, the Internet as a social arena and the development of self-feeling in online communities.

THE EVERYDAY LIFE ONLINE

Online Time and Net Activities

The Internet is the common activity of the group of eleven interviewees and the 'net' becomes a place where they spend a large proportion of their everyday lives. At first glance the vast time spent online is not seen as problem by the interviewees. However, in some of the narratives the interviewees mention deficient success at school and insufficient commitment in other communities as a problem due to time spent on the Internet. The lack of commitment is something one regulates by decrease the online activity or by cutting down the number of active online communities.

Unfolding the question of time spent online a number of different activities were mentioned. The interviewees pronounce reading news, chatting, gaming and activities related to blogs such as writing, reading and commenting. The Internet was used both for work in school and more social activities at leisure time. The former was also the factor that regulated the latter. If one had much work to manage in school one had to stay away from social activities online. On the other hand there also existed a common understanding that school was an acceptable excuse for being absent from online communities. It can be argued that plenty of work in school is carried out online or at least in front of a computer, which makes the balance between times spent with school related work and free time more interesting to our study than time spent online versus time spent apart from the Internet.

The interviewees seem to keep different kinds of activities separated. The major division in the study arises between life apart from the Internet and the life online. Notably, some interviewees found it difficult to define what activities belonged to which category due to the fact that many activities were ongoing parallel to each other. The topic of defining had to be abandoned since the distinction between their life apart from the Internet and the life online was somewhat blurred. The interviewees preferred to discuss in what different contexts the Internet was used. In the discussion, focusing the division between online activities and life apart from the Internet, the interviewees touch upon an additional subject of interest to the study. What is the difference between the two worlds and is not the Internet part of the real world? The question is intimately attached to the apprehension of the Internet as a second form of reality. On the one hand the interviewees see communication taking place in different Internet communities and chat services as real. On the other hand the interviewees differentiate between the real world and the online worlds. One interviewee states he has a personal scale of priority in which he promotes the real world apart from the Internet above the online world. To prioritise the world apart from the Internet is not to neglect the Internet as less real, but rather to see the world apart from the Internet and the Internet as a complementary, both promoting social interaction. As it seems it is a challenge to balance the better of two worlds, not neglecting any of them as less real.

I have to say that I really have been able to balance my online gaming with the real world off line. I have tried to prioritise the real world above of the virtual. (...) Since I play so much I try to meet people offline whenever it is possible. It sort of a personal guideline, that that reality goes first. Then again, the only reason for me to play World of Warcraft is due to all the social possibilities. I have met quite a few really nice people online, some of them have become real friends... I mean offline friends. [Excerpt; IP11]

One of our interests in the study concerned in what different contexts the Internet was used among our interviewees. The residence on the Internet was not made clear by the interviewees. In fact they seemed scattered all over the Internet and did not favour a certain site or service exclusively. It can be argued that the Internet is an arena in itself, constantly changing of features, which make it problematic to the interviewees to depict favourable sites. However, our data show kinds of digital preferences on the Internet in general. The interviewees explain blogs and huge net communities as more distant to them than minor net communities. Above all chat programs such as MSN Messenger were experienced as personal and intimate. Further more the contacts made at MSN Messenger had a higher priority than contacts made on Facebook. MSN Messenger contacts are also more likely to be maintained apart from the Internet. For instance the interviewees found new acquaintances by means of friends they already know or by seeking potential contacts within communities based on common interest.

Setting out from the narratives of the interviewees the Internet is not an arena young people visit in order to meet new friends by random. On the contrary the Internet seems to create new opportunities to stay in touch with contacts already established. In case of new contacts were made they were based on common friends or common interests. Some interviewees explain that if they have made new friends in environments such as online games or chat forums there are persons they meet offline later on in the world apart from Internet.

Inclusion and Exclusion

So far we may come to the conclusion that the Internet is used excessively by the interviewees in a relatively small number of different online communities. In the interview material we can see at least three aspects that lie near at hand to explain why spending time online; *to prevent loneliness, to create opportunities of inclusiveness* and *to avoid exclusion*. All categories are interlinked, but also display differences among them; to prevent loneliness and to be part of something is made by free will whereas to avoid exclusion becomes a kind of social pressure or social control in order to come up to the expectations raised by oneself and the online community.

The interviewees argue the main reason for being active online is that it creates a sense of presence and a chance to interact with others. In the interviews it seems important to explain that the Internet provides an interactive arena that is very much social in every aspect. To have someone to talk to and not be left alone appears as main argument why the use the Internet.

I'm never alone in front of the computer. It's always someone out there in cyberspace to talk to. [Excerpt; IP10]

It has been argued that loneliness is a common problem of the younger digital generation in our society today, seeing solitude in a strictly physical sense. However, on the Internet our interviewees express a choice of being lonely or not. Staying online they can choose whether be left alone or take part in an Internet community actively. Loneliness is never inflicted on them since the online community always offers someone to talk to and an opportunity to stay out of solitude. Notably is that the presence of the interviewees can be more or less anonymous on the Internet.

The interviewees return to the question of the inclusiveness character of the Internet and regard the Internet as a social arena always open for participation. The Internet seems to provide a place of social belonging which opposes loneliness and exclusion. Cooley (1922) argues the coming into existence of the individual involves the presence of others. The individual needs a social context in which he or she can interact with others to exercise one's individuality. We believe the Internet can make such social context significant to young peoples' opportunities to develop inclusiveness and adherence into a larger community. In our interviews we also found an opposing argument towards the two already mentioned above why stay online, namely the social pressure. The Internet is constantly changing and the interviewees feel a risk to miss what is happening if not stay online all the time. From the narratives we interpret a 'social pressure' to stay uplinked and updated. Some of the interviewees also use mobile Internet connection in order to keep control of the course of events.

There's something compulsory with all the communities... I've to stay there because all the others do. There is a social standard that orders you to stay updated and keep things under control. [Excerpt; IP7].

In the narratives we interpret a fear to be excluded if not being able to stay in command of the flow of information. Cooley (1922) refers to the activity as main source of developing the individuality as a reflection of the community. Action is important in order to take part in the happening of the community and to be involved and updated in current discussions and topics. To be active online is to take part in all the interactions of the community and a way to stay with your friends, or simply to socialise.

MAKING FRIENDS IN A PERSONAL SPHERE

As have been mentioned earlier, young people spend at lot of time on the Internet in different online communities. The key factor seems to be the opportunity to communicate with online contacts, wherever they can be found, not to visit a specific site. The opportunity to interact is a main concern of the interviewees seeing the Internet mainly as a media of a two-way communication. The interviewees are attracted to people that make comments, evaluate and reply to their presentations. In our data we see two different reasons to why young people stay active on Internet. Young people are active because of the actual content, but more importantly due to the opportunity of making contacts and responses. It seems significant that interaction is essential to the experiences of the interviewees and that they are positive to overall online activity.

It's fun if people comment. It would have been boring if no one commented or if it was only bad comments. (...) If no one wrote to me or wanted to comment my things I would have experienced it in a bad way, like no one cared. [Excerpt; IP 5]

The interviewees express a wish for that people care and that they being noticed. The sense of being observed in a positive way is mentioned as vital in several of the interviews. We can also interpret a distinction between really being noticed and merely draw attention. The interviewees expel that online presentations is not only a question of attention but to be acknowledged. The interviewees separate those two dimensions carefully and explain that they are not generally looking for responses; rather they are trying to maintain an ongoing dialogue, to be part of something that continuous or a forward-looking communication. The balance between the input and the output of the ongoing interaction are significant to the continuing interaction process. If response fail to come off the interviewees feel excluded and lonely. We may interpret a strong urge to communicate and willingness to interact with others, no matter how. We can also see contradictions in the material to such interpretation of the online user's wish to communicate freely. Some interviewees are not interested in interaction in a broad sense. The interviewees in this case feel it would be a conflict of integrity if others could take part in their lives and the things they were doing. These interviewees were not less active online than the others in the sample, but more careful in their selections of who will have the opportunity to observe the photographs or texts they presented on the Internet.

A major part of the images are quite personal, [I'm] trying to stay anonymous on the Internet (Excerpt; IP 2)

The cautiousness of the interviewees might not be a question of integrity primarily. Young people may try to keep a personal sphere. One interviewee explained that social networks are more like contact surfaces without actual content. The interviewee did not look for new contacts, merely maintaining old ones in already established networks, and interaction in a general sense was not an option preferring direct contact with a group of chosen contacts.

I do not accept anyone I don't already know as a friend. I'm not comfortable with letting people I don't know looking at pictures of me and save them for instance. I'm not interested in writing to people I don't know and never will meet. (Excerpt; IP8)

Somewhat contradictory the interviewee in the excerpt above is looking for a personal sphere in a public network trying to protect personal matters from observation. It seems as the interviewee fear visibility on the Internet and that the interviewee expresses a desire not being exposed publicly. The exception is contacts of interest and consequently the interviewee blocks unapproved visitors on a regular basis. The blocking is also seen as a successful way to cope with the problem. The quest for acceptance and inclusion is made through already established social networks. We may argue the restriction of the interviewee somewhat problematic, since the online communities are designed for expansion of social networks. To stay out of attention from others is inconsistent with the idea of many community services.

SELF-PRESENTATION AND FEEDBACK

Online interaction is a matter of exchange information and to spend time with friends. Online communications is also an issue of presenting oneself and receive feedback on that presentation. One of the most important things mentioned by the interviewees is the response or feedback to one's presence on the Internet. To publish material without receiving feedback is experienced negatively and may evoke feelings such as 'one does not exist' or as if 'no one cares' out there. The character of feedback is an immediate matter pointed out by the interviewees. In general, one is looking for a certain kind of feedback that favours one's presentation. However, response to presentations may not always be constructive and according to the interviewees feedback does not have to have a positive character. The interviewees argue; if one receives 'misguided' responses one can regard such feedback as peculiar and negligible.

It gets so awkward if what you try to communicate gets really misunderstood. Then I rather don't publish anything at all, if I'm not sure of getting the comments I plan. [Excerpt; IP9]

We may interpret negative feedback as something the interviewees can ignore. Consequently, negative response should not interfere with one's self-feeling. The opportunity expressed by the interviewees to choose how one will react to feedback contradicts the fundamentals of the interpretative theory. Cooley (1922) argues the self-feeling is an emotion apprehended by ourselves, but confirmed by others. We believe that all kinds of feedback are important to how we experience the self and feedback may support or damage an ideal image. The self-feeling is 'co-constructed' and thus sensitive to all forms of feedback. The interviewees argue one can ignore feedback that is not satisfying one's self-presentation. The selective realisation of misguided information described by the interviewees raise an additional question; does one have to stay truthful on the Internet? The intention of communication is part of what we may understand as a 'selective honesty'; that is to say, you may choose to display honest parts of yourself to a certain audience.

I present different images of my self at different sites and receive alternate responses. At personal sites, such as MSN I believe people apprehend me as in reality. (...) We all have frontages. But I'm quite honest presenting my personality. (Excerpt; IP4)

As it seems from the interviews we can attach a sense of honesty to online publication. The interviewees try to make truthful presentations. In case one likes to withdraw certain kind of information it is not regarded as being dishonest. The online publisher only uses material that he or she thinks can describe a person adequately. However, it is important that others do not misunderstand or misinterpret information that describes the publisher. Consequently, you are honest with what you show to others, but you are not dishonest due to the fact that you keep things to yourself. The image one present is truthful although not necessarily complete.

I often feel I can't show who I really am. I have to hide all my bad qualities even if they are a part of me. Nobody really knows me, that am why I feel lonely. [Excerpt; IP7]

One expression that returns in the interviews is the term 'reflection'. The notion is used in a sense that makes us think of the published material as a mirror image of the publisher. The material, such as photographs is to suppose to express a reflection of the intended meaning you are willing to send.

You always have a sense of how you think people will react on the images (...). I don't contemplate enormously on receiving a positive reflection on my self. It is more like the material shall reflect me. (...) I do not publish photographs that do not reflect my type of personality (Excerpt; IP 5)

In the excerpts above the interviewee display that photographs are meant to reflect the personal interest of the publisher. The photographs may also attract deeper qualities and emotions decisive for one's comfort. The publication of delicate material can be seen as a sort of test in order to find out if others acknowledge and enhance the qualities one tries to communicate. The examination process makes some of the interviewees careful of what they are willing to publish on the Internet. The fundamental idea of the theory of the looking-glass self is how we come to realise ourselves as a reflection confirmed and proved by others. In the interview material we can see how the publication of photographs and comments are intimate connected to the image one like to present. A presentation or an image that fails to make the intended expressions will damage the self-feeling. The interviewees do not want to enhance certain perspectives of the self since an unsettled impression may end up in a false image.

Mainly I present images of myself from a party being happy. I never publish an image in which I have a hangover or is depressed. (...) What I'm trying to say is that you can create a good or bad impression by means of photographs and I choose to expose my good sides. (Excerpt; IP3)

I judge myself very much from the perspective of others. (...) My confidence is based on what I experience that others think of me. (...) If people do not comment I feel confident (...) not good enough. (...) I feel inadequate if others do not prove me wrong. (Excerpt; IP7)

Is not information online distorted and risk creating a false image of people? The interviewees state there exist a form of equality between all publishers on the Internet; the same goes for everyone. Zhao (2007) argue there is a tacit agreement concerning the development of one's identity in the network online, namely you must accept what has been published. The interviewees select carefully the type of image they like to present and likewise they are aware that others do the same thing. The interviewees argue the distortion of reality makes it possible to hide characteristics that are negative, that we are not proud of and do not wish to make visible. Depending on what platform being used and the sort of relation one has to others on that particular platform one may receive different sorts of images. Information runs through alternate forms of filters that 'interpret' and deals with information in a certain way. The filter processing is not seen as a problem by the Interviewees. The Internet is simply working this way. One has to do with multiple kinds of self presentations, which depends on a particular situation and the particular interest one has for the moment.

CONCLUSION

In this study we have outlined a sociological perspective on online interaction and net cultures in a Swedish context. The aim of the study was to discuss how young people were engaged in social interaction in the context of online communities. Three questions were at stake in the study. The first question concerned the time spent online by young people. Our study shows that the interviewees spend a large proportion of their time on the Internet. Actually some spend a major part of their everyday lives on the Internet in different forums and online communities. The contexts favourable by the interviewees were mainly minor personal online communities together with services characterised by a high degree of social interaction.

The second question handled the character of interaction of young people in online communities. In the study we could see how the interviewees wished to stay in contact with friends as the main reason spending time on the Internet. The online communities appeared to be arenas for social contacts providing a place to meet, talk and exchange information, stories and photographs. The character of communication could be described as intimate and sincere, although blended with a pinch of suspicion. In some online communities it is optional to give access to material one has published and the interviewees tends to make photographs and comments exclusive for others than friends. A form of duplicity concerning interaction in the online communities appeared in the study. The online communities created both a chance of inclusion; a sense of belonging by staying updated as well as exclusion; a fear of being put aside not being committed sufficiently.

In the study we raised the question of how young people construct self-feeling in online interactions. Online publishing was common among our interviewees and it was also considered to be of great importance to them. The published material worked like a 'looking-glass' generating feedback and the material was published in order to generate responses. The existence of groups of acquaintances was based on the transaction of information. Our study shows that dialogue is important and that it is equally important to present oneself in an acceptable manner. For instance photographs are not important if they cannot evoke some kind of response. It becomes crucial to the self-feeling that the online communities confirm the image one tries to mirror. The sense of being 'good enough' is an example of such construction of self-feeling. Out of presentations the self-feeling is created. The process may be limited to a single community or occasionally spread in a broader perspective. The interviewees in the study express the importance to belong to a larger social group in order to develop self-feeling.

Socialisation in Digital Neighbourhoods

Can we define the Internet as a social arena? The question seems almost absurd in the context of this article and undermines the debate of 'social media' in Sweden. The activities within the framework of social media suppose to combine technology and social interaction in order to create and share information with others. It can be argued that the online communities used by the interviewees in the study appear to be true social arenas and that the relation between online activity, interaction and socialisation becomes significant. The idea of the Internet as a social arena and the notion of socialisation is a particularly interesting and it is a relation that we like to pick up in this section.

Cooley (1922) claims the society as living whole made up of differentiated persons. In the society an active person will develop an identity based on the interplay made with others in a number of social institutions; such as the family, school and working life. We may argue that the Internet is a social institution in several ways; first, as a technical structure set up for communication in the society similar to telephones and railroads and second as a forum or a social arena for the interplay of people in situations of communication.

The interviewees interact with friends sharing personal information and intimate feelings. The interviewees return to the importance of interaction and even feel set aside if not perceive feedback. The lack of feedback evokes feelings of loneliness. Communication online is personal and intimate and in every sense similar to interaction in other institutions of the society. We cannot disregard the online interaction as less real or less significant to how people develop different personalities. It has been said that young people spend a vast of time online and we believe the Internet may affect young people in the same way as interaction in other parts of everyday life. In fact online communities permeate the social institutions of the society being part of education, work and leisure time.

We argue the Internet communities are fundamental institutions of socialisation fostering a young generation into the society. Cooley (1922) sees the importance of socialisation in primary groups such as the play-group of the vicinity. Similar to the idea of local play-groups we believe the context of socialisation has been transformed, as the communications of the society has been transformed, and made new arenas. The physical play-group of the neighbourhood has simply moved to the Internet and by doing so they have became digital neighbourhoods. The function of the collective remains the same primarily to develop individual and collective citizenship and self-feeling. The connection between activities on the Internet and the socialisation process is strengthened by Tapscott (1998) arguing the Internet is a specific culture rather than a tool of communication. The results of the study support the idea of the Internet as a culture and a social arena in which people express themselves similar to everyday communications beyond the network.

In conformity with Cooley we see an intimate relation between the individual and the society. Unquestionably, the online communities are part of the wider society and to interact in online communities is to interact with the society.

REFERENCES

Berger, P., & Luckmann, T. (1966). *The Social Construction of Reality. A Treatise in the Sociology of Knowledge*. New York: Anchor books.

Blumer, H. (1969). *Symbolic Interactionism. Perspective and Method*. Berkley, CA: University Press.

Bratton, J. (2007). *Work and Organizational Behaviour*. New York: Palgrave Macmillan.

Buckingham, D. (1999). Children, Young People and Digital Technology. *In Convergence*, *5*(4).

Buckingham, D. (2008). *Youth, Identity, and Digital Media*. Cambridge, MA: Massachusetts Institute of Technology.

Castells, M. (1996). The Information Age. Economy, Society and Culture.: *Vol. 1. The Rise of the Network Society*. Oxford, UK: Blackwell.

Charon, J. M. (2007). *Symbolic Interactionism. An Introduction, an Interpretation, an Integration*. London: Pearson Prentice Hall.

Cooley, C. H. (1909). *Social Organization*. New York: Charles Scribner's Sons.

Cooley, C. H. (1922). *Human Nature and the Social Order* (Revised Edition). New York: Charles Scribner's Sons.

Dunkels, E. (2007). *Bridging the Distance*. Umeå, Sweden: Umeå University.

Dunkels, E. (2008). Children's Strategies on the Internet. In *Critical Studies in Education*, vol. 49 no. 2 Sep. 2008.

Dunkels, E., & Enochsson, A. (2008). Interview with young people using online chat. In Quigley, M. (Ed.), *Encyclopaedia of Information Ethics and Security*. Hershey: Idea Media Group.

Kvale, S. (1997). *Den kvalitativa forskningsintervjun*. Lund, Sweden: Studentlitteratur.

Masuda, Y. (1983). *The information Society as Post-Industrial Society*. Bethesda,MD: World future society.

Mead, G. H. (1967). *Mind, Self and Society*. Chicago: The University of Chicago Press.

Mediarådet (The Swedish Media Council). (2008). *Ungar och medier 2008*. Elanders.

Patton, M. Q. (1990). *Qualitative Evaluation and Research Methods*. London: Sage.

Prensky, M. (2001). Digital Natives, Digital Immigrants. In *On the Horizon*, 9 (5), October, 2001.

Sveningsson, M., Lövheim, M., & Bergquist, M. (2003). *Att fånga nätet – kvalitativa metoder för Internetforskning*. Lund, Sweden: Studentlitteratur.

Tapscott, D. (1998). The Rise of the Net Generation. In *Meridian, January, 1998*. Growing Up Digital.

Tapscott, D. (1999). Educating the Net Generation. *Educational Leadership*, (Feb): 1999.

Toffler, A. (1980). *The Third Wave*. New York: Collins.

Webster, F. (1995). *Theories of the Information Society*. London: Routledge.

Winther Jorgensen, M., & Phillips, L. (2000). *Diskursanalys som teori och metod*. Lund, Sweden: Studentlitteratur.

KEY TERMS AND DEFINITIONS

Functionalism: the notion of functionalism is chiefly used in social science to describe the exercise of effects of customs, behavioural patterns and institutions.

Interaction: refers to communication and interplay between people.

Interpretativism: is a concept that involves an understanding of meaning making through interpretations.

Semi-Structured Interviews: refers to a guide in which questions are organised in a structure of coherent themes in order to increase flexibility.

Social Institutions: a number of social relations based on conscious or unconscious rules and standards.

Socialisation: a process in which proficiency is mediated and learned. Socialisation involves the forming of individual personality.

Sociology: is the science of the social. Sociology involves a wide range of objects of study from the interplay between people to global structures and processes (macro-sociology). Sociology often analyses phenomena such as institutions, organisations and collective behaviour (micro-sociology).

Symbolic Interactionism: is a sociological perspective that focuses on interaction and communication.

Primary Groups: refers to a group formation in which the members have a frequent and lasting relation face-to-face.

Chapter 7
The Use of Interactive Media in Identity Construction by Female Undergraduates in a Nigerian University

Oyewole Jaiyeola Aramide
Bowen University, Nigeria

ABSTRACT

The world has become a global village with the aid of the internet whose attributes and capabilities are constantly surpassing all other sources of information that existed before it. The internet facilitates accessibility and availability. Distance in communication has evaporated, making interpersonal and group communication across continents possible and easy. However, a more outstanding use is recorded in its interactive segment. These interactive media come in the forms such as Chat rooms, Facebook, and Email. The chapter examines the uses of the internet interactive media by a selected group of Nigerian female undergraduates in identity construction. Results show sample population prefer and use the facebook for internet interactions due to its affordances. The e-mail and chat room media followed closely in order of preference. The findings support the uses and gratification theory which holds that people manipulate the media for self gratification.

INTRODUCTION

Marshall McLuhan a mass communication theorist (Uche 1996) first theorized that the world is fast becoming a global village. This speculation has become a reality through the internet whose qualities and facilities constantly surpass all other sources of information that have existed before it. The information age has produced a wealth of information which is supposed to be readily available to anyone who wishes to use it. Indeed with increased popularity of the number of on-line services, people having access to information appear to be growing at an exponential rate. The uses of the internet are many and this chapter seeks to examine its interactive segment as employed by a selected group of Nigerian female undergraduates.

Communication is life and the essence of living is justified with the ability to interact with one's immediate environment especially with

DOI: 10.4018/978-1-60960-209-3.ch007

other human beings. In the immediate past, this was possible through what is referred to today as the old media of communication, such as the broadcast and print media: the radio, television and newspapers.

The twenty first century story is different, with globalization. The internet has evaporated distance in communication. Bell and Gray projected that by the year 2047…..all information about physical objects, including human beings, buildings, processes and organizations will be online. This is both desirable and inevitable. Globalization or Internet Communication technology in the words of Brown and Duguid (2002), has replaced old media of communication since where old technologies inherently forced people together in factories, offices, schools, libraries, for interaction and relationship, new ones tempt them to stay apart while providing all of these on-line with tremendous freedom. Consequently, centripetal social needs, which call people together, compete with centrifugal technologies that allow them to move apart but closer with the click of the button. The world is connected with technology and space is no longer a barrier to interpersonal or group communication.

This chapter examines the uses of the internet interactive media employed by female undergraduates of Bowen University. Bowen University is a private Christian University owned by the Nigerian Baptist Convention, located in Iwo town in south-western Nigeria. Michener, Delamatar &Myers (2004) observed that some people pursue the search for self-knowledge and for a meaningful identity eagerly, others pursue it desperately. College students (undergraduates) in particular are often preoccupied with discovering who they are. A few, however, have experienced existential uncertainty.

Basically, females are particular about their person, class, status and others' perceptions of who they are, and the factors responsible for their societal acceptance. The availability of electronic media especially the internet provides these young females the opportunity to avail themselves of different websites. The internet provides strong influences that shape their perception of reality and means to project their perception of reality of the ideal female. In examining various media; the uniqueness of each will be appraised and the reasons for preference of one over the others will be highlighted as well as the advantages and disadvantages of each.

The main objective of this study was to examine how these Nigerian female undergraduates construct identity by the use of internet interactive media (IIM).

Other objectives are:

- To explain identity construction among these female undergraduates and propose reasons for it.
- To ascertain why these young females use the interactive media available on the internet in identity construction.
- To determine the features and affordances of each internet interactive medium that makes it appropriate for young female identity construction.
- To determine the most suitable of all available internet interactive media for these young females.

BACKGROUND

Communication Defined

Disciplines such as Psychology, Anthropology, and Human communication have all given a variety of perspectives to the definition of communication. Eyre, Chapell and Read, as well as Ugboajah cited in James et al (1990) gave similar definitions of communication as follows:

Communication is not just giving information but understandable information, receiving and understanding the message. Communication is the transferring of a message to another party so it can be acted upon. (Eyre, 1983)

Communication is any means by which thought is transferred from one person to another. (Chappell and Read 1984) and

Ugboajah (1985) posits that the communication process involves all acts of transmitting messages to channels which link people, languages and symbolic codes which are used to transmit messages, the means by which messages are received and stored. These definitions encompasses rules, customs and conventions which define and regulate human relationships and events.

Communication includes: verbal communication, non-verbal communication, visual communication and electronic communication. It is vital to personal life and is also important in business, education and any situation where people encounter one another. It touches all spheres of human activity. It informs all of man's actions because it is occasioned by his need to interact with his fellow man. In every communication process the following elements cannot be neglected:

The Source: This is the originator or encoder. The communication process begins with the sender or source that transmits the message.

The message: The content or substance of a communication act. The message is a signal that triggers the response of the receivers.

The channel /medium: The medium could either be verbal or non verbal while the channel is the method through which the message is converted or delivered.

Receiver: This is the person to which the message is targeted. A receiver notices or attaches meaning to the message. The receiver may also be known as a decoder or recipient.

Noise: This is any unwanted signal or distortion that leads to the ineffective flow of the message or which deviates from the fidelity of a communication message.

Feedback: This is the verbal or non-verbal acknowledgement of the message received. It is vital to communication because it determines the success or failure of the message sent.

THEORETICAL FRAMEWORK

Identity Construction for Gratification

Uses and Gratification Theory

The uses and gratification theory is about what people do with the media, rather than what the media does to people. In essence, the audience or receiver uses the media to gratify needs and desires such as entertainment, relaxation, information and many more.

Different from the Bullet theorists, who did not base their theory on research, the uses and gratification school of thought used empirical research and found that people only listened to what actually interested them. This in contrast to critical theory which says that the masses are easily swayed with media messages which shape their attitudes and beliefs. Uses and Gratification theorists including Dennis McQuail and Jay Blumler (Watson 2003) see the audience as active participants in media communication, using the media to gratify their own needs and purpose. In essence, people may tune in to their own radio or television stations and read newspapers or magazines for a number of reasons such as entertainment, escapism, relaxation, news as well as source of information. They make further choices about when, where, how often and how long they will engage with the media. They consume the media because they like what they are getting. In a way, the audience determines the media content. In television and radio ratings, the media have to give the masses what they want, otherwise the people switch off and the program dies. It is all about what the viewers want to see or hear.

Uses and gratification theory challenged the idea that the masses are 'passive' or 'careless' 'sheep' who all read media texts or messages and that the audience does not necessarily buy into the message of the media product. In effect, it means the audience controls the media and not vice versa. Blumer and Katz cited in Watson (2003) concluded that audience's fulfillment of needs come within the broad generalization of four desires diversion, personal relationships, personal identity and surveillance. This is achieved essentially from media characters on radio, television and the like. However, these four desires could be adapted to mean:

Flight: a voyage from routine pressures of life, release from physical, emotional and intellectual nervousness.

Friendship: It grants audience the liberty and capacity to analyze their lives vis-à-vis their preferred character and the privilege to make invisible friends.

Fix identity: *This* gratifies the audience's need to situate his/her identity in the light of the characters, individual experiences of life such as fear, joblessness, loneliness, illness disappointments, pleasures and success.

Focus: the media with an overwhelming force influence and create in the audience the ability to focus on the important issues of life through the means of appealing to their desires and needs. This position must be examined in contrast with the notion that 'everyone is free' meaning that the mass media's audience can resist being influenced, simply through choice. An individual makes the choice, and the selection that is made will merely underpin the views and influences they already have.

The internet is deployed by different people for different purposes. Everyone seems to have something of value to do on the internet. However, the interesting thing is that a bee-hive of activities is constantly taking place on the internet which means the internet enjoys active audience most of whom are youths. These youths utilize the internet for various reasons but of interest in this chapter is how female undergraduates construct identity while using the internet. What is identity construction and why identity construction for female youths?

Identity Construction

The issue of identity construction is central to human existence and is as old as humanity. Identity presents itself in several ways, such as individual identity, cultural identity, collective identity, religion identity, national identity. Defining one's identity will determine the kind of complex (superiority or inferiority) that such an individual will grapple with in life. Identity is the way self is construed, the portrait of one's self painted by one and viewed by others. Identity is conceived of as a very peculiar set of opinions, judgments, evaluations, attitudes, manifested by a person towards himself or herself (Doise 1998). It is about "Self perception" and "Others perception". Self perception refers to the way we view ourselves, our impression, outlook, perception and categorization of who we are. 'Others perception" is the way we conjecture other human beings in our thoughts, minds and in real life relationships and interactions. In identity construction, there has to be a hierarchy of norms, as well as lateral, reciprocal relations which require people to be "judgmental" while excusing, approving or disapproving others. Schopflin (2001)

Regarding self perception, it is a search to discover who we are to ourselves, what we are to others and what we wish we are. Who we are to ourselves is usually different in most cases to what others view us to be, George Herbert Mead(cited in Ritzer 2000) categorized identity into two.

The "I" and the "Me". The "I" is the immediate response of an individual to others. It is the incalculable, unpredictable and creative aspects of the self. Mead lays great stress on the "I" for four reasons:

- It is a key source of novelty in the social process

- It is where our most important values are located
- It constitutes something that we all seek- the realization of the self.
- It permits us to develop a "definite personality".

Mead says the 'me' is the organized set of attitudes of others which one himself assumes. 'Me' is equal to conformity. Doing and accepting what others want.

Imagine the following scenario:

A pretty young girl of eighteen, engaged to be married to a young handsome Romeo, weighed 45kg before marriage and was a delight to all who saw her beauty. However, after two years of marriage and a set of twins, she added on 30kg to make 75kg. Yet her view of herself as being a slim beautiful queen did not change. Whereas, she had changed in reality from her previous description, in her thoughts, mind and imagination, she has a false identity of who she is. The way she is viewed and may be addressed and treated is different from the picture she has stored up. In essence, there will be a conflict of identity when she receives a reception contrary to her perception.

The 'ME' is the quest to be what we think is acceptable to others and by extension ourselves. That is the wish to be something and someone else creeps into our desire and pursuits unknowingly.

In social psychology, identity is a central concept which has been the subject of long standing debates regarding the conflict of affirmation and individual necessity.

It is imperative to admit that there is what is called personal identity and social identity. Deschamps and Devos (1998) clarified these two concepts. According to them, these concepts are based on the idea that every individual is characterized by social features which show his or her membership of a group or a category, on the one hand, and by personal features or individual characteristics which are more specific, more idiosyncratic on the other hand.

Simply put, personal identity refers to a feeling of difference in relation to some others while social identity refers to a feeling of similarity to some others.

Schopflin (2001) observes that Identities are anchored around a set of moral propositions that regulate values and behaviour, so that identity construction necessarily involves ideas of 'right' and 'wrong', desirable/undesirable, unpolluted/polluted. Identity construction includes inclusion and exclusion. Inclusion may be a synonym for acceptance, that is, you create or construct an identity that will make you acceptable first to yourself since this is important to your self esteem and complex formation. When you accept yourself in your own estimate, then presenting yourself as acceptable may not be a real problem. However, if the acceptance or inclusion factor is not thoroughly dealt with and you have a struggle accepting your identity, selling it to others becomes a herculean task.

Exclusion is when you wish your name expunged from a particular list, you worry about others' perception of your identity and you wish to have your portfolio deleted from a particular group or category. Categorization according to Deschamps and Devos (1998) refers to psychological processes which tend to organize the environment into categories or groups of persons, objects, events (or groups of some of their characteristics),as well as according to their similarities, their equivalences concerning their actions, their intentions or behaviour.

Categorization as regards identity construction in effect translates to definite groupings in which people locate themselves and emphasis is in the differences between categories (contrast or cognitive differentiation effect). Thus, a category is a group of elements which have in common one or several features.

Gender and Identity Construction

There has always existed dichotomy between masculinity and feminity. It was this differentiation and categorization between male and female, based not on physical features alone but beyond that and more, that feminism rose to correct. This anomaly calls for the need for females to examine self identity. Female identity construction becomes a crucial and critical issue among feminist, philosophers and psychologists. Women consider it imperative to determine their identity by constructing one. According to Giddens as cited by Tucker (1998), with feminism, the body itself became subject to reflexive awareness as people became aware that the self is socially constructed as well as biologically given.

Thus, identity construction has a lot to do with gender demarcation and body design. Simply put, identity construction originates from formation of an individual, personality and entity. It establishes itself through physical appearance, psychological endowment and intellectualism and sexuality.

Physical Appearance (Self Esteem and Body Language)

Identity construction has some business with how beautiful or ugly a personality is. This becomes necessary especially for a female. Female perception of who they are starts from childhood. Thio, (1989) remarked that as soon as a baby is born, parents are likely to ask Is it a boy or girl? This one million dollar question sets the stage for gender differences since the sex of the newborn determines its gender classification and subsequently determines what it would be in future. The gift given to a new baby girl is that of a beautiful pinky flowered doll while a baby boy gets a ball, or toy gun. In Nigeria, colour differentiates; a baby girl receives a pink coloured gift while a baby boy receives a blue coloured gift. Both gifts communicate non-verbally. The doll shows that a girl must be beautiful and bright while the gun expects the boy to be warlike (an attacker of some sort) macho and sporty. He must know and learn how to shoot at the enemy. (perceived and real) Thus, from the beginning of life, a female is concerned about her looks. This sets the female on her toes to construct an acceptable identity for herself in the society. She develops her self esteem from what she wears, how she looks and what adornment she could manipulate to present herself acceptable to herself as well as the world around her.

Psychological Endowment and Intellectualism

Women like men aim high in life and desire to attain the highest pinnacle possible. It is not surprising to have a few of them as Governors, Ministers, Senators and even Presidents in the world. The emergence of notable women of integrity in the political and policy-making scene of Nigeria democratic system in the immediate fourth republic and the present fifth republic attests to women's capability to perform credibly well in governance and policy making. In 2007 election in South western, Nigeria, three female emerged as deputy governors. Sarah Sosan in Lagos state, Olusola obada in Osun state and Salmot Badru in Ogun state. However, the desire to become relevant and significant in life has to do with certain psychological processes. One is the goal or vision of such females. It has to do with what a person is thinking. At the thought and emotion level, there is a lot of construction and imagination, beginning from patterning of life after role models who are successful as well as the attempt to carve a niche for oneself on the world scene. This innate desire to become important and acceptable is closely connected to the reason for identity construction especially on the net.

Sexuality

The internet is filled with sex and sex-related pictures and articles. Websites exist that are full of pornography and some females' construct their identity from such images. Identity construction sometimes erupts from imitation of what is considered the ideal. Hence, a visual expression of a particular action, ideology or picture may send a massage to a female about the "current acceptable way of constructing identity". Freud, the great philosopher, in his writings in the late nineteenth and early twentieth century argued that boys and girls grow up and mature in different ways.

Thus, he tended to see females as more or less stunted males, who, because they do not undergo the oedipal transition do not develop strong egos, have little capacity for holding strong values and tend to be more emotional and sensual but more sexually passive than males.

All of Freud's assertions were dispelled by feminism. For feminism contends that what a man can do a woman can do (better). Choldrow came to the rescue of the females with his analysis that girls find it easier to develop a gender identity through a continuous relationship with their mother. Thus, girls have an ambivalent sexuality than do males. Girls during Oedipal phase, crave an identity with a male (often their father) and later in life continue and maintain a close attachment to other women throughout their lives for only females can understand the true cravings of another female.

Why Young Female Use the Internet Interactive Media (IIM)

For young females the importance of communication as emphasized by Judy et al (2003) reveal that communication can improve the way others see a female, increase what she knows about human relationships, teach her important life skills, as well as permit her to exercise her constitutionally guaranteed freedom of speech. The above suggests the main reason why the choice of the internet for interpersonal and group communication is preferable. According to Ramanujam (2008) the turn of the 21st century signaled the birth of a new medium of communication, the World Wide Web with a decentralized network involving hundreds of millions of people and over a billion pages. The web is a personal medium allowing one-on-one conversations, support groups, virtual communities and individualized access to information hence the high patronage by all.

Female undergraduates are usually in transition from adolescents to young adults with the age range of fifteen to twenty five. These groups of people tend to be hungry for information and therefore explore all available means of acquiring one. It is therefore likely for the search to be extended to the internet. Literature reveals that this age group seeks Identification and appreciation. The search for relevance and importance seems on-going. The interactive formats and pages available on the web permit its use for identity construction by young adults.

Interactive Media

Communication is all about human interaction verbally or visually. Any media selected to carry out this interaction between two or more people is referred to as interactive media or platform. Specifically, communication forms encompass interpersonal, group and mass. Interpersonal medium is face-to-face or one- on-one relationship. Group communication involves many people connected with a purpose.

Some Interactive Media

The internet is filled with media interaction opportunities and websites, the following are a few commonly used by the sample population:

Facebook

Facebook is a global social networking website, that is operated and privately owned by Facebook, Incorporated. Users can add friends and send messages, and update their personal profiles to notify friends about themselves. Additionally, users can join networks organized by city, workplace, school and region. Open to anyone with a valid email address, the site includes members' pictures; biographies, interests and messages, and members can browse freely through open profiles. Members may choose their personal levels of security to prevent strangers from accessing their personal information. The platform enables anyone, anywhere, to build complete applications that members can choose to use. Applications range from photo sharing to graffiti walls.

Chat Room

Electronic chats are a synchronous form of communication, closely resembling actual, real-time conversations. Individuals' log-into a system and can communicate directly with anyone else logged into the same system. Within seconds of writing a comment online, everyone else logged into the system can view and immediately respond to these initial remarks. Electronic chats are synchronous or 'live'. As such two or three students can effectively communicate at any one time.

The primary use of a chat room is to share information text with a group of other users. New technology has enabled the use of file sharing and webcams to be included in some programs and almost all internet chat or messaging services allow users to display or send to each other photos of themselves. With a prolific rise in the activity of 'Sexting', where people send digitized nude or otherwise sexually explicitly photos of themselves electronically to be viewed by others, it is becoming increasingly common for chat room users to exchange nude or sexual photos.

Some people who visit chat room use them as a place to experience online sex, also known as cybersex or computer love. While not physically able to see their partner, cyber-ers get stimulation by reading x-rated quotes. While many in the media focus on this aspect of chat rooms as it certainly boosts their ratings, it is by no means the only thing chat rooms are used for. While many people engage in "cybersex" for many reasons, it is also true that sexual predators use cybersex conversations as a means of identifying potential victims.

Electronic Mail

Electronic mail popularly written in its short form email is a form of digitized typewritten letter sent online to the receiver. It enjoys immediacy of feedback if the receiver is online as at the time such message is sent. It may be forwarded to as many e-mail addresses as possible and may be stored and retrieved for use as at when needed.

Badoo

This is a multi-lingual, social networking base which allows users to share photo and videos, create reportages of their lives and promote themselves and their work.

Bedo

This is a social networking website created to enable friends to stay in contact. It has developed into an online community where friends can post pictures; write blogs and send messages to one another.

Friend Finder

It is an interactive medium where members are connected globally. Subscribers are admitted if aged eighteen and above.

Perfspot

It is an interactive medium in which photographs, images, graphic video or other such contents may be posted by users and these becomes published news or informational contents and are no a longer personal property.

Research Questions

- How do the selected female undergraduates construct identity by the use of internet interactive media (IIM)?
- What does identity construction mean to these female undergraduates?
- What are the reasons given by female undergraduates for identity construction?
- Why do young females use the interactive media available on the internet in identity construction?
- What are the features and affordances of each internet interactive medium that makes it appropriate for these young females for identity construction these young females?

Method of Study

The study examines the uses of the internet as an interactive media among Bowen University female students who used the internet regularly. The purpose was to determine the relevance of the internet as an interactive media among youths and to discover the frequency of use and reasons for use in identity construction.

The study employed the survey technique for generating data. Purposive sampling of all levels of female undergraduates of Bowen University, Iwo was used. Only female students who were internet users at all the four levels of study were selected for the research. A total of 100 questionnaires were given out to them each level got 25. Questionnaires were administered by the researcher herself and all were returned, signifying one hundred per-cent return rate.

Results

The age brackets of respondents were between 16 and 25 years. All respondents were familiar with the internet and Daily use was sixty percent because of the irregularity of energy supply to activate the internet at the available cybercafés. Ninety percent of the respondents used the internet for identity construction explicitly and implicitly. Specifically, the students went on the internet for asynchronous reasons. They utilized the face book, e-mail and chat rooms in order of preference to communicate with peers, friends, family and loved ones globally.

Of the respondents, ninety percent interact with at least a male acquaintance each time they went on the net. Also, ninety percent had their photographs on the internet. The uploaded pictures on the internet serve as a non-verbal communication to communicate identity and to solicit recognition, relevance and acceptance. The respondents said that the reason for uploading their pictures, in addition to the fact that the media is called Facebook (a medium to show your face and by extrapolation your person) was that it advertises them to the entire users who may become prospective partners. When asked whether internet interaction wastes time, ninety percent of the respondents disagreed and applauded the internet as a place for relaxation and entertainment.

Discussion

The sample population revealed that although written messages accompany the visual communication on the net, essentially, the messages were passed non-verbally through the upload of pictures to communicate with other users. The messages range from demographic information to academic achievements to sexual preferences.

Ways Undergraduate Students Use Internet Interactive Media (IIM) to Construct Identity

The female undergraduates use the net to construct identities for themselves by utilizing the internet facilities in many ways than one. This self disclosure agrees with Seabrook's view that …"the home in the real world is among other things, a way of keeping the world out"… and "an online home, on the other hand, is a little hole you drill in the wall of your real home to let the world in". The motive of pasting ones personal details online in a facebook page for instance therefore is to let in others into a personal world. In addition, the types of pictures posted determine and boost the type and number of visitors to the site. Being non-verbal communication, the receiver of the message is at liberty to construe his or her own meaning which may be in consonance or not with that of the sender. In actual fact, most respondents wished to communicate sex or sexually related messages to their receivers most of were male. The study's result endorsed the fact that dressing or physical appearance is a major way of constructing identity by young females. Sometimes some females change their pictures several times in one month just to construct the identity of a beautiful woman. The pictures speak and the subsequent comments of each visitor to the site determine whether such pictures are replaced or not.

Meanings of Identity Construction to the Female Undergraduate

The results shows that a young female undergraduate finds fulfillment in others approval of her person especially males. Her effort to construct the identity of a female with beauty and brain is mainly for acceptance reasons. In some cases many females have inferiority complex because they grew up with criticisms and some with utter condemnation. This accounts for the search for approval and acceptance from a neutral person. Previous assessment of who she is by close associates is then compared with fresh comments from visitors to her site who may see her differently and give soothing assessments pleasurable to her and helpful for her identity.

Reasons for Identity Construction

From the findings of the study, several reasons were proposed for engaging in identity construction on the net. Some of these were:

- Peer influence, because it is the "in thing" to have a web page where your voice is heard. In sum, these young females were easily influenced by what their contemporaries were doing especially if there was societal acceptance of such actions.
- For self actualization, some of these undergraduates considered males' opinion of who they are as very important. As such, when positive comments were posted on their web page it was confirmatory of their opinion of themselves. Sometimes such positive and encouraging comments were copied and placed in conspicuous places where they and others could view it. This is a way of creating identity.

Reasons for Using the Internet Interactive Media (IIM) in Identity Construction

The study results show that the internet is the fastest and quickest way to reach the world; hence accessing its facilities in identity construction by young females is inevitable. Technology has enhanced communication; therefore young people spend more time on the internet than with the traditional media of communication. Even in developing countries female undergraduates spent more time on the internet than they do with television. Therefore considering the traffic on the net these female undergraduates prefer to interact on the net more than any other way. The closest way of interaction after the net is through the mobile phones. Both means of communication shared the same advantage of immediacy of feedback.

Feature and Affordances of Each Internet Interactive Medium for Young Female Identity Construction

Affordance is a concept in perception coined by Gibson (1979) to refer to the features an environment or an object offers or provides the person in that environment or using the object. It emphasizes what the person recognizes or perceives as useful for her purpose rather than what is inherent in the environment or object. So though a facebook may afford to a politician a way to mobilize a following, it affords a female undergraduate a different use. Different motives therefore determine different gratification for each category of users.

This study illuminates the motives of the female undergraduates and the affordances the most widely used interactive media furnish them.

Facebook

- It enables these female undergraduates to find friends and send messages easily.
- It enables them to update their personal profiles and to inform friends about themselves.
- It enables them wherever they are to build personal web pages using an array of application from which they can choose. These applications allow them to upload photos and pictures, use emoticons and graphics.
- Facebook allows these female undergraduates to interact with others without restriction without being shy. This is evident in the photos of different degree of nudity which they upload.

Chat rooms pride itself in the following affordances:

- It is used by the female undergraduates to carry out discussions about themselves, others and current events of interest to them and others in their virtual world.
- Female undergraduates frequently use this medium for sexting with males in the chat room that sometimes lead to cybersex
- Can be used by students to present and describe their work to others; feedback or advice can be sought from other participants.
- Female undergraduates use chat room as a community meeting place where the 'I' is projected or constructed.

Most Suitable of All Available Internet Interactive Media for Young Females

The findings of the research showed the facebook as the most suitable of all interactive media appraised in the study followed closely by the chat room and e-mail. Essentially, the ability to upload pictures is a major advantage of the facebook since youths are fascinated by looks. Moreover, the IIM allows free expression of "I" and "ME" as explained by Mead self identity.

Identity Construction and Gratification

These female undergraduates used the IIM to meet certain needs that relate to constructing acceptable identity engendered by normative influence. These were portrayed in physical appearance (self esteem and body language), psychological endowment and intellectualism, sexuality and relationships.These needs are projected through pictures, text messages (sexting) invitation to join pages, as well as engaging in discourse with others about a chosen topic of interest. The IIM affords female undergraduates these gratifications.

Internet Interactive media abound on the internet but the two most widely used by the respondents in this study are the Facebook and the Chart room.

CONCLUSION

This study affirms the "uses and gratification" theory that media users or the audience engage the media actively by utilizing it for gratification purposes to suit their desires. In this regard, the active users of the internet were primarily youths who sought affirmation for their self identity on these interactive media. However, it is imperative to be conscious of the merits and demerits of internet interactive media use. While it provides a way of escape for youths to explore, discover, assess, achieve and construct a concrete identity for themselves, a major demerit of the interactive media in identity construction is that it may encourage subjective assessment of young females and contribute adversely to their identity construction. Others are that it may also affect complex formation negatively, if mischievous comments are made. It may also encourage indecency on the net. Although most respondents denied the fact that internet surfing wastes time it is a known fact that sometimes, youth waste time on the net while engaging in cybersex.

On the whole, the use of the internet for identity construction can not be ignored and it is unavoidable for youths to employ the internet interactive media for self gratification. Since the internet has come to stay, the youth workers and counselors must not dismiss the long term effect of these media on their perception of themselves.

SUGGESTIONS FOR FURTHER RESEARCH

Further research may be conducted to appraise the way male undergraduates construct identity through the internet and what identity construction means to males.

A comparative analysis of findings can thus be carried out and a position paper published to endorse or contest existing literature on differences and similarities between young male and females.Moreover, it may, also be worthwhile to do a content analysis of the actual comments available on selected facebook page to appraise its appeals.

REFERENCES

Abanihe – isiugo, U.C. (2002). *Currents and Perspectives in Sociology*.

Brown, J. S., & Duguid, P. (2002). *The Social Life of information*. Boston: Harvard Business School publishing Corporation

Deschamps, Y. C., & Devos, T. (1998). Regarding the Relationship between social identity and Personal Identity. In Worchel (Ed.), *Social Identity: International Perspectives* (pp1- 12). London: Sage Publications

Doise, W. (1998). Social Representations in Personal Identity. In Worchel, Y. (Ed.), *Social Identity: International Perspectives* (pp. 13–23). London: Sage Publications.

Gamble, T. K., & Gambel, M. (2002). *Communication works* (7th ed.). Boston: McGraw Hill.

Gibson, J. J. (1979). *The Ecological Approach to Visual Perceptions*. Boston, MA: Houghton Mifflin.

http://ww. Oefg.at/text/veran staltungen / wissenschaftstag/ Beitrag –

Hybels, S., & Weaver, R. L. (2001). *Communicating Effectively* (6th ed.). Boston: McGraw Hill.

James, S.L., Ode, I., & O., Soola (1990).*Introduction to Communication: For Business and Organisations*. Ibadan: Spectrum Books Ltd.

James, W. (2003). *Media Communication: An Introduction to Theory and process* (2nd ed.). New York: Palgrave Macmillan.

Keesing, R. M & Keesing, F.M. (1971). *New Perspectives in Cultural anthropology*. New York: Holt, Rinehart and Winston.

Knowledge-based webliography. Retrieved January 13, 2010, from

Lagos, N. Malthouse Press Limited Adedayo, O. (2009). *The internet as a medium of communication: A uses and Gratificationtheory amongst* Bowen University Students. Unpublished Undergraduate Long essay, Bowen University, Iwo Nigeria.

Lamy, M. N. Hampel. R. (2007). *Online communication in language learning & teaching.* London: Palgrave Macmillan.

Lister, M., Dovey, J., Giddings, S., Grant, I., & Kelly, K. (2009). *New Media: A critical Introduction* (2nd ed.). London: Routledge.

Pearson, J., Nelson, P., Titsworth, S., & Harter, L. (2003). *Human Communication*. Boston: McGraw Hill.

Ritzer, G. (2000). *Classical Sociological Theory* (3rd ed.). Boston: McGraw Hill.

Schopflin, G. (2001). *The construction of Identity: Learning-Theories*. Communication.

schopflin. pdf.

Thio, A. (1989). *Sociology an Introduction* (2nd ed.). New York: Harper & Row Publishers.

Thomas, E. K., & Carpenter, B. H. (Eds.). (2001). *Mass Media in 2025: industries Organizations, People and Nations Westport*. Santa Barbara, CA: Greenwood Press.

Tucker, K. (1998). *Anthony Giddens and Modern Social Theory*. London: Sage Publications.

Uche, L. U. (1996). *North-South Information Culture: Trends in Global Communications and Research Paradigms*. Lagos, Nigeria: Longman Plc.

Uduoakah, N., & Iwokwagh, N. S. (2008). Communication and HIV/AIDS Prevention among Adolescents in Benue State. *The Nigerian Journal of Communications, 6*(1 & 2), 44–58.

ADDITIONAL READING

Aftab, P. (2000). *The Parents Guide to protecting your children in cyberspace*. London: McGraw-Hill.

Albarran, A.B. (200). *Management of Electronic Media*. (2nd ed.) Australia: Wadsworth.

Asika, N. (1991). *Research Methodology in Behavioral sciences*. Ibadan: Longman.

Barran, S. J., & Davis, K. D. (1995). *Mass Communication Theory: Foundations, ferment and future*. Florence, KY: Wadsworth.

Barran, S. J., & Davis, K. D. (2003). *Introduction to Mass Communication: Media literacy and culture* (3rd ed.). Boston: McGraw-Hill.

Belmont, CA: Wadsworth.

Blumer, J., & Katz, E. (Eds.). (1974). *The Uses of Mass Communication*. Beverly Hills, CA: Sage.

Cobb, N. J. (2000). *Adolescence: continuity, change and diversity* (4th ed.). New York: Mayfield Publishing Company.

DeFleur, M. L., & Ball-Rockeach, S. J. (1989). *Theories of Mass communication* (5th ed.). London: Longman.

Disch, E. (1999). *Reconstructing Gender: A Multicultural Anthology* (2nd ed.). New York: Mayfield Publishing Company.

Dominick, J. (1999). *The Dynamic Mass Communication* (6th ed.). New York: McGraw-Hill.

Ehikhamenor, F. A. (2003). Internet Facilities: Use and non-use by Nigerian University Scientists. *Journal of Information Science*, *29*(1), 35. doi:10.1177/016555150302900104

Hancock, P. (2000). *The Body, culture and society. An Introduction*. Buckingham, UK: Open University press.

in Nigeria: A Wake-Up Call. Abuja, Nigeria: UNICEF and NPC Publishers.

Internet Society Site. (2010).Retrieved from http://www.isoc.org/internet/history/brief.shtml

Lee, N. (2001). *Childhood and Society. Growing up in age of uncertainty*. Maidenhead, UK: Open University Press.

Littlejohn, S. W. (1992). *Theories of Human communication*. Belmont, CA: Wadsworth Publishing Company.

Lopez, J. J., & Scott, J. (2000). *Social Structure: Concepts in Social Sciences*. Buckingham, UK: Open University Press.

McQuail, D. (1987). *Mass Communication Theory: An introduction*. Thousand Oaks, CA: Sage.

McQuail, J. D. (2005). *McQuails Mass Communication Theory* (5th ed.). London: Sage.

Media. New York: Longman.

Pfohl, S. (1994). *Images of Deviance and Social Control: A Sociological History* (2nd ed.). New York: McGraw-Hill.

Severin & Tankard J. (1997). *Communication Theories*. Origins, Methods and Use in the Mass.

UNICEF & NPC (2001). *Situation Assessment and Analysis on Children's and Women's Rights*

Wimmer, R. D., & Dominick, J. R. (1989).*Mass Media Research: An Introduction,* (2nd ed)

Worchel (Ed.) (1998). *Perspectives Social Identity:International Perspective*. London: Sage

Zinn, M. B. (2000). *Gender through the prism of difference* (2nd ed.). Boston: Allyn and Bacon.

Chapter 8
The Representation of Female Friendships on Young Women's MySpace Profiles:
The All-Female World and the Feminine 'Other'

Amy Shields Dobson
Monash University, Australia

ABSTRACT

This chapter examines the representation of female friendship on MySpace, based on a sample of 45 public MySpace profiles owned by young Australian women, aged between 18 and 21 years old. Two prominent constructions of female friendship on this social network site are outlined: firstly, female friendships as idealistically party-oriented, 'wild', and rowdy; and secondly, female friendships as close, loyal, and intimate — comparable in the depth of feeling and connection expressed to romantic partnerships or family ties. These idealised, performative constructions of female friendship, in the context of online self-presentation, also seem to rely on exclusivity, and opposition of selves and friendship groups to a feminised outsider/'other'. Some of the political implications of such representation are discussed from a feminist perspective. I suggest some ways in which ideals and goals of female representation to emerge from second-wave feminist media and performance critique might be said to have actualised and failed to actualise in these online performances of friendship and identity created by young women.

This chapter contains explicit content

INTRODUCTION

This chapter examines the representation of female friendship on online social network sites, based on a sample of 45 public MySpace profiles owned by young Australian women, aged between 18 and 21 years old. This chapter draws on research conducted for my doctoral thesis, which examines the performance of popular modes of femininity by young women on MySpace, employing media and performance theory. Perhaps unsurprisingly, a display of sociality and friendship with other young women is one of the most prevalent aspects of these young women's self-presentation, a finding in line with ethnographic work which documents the central importance of friendships with other

DOI: 10.4018/978-1-60960-209-3.ch008

females in the lives of teenage girls and young women (McRobbie and Garber, 1976; Hey, 1997). On 40 of the 45 public profiles examined, young women depict themselves with female friends, often displaying several albums with dozens of images of themselves and their friends in each album. All 45 of the profile creators in my study centrally reference time spent with friends, or the importance of their friends in text displayed on their profiles. On some of the profiles examined, much space is devoted to lengthy descriptions of a profile owner's close friends, or to textual dedications to friends.

In my examination of the data contained in the 45 MySpace profiles, two key constructions emerge in the presentation of close female friendships. Friendships are presented as idealistically party-oriented and rowdy. They are also presented as idealistically close, loyal, and intimate—comparable in the depth of feeling and connection expressed to romantic partnerships or family ties. Both of these constructions lead to an overall image for viewers of an idealistic female-centred world which is all-encompassing and self-enclosed, and able to satisfy almost all of the young women's social and emotional needs. Close, loyal, and exceptionally fun friends are constantly present, and time with friends and wild parties together abound.

The friendships that occupy most of the photo imagery and text examined from the profiles are most commonly between two or more young women, with male friendship, and in many cases, even romantic relationships with males, playing a significantly smaller role in the self-presentation examined. Of course, several young women display close friend relationships with young men. Also, in several cases, the boundaries between non-sexual and sexual relationships with other young women are blurred on these profiles, as I discuss. However, in none of the profiles examined are sexual relationships with the close friend in question claimed or directly named as such by the young women involved. Rather, displays of sexual affection and homoeroticism between female friends are depicted with undertones of 'tongue-in-cheek' and seduction of a (male) viewer, as I explain. Henderson (2008) and McRobbie (2008) have both noted a similar phenomenon in their recent writings on young women's culture in the West.

In this chapter I do not attempt to delve into the actual practice or 'reality' of these relationships by examining communication between friends or comments displayed on each other's profiles. My primary concern is with the *representation* of female friendship as part of an individual's overall identity performance in this public forum. I textually analyse female friendships on the MySpace profiles by examining the images and textual self-descriptions posted by the profile owners. I employ a feminist framework to do so, with particular reference to feminist representational theories from media and performance studies. I suggest that the representation of friendship on these young women's MySpace profiles depicts young femininity as active, pleasure-seeking, and sometimes also more typically 'masculine' in the type of hedonistic and rowdy behaviour shown to be engaged in by young women. This contributes to a construction of female subjectivity often called for by second-wave feminist representational critics and scholars. However, such presentation of female friendships is also quite romanticised, perhaps serving to contribute towards the construction of a kind of idealised, young, 'new femininity' (Tincknell, Chambers, Van Loon, and Hudson, 2003; Gill, 2006). Whilst self-alignment with this dominant construction of a kind of post-feminist 'new femininity' perhaps allows young women greater freedom in terms of the type of behaviour they can be publicly seen to participate in, it also seems reliant on a typically 'masculine' performativity, or 'laddishness', as I discuss. Further, I argue, these close female friendships, in their performance on MySpace, seem to rely on exclusivity, and opposition to a stereotypically feminine 'other', in the same way

Simone de Beauvoir argues that any dominant group automatically and often unconsciously opposes itself to an 'other' (de Beauvoir, 1949/1972). On MySpace, this 'other' is often constructed as egotistical, superficial, nasty ('bitchy') and/or ugly — in other words, as a stereotypically 'bad' female. I argue that such constructions of female friendship and subjectivity then leave stereotypes of femininity intact, simply applying them to 'other' young women and distancing them from self, and thus, somewhat eluding a deeper deconstruction of gender binaries.

THE SAMPLE OF PROFILES

The 45 MySpace profiles examined in the doctoral research from which this chapter draws were selected from a viewing of over 300 public profiles on this social network site, maintained by Australian 18-21 year-old women. Social network sites are websites that allows users to create personal profiles to represent themselves, and connect with other users — most typically, users who are part of their offline social network. Social network site members can chat, share media, post blogs, expand their social networks by connecting with new people, and see the connections between other members of the network. The viewing and selection of the social network site profiles in my sample was conducted between September 2007 and February 2008. Profiles were included in this study based on the richness of expressive data they contain, and also, because of the way these young women perform their identities within current conventional models of heterosexual femininity in Western popular culture. In this study I have not looked at self-representations on MySpace which are overtly subcultural, or which sit radically 'outside' the current normative models for femininity in the aesthetics used and the identity presented. Rather, just as Janice Radway (1984) and Ien Ang (1985) were interested in the way female consumers consumed 'mainstream' and gender-typical media such as heterosexual romance novels and soap operas, I am interested here in the cultural productions of young women who, in their own identity constructions, engage with gender-typical aesthetics and current heterosexual paradigms of feminine performativity. Female friendship is one of four aspects of self-presentation that emerge as commonly important across the sample of profiles examined. The other three prevalent aspects of self-presentation on MySpace I identify in my doctoral research are: hetero-sexy profile decorations and images (see 'Key Terms' for definition of hetero-sexy); crass, vulgar and 'laddish' self-imagery; and a kind of pedagogical and individualistic discourse that comes across in the self-descriptive texts and mottos which these young women commonly display (Dobson, 2010).

The profiles selected as part of this sample align, in their visual appearance, with prevalent modes of feminine performativity in recent popular Western media culture. However, it should be noted that the young women whose profiles are examined here are not representative of a *majority* group more broadly. Precise and up-to-date figures for social network site use by Australian young women are difficult to obtain. Writing in 2007, Kennedy, Dalgarno, Gray, Judd, Waycott, *et al.* report that more than 50% of the Australian undergraduate university students in their study (which involves multiple universities across Australia) had never used a social network site (Kennedy et al., 2007, p. 520). Just three years later, this figure is likely to be much smaller. Comprehensive demographic research on Internet use has been conducted in the USA by Pew Internet, who suggest that 74% of Americans over 18 use the Internet (Rainie, 2010), and of these users who are aged between 18 and 24, 75% have a profile on a social network site. It is important to keep in mind, however, that, as Murthy's (2008) research indicates access to the Internet and to social network sites 'remains stratified by class, race, and gender' (Murthy, 2008, p. 837).

The names of all the profiles owners and their friends have been changed; any identifying photographs shown have been de-identified; and any identifying information has been changed or removed from material discussed here to protect the privacy of the profile owners and their friends. The pseudonyms I use have been constructed to give readers a sense of the kind of names chosen by the creators. They are in keeping with the spirit of the names commonly used by young women, but do not reveal the individuals' MySpace alias. It can get confusing at times when trying to decipher in some of the material found what spelling and grammar mistakes are intentional 'expressions' on the profiles and what are genuine errors, because of the way that language, capitalisation and grammar are used in these young women's textual expressions. For the purposes of clarity in this chapter, what appear to be obvious 'typos' in text from the profiles have been fixed, but I have left grammar, capitalisation and 'loose' forms of spelling and abbreviated language alone, and displayed most of the text quoted as it appears on the public profiles. As said, all of the profiles were initially accessed and textually analysed between September 2007 and February 2008, and screen-grabs of the profiles were collected during this period. Many profiles have been subsequently accessed, and the access dates given in reference to material shown from the profiles reflect the date on which specific material from the profile was last accessed.

BACKGROUND OF THIS RESEARCH

My lens of analysis considers the phenomenon of self-production and identity performance by young women on MySpace in relation to some central objectives set out by feminist media and performance theorists. Feminist scholars from across a range representational disciplines including theatre, film, and television, have expressed as a goal the need for 'better', more 'authentic' female subjectivities in the representation of women. They have called for subjectivities which are multifarious, diverse, complex, and strong, rather than one-dimensional, 'decorative', and passive, as it has been argued so much representation of women is, across a range of media, narrative forms and representational genres (Friedan, 1963; Greer, 1970; Butcher, Coward, Evaristi, Garber, Harrison, and Winship, 1974; Edgar and McPhee, 1974; Busby, 1975; Mulvey, 1975/1989; Betterton, 1987; Case, 1988; Ferris, 1990; Faludi, 1992). There is an assumption underlying much early second-wave feminist film, television, and theatre critique that if and 'when' more women are themselves *producing* cultural representation, that images of women in the media will, as a matter of course, give rise to more diverse, authentic and multi-dimensional female subjectivities. I explore how this feminist ideal is playing out today in the self-representations of young women on MySpace — the different ways in which, in relation to this early stage of feminist media and theatre theory, feminist ideals might be said to have actualised and failed to actualise.

This lens of analysis contributes a significant, and experimental, change of approach to the questions posed about young women's self-made media, by examining social network site profiles primarily as a form of cultural production by young women. Much literature on young women's Internet use asks how young women interact with peers and express identity in various online environments (Stern, 1999, 2002; Takayoshi, Huot and Huot, 1999; Davies, 2004; Reid-Walsh and Mitchell, 2004; Bortree, 2005; Grisso and Weiss, 2005; Schofield Clark, 2005; Thiel, 2005; Kearney, 2006; Kelly, Pomerantz, and Currie, 2006; Dobson, 2008). In this literature, the Internet is often positioned as a 'safe space' for young women to express themselves, experiment with identity, and communicate more freely than they may be able to in 'offline' environments. Several studies on girls' and young women's media use flag the importance of addressing girls and young women as media *producers* as well as media consumers (Bloustien, 2003; Kearney, 2006; Tsoulis-Reay, 2009), which this chapter does. By locating my examination of MySpace self-representation tentatively within the field of feminist media and performance theory, my aim is to move beyond the scope of exploring the personal, social, and perhaps political opportunities and risks this kind of creative, digital-media identity production may offer young women as individuals, and into the political implications that these expressions may hold as public, mediated images of (young) women.

FEMALE FRIENDSHIP AS ROWDY AND 'LADDISH'

In their photo galleries the young women whose profiles I have examined create an image of themselves and their social worlds as jubilant, rowdy, adventurous, and party-oriented. The profile producers show themselves and their friends playing, laughing, dancing, drinking, socialising and creating poses together. They pose sometimes in exhibitionistic and sexually provocative ways, as well as in silly ways, such as with wide open mouths and protruding tongues, and in vulgar and crude fashions that may be seen as more typically 'masculine' or 'laddish'. The representation of young feminine sociality seen on the MySpace profiles examined is in line with that presented in popular reality TV shows such as *Girls Gone Wild* (1998-present) and *Ladette to Lady* (2005-present). This amounts to a performance of young feminine sociality centred around debauchery and 'laddishness'. Several scholars and journalists have noted the take up, as well as the media representation, of typically 'masculine' and 'laddish' behaviour by girls and young women (Whelehan, 2000; Winship, 2000; Tincknell *et al.*, 2003; Verghis, 2003; Jackson, 2006; Jackson and Tinkler, 2007). Jackson describes the way the term 'ladette' is used in the mainstream UK press, 'to refer to girls that are a bit "laddish" in terms of being a bit loud, maybe a bit boisterous, a bit jokey, they go out drinking and things like that' (Jackson, 2006, p. 345).

As Figures 1-3 exemplify, there are numerous images in the MySpace photo galleries of young women being crass and vulgar in typically 'unfeminine' ways: for example, sticking their tongues out, showing wide open mouths and depicting (often excessive) alcohol consumption. By displaying images of themselves and their friends in such poses and positions the subjects present themselves as wild, uninhibited, licentious, and 'laddish' or 'phallic' in the way that McRobbie has argued young women now increasingly behave (2007), and also in a way that subverts traditional notions of young femininity. These images of young women playing together, drinking, and generally having fun, depict, as Stallybrass and White note of Bakhtinian (1968) 'grotesque' body imagery, 'a subject of pleasure in processes of exchange' (Stallybrass and White, 1986, p. 22). Young women are seen on the profiles examined jumping on trampolines, poking their tongues out and laughing with friends, stuffing food into their mouths, and grasping each other's limbs in playful exchanges and displays of carefree abandonment. Posing with protruding tongues and wide open mouths is a particularly common mode of self-presentation, demonstrated in Figures 1, 2 and 3. So too are photos of young women drinking together, exemplified by Figure 2. These young women depict themselves as active, pleasure-seeking subjects, rather than inert, passive objects of other's pleasure, whether it be through play, participation in social festivities, or the consumption of alcohol or food.

The following examples of text displayed on young women's profiles — often in large, decorative fonts or graphically designed frames, which I refer to as graphic text (defined in 'Key Terms') — extend on this notion. The graphic text furthers the construction of young feminine sociality as similar to that of conventional young masculine

Figure 1. Harriet, accessed 03/03/08 (Used with permission)

sociality in its level of crassness, unabashed hedonism and *public* pursuit of — often somewhat licentious — entertainment and gratification. By extension, perhaps a *feminine* individual who is herself crass, vulgar, and a pleasure-seeking *subject* in her own right becomes possible.

Yeah, we LAUGH way TOO HARD,

and act TOO imamature, BUT i

wouldn't have it any other way

(Georgie, accessed 03/10/07)

Everyone Has A Wild Side

BUT ME AND MY FRIENDS JUST PREFER TO MAKE OURS PUBLIC

(Micki, accessed 17/09/07)

The above quotations and the images and graphic text in Figure 4 indicate that notions of laddishness are being readily taken up by these young women in their own self-presentation. McRobbie (2007) refers to this emergent construction of girlhood as 'phallic girlhood'. The raucous, hedonistic style of self-presentation and the drinking and sexual leisure culture that accompany it was once seen exclusively as the domain of young men. McRobbie argues that some limited privileges of masculinity have been afforded to girls and young women as a way of deflecting or silencing critique of hegemonic masculinity (McRobbie, 2007, p. 718). In other words, the social acceptability of 'laddish' behaviour — which often involves excessive alcohol consumption, rudeness, aggression, sexual objectification and chauvinism — goes unquestioned as an acceptable paradigm for masculinity, so long as girls are allowed to partake too. According to McRobbie, girls' participation in sexual promiscuity and binge

Figure 2. Alexx, accessed 15/12/09 (Used with permission)

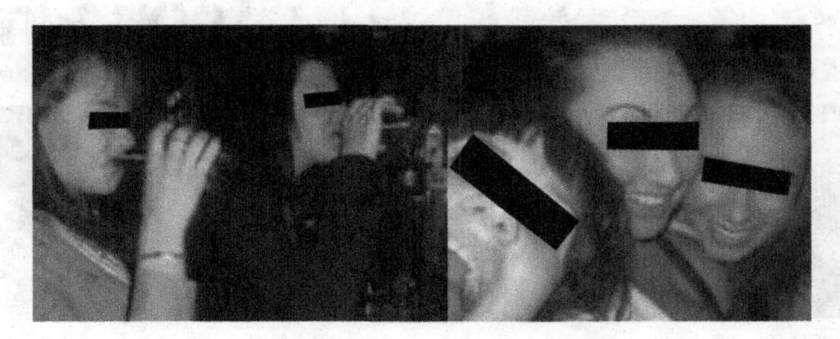

Figure 3. She's infamous, accessed 15/12/09 (Used with permission)

drinking culture is a central part of what she refers to as a 'new sexual contract':

On condition that she does not reproduce outside marriage or civil partnership, or become the single mother of several children, the young woman is now granted a prominence as a pleasure-seeking subject in possession of a healthy sexual appetite and identity. Consumer culture negotiates this complicated terrain by inviting young women to overturn the old sexual double standard and emulate the assertive and hedonistic styles of sexuality associated with young men, particularly in holiday locations and also within the confines of licensed transgression of, for example, weekend heavy drinking culture. (McRobbie, 2007, p. 732)

The quotations given above from Georgie, Micki and She's infamous contain expressions of the same kind of rowdiness, assertiveness and risky behaviour to which McRobbie is referring in the context of sexual hedonism. This chapter focuses on the presentation of female friendships on MySpace, not on sexuality. However, I reiterate McRobbie's points here because I would extend her assertion to point out that in their self-presentation on MySpace, these young women not only emulate the 'assertive and hedonistic styles of *sexuality* associated with young men', but the brazen and hedonistic styles of drinking and general social leisure behaviour typically associated with young men. Of course, as McRobbie implies, these elements of leisure culture go hand in hand. As Figure 4 demonstrates, female sexual promiscuity is often represented by young women themselves in a much less damming light than it traditionally has been. Whether or not the kind of behaviour associated with 'whores' and 'sluts' is actually acceptable amongst young women in 'real life' contexts, such behaviour is often joked about light-heartedly, or even presented in a posi-

Figure 4. She's infamous, accessed 06/08/08 (Used with permission)

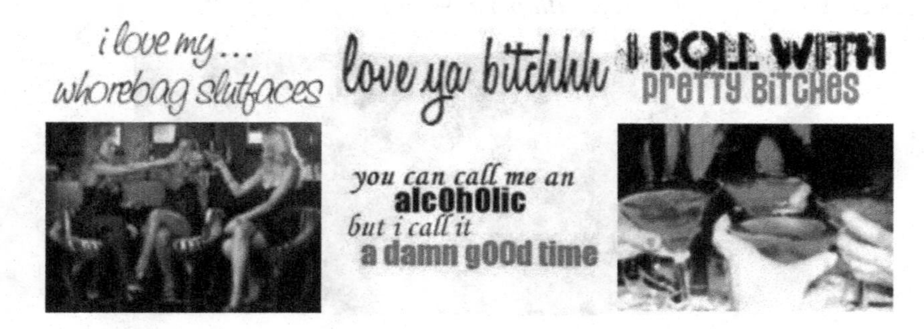

tive light when referring to the actions of one's friends in the MySpace representations examined.

The use of the phrase 'whorebag slutfaces' to refer to one's close and beloved friends exemplifies the way in which derogatory, gendered and offensive terms to describe female sexual promiscuity have been reconfigured in the representations examined, and are now often used affectionately between friends, sometimes in a context of pride in an individual's unabashed sexual pursuit. Terms like 'bitches', 'ho's', 'wifey's' and 'sluts' have been used in rap and hip-hop music, amongst other forms of popular culture, to describe women since at least the 1980s. Gill (2003) has argued that in a post-feminist cultural and political context, these terms seemed to have gained some new credence, particularly in popular youth culture. Whether young women simply imitate popular musicians and men to show their 'cool' complicity, as McRobbie has suggested (2004, p. 9), or adopt these terms consciously in a similar way to the adoption of the term 'nigger' in some black American cultures, cannot be determined from the representation in this sample alone. What the MySpace profiles do demonstrate, however, is that, as I later discuss, these terms are still used by young women negatively, and invested with their original derisive meanings and venom when applied to *other* females outside the immediate friendship group.

Much more could be said of these displays of rowdy 'laddishness', particularly in regards to issues of alcohol consumption and the public display of such. It is beyond my scope here to comment on the sociological concerns associated with the public display of heavy alcohol consumption that seemingly accompanies this kind of 'laddish' performativity (Jackson, 2006; Jackson and Tinkler, 2007). Such an issue deserves more detailed attention and warrants further research. McRobbie sees constructions of femininity in girls magazines as places where girls are encouraged to be 'loutish' and behave like boys are traditionally seen to behave, and she poses the question, 'What kind of confidence and enjoyment does it give girls to have won the freedom to act like boys' (McRobbie, 1999, p. 51). In this space of new-media *self*-representation, not only do the young women in question depict themselves as pleasure-seeking subjects, who behave loutishly 'like boys', but significantly, as subjects who are able to garner immense pleasure and enjoyment purely from their own, autonomous and somewhat-closed social groups, from which males seem largely absent, as the image and textual examples discussed in this chapter demonstrate.

LOVE ALL MY DANCING GIRLS! and give me a D floor, my gals, some tunes..the occasional drink and im set for the long night/morning ahead! ohh and throw in an early chunder in there and im definetely set!! ;-

(Lexxi, accessed 03/09/07.)

I LOVE GOING OUT AND DANCING WITH MY GIRLS!!! There's nothing i wouldn't do for my friends, (even if they've pushed me past my limits, no matter what, IM STILL THERE FOR THEM!!!) I love my girlies, without them im nothing!! (Bitch Cunt/Janine, accessed 06/11/07)

I'll always live for the nights

I can't remember

And the friends

I'll never forget

(Fee, accessed 29/10/07)

You're only as **strong** *as The tables you* **dance** *on. The drinks you mix & the* **friends** *you roll with.*

(Anna, accessed 20/02/08)

A Closed and Idealised World

This performance of an apparently exclusive 'female world', where young women freely participate in pleasurable, hedonistic and impious behaviour, is significant in relation to the history of traditional and stereotypical feminine imaging on screen and stage. Females have traditionally been seen as passive, inert 'backdrops', whose presence onscreen in much cinema simply provides a pleasurable visual spectacle for consumption by the (masculinised and hetero-sexualised) viewer (Mulvey, 1975/1989). In canonical dramatic narratives and myths, it has been suggested by feminist representational theorists that the role of the female has not been to seek her own pleasure or gratification, but to aide the male subject in finding his, if not to be his source of pleasure herself (Edgar and McPhee, 1974; Mulvey, 1975/1989; Doane, 1987; Case, 1988; Ferris, 1990; Faludi, 1992; Baum, 2003). These theorists have suggested that females have rarely been the subject in dramatic narratives. That is, they are rarely the driving force of the action, the protagonist, or, the one with whom the audience is meant to identify and sympathise. Instead, in dramatic narratives, the presence of the female is usually dependent on the male and his use for her/what he does with her: whether it be to love her, save her, kill her or sexually objectify/idealise her. Particularly in fairy tales the 'pleasure', or more accurately, the salvation, of the female is most often reliant on the male protagonist.

Even in more modern, popular narratives on screen in which women are undeniably the subjects of the dramatic action, and which also depict the kind young feminine party culture and 'phallic behaviour' to which McRobbie (2007) and others refer, such as *The Sweetest Thing* (2002), and *Sex and the City* (1998-2004), rarely is there *not* a man waiting by the end of story to put an end to the female subject's party-centred lifestyle and the sexual freedoms associated with her singlehood. In the images and text from MySpace discussed above, young women demonstrate and voice their pride and appreciation of their female friendships, and express the pleasure and gratification gained from time spent with their (almost exclusively female) friends. That is, often in the absence of males. They present themselves as wild and hedonistic when together with other female friends. In relation to some fundamental feminist representational critiques, this kind of representation may be considered significant. It forces a consideration of the ways in which some feminist representational goals are met in such image production, as well as the ways in which gender binaries may still be left intact. In fact, not only do the profiles examined depict a pleasurable social world largely autonomous from males, but from any 'others'. As I later discuss, this leads, almost by default, to the construction of an exclusive social world, and the 'othering' of females who are not part of the immediate friendship group. So whilst feminine subjectivity portrayed onscreen here may be more in line with the kind of active, desiring subject that emerges as preferable in the work of scholars such as Mulvey (1975/89) and Case (1988), the kind of recognition of gender as a source of inequality, on which feminism is founded, seems largely absent from these displays of strong female friendship on MySpace.

The text on the profiles which references friendship and sociality depicts quite a closed social world. One may assume from the sentiments quoted so far that the profile creators go out not necessarily to meet new people or pursue courtship of the opposite sex (flirtation or sexual/romantic encounters) but primarily to have fun with their close friends. I do not mean to imply that males, romantic partners, and heterosexual sex are irrelevant to these young women. However, I suggest that the profile owners are actively constructing an image of their social world as exciting and gratifying in itself, self-contained and self-sufficient in regards to the presence of the opposite sex, or anyone else. This could perhaps be due to the 18-21 year-old age range I have examined; perhaps slightly older women would demonstrate more focus on sexual pairing and relationships in their self-presentation. Nonetheless, the romantic ideal is a strong one in popular culture, and one that is markedly absent from the profiles examined. Hopkins suggests that fame has replaced romance as the dominant fantasy for this generation of young women (Hopkins, 2002, p. 4). Perhaps the representations examined, not to mention the MySpace phenomenon itself, are indicative of this — '*acting* like a rock star' does commonly seem to involve endless, wild partying and unashamed sexual promiscuity.

The text and imagery on these profiles constructs a somewhat idealised world, not unlike that perhaps fantasised or imag(in)ed as accompanying fame and celebrity-hood. In the imaged/imagined world of dancing, drinking, and partying with friends, there is little to indicate any kind of conflict or betrayal between friends. Whilst a closer inspection of, for example, the comments between friends on these profiles may reveal dramas, fights, insecurities, and jealousies that Hey's (1997) research suggests often abound in young women's close relationships, it is not my purpose here to examine the profiles in terms of the 'real lives' they reveal. My purpose is to examine the immediately accessible/overall image being presented to a public audience — in other words, how female friendships are *performed*. What is constructed for viewers is a positive, idealistic, and somewhat fantastical or escapist social world, one which is also closed off to some degree from males, from others outside the friendship group, and from the everyday, mundane life-worlds of school, work, and the day-to-day dramas of young women's lived social lives.

CLOSENESS, LOYALTY, AND INTIMACY IN FEMALE FRIENDSHIPS

So far I have discussed the first of two main constructions of female sociality on the profiles examined: that of rowdy and 'laddish' behaviour in which female friends joyfully engage together. Now I discuss the second main construction of female friendship: that of friends as extremely close, loyal and devoted to each other in a way comparable to romantic relationships or family. On many profiles, text about friends is posted which describes an individual friend and the relationship between her and the profile creator. Sometimes such text is written as if being directed to the individual friend herself, but it is posted publicly as a blog, bulletin, or self-description, leading me to read such texts more as public *tributes* than as personal communications between friends. What is communicated very strongly to a public audience is the significance of the relationships, and the degree of loyalty upon which the relationships are based. As the following text shows, often female friend relationships are described as being of vital importance, with the profile owners expressing a willingness to go to great lengths to serve and protect their friends and their friends' interests, and praising the dedication and loyalty reciprocated by friends.

Mel and Kate.. 2 girls who have seen the worst & best of me & have stuck by me through everything... Mel has sculpted the person i am & Kate has saved my life!! I LOVE MY GIRLS & no matter what happens... I will never forget what you did for me... Words can not describe how much i love my baby girl liz... She is (AND THIS IS TOTALLY CORNY) but truly she is my other half... There is no other person who can make me laugh, cry & piss me off more than her... I LOVE YOU BABY

GIRL, thankyou for reminding me who i am & what i can be....

(Bitch Cunt/Janine, accessed 03/03/08)

..:: Friends are like family... only better::.. Belinda (left) I LOVE LOVE LOVE LOVE LOVE this girl to death! no words can describe the way i feel about her, we just have this magical bond that i dont have with anyone else.. shes fucking funny as!! and we can sit there for days just goin on about random shit and personal jokes..she is just truly a best friend and im going to grow old with this girl!! Crystal. This girl is gorgeous, shes such a sweetie and im so glad i got to know her better as shes has become the apple of my eye:) we laugh, we dance we go home with each other but most of all we have fun! we look out for each other and talk everyday.. shes my hot wifey!!♥LOVE U BITCHES♥

(lil' freakshow Erica, accessed 03/03/08)

Karen you are my life i love you with everything i have your my best friend and my sister you have always been there for me I would do anything for you your motherfucking RAD FUCK SHIT CUNT WHORE..as always Manson whores for life! your the greatest and you laugh funny friends for life babe!! ily.. (Amanda/Fuckhead, accessed 03/10/07)

....bRianNa....

*what can i say about this chick? i absoultely u her with all my heart! you have been there for me through thick and thin, you have sat there and listened to me when i have cried my eyes out when you could have said i told u so as that was probly the only thing to say in some cases!! you are the best! the 14 years i have known u have been the best, we have grown up together and i couldnt ask for anything better!!! love you with all my heart xoxoxoxoxoxox STAY BEAUTIFUL ** LICK** haha you know what im talking about hahahahaha!!!!*

(Danirocks, 29/11/07)

Metaphorical Language

Some of the significant semantic themes in this text include the use of lover/mother metaphors to describe friends; the way language is used in a kind of one-up-man-ship style; and the use of obscenities and defamatory terms in a positive context. For example, Bitch Cunt/Janine refers to Bobsi as 'beautiful, 'my angel' and 'my saviour', and says of Liz, 'truly she is my other half'; Amanda/Fuckhead says of Karen 'you are my life I love you with everything I have'; and lil' freakshow Erica tells us that she 'LOVE LOVE LOVE LOVE LOVE'S' her friend Belinda 'to death', and that 'no words can describe' how she feels about this girl. These young women are utilising a kind of romanticised discourse usually reserved for sexual partners. Another heterosexual metaphor is the term 'hot wifey' ('wifey' is a term from American rap culture), which lil' freakshow Erica uses, perhaps to connote the marriage-like relationship she has with her friend. Similarly, Bitch Cunt/Janine refers to several friends as her 'baby girl', in a way that evokes both parental and sexual-partner metaphors. That is, romantically involved partners often use the term 'baby' towards each other metaphorically, to connote affection, and a protective, caring attitude towards the person, which is how I would suggest the term 'baby' is being used here.

As in the above section describing the performance of laddishness in young women's friendships, terms such as 'bitch', 'cunt', and 'whore' are used in a positive context when directed towards one's closest friends, as for example, when lil' freakshow Erica signs off with 'Love U Bitches', and Amanda/Fuckhead calls Karen a 'rad fuck shit cunt whore'. In this example there also seems to be a kind of 'one-up-man-ship' with language taking place. The writers often seem to go to greater and greater extremes of offensive language in order to express the gravity of emotions they feel for close friends. When Amanda/Fuckhead refers to her friend as a 'motherfucking rad fuck shit cunt whore', the combination of traditionally derisive terms like 'motherfucker', 'cunt', and 'whore' with the complimentary term 'rad' is not meant to contradict the positiveness of being 'rad', but rather, enforce and build upon the significance of the compliment. The more offensive the language seems, the greater the impact and significance of the dedication. When Danirocks writes to her friend Brianna 'love you with all my heart xoxoxoxoxoxox STAY BEAUTIFUL ** LICK**', once again, viewers might perceive the 'lick' as a parody sexual innuendo, one that is perhaps meant to shock. A lick is conceivably one step up from a metaphorical 'kiss' in terms of explicitness (whilst also being a private joke, as indicated by the text 'you know what I'm talking about'). Coarse language is used to shock, as well as to emphasise importance and gravity in these text-based displays of loyalty and closeness between friends.

There is an argument to be made about whether terms such as 'bitch', 'slut', 'whore', and 'cunt' can actually be 'reclaimed' or not, and about the political implications of adopting such heavily invested and sexist terms for use by those who have traditionally been the objects of their derision. Further, as I discuss later, such terms are also used in more traditionally derisive fashions in relation to 'other' women outside the friendship group. However my point here is primarily to describe and analyse the way I perceive such language being used by profile owners in the data, and the way it contributes to the constructions of both crudeness and loyalty between female friends. I am suggesting that this construction of closeness and loyalty is slightly different in flavour to the more outward-focused, public image of young women together as rowdy, boisterous 'party animals' described above. However, the two modes of friendship performance often go together. Further, these textual constructions of young women as loving and loyal to one another do not contradict the image of young women when together as raucous, licentious and even vulgar. Instead, the appropriation of offensive language and derisive terms as extremely complimentary and affectionate descriptions of close friends serves to uphold this image, to tie or associate vulgarity towards friends with emotional outpourings of love and care. The textual dedications quoted further contribute towards the overall world being constructed/imag(in)ed here: a world in which the female relationships in question are all-encompassing, and meet all of an individual's emotional and social needs. The profile owners construct an almost exclusively female social world of inter-dependant, and at the same time, self-reliant individuals, which is in itself wholly satisfying, enclosed and complete.

Female Friends and Homoerotic Display

Photo images too play a large part in the construction of close, intimate and all-encompassing female friendships on these profiles. Figures 5 and 6 represent just some of the copious displays of physical affection between female friends displayed on the profiles examined. These images indicate for viewers the level of intimacy in these friendships, whilst also sometimes depicting a performance of homoeroticism. Whether lesbian tendencies or desires underlie such images somewhat beside the point here. What can be said about the majority of these images is that the poses and tone of the photos suggest that these friends wish to perform their intimacy, and often also display eroticism for viewers. The young women look as if they are conscious of their construction, and purposeful in their exposure of 'intimate' poses and homoerotic scenarios. They are performing for the camera, and many of the subjects in such images look to be creating poses and image sequences as a leisure activity in itself. This may be done in order to demonstrate to other friends and viewers the depth of feeling, the level of connection and intimacy present in the friendship. Other images look decidedly 'posey' and purposefully provocative, or perhaps as if they may be intended to convey eroticism to viewers. Some of the images found in my sample explicitly work sexuality into the level of intimacy staged for viewers. Some young women engage playfully with the borders that lie between friendship and sexual intimacy in these close relationships. On several of the profiles examined there looks to be a playful dance around the boundaries of friendship and sexuality, a production of eroticism between friends that fits precisely into the mould of what is seen as seductive, titillating and hetero-sexy in popular raunch culture over at least the last ten years.

Out of the 45 profiles examined, only a handful of the owners claim a sexual identification other than 'straight' (heterosexual). Two claim bisexuality; three claim as their sexual status 'not sure'; and only one claims to be 'lesbian', however, appears to be in a relationship with a male who is displayed and discussed prominently on her profile. Viewers and friends may well assume the 'lesbian' sexual status listed is intended humorously. All 39 other young women claim heterosexual status on their profiles. An excerpt from the blog of one of the young women who claims her sexual status as 'not sure' exemplifies the way in which a performance of homoeroticism is often engaged in with perhaps a male heterosexual audience in mind. Much the same way, perhaps, as two (feminine looking) females kissing has become a popular motif in both heterosexual pornography, and on shows like *Girls Gone Wild* (1998-present) and *Ladette to Lady* (2005-2009), following the 'pornification' of mainstream culture (Levy, 2005; Paul, 2005; Henderson, 2008; McRobbie, 2008). Bec writes,

23 Nov 2007.

Well Im off 2 [name of dance party removed] 2night cant wait it the one thing that will make this week better i hope...

Ill be going with Lucy & Alison and hope 2 have a blast ☺ hehe

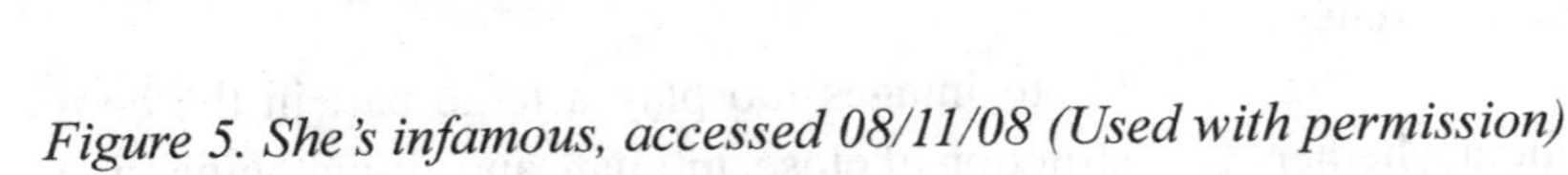

Figure 5. She's infamous, accessed 08/11/08 (Used with permission)

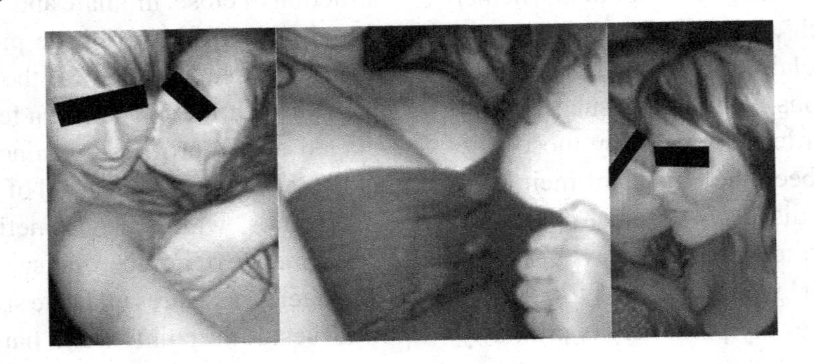

Figure 6. Katy/Kuntastic Bitch, accessed 18/03/08 (Used with permission)

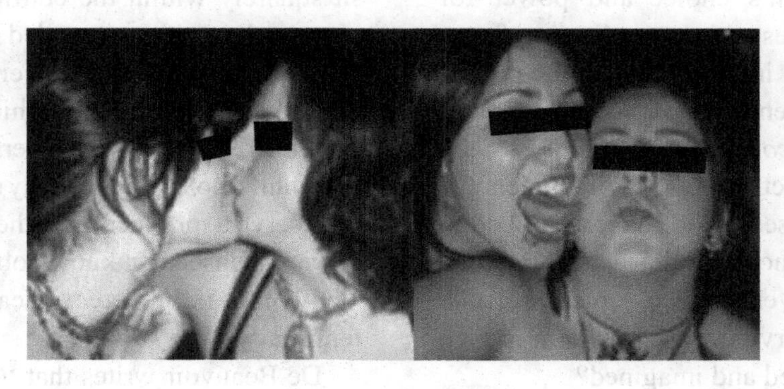

Dance, Dance, Dance will be the theme of the night make the guys pass out from lack of oxygen as 2 hot chicks get it on lol

26 Nov 2007

Lucy and I got closer and all guys watching got a free show... not that anyone complained poor Max I think we shocked him hehe...

(Bec, accessed, 29/11/07)

As McRobbie notes,

The taking up of the position of phallic girl bears superficial marks of boldness, confidence, aggression and even transgression [...]. However this is a licensed and temporary form of phallicism [...] predicated on the renunciation of the possibility of critique of hegemonic masculinity, for fear of the slur of feminism or lesbianism. Indeed lesbianism is re-configured as a popular (rather than pornographic) space of activity for phallic girls within circumscribed scenarios of male pleasure. (McRobbie, 2007, pp. 732-733)

McRobbie (2008) addresses the phenomenon of homoerotic display in popular culture and youth culture by young women who often defy the label 'lesbian' or 'queer'. She critiques these cultures in which, as she notes, lesbianism itself is still 'aggressively disavowed' (McRobbie, 2008, p. 235). However, she also suggests that in a culture where, as Harris argues (2004), young women's lives, feelings, habits, education, desires, and sexuality are under intense scrutiny from a variety of different angles and institutions, homoerotic display by 'straight' women may be a way of eluding complete comprehension. In other words, 'it could be understood as a field of female desires that shape up and take place outside and against the normative requirements to conform to specifically readable modes of femininity. [...] Taking part in same-sex relations without "coming out" is also a way of remaining illegible' (McRobbie, 2008, pp. 231-232).

Again, whether Bec actually gains sexual pleasure from the acts she describes is another matter, somewhat beside my central concern here. Rather, it is these performances and self-representations and what they might signify, or how they may be read by an audience that is my primary concern. As Bec's blog excerpt may be said to reveal, this type of performative homoeroticism is often a mimicry of heterosexual pornography. It is important to ask what types of pleasure, power, and agency young women gain from these performative displays of homoeroticism. But it is also important to question the extent to which the, now generic, pornographic performative framework associated with such display overshadows discourses of women's 'choice' and 'power' for viewers. Such images add a complicating factor of sexuality into any interpretation of the performance of female 'friendship' on MySpace, and as McRobbie notes, a *complex* sexuality it is. They also serve in constructing the kind of 'all-female' autonomous and closed world I have described so far. The displays of homoeroticism also provoke the question, however, of whose pleasure such performances are serving, and for whom is such a world being imaged and imagined?

SUPERIORITY OF FRIENDSHIP GROUPS AND CONSTRUCTION OF THE FEMALE 'OTHER'

Both dominant constructions of female friendship on the MySpace profiles examined in this study — as 'laddish' and as intimate — serve to create for viewers an idealised image of a female-centred world, one which is self-contained, self-satisfying and somewhat exclusive, as I have argued. An unknown viewer (defined in 'Key Terms'), that is, a viewer who is not part of the immediate friendship group in question, is positioned as an outsider. In this final section I discuss examples of 'hostile' graphic text statements directed towards unknown viewers. These, I contend, firstly, serve to further produce exclusivity amongst friendship groups through discourses of superiority and inferiority. Secondly, these statements construct an 'other' who is imagined as possessing the negative qualities stereotypically associated with femininity. As the quotations below demonstrate, the negative qualities associated with traditional femininity have not changed significantly as they are communicated in these texts by young women themselves, directed towards 'other' female viewers from which the profile owners consciously and clearly distance themselves. I am not arguing that these profiles are created with only a female audience in mind. As discussed above, I suggest that parts of the representation specifically appeal to a masculinised, hetero-sexualised gaze, and sit squarely within the confines of hetero-sexy visual culture, if not intended specifically to titillate male heterosexual viewers. However, in this next section I discuss text which is hostile in tone and aimed at unknown viewers, arguing that such text, whilst not always overtly directed at females, does serve to produce an/'other' feminine object: that is, an unseen, unknown object who possesses many of the traits stereotypically associated with femininity.

De Beauvoir writes that 'Otherness is a fundamental category of human thought', and as she argues, 'no group ever sets itself up as the One without at once setting up the Other over and against itself' (de Beauvoir, 1949/1972, p. 17). The presentation of one's close girlfriends' and friendship groups as superior (to 'outsiders', unknown viewers, or any individuals who are not part of the immediate friendship group in question) is a striking feature of friendship construction and female relationships in the profiles examined. Demonstrating performatively the closeness and importance of one's friendships, and building a sense of exclusivity in relation to one's friends also seems to involve expressing meanness and scornfulness towards other young females. Along with the construction of one's own friends as fun, loyal and extremely close and important to a profile owner, there seems to go a discourse strongly reliant on binary opposition. This discourse creates self-superiority through the construction of an 'other' as inferior. I am suggesting that the 'other', in this context of *friendship* performance at least, is not men, but those who are peers of the subject — other girls and young women outside the friendship group being represented. This potentially negates the possible feminist implications of the 'all-female world' that I have been describing so far, because when seen in light of the following examples, it is not the 'femaleness' of the world which is perhaps of greatest significance, but the general 'exclusiveness' of the world. Solidarity is with particular, known females, not with women generally. Slogans and sentiments such as those below are found on many of the profiles examined, often in very large, imposing graphic text, decorated with traditionally feminine colours such as bright pink and purple, bold fonts, and glittering, sparkling or flashing letters.

girls like us don't come around too often

yOu'll LeArn To LOVE uS.

hate us and your simply wasting your time

because you can't BEAT us

and you sure as hell can't join us

(Mz_Princess, accessed 18/03/08)

just when you thought

you were the centre of

ATTENTION

[[we showed up]]

(Rochelle, accessed 30/10/07)

OOPS!

DID MY SARCASM HURT YOUR FEELINGS?

get over it.

(SheZzaCat, accessed 06/12/07)

hate it or love it

ya whispers r endless

KEEP TALKING bitches

(SheZzaCat, accessed 06/12/07)

Jealousy is a

terrible sickness

get well soon

whore

(MZ KitTy, accessed, 17/09/07)

we're not sarcastic — we're hilarious

we're not annoying — we're just cooler than you

we're not bitches — we just don't like you

and we're not obsessed — we're just best friends.

(MZ KitTy, accessed, 17/09/07)

A common mode of addressing unknown viewers, as the above quotations show, presumes them to be jealous, judgmental, 'bitchy', unliked by the profile owners, rivals, or in 'doomed competition' with the writers (i.e., they cannot compete or do not match up in seeking attention, sexual partners, or friends). There may be no text on a profile that directly specifies the gender of viewers. However, when allusions are made to bitchiness, jealousy and competitions for attention, I would suggest that it is a female, and stereotypically feminine, rather than male and/or masculine reader to which such hostile text is directed. The name 'whore', at least, in MZ KitTy's statement, is clearly indicative of the female gender. Above I have discussed the way this name is used in an affectionate manner towards close friends. However, this quotation demonstrates the way that 'whore', like other similarly derogative terms, is still used negatively and contemptuously when the female in question is not a close friend, but an outsider. In the other quotation from MZ KitTy's profile, there is no such clear indication of gender as the name 'whore'. However, the tone of the text combined with its location in close proximity on her profile to the other graphic text seems to suggest that it is also directed towards an unknown viewer who is imagined as female. MZ KitTy creates an 'us and them' tone through the speaking position of 'we' rather than 'I', as is also the case in the quotations from Mz_Princess and Rochelle. This competitive and defensive language is more commonly thought of in relation to, and thus indicative of, female-to-other-female discourses, as opposed to, for example, one heterosexual couple addressing another, or two or more girlfriends addressing a group of boys. In discourses of competition and superiority in which the *subject* position is female, the viewers addressed, I contend, are usually also (subtly or openly) engendered as female.

De Beauvoir argues that women themselves use men's terms when referring to other women. She writes,

Proletarians say 'We'; Negroes also. Regarding themselves as subjects, they transform the bourgeois, the whites, into 'others'. But women do not say 'We', except at some congress of feminists or similar formal demonstration; men say 'women', and women use the same word in referring to themselves. They do not authentically assume a subjective attitude. (de Beauvoir, 1949/1972, p. 19)

In the example from MZ KitTy's profile, 'we' is not being used to differentiate between females as a unified group and males, but rather between 'we the female *subjects*', and 'other' female *objects*. Thus the implication of de Beauvoir's argument, that women perhaps imagine other women as different from, rather than similar to themselves, holds true in this example, and also possibly more broadly too, in the post-feminist, individualised political and social context in which such self-representation sits.

Surely, the second wave of feminism in the 1970s broke new ground on a fundamental level, in terms of fostering even a mere *capability* in women to perceive themselves as *subjects*. In some respects second-wave feminism did much to change the way women perceive themselves personally and as a group with common experiences and knowledge different to those of men. However, data discussed in this chapter indicates that young women today are still using 'male' terms to describe themselves, such as 'bitch', 'whore', and 'cunt', even if they are also appropriating such terms to imply a sense of strong, aggressive, and perhaps more 'masculine' subjectivity. As the quotations displayed above demonstrate such negatively gendered terms are still used in their originally derisive fashion when referring to 'other' women. Further, such terms are still thought of and used almost exclusively in reference to femininity and females, rather than masculinity and males. That is, 'bitchiness', for example, whether positive or negative, is still in this context being engendered as a 'naturally' feminine trait: A(mmm)y writes in her self description, 'Girls are bitches, but we were born to be' (accessed 11/12/07). When this term is used to refer to males, it is usually to imply a degrading feminisation of the male in question.

In summation, two consequences can be perceived here regarding derisive feminine terms. Firstly, formerly negative terms for females have been re-appropriated and used affectionately by the profile creators in relation to their own exclusive circles, but are also becoming perhaps more naturalised as 'female' in the process. Secondly, these same terms are applied in the old derisive ways to 'others' by the profile owners. Thus, it may be seen that young women are still investing in, and propagating some negative stereotypes of femininity, not surprisingly in relation to 'others' rather than self. Some more examples of addressing an unknown female, or perhaps unconsciously *feminised* and 'othered' viewer follow. As they demonstrate, there are numerous references on the profiles examined to 'fakes', 'bitches', 'backstabbers', 'hos' and 'plastics' as nouns to describe people, qualities and traits unliked by the profile owners. The term 'plastic' as a noun seems to be used to describe superficiality, fakeness, and also a type of typically 'girly' femininity embodied by, for example, a plastic Barbie doll. It may be a reference to the film *Mean Girls* (2004), in which the clique of 'means girls' around which the film centres call themselves The Plastics, and who embody these qualities.

your young have some fun go live your life just remember you cant make a HO A HOUSEWIFE!

[...]

names cinderella my friends and my family are my life fucking run your mouth about me and your dead to me I have no time for fakes or hy-procrites (Cinderella, accessed 29/10/07)

Dont add me if your anything less then me. If your profile takes 10 minutes to load, it will take me 10 seconds to delete you. Im hard to please. Before adding me, write me a msg or i WILL NOT accept you. Much Love to all Bitches. Penny_XOX

(Penny Darling, accessed 09/10/07)

.. if you piss me off or get on my bad side you will F^#ing know about it..i hate StuCK up Bitches, liars and backstabbers.*

[...]

I Hate Fakes and Plastics

(Georgie, accessed 03/10/07)

Im over the fake attention seekin drama. thank you. Getta hobby.

(Catiaaah, accessed 06/12/07)

I am who I am

and Sweetie your approval

isn't needed; no matter how

much you think it is!

(Violet, accessed 06/12/07)

I HATE YOU

(Penny Darling, accessed 06/11/07)

BITCH CLUB

BECAUSE

WE'RE BETTER

THAN YOU

(RadBitch, accessed 29/10/07)

BITCH PLEASE

(SheZzCat, accessed 06/12/07)

The nouns 'bitch', 'ho', 'fake', 'plastic', 'backstabber' and 'drama queen' are used to describe traits such as meanness, promiscuity, in-authenticity, hypocrisy, superficiality, and the kind of gossip-driven and overly emotional sociality associated with a typically 'feminine' personality or character. In this way, a binary is constructed between the female 'subject' and the often stereotypically feminine 'other'. When profile owners address viewers as a bitch, a whore so on, or refers to their dislike of bitches, ho's, fakes, drama queens, and so forth, they are evoking gender stereotypes long propagated and circulated throughout the history of Western female representation on stage and in the mass media. So too, when profile owners imply that others are or should be jealous of their physical appearance, friends, boyfriend or life in general, they are invoking gender in relatively traditional and stereotypical ways.

This stereotypically feminine 'other' is imaged in her absence and through the negative mode. That is, an image of the 'other' is formed in the minds of viewers through these descriptions of what the subject herself is *not*. Thus, the tie is maintained, and perhaps even strengthened, between femininity and the ever-evasive, never possess-able object, and a deeper disruption of representational gender binaries continues to elude us. The propagation of stereotypical female 'others' and 'objects' on these profiles perhaps then happens similarly to the way de Beauvoir argues it long has in Western society more broadly: femininity is 'othered', even by women themselves, and by default pertains to all the negatives in a (patriarchal) system of thought, knowledge and culture restricted by, and limited to, binary oppositions (de Beauvoir, 1949/1972). As Callaghan (1989) and Baum (2003) argue of Western theatre throughout history, the female object/'other' is imaged in the minds of audiences despite, and because of, her *absence* as a subject in the narratives, and also because of the absence of actual female bodies on the stage itself (Callaghan, 1989; Baum, 2003). Just as Schneider describes idealised female objects onscreen and in advertising particularly as forever just out of reach, beyond the grasp of viewers and consumers (Schneider, 1997, pp. 89-93), the unknown and/or unliked feminine other in the MySpace context remains forever an object of deferral — never specified, never inhabited, never actually present in the representation, and never of course, the *self*, but rather, constructed in and through her absence.

CONCLUSION

Female friendships are performed on MySpace through the presentation of an idealised subjectivity aligned with 'laddishness' as well as intimacy, and reliant on the exclusivity of close ties. I have described the way female friendships are represented on the profiles examined as fun, wild, strong, and close, and the way the representation of friendship in the sample examined also constructs a world for viewers which seems largely closed, and completely satisfied by almost exclusively female company. Further, as I have argued, friends also remain opposed to a stereotypical *feminine* 'other' through the MySpace performance of friendship.

There is much scope within the MySpace self-presentation I have examined for a radical disruption of gender boundaries and binaries. Young women are key participants in the online self-expression phenomenon (Orlowski, 2003; Lenhart, Madden, Smith, and Macgill, 2007; Thelwall, 2008). As I argue in my doctoral thesis (Dobson, 2010), high participation in this public form of self-expression by young women is itself a sign of significant change from a recent past where educational scholars and psychologists saw the need to encourage girls to 'speak up' and to express themselves more openly and publicly (Tsolidis, 1986; Kenway and Willis, 1990; Orenstein, 1994; Pipher, 1994). Several aspects of the self-presentation examined can be seen as in line with some prominent goals of feminist representational theory and critique. These young women depict themselves and their friends as active, pleasure-seeking *subjects*. They often present a defiant, multi-dimensional, and strong, uncompromising subjectivity. Unashamed performances of vulgarity, exhibitionism, and debauchery often seen in the photo galleries of these profiles contradict more idealised, normative, and traditionally 'feminine' imagery, which has been heavily critiqued by feminist representational scholars. Such performances also perhaps create space for new modes of female performativity which incorporate certain freedoms previously held exclusively by young men, as I have argued.

However, the construction of the feminine 'other' on the profiles examined, through its opposition to a more 'laddish' and crude 'new feminine', 'post-feminist' individual *subject*, reveals that stereotypical and traditional notions of femininity have not actually shifted dramatically in the eyes of these young women themselves. 'Other' women are still associated with such stereotypical traits as superficiality, bitchiness, and vanity, or alternatively, they are excluded because of ugliness. (Remember phrases such as 'I roll with pretty bitches', 'we're better than you', 'just when you thought you were the centre of attention…', and 'you sure as hell can't join us'). Whilst the profile owners are free to adopt certain hedonistic freedoms and behaviours such as excessive drinking, partying, and rowdiness with their friends, as well as to display the 'private' and intensely personal subjectivity associated with very close and loyal female friendships, 'other' young women are perhaps seen as stereotypically feminine and not in possession of the same subjectivity and freedoms as they are. The negatives associated with being feminine have not changed — it is simply that these young women do not associate such qualities with their own subjectivity. When the negative and repudiated qualities of traditional femininity are 'othered' onto unknown, unseen 'females', rather than facing interrogation and deconstruction, they remain chiefly intact.

Further, as McRobbie suggests, young women themselves engage in a performance of 'phallic girlhood', and show themselves wholeheartedly enjoying the excesses of pleasure, hedonism and abandonment that young men have long enjoyed, but this also means that the paradigm of hegemonic masculinity is itself left intact (2007). As mentioned, the kind of behaviour and/or public display, particularly of alcohol consumption, that accompanies this 'phallic' paradigm is itself perhaps problematic, and presents an area for further research. Other important areas for future research include the kind of hetero-sexy appeal commonly constructed on the profiles through decoration and self-imaging; political individualisation and its connection to the beliefs, philosophies and moral code commonly expressed in this mode of identity performance; the shifting nature of 'production' and 'consumption' within this new-media environment; and of course, the shifting and complex nature of notions of 'public-ness' and 'private' information that accompany social network site practices. I have raised the question of who might be viewing these public social network site profiles, besides the friends of owners. This question too presents an interesting avenue for future research. Such research may productively ask not only who is watching, but how viewers are interpreting these performances of identity.

In this new-media landscape, the representation of women is changing in ways that feminists could not have predicted twenty, or even ten years ago. These young women on MySpace are producing strong, female-centred onscreen subjectivities, perhaps in ways feminists may never have imagined, and yet *re*-producing rather tired sexist discourses at the same time. Thus, some vigorous re-evaluation of feminist representational theories, as well as further interrogation of young women's online self-productions is warranted.

ACKNOWLEDGMENT

I am most grateful to the two anonymous reviewers for their useful comments and suggestions. This research is also greatly indebted to Professor Denise Cuthbert, who has provided rigorous intellectual feedback, suggestions, and support on all fronts throughout the project.

REFERENCES

Ang, I. (1985). *Watching Dallas: soap opera and the melodramatic imagination* (Couling, D., Trans.). New York: Routledge.

Bakhtin, M. M. (1968). *Rabelais and his world*. Cambridge, MA: M.I.T. Press.

Baum, R. (2003). *Female absence: women, theatre and other metaphors*. New York: P.I.E.-Peter Lang.

Betterton, R. (Ed.). (1987). *Looking on: images of femininity in the visual arts and media*. New York: Pandora Press.

Bloustien, G. (2003). *Girl making: a cross-cultural ethnography on the processes of growing up female*. New York: Berghahn Books.

Bortree, D. S. (2005). Presentation of self on the Web: an ethnographic study of teenage girls' weblogs. *Education Communication and Information, 5*(1), 25–39.

boyd, d., & Ellison, N. (2007). Social network sites: definition, history and scholarship. *Journal of Computer-Mediated Communication, 13*(1), 210-230.

boyd, d. (2008). Why youth (heart) social network sites: the role of networked publics in teenage social life. In D. Buckingham (Ed.), *Youth, identity and digital media* (pp. 119-142). Cambridge, MA: MIT Press.

Busby, L. (1975). Sex-role research on the mass media. *The Journal of Communication, 25*(4), 107–131. doi:10.1111/j.1460-2466.1975.tb00646.x

Butcher, H., Coward, R., Evaristi, M., Garber, J., Harrison, R., & Winship, J. (1974). *Images of women in the media*. Birmingham, West Midlands: Centre for Contemporary Cultural Studies, University of Birmingham.

Callaghan, D. (1989). *Woman and gender in Renaissance tragedy: a study of King Lear, Othello, The Duchess of Malfi and The White Devil*. New York: Harvester Wheatsheaf.

Case, S.-E. (1988). *Feminism and theatre*. New York: Routledge.

Darren Star Productions, D. Star, M. P. King & S. J. Parker (Producer). (1998-2004) *Sex and the City*.New York: Home Box Office.

Davies, J. (2004). Negotiating femininities online. *Gender and Education, 16*(1), 35–49. doi:10.1080/0954025032000170327

de Beauvoir, S. (1949/1972). *The second sex* (Parshley, H. M., Trans.). Harmondsworth, UK: Penguin.

Doane, M. A. (1987). *The desire to desire: the woman's film of the 1940s*. Bloomington, IN: Indiana University Press.

Dobson, A. S. (2008). Femininities as commodities: cam girl culture. In Harris, A. (Ed.), *Next wave cultures: feminism, subcultures, activism* (pp. 123–148). New York: Routledge.

Dobson, A. S. (2010). *Bitches, Bunnies and BFF's: a feminist analysis of young women's performance of contemporary popular femininities on MySpace*. Unpublished doctoral dissertation. Monash University, Melbourne.

Edgar, P., & McPhee, H. (1974). *Media she*. Melbourne, Victoria: Heinemann.

Faludi, S. (1992). *Backlash: the undeclared war against women*. London: Chatto & Windus.

Ferris, L. (1990). *Acting women: images of women in theatre*. Basingstoke, UK: Macmillan.

Francis, J. (producer) (1998-present) *Girls Gone Wild*. Los Angeles, CA: Mantra Entertainment.

Friedan, B. (1963). *The feminine mystique*. London: Gollancz.

Gill, R. (2003). From sexual objectification to sexual subjectification: the resexualisation of women's bodies in the media. *Feminist Media Studies, 3*(1), 100–106.

Gill, R., & Arthurs, J. (2006). Editors' Introduction–New Femininities? *Feminist Media Studies*, *6*(4), 443–451.

Goffman, E. (1969). *The presentation of self in everyday life*. London: Allen Lane.

Greer, G. (1970). *The female eunuch*. London: MacGibbon & Kee.

Grisso, A. D., & Weiss, D. (2005). What are gURLS talking about? Adolescent girls' construction of sexual identity on gURL.com. In Mazzarella, S. R. (Ed.), *Girl wide web: girls, the Internet, and the negotiation of identity* (pp. 31–49). New York: Peter Lang.

Harris, A. (2004). *Future Girl*. New York: Routledge.

Henderson, L. (2008). Slow love. *Communication Review*, *11*(3), 219–224. doi:10.1080/10714420802306650

Hey, V. (1997). *The company she keeps: an ethnography of girls' friendships*. Buckingham, Milton Keynes, UK: Open University Press.

Hopkins, S. (2002). *Girl heroes: the new force in popular culture*. Annandale, Australia: Pluto Press.

Jackson, C. (2006). Wild girls? An exploration of ladette cultures in secondary schools. *Gender and Education*, *18*(4), 339–360. doi:10.1080/09540250600804966

Jackson, C., & Tinkler, P. (2007). 'Ladettes' and 'Modern Girls': 'troublesome' young femininities. *The Sociological Review*, *55*(2), 251–272. doi:10.1111/j.1467-954X.2007.00704.x

Kearney, M. C. (2006). *Girls make media*. New York: Routledge.

Kelly, D. M., Pomerantz, S., & Currie, D. H. (2006). "No boundaries"? Girls' interactive, online learning about femininities. *Youth & Society*, *38*(1), 3–28. doi:10.1177/0044118X05283482

Kennedy, G., Dalgarno, B., Gray, K., Judd, T., Waycott, J., Bennett, S., Maton, K., Krause,

Kenway, J., & Willis, S. (1990). *Hearts and minds: self-esteem and the schooling of girls*. New York: Falmer Press.

K.L. Bishop, A., Chang, R. & Churchward A. (2007). The net generation are not big users of Web 2.0 technologies: preliminary findings. In *ICT: Providing choices for learners and learning. Proceedings ascilite Singapore 2007*. Available from http://www.ascilite.org.au/conferences/singapore07/procs/kennedy.pdf

Kumble, R. (Director)(2002). *The Sweetest Thing.*. New York: Sony Pictures Entertainment.

Lenhart, A. (2009). *Adults and social network websites*: Pew Internet. Available from http://www.pewinternet.org/Reports/2009/Adults-and-Social-Network-Websites.aspx

Lenhart, A., & Madden, M. (2007). *Social networking websites and teens*: Pew Internet. Available fromhttp://www.pewinternet.org/Reports/2007/Social-Networking-Websites-and-Teens.aspx

Lenhart, A., Madden, M., Smith, A., & Macgill, A. (2007). *Teens and social media*: Pew Internet. Available from http://www.pewinternet.org/Reports/2007/Teens-and-Social-Media.aspx

Levy, A. (2005). *Female Chauvinist Pigs*. Melbourne: Schwartz Publishing.

Liu, H. (2007). Social network profiles as 'taste performances'. *Journal of Computer-Mediated Communication*, *13*(1), 252–275. doi:10.1111/j.1083-6101.2007.00395.x

McRobbie, A. (1999). *More*! New sexualities in girls and women's magazines. In McRobbie, A. (Ed.), *In the culture society: art, fashion, and popular music* (pp. 46–61). New York: Routledge.

McRobbie, A. (2004). Notes on postfeminism and popular culture: Bridget Jones and the new gender regime. In Harris, A. (Ed.), *All about the girl: power, culture, and identity* (pp. 3–14). New York: Routledge.

McRobbie, A. (2007). Top Girls? Young women and the post-feminist sexual contract. *Cultural Studies*, *21*(4-5), 718–737. doi:10.1080/09502380701279044

McRobbie, A. (2008). Pornographic Permutations. *Communication Review*, *11*(3), 225–236. doi:10.1080/10714420802306676

McRobbie, A., & Garber, J. (1976). Girls and subcultures: an exploration. In Hall, S., & Jefferson, T. (Eds.), *Resistance through rituals: youth subcultures in post-war Britain* (pp. 209–222). London: Hutchinson.

Media, R. D. F. (Producer). (2005-present) *Ladette to Lady*. UK: ITV (UK) and Channel 9 (Australia).

Mulvey, L. (1975/1989). Visual Pleasure and Narrative Cinema. In Mulvey, L. (Ed.), *Visual and other pleasures* (pp. 14–26). Houndmills, UK: Palgrave.

Murthy, D. (2008). Digital ethnography: an examination of the use of new technologies for social research. *Sociology*, *42*(5), 837–855. doi:10.1177/0038038508094565

Orenstein, P. (1994). *Schoolgirls: young women, self-esteem, and the confidence gap* (1st ed.). New York: Doubleday.

Orlowski, A. (2003). Most bloggers 'are teenage girls' – survey. *The Register*. Retrieved 23/04/08 from http://www.theregister.co.uk/2003/05/30/most_bloggers_are_teenage_girls/

Paul, P. (2005). *Pornified: how the culture of pornography is changing our lives, our relationships, and our families*. New York: Times.

Pipher, M. B. (1994). *Reviving Ophelia: saving the selves of adolescent girls*. New York: Putnam.

Radway, J. A. (1984). *Reading the romance: women, patriarchy, and popular literature*. Chapel Hill, NC: University of North Carolina Press.

Rainie, L. (January 5, 2010). *Internet, broadband, and cell phone statistics*. Pew Internet. Available from http://www.pewinternet.org/~/media//Files/Reports/2010/PIP_December09_update.pdf

Reid-Walsh, J., & Mitchell, C. (2004). Girls' web sites: a virtual "room of one's own"? In Harris, A. (Ed.), *All about the girl: power, culture, and identity* (pp. 173–182). New York: Routledge.

Schneider, R. (1997). *The explicit body in performance*. New York: Routledge. doi:10.4324/9780203421079

Schofield Clark, L. (2005). The constant contact generation: exploring teen friendship networks online. In Mazzarella, S. R. (Ed.), *Girl wide web* (pp. 203–221). New York: Peter Lang.

Stallybrass, P., & White, A. (1986). *The politics and poetics of transgression*. London: Methuen.

Stern, S. (1999). Adolescent girls' expression on web pages: spirited, sombre and self-conscious sites. *Convergence (London)*, *5*(4), 22–41. doi:10.1177/135485659900500403

Stern, S. (2002). Virtually speaking: girls' self-disclosure on the WWW. *Women's. Studies in Communications*, *25*(2), 223–252.

Takayoshi, P., Huot, E., & Huot, M. (1999). No boys allowed: the World Wide Web as a clubhouse for girls. *Computers and Consumption*, *16*, 89–106. doi:10.1016/S8755-4615(99)80007-3

Thelwall, M. (2008). Social networks, gender, and friending: an analysis of MySpace member profiles. *Journal of the American Society for Information Science and Technology*, *59*(8), 1321–1330. doi:10.1002/asi.20835

Thiel, S. M. (2005). "IM me": identity construction and gender negotiation in the world of adolescent girls and instant messaging. In Mazzarella, S. R. (Ed.), *Girls wide web* (pp. 179–201). New York: Peter Lang.

Tincknell, E., Chambers, D., Van Loon, J., & Hudson, N. (2003). Begging for it: "new femininities," social agency, and moral discourse in contemporary teenage and men's magazines. *Feminist Media Studies*, *3*(1), 47–63. doi:10.1080/14680770303796

Tsolidis, G. (1986). *Educating Voula: prepared by Georgina Tsolidis for Ministerial Advisory Committee on Multicultural and Migrant Education*. Melbourne, Victoria: Ministry of Education, Victoria.

Tsoulis-Reay, A. (2009). OMG I'M ONLINE... AGAIN! MySpace, MSN and the everyday mediation of girls. *Screen Education* (53), 48-55.

Verghis, S. (2003). Girls stalk. *The Sydney Morning Herald,* Wednesday, January 8, p. 19.

Waters, M. (Director)(2004). *Mean Girls*.. Los Angeles: Paramount Pictures.

Westlake, E. J. (2008). Friend me if you Facebook: generation Y and performative surveillance. *The Drama Review*, *52*(4), 21–40. doi:10.1162/dram.2008.52.4.21

Whelehan, I. (2000). *Overloaded: popular culture and the future of feminism*. London: Women's Press.

Winship, J. (2000). Women outdoors: advertising, controversy and disputing feminism in the 1990's. *International Journal of Cultural Studies*, *3*(1), 27–55. doi:10.1177/13678779000300103

ADDITIONAL READING

Beer, D. (2008). Social network(ing) sites. Revisiting the story so far: a response to danah boyd and Nicole Ellison. *Journal of Computer-Mediated Communication, 13*(2), 516–529. doi:10.1111/j.1083-6101.2008.00408.x

boyd, d. (2007). Social network sites: public, private or what? *The Knowledge Tree: An e-Journal of Learning Innovation, 13*. Retrieved 17/7/08 from http://kt.flexiblelearning.net.au/tkt2007/edition-13/social-network-sites-public-private-or-what/

(2007). boyd, danah (2007). Creating culture through collective identity performance: MySpace, youth, and DIY publics. *International Journal of Communication, 1*, 91–92.

Bryant, E. (2008). A Critical Examination of Gender Representation on Facebook Profiles. *Conference Papers – National Communication Association*.

Bryant, J. A., Sanders-Jackson, A., & Smallwood, A. M. K. (2006). IMing, text messaging, and adolescent social networks. *Journal of Computer-Mediated Communication, 11*, 577–592. doi:10.1111/j.1083-6101.2006.00028.x

Buckingham, D. (Ed.). (2008). *Youth, identity, and digital media*. Cambridge, MA: MIT Press.

Coleman, R. (2008). The becoming of bodies: girls, media effects, and body image. *Feminist Media Studies*, *8*(2), 163–179. doi:10.1080/14680770801980547

Day, K., Gough, B., & McFadden, M. (2004). Warning! alcohol can seriously damage your feminine health. *Feminist Media Studies*, *4*(2), 165–183. doi:10.1080/1468077042000251238

Diamond, E. (1997). *Unmaking mimesis: essays on feminism and theater*. New York: Routledge. doi:10.4324/9780203358900

Donath, J., & boyd, d. (2004). Public displays of connection. *BT Technology Journal*, *22*(4), 71–82. doi:10.1023/B:BTTJ.0000047585.06264.cc

Drotner, K. (2005). Media on the move: personalised media and the transformation of publicness. *Journal of Media Practice*, *6*(1), 53–64. doi:10.1386/jmpr.6.1.53/1

Durham, M. G. (2009). *The Lolita effect: the media sexualisation of young girls and what we can do about it*. New York: The Overlook Press.

Ellison, N., Steinfield, C., & Lampe, C. (2007). The benefits of Facebook "friends": exploring the relationship between college students' use of online social networks and social capital. *Journal of Computer-Mediated Communication*, *12*(4), 1143–1168. doi:10.1111/j.1083-6101.2007.00367.x

Gill, R. (2007). *Gender and the media*. Cambridge, UK: Polity Press.

Harris, A. (2008). Young women, late modern politics, and the participatory possibilities of online cultures. *Journal of Youth Studies*, *11*(5), 481–495. doi:10.1080/13676260802282950

Hodkinson, P. (2007). Interactive online journals and individualisation. *New Media & Society*, *9*(4), 625–650. doi:10.1177/1461444807076972

Hodkinson, P., & Lincoln, S. (2008). Online journals as virtual bedrooms? Young people, identity and personal space. *Young: Nordic Journal of Youth Research*, *16*(1), 27–46.

Kelly, D. M., & Pomerantz, S. (2009). Mean, wild, and alienated: girls and the state of feminism in popular culture. *Girlhood Studies*, *2*, 1–19. doi:10.3167/ghs.2009.020102

Livingstone, S. (2008). Taking risky opportunities in youthful content creation: teenagers' use of social networking sites for intimacy, privacy and self-expression. *New Media & Society*, *10*(3), 393–411. doi:10.1177/1461444808089415

Lundby, K. (2008). *Digital storytelling, mediatised stories: self-representations in new media*. New York: Peter Lang.

Magnet, S. (2007). Feminist sexualities, race and the Internet: an investigation of suicidegirls.com. *New Media & Society*, *9*(4), 577–602. doi:10.1177/1461444807080326

Magnuson, M. J., & Dundes, L. (2008). Gender Differences in "Social Portraits" Reflected in MySpace Profiles. *Cyberpsychology & Behavior*, *11*(2), 239–241. doi:10.1089/cpb.2007.0089

Manago, A. M., Graham, M. B., Greenfield, P. M., & Salimkhan, G. (2008). Self-presentation and gender on MySpace. *Journal of Applied Developmental Psychology*, *29*(6), 446–458. doi:10.1016/j.appdev.2008.07.001

Mazzarella, S. R. (2005). *Girl wide web: girls, the Internet, and the negotiation of identity*. New York: Peter Lang.

Pearson, E. (2009). All the World Wide Web's a stage: The performance of identity in online social networks. *First Monday*, *14*(3).

Phelan, P. (1993). *Unmarked: the politics of performance*. New York: Routledge. doi:10.4324/9780203359433

Pitcher, K. (2006). The staging of agency in *Girls Gone Wild*. *Critical Studies in Media Communication*, *23*(3), 200–218. doi:10.1080/07393180600800759

Railton, D., & Watson, P. (2005). Naughty girls and red blooded women: representations of female heterosexuality in music video. *Feminist Media Studies*, *5*(1), 51–63.

Ringrose, J. (2008). Every time she bends over she pulls up her thong Teen Girls Negotiating Discourses of Competitive, Heterosexualized Aggression. *Girlhood Studies*, *1*, 33–59. doi:10.3167/ghs.2008.010104

Stern, S. (2008). Producing sites, exploring identities: youth online authorship. In Buckingham, D. (Ed.), *Youth, identity and digital media* (pp. 95–117). Cambridge, MA: MIT Press.

Subrahmanyam, K., & Greenfield, P. M. (2008). Online communication and adolescent relationships. *The Future of Children*, *18*(1), 119–146. doi:10.1353/foc.0.0006

Sveningsson, Elm, Malin. (2007). Young people's presentations of relationships in a Swedish Internet community. *Young*, *15*(2), 145–167. doi:10.1177/110330880701500203

Thelwall, M. (2008). Fk yea I swear: cursing and gender in MySpace. *Corpora*, *3*(1), 83–107. doi:10.3366/E1749503208000087

Thiel-Stern, S. (2009). Femininity Out of Control on the Internet: A Critical Analysis of Media Representations of Gender, Youth, and MySpace.com in International News Discourses. *Girlhood Studies*, *2*, 20–39. doi:10.3167/ghs.2009.020103

Thornham, S. (2007). *Women, feminism and media*. Edinburgh, UK: Edinburgh University Press.

Van Doorn, N. (2009). The ties that bind: the networked performance of gender, sexuality and friendship on MySpace. *New Media & Society*, *11*(8), 1–22.

KEY TERMS AND DEFINITIONS

Social Network Site (SNS): A social network site is a website that allows users to join, create a personal profile and connect with other users (most typically, users who are part of their offline social network) by requesting to be 'friends' with them online. SNS members can chat, share media, post blogs, expand their networks and see the connections between people. This is because icons representing all of the other users whom one has marked a connection with by 'friending' them are displayed on one's profile. boyd and Ellison (2007) argue that this visible layout of one's social network is the defining feature of SNSs.

Social Network Site Profile: A SNS profile is like a personal home-page that users can create using simple, pre-determined settings, features and layouts on SNSs such as MySpace and Facebook. On a MySpace profile, users are asked to fill in a short self-description, list their favourite books, music, movies and television shows, and so on. They may provide their sex, race, sexual preference, marital status, birthday, star sign, school or work, location, and other details in a section called 'My Details'. MySpace users can decorate their profiles by selecting or creating their own background page themes and designs, and also by pasting graphic decorations onto their profile by copying web code into different sections of their profile. At the time this research was conducted, users could choose to set their profile to 'private', in which case it could only be viewed by those marked as 'friends', or they could make it publicly accessible to anyone browsing MySpace. More numanced privacy settings are now available

'Friends' and 'Friending': On SNSs, users may make a visible connection with other users, and share information and data with them by 'friending' them. A request is sent from one user to another to be added to each other's network as 'friends'. A 'friend' on a SNS may be someone the user has never met offline, but for most users, the vast majority of their SNS 'friends' are people from their offline social network (Lenhart and Madden, 2007; boyd, 2008; Thelwall, 2008).

Unknown Viewers: I refer to 'unknown viewers' of the MySpace profiles in my sample, as the sample consists only of profiles which were publicly accessible at the time of my data collection. Whilst most of the communication and interaction that takes place through SNSs may be between friends that know each other offline, a public profile may be viewed by anyone, thus by individuals who are unknown to the profile owner. In my sample of profiles, self-descriptive text often seems directed in tone to an audience of people who are unknown to the writer.

Performative: In this chapter I refer to identity construction on SNS as 'performative', as do other SNS site scholars (boyd, 2008; Liu, 2008; Westlake, 2008) I see these identities as performative, perhaps to an even greater degree than that described by Erving Goffman in his writings on 'the presentation of self in everyday life' (1969). The presentation of self on a MySpace profile is performative, perhaps more so than in everyday life, in that it is primarily visual; consciously planned and constructed; and on public profiles, self-presentation is often devised with an audience of both friends and unknown viewers in mind.

Graphic Text: I use this term to describe a particular kind of profile decoration common in the sample of MySpace profiles. Graphic texts are often small boxes or frames pasted onto a MySpace profile, which often contain some text and graphic decoration. For example, graphic text may consist of a motto or aphorism, a short statement or a quotation, decoratively written and framed with some graphics or small images such as hearts and stars.

Hetero-Sexy: In this chapter I use the term 'hetero-sexy' instead of more general terms such as 'sexy' or 'sexualised' in relation to contemporary girl culture and the representation of young women's bodies in the mass media and visual culture. I use this term to draw readers' attention to the fact that it is a specific type of sexual appeal that is commonly evoked when notions of 'sexiness' are discussed in relation to contemporary visual and media culture.

Chapter 9
YouTube as a Performative Arena:
How Swedish Youth are Negotiating Space, Community Membership, and Gender Identities through the Art of Parkour

S. Faye Hendrick
Umeå University, Sweden

Simon Lindgren
Umeå University, Sweden

ABSTRACT

The video sharing site YouTube is used by huge numbers of young people in the roles of consumers and producers of content and meaning. The site hosts more than 120,000,000 video clips, and its users represent a wide variety of nationalities, religions, ethnic backgrounds, identities and lifestyles. Due to the scale of YouTube it is hard to see how a tangible sense of actual community could be created within the site. Using on- and offline ethnographic data in the form of footage, interviews and patterns of community interaction (favoriting, subscribing, commenting, rating, and video replies), this chapter presents the results of a case study that aims to analyze how a specific interest group with a certain national anchoring (Swedish parkour youth) deal with the vastness and complexity of YouTube in creating a sense of identity and community in relation to their specialized interest.

NEGOTIATING "THE YOUTUBE SUBLIME"

Using the concept of "the YouTube sublime", Grussin (2009) describes the immensity and complexity of the YouTube site. The number of videos in its online repository is so large that it is nearly impossible to comprehend, and it is definitely impossible for any single user to get a comprehensive overview of them all. Furthermore, the YouTube user base is massive, global, and mixed to the extent that is seems hard to conceive of any real sense of close community being created or experienced here. The aim of this chapter is to analyze the strategies employed by individual young YouTube users to negotiate and

DOI: 10.4018/978-1-60960-209-3.ch009

the size and scope of YouTube. This will be done using the specific case example of youth engaged in the activity of parkour in Sweden.

Parkour involves free running and acrobatics in urban environments. Swedish youth participate in this alternative sport often film their parkour runs and upload them to YouTube, creating an arena in which they perform aspects of identity and membership, and connect with an international group of youth who share a common interest in parkour. By analyzing the creation, composition, consumption and cultural context of these video clips, one may gain an insight into the conditions and the potential for constructing community in the face of the reality of "the YouTube sublime".

YouTube has become a platform where commercial and amateur videos share the same hybrid media space as artful productions and documentaries, video diaries and powerful physical performances (Benkler, 2007). As Jenkins (2007) puts it, the site is "a space where commercial, amateur, nonprofit, governmental, educational, and activist content co-exists and interacts in ever more complex ways". YouTube users represent a wide demographic range, and an even broader range of purpose as the affordances of the platform lend itself to the convergence of different cultures in an increasingly complex way (Green & Jenkins 2009). For example, globally, every minute 13 hours of video material is uploaded by YouTube's 3,75 million users, and 25% of these users are young people between the ages of 12 and 17 (Wesch, 2008). However, despite its large potential for providing an arena for networked participatory culture (Jenkins 2006, Rheingold 2003), the discourse surrounding YouTube in the Swedish mass media has centred on the platform's "dark side" – not least when the participation has involved youth-generated videos. Examples of this debate in the Swedish media can be seen through the discussion of 'warning films' created by school shooters (Lindgren 2009), or in the considerable number of mediatised discussions surrounding cyber bullying and over-sexualized teen videos.

At every historical moment when a new medium enters the stage, debates occur that are strikingly similar to those about YouTube which revolve around issues of basic social and cultural norms (Drotner 1999). These discussions are always simplistic and tend to become very clearly polarized. While the new medium is demonized by some, it is at the same time celebrated and dedramatized by others. Historically, and YouTube is no exception, it is often the negative side that has become the most visible in the news media (Springhall 1998). The debate tends to play out among "responsible" adults: Teachers, librarians, cultural critics, politicians and researchers – each and everyone with their own interests and stakes in the issue – produce various diagnoses and offer a range of solutions directed towards children and young people. The new medium under discussion becomes a fitting rhetorical device in discussions that are often about something completely different.

These "media panics" (Drotner 1999) tell us very little about the actual use of the discussed mediums, but more about social and cultural dilemmas – relating to issues of socio-cultural change, education, gender roles, parental control etc. – that are of a much wider character. The moralizing discourse on YouTube, for example, directs attention away from much of what is actually happening on this site in terms of creativity, participation and knowledge exchange. Because of this, we want to make a contribution by backgrounding these "effects" debates and instead bring the issues of young people's actual YouTube use to the fore. Notwithstanding the negative picture in the Swedish media of youth on YouTube, Swedish youth are well represented on the site, and are creating and making accessible films that run the gamut between diary-like video diaries, to mash-ups, and to innovative, experimental films that blend both an awareness of the physical space, as well as the digital – a hybridization of online and offline space.

In this chapter, we present an analysis of three interrelating dimensions of how youth are negotiating "the YouTube sublime" to construct a creative and meaningful arena through physical and artistic performance. As stated above, the YouTube sublime is a term used by Grussin (2009) to describe the enormity and promiscuity of YouTube. While the first of these terms obviously refers to the size of the YouTube user base and the volume of its ever growing movie archive, the latter term refers to the embeddedness of the site.

READING THE PHYSICAL GRAFFITI

Parkour is essentially the art of moving oneself from point A to point B in the quickest, smoothest, and the most efficient way possible (Apell, 2008). Being a form of street art performed with the body, Parkour is often described as "physical graffiti". It is a combination of free running and acrobatics, most often practiced in urban environments. It is also a non-competitive sport and artform based on the premise that the practitioners, traceurs, or the feminine traceuses, can overcome any mental or physical obstacles (Le Parkour Sweden). Parkour's popularity in Sweden may be due in part to a culture of groups and non-profit clubs. There are many parkour clubs in cities throughout the country, and by joining members can practice with experienced trainers, have access to indoor facilities, and not least meet others with the same interest.

In Figure 1, which is an illustration of how parkour comes into expression on YouTube, the most recent comment on the clip is a request from two 15-year-old parkour runners who are looking for others to join in on their training sessions. Here, they state their ability level, the amount of time they have been practicing parkour, and ask if anyone in their region would like to run together with them.

Filming and uploading runs is what connects parkour runners, both in and out of clubs, together. At the time of writing, a search for the term "parkour" on YouTube results in 187,000 hits. Narrowing the search down to "Swedish parkour" returns a list of 821 hits. While the number of parkour videos on the site is relatively large, it is, as Grussin states, the promiscuity of

Figure 1. Parkour in Swedish urban space

the platform which simultaneously fosters social networking, but also limits the sense of community on the platform itself. YouTube videos allow for embedding on websites, discussion forums, blogs and social networking services - nearly anywhere online. And while videos gain in popularity on YouTube itself through user generated ratings and comments, videos reach a viral level of popularity because users email links and embed videos in a variety of social platforms. This practice was pointed out by several of our informants in this case study.

The informant group was comprised of twenty Swedish youth, ranging from the ages of 15 to 25, who participate in the art of parkour, film their runs, and upload their own original content to YouTube. All informants in this study actively participate in parkour either through a parkour club or on their own, and they have all filmed and uploaded a run within the last 6 months. Furthermore, each one of the informants has his or her own channel on the YouTube site. While language was not a consideration for inclusion in this case study, all informants did use Swedish in their online descriptions, as well as in the spoken language of all of their videos. The videos that we studied were uploaded during a period of one year stretching from August 2008 to August 2009.

We used the full content of these audiovisual data – together with the complimentary textual data in the form of tags, comments and descriptions – as the basis for a qualitative content analysis which was, to use the terminology of Krippendorff (1980, pp. 87-89), "open-ended" and "tentative". In order to stay open to several possible interpretations of the material, we used a strategy entailing re-reading and re-contextualizing it several times. More concretely, we employed the so-called constant comparative technique (Glaser & Strauss 1965; Lincoln & Guba 1985; Wimmer & Dominick 2003, pp. 86-88). This approach means that important themes and patterns in the data are initially assigned to provisional analytical categories. These categories are then renegotiated, refined and gradually abstracted as the material is being read and re-read iteratively. All films on each one of the channels were watched in order to determine which offline spaces participants were using. After compiling a list of physical spaces, the films were re-watched and coded for offline/physical spaces. The films were also examined and coded for physical behaviours/actions, and video editing techniques. The aim throughout was not simply to boil everything down to the lowest common denominator but also to stay sensitive to nuances and contradictions in the material.

In a first analytical step we went through all of the data – simultaneously creating preliminary codes and taking notes. Borrowing a term from Althusser (1965), and reaccommodating it somewhat, we have strived to discern the "problematic" of our material. This means that we have tried to map out its "unconscious infrastructure" (Resch 1992, p. 177). By looking at the

creation, composition, consumption and cultural context of the parkour videos, we have strived to find the dominant themes and patterns of meaning in our data. Furthermore, this means that we have an idea that all of the online and offline parkour related material that we have used can, in some sense, be perceived as one single, and relatively consistent, expression. Using a concept from van Dijk (1991, p. 46) we look at the data as "one overall, macro-speech act" that has a common "structure", "thematic organization" and "style" throughout that holds it all together as a meaningful unit (van Dijk, 1991, p. 41). Throughout the rest of this chapter, the three dimensions that emerged in our analysis of the material as vital for understanding parkour culture, online and offline, will be addressed. We will deal with issues relating to, first, space; second, community membership; and third, gender identities.

Although the young people in this case study have openly published their content to YouTube, we have attempted in this chapter to keep their names, locations and appearances as anonymous as possible. No identifying characteristics, such as screen names and video titles, have been used in the chapter. However, screenshots where faces are not shown, or their faces are only shown in profile, have been used. It still remains theoretically possible to identify traceurs in these films, although it has been made more difficult by minimizing identifying characteristics. In addition, informants who provided identifying information in their channels, such as full names or addresses, have been excluded from the study in order to obscure any direct links between published results and the informants' channels.

DIMENSION 1: SPACE

Parkour clips hosted on YouTube function as static artefacts, maintaining an archived permanent, if not arguably somewhat ephemeral, place on the site. Additionally, they represent a blend of the physical and the digital by mediating the physical urban environment into the virtual environment. Furthermore, these uploaded runs connect an international group of youth who share a common interest in the sport by creating digital links between both local and international nodes of Parkour practitioners. Space is in other words of key importance when reading parkour films posted to YouTube. And spaces are always defined by the actions performed within them (Lemke 2005). The social and the spatial are interlinked, as space is constantly structured by those who occupy it. Space is therefore always a social product (Lefebvre, 1974) marked by various spatial practices that can be read and interpreted.

Offline public spaces are functional because they tend to have sanctioned behaviours linked to them that are culturally agreed upon. People sit on benches, walk down pathways, and hold handrails. Occasionally these behaviours are extended as a part of dramatic performances such as street theatre or demonstrations of protest. Parkour, however, systematically uses objects in public spaces to run, jump, bounce, and flip from. While these acrobatic performances may seem like deviant public displays to simply show off strength and agility, parkour is considered an art by its practitioners and is taken very seriously, and much emphasis is placed on proper training and thoughtfulness during runs. In the Swedish parkour videos examined in this case study, three main offline spaces were mapped: the urban space, the training space and the rural space. In addition to this, we also analyzed the workings of the online space of parkour.

The urban space is the most common space in which parkour is performed and filmed. Runners use physical objects as impromptu obstacle courses by running through spaces, jumping between rooftops, hopping over handrails, etc. Interestingly, it was very rare to see non-parkour people present in the background of the footage – even though runs were undertaken during both day and night, and in what appears to be city and town centres.

Figures 2 and 3 depict parkour performances in urban spaces. Traceurs re-appropriate and reconstruct social, public spaces by performing physical graffiti against concrete, 'adult'-driven city-planning. The traceurs are, in effect, making a political statement through finding "new and playful uses for redundant architectural features" (Buckingham, 2009: 134). Similarly, research on skater communities report that skating in public spaces is a response which is "often provoked by the... punitive responses of adult authorities" (Buckingham 2009: 134), or even as an overtly political statement in critique of "the routinised efficiency of the modern capitalist city, with its 'zero degree architecture' " (Borden 2001: 206).

The second most common space where parkour is filmed is the training space. The training space, usually a gymnastics hall, is an offline space in which training sessions are documented. These films are usually edited cuts of traceurs practicing or performing tricks.

The training space is interesting for several reasons, and most generally because training is an important facet of parkour. Most Swedish

Figure 2. Parkour in Swedish urban space

parkour clubs hold regular workshops on tricking, weekly training sessions with a coach, and stress knowing your limits before pushing through them (Le Parkour Sweden, 2009). The filmed spaces in which practitioners do their training are located both indoors and outdoors, but often in sporting arenas. Looking at parkour videos from outside of Sweden, training videos still focus on the movement or trick, but are often shot in outdoor, urban areas. One possible reason for this difference is the aforementioned culture of non-profit clubs. In Sweden, a group of people can register as a non-profit organisation and apply for funding from the national or local government, or from other funding agencies. This club status can help the group gain access to local sport arenas where they can practice parkour. These sport arenas are common both in larger and rural areas in Sweden.

The rural space, the third most common parkour performance space, offers different running and tricking possibilities from the urban space. Open fields and fewer structures allow for longer free running and flips off the ground, rather than the urban runs which involve more start-and-stop running.

Figure 6 is a screenshot of a run that was filmed on Midsummer's day. Most Swedish towns have a common public gathering place where a maypole is erected and danced around. What is

Figure 3. Parkour in Swedish urban space

Figure 4. Parkour in Swedish training space

interesting about this run is that in this more rural setting there are many people passing by – but no one seems to stop to watch the parkour display. The only people watching are the two boys who are waiting for their turn to trick in the field. In this space, the actions performed have become the background noise to the main event of the midsummer celebration.

THE ONLINE SPACE

Initially the online space only consisted of the YouTube platform where parkour videos are uploaded and shared. However, during the interview process it became clear that much of the value of online interactions occurred in spaces other than individual YouTube channels. Grussin (2009) describes the interaction outside the bounded spaces of the YouTube collective as the promiscuity of YouTube. YouTube allows for embedding of videos in nearly all HTML spaces such as blogs, webpages and discussion forums. Additionally, permanent links (permalinks) are assigned to all YouTube videos, and these links can be sent through emails and microblogging services such as Twitter. While the ability to easily spread video is arguably an important part of the success of YouTube, it has also made tracking and participating in social interactions between videos very difficult. Many of the spaces in which these videos are embedded also allow for commenting

Figure 5. Parkour in Swedish training space

Figure 6. Parkour in Swedish country space

and linking, which makes visiting the YouTube page on which the video is hosted unnecessary.

The YouTube platform allows the user to get to know the space through various levels of participation. Users can watch, link, embed, favorite, rate, and comment on videos. Users can also become producers by uploading video comments, putting together a mash-up of favorite videos, or creating and uploading original content.

On the left hand side of Figure 7, we see an example of a parkour video hosted on YouTube. On an individual video page, viewers can rate, favorite or comment on a video. On average, Swedish youth parkour videos received a few dozen comments, around the same amount of ratings, and a few hundred views. As with any community of practice (Lave & Wenger, 1991), there are central figures who receive thousands of hits, and who are seen as the 'celebrities' of the community. These members tend to be the first with new tricks, advanced video editing skills, and are often prolific uploaders of videos.

On the right hand side of Figure 7, we see a screenshot of a Swedish parkour channel. Each user, even if they only registered with the site in order to comment on another's video, has his or her own channel. The channel is the profile of the user. Here, the user can give information about him- or herself. The user can set different levels of privacy, choosing to allow others to see his or her subscriptions, or recent activity. The channel also contains much information outside of the control of the user. For example, anyone viewing the channel can see how many videos the user has watched or favorited or how long the user has been registered on YouTube.

Social networking affordances are present on both the individual video pages, and on the user channels. Viewers can subscribe by clicking a 'subscribe' button on either the individual video page or the channel. Viewers can also spread a video by sending it to their social networks in various ways. When interviewing a northern Swedish parkour group, the informants described using YouTube as a search engine rather than a social networking platform. Value was found in finding new tricks and places in which to practice, rather than in finding new traceurs with whom to run.

Just on this site YouTube, it is not there that we meet. It is not just there that we find the communities and it is not there we get the first feedback. I find more if I go to their websites and their forums. I mean that there is not the same feeling of togetherness that you find on other (Parkour) sites (Informant 2).

I agree completely (with Informant 2), but that is not the point of YouTube. YouTube is like 'Spre-

Figure 7. Examples of parkour hosted on the YouTube platform. On the left, a parkour film on the film's hosted, permanent location (permalinked page). On the right, the YouTube channel of a Swedish traceur.

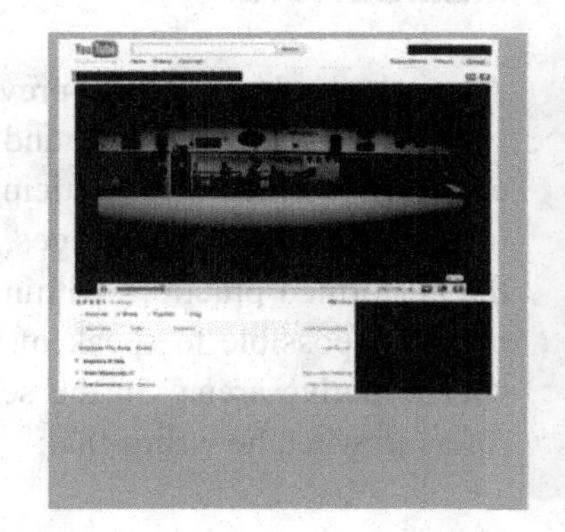

eeeeead it!" There it is like get so many views as possible, so many rates as possible, so many favorites as possible. When I go there, it is to find new moves. There are tons of new tricks you can learn (Informant 3).

On YouTube I get a lot of hits and a lot of comments. They are never like, 'you suck'. They are more like, 'sick! WOW!' or good critique on how to make a move better (Informant 3).

Rather than finding social value on YouTube itself, the informants in this group reported finding it on specialized Parkour websites. It should be noted, however, that this group is an organized, official club who has a professional and very active web presence. On their site a discussion board is separated into different topics, two of which are dedicated to videos – one to their own productions and one to other's productions. The board which lists their own videos has 95 different threads, on which they have received between 22 and 150 views. The board which contains others' videos has 333 threads, but a much lower rate of views (the average at around 45 per thread). The rate of views per thread on their own videos versus others' videos suggests a high level of interactivity and commitment to the local club's members. With an active web presence already established, this group uses YouTube as a place from which to host videos rather than as a place to establish connections with other traceurs. Two informants in this group commented on how their club used YouTube. Informant 1 was a part of the group before they became an official club, and stated that YouTube was a way to identify themselves as legitimate members of the Parkour community.

Before we were a club we were just a group of 10 or so and then we made more videos because we wanted to show people what we could do but now that we have so many classes and workshops, it does not feel that we have to make videos anymore (Informant 1).

Informant 4, on the other hand, experienced YouTube as an added identifying element to an already established local group identity.

The people in our club found us through our website after they saw us performing in town. They find our YouTube films after they found us (Informant 4).

When examining comments on traceurs who do not have a strong localized group, YouTube is used to a much larger extent as a social networking platform. As was shown in Figure 1, two boys in Stockholm were using the comments to find others with whom to run. The affordances of the YouTube platform in such cases are important in creating an arena of participation, however the affordances of the community in which you are performing appears to be equally important. The distinction between offline and online spaces in regards to maintaining a community of practice is not clear-cut.

Thomas (2007), in her work with youth and digital illiteracies, discusses the seamlessness of youth's experiences connecting online and offline worlds as a consequence of this generation being born into a time where the digital is so interwoven into every part of their lives. Thomas suggests that there is no distinct demarcation between how youth interact in their offline and online worlds. Scollon (2005) comments on space by stating that it "is always more complex than a physical location, consisting of the interaction among the built environments, the relationships among participants and the meanings assigned to the various semiotic tools introduced into the space (Jones, 2005:141). Further, Scollon uses the concept site of engagement to describe the point at which mediated action happens and how these points foster the construction of social identities, communities and practices.

To conclude the discussion about the spatial dimension, we argue that the use of the content on YouTube is equally important for the parkour youth as the social networking afforded on the site. In fact, it is the interplay of the video and networking capabilities that make participation in the network meaningful. This is similar to what Burgess and Green (2009) noted when examining YouTube as a social network: the content created was less significant than the uses of that content by the social network. Not only do Swedish youth parkour runners mark aspects of their identity within the videos themselves, but by linking to each other, watching each other's videos, sharing and embedding videos on their club websites, and not least by commenting on both the quality of the parkour and the quality of the video, these youth are creating an online space that is very real and very active.

DIMENSION 2: COMMUNITY MEMBERSHIP

As we have shown in the previous section, the ways in which users collect and subscribe to videos, and in which they participate and interplay on each other's channel pages, give meaning to the performed practices within the videos. This makes it possible to speak of this context as a "performative arena" in the sense that LaBelle refers to when he writes that:

(...) in underscoring the art object and the art viewer as interwoven into a conversational exchange in which the object produces the looking/listening, and the looking/listening produced the object, comes to suggest the field of attention as a performative arena. Thus, art objects do not so much contain or embody meaning but rather are given meaning through a performative exchange (LaBelle, 2006, p. 101).

This field of attention that LaBelle writes of comes into expression in the interaction of communities of practice among Swedish youth parkour runners. This group has established practices for both their offline performances and their online interactions, and it is the reciprocity between these spaces and interactions that give rise to an arena which is simultaneously meaningful, shared and experienced. Showing proficiency in parkour does not necessarily mark one as a member of the YouTube community, however. Many videos within this online community were in fact made by beginners.

In order to become part of this community, as indicated by reciprocal linking and commenting, these beginners had to participate through both training and showcase videos. These videos not only distinguish the user as a legitimate participant of the Swedish parkour community, they also serve to give this community status as a legitimate part of the larger, international parkour community and act as a link connecting Swedish youth to a vibrant, global group of youth who share common a interest. Although YouTube has a potential for community-building through its social networking capabilities such as commenting, subscribing, and the possibility to rate and favourite videos, community is experienced differently on YouTube by different communities of practice. The parkour group in this study reported a decrease in interactivity online which coincided with an increase in physical interactions. As we touched upon in the previous section, several informants in this study described no longer feeling a need to post films once their identity as a Swedish parkour runner was established. However, looking at the rate of interactivity, online participation in this community seems to be quite high.

Several markers were present in all films in this case study which serve to signal the uploader as a legitimate member of the parkour community by performing the traceur identity. These markers perform the identity of parkour runner through the enactment of skill, movement, and video editing proficiency. Swedish youth parkour films enact the identity of parkour runner through the skill portrayed in the run itself. Films by experienced parkour runners focus on the terminology of the tricks in the film's description, or by enacting the tricks in a teaching or instructive capacity. Conversely, films by beginner runners often linguistically hedge the skill level in the description by labelling them 'quickly made' or 'made when slippery outside'. Another marker of beginner versus more experienced parkour runners can be seen in the speed of the film. Less experienced runners, although possibly more experienced video editors, would slow down the motion of the trick and ask for commentary on the performance of the trick. One runner even entitled his video 'What did I do wrong?'.

As mentioned previously, training is an important part of parkour. There are an equal number of videos that show training in session as there are videos that showcase the runners' skills. Again, in contrast to international training videos, of which there is also a fairly equal representation to showcase videos, Swedish youth often train in either indoor or outdoor sport arenas.

Digital remixing in order to create videos which show the flow of the run, or a flow of tricks also proved to be an identifying marker of parkour runners. This is common both in amateur videos, but also in international, professionally made parkour YouTube videos. This is not to say that the Swedish youth are copying or mimicking international videos, rather that there seems to be implicit, agreed upon ways of editing a video together. Several informants in this case study termed these videos 'flow films', which may be perceived as a mediated edit of a central concept in parkour. Flow, in parkour, refers to the smoothness of transitions between moves. According to howtoparkour.net, the quality of the flow of a run is a marker of the runner's skill. The videos, which used the term flow, were highly edited, thus it is unclear if the uploaders were referring to the flow between the edited tricks, or flow as a parkour aesthetic. Background music is also an important part of the digital remix. Hard beats seemed to be the preferred style, although this varied from user to user. Interestingly, slower music was often chosen to highlight videos in which tricks were asked to be evaluated for technique.

DIMENSION 3: GENDER IDENTITIES

Gender appeared as an increasingly important theme throughout the course of this study due to the obvious lack of girls represented in the YouTube parkour community. Initially we assumed that this is a male-dominated sport and that girls were not participating, thus there were no girls to film. However, girls were found on the websites of the Swedish parkour clubs. Not only were girls pictured in equal numbers as the boys, workshops geared towards girls' participation in parkour were featured on at least two of these websites. Girls are clearly participating in parkour, at least according to published club pictures. However, after an exhaustive search of Swedish parkour videos on YouTube, only three individual videos containing girls were found. One film was a training video with one girl in a large group of boys. The second film was of two girls trying out parkour for the first time, and the third was of a girl who would run, stop, giggle, etc.

This made us reconsider the idea of parkour being marked by male dominance. It is possible that parkour is dominated by boys to a larger extent than what seems to be the case on some club websites. And maybe the more exhibitionistic way of performing parkour – that is, through publishing one's skills on YouTube – is a more masculinely coded activity. Writers such as Robert Brannon (1976) have shown how the socially and culturally established ideas about masculinity encourage men to seek out adventure and danger. In line with this, a clear element of risk taking – risking breaking ankles or wrists, or getting injured while taking high falls on concrete or asphalt – characterizes the practice of doing parkour. The fact that risk taking plays such an important part no doubt makes it possible to see parkour culture in relation to "hegemonic masculinity", as defined by Connell (1995). This concept refers to a form of male identity which is based on a goal oriented model of thought typical of the hunter/leader stereotype. Furthermore, it has obvious elements of competitiveness – of "sports as embodiment of masculinity" (Connell, 1995, p. 54), and also of homosociality (Sedgwick, 1985), that is of "boys being boys" in gender exclusive peer groups.

In the male-dominated videos, discussion in the comment section ranged from commenting on the skill of the trick performed to looking for others with whom to train. Participation relating to the girls' videos was mainly about the girls' appearance or commenting on the lack of skill portrayed. It is difficult to say if the lack of parkour videos featuring girls is due to the negative comments received in the few girl-only examples, or if there are other factors at play. Trivialising a girl's performance through negative remarks, however, is not restricted to the parkour community, but can also be seen in other extreme sports labelled 'gender-neutral' (Laurendeau, 2003). The gap between how female performance is experienced in the physical parkour community, which is reported as friendly, non-competitive and supportive, and how it is experienced in the YouTube community is explored through the practices and perceptions of community in the two spaces.

As has been shown by other researchers (Lindgren & Lélièvre 2009), alternative sports – a category where parkour would fit in – have a certain connection to masculinity. Loret (1995) describes how a number of alternative sports based on different forms of appropriation of the environment or of public space were emerging during the 1980's and on. Examples of these sports are skateboarding, snowboarding, surfing, inline skating, BMX cycling and extreme climbing. Similar to parkour, all of these have as their aim to try to make more and more technically advanced performances that often are spectacular and have elements of individualism, aggressiveness and risk taking (Loret, 1995). In other words, parkour belongs to a group of sports that are about trying to stretch the limits of one's physical abilities as well as of how social space is used and appropriated. The focus on girls' appearances, and the trivialization of their achievements can therefore be understood in terms of parkour being constructed in some key senses as a "boys-only" activity.

CONCLUSION: ESTABLISHING A PERFORMATIVE ARENA

We argued in the beginning of this chapter that the vastness and embeddedness of YouTube make it hard to measure how, and the extent to which specialized communities with specific norms and boundaries can be launched here. Our aim has been to analyze the creation, composition, consumption and cultural context video clips posted by Swedish parkour youth. The goal of this case study was to gain an insight into wider issues regarding the possibilities for forming community in spite of the size and complexity of this environment.

In a qualitative content analysis of on- and offline ethnographic data (uploaded footage, face-to-face interviews, and patterns of community interaction online) we found three interrelating dimensions that shed light upon how Swedish parkour youth are negotiating "the YouTube sublime". The first of these was the spatial. In the analyses relating to this dimension we showed how, in parkour youth culture, a set of offline spaces (the urban space, the training space, and – to some extent – the rural space) is mediated in to an online space where YouTube functions as one of the key tools and platforms. This leads to parkour being acted out on an arena of participation where online and offline activities and spaces are interlinked. This arena can be conceived of in terms of Scollon's concept of "sites of engagements". This is a place where the users are involved in mediated action, the construction of social identities and of community. Much of this constructive work comes into expression in the uploaded parkour clips themselves, as well as in the practices of linking, watching, sharing, and embedding this content elsewhere (e.g. on club websites).

The second dimension that we analyzed related to constructions of community membership among parkour youth. We found that there is a high degree of interactivity, in terms of online participation, within this group, and that much of the interaction revolves around the negotiation of status and membership. To be a skilful traceur does not seem to be enough, and sometimes not even required, in order to be regarded as a member of the parkour community on YouTube, To be regarded as legitimate one needs to display offline skills (parkour proficiency) as well as online skills (knowledge of video editing, being active in terms of linking, commenting, etc.). Borrowing a concept from LaBelle, we argued that YouTube can be recognised as a performative arena due to the interplay between the users who perform aspects of community membership and identity through the blend of offline and offline practices, but also due to the social networking affordances allowed by the YouTube platform (Cheng, Dale, & Liu, 2008; Lange, 2008; Santos, Rocha, Rezende, & Loureiro, 2007).

Jenkins (2007) writes, in line with what we have found, that hybrid media spaces such as YouTube are very powerful for building community, but also stresses that there are always power structures involved. Any space of this kind "potentially represents a site of conflict and renegotiation between different forms of power". The third and final dimension that we addressed in our analysis has to do with gender identities, but our results in this case were not fully conclusive. On the one hand, many girls were seen on some of the club websites. But on the other hand, comments to girls' parkour videos on YouTube were often bordering on sexism. Furthermore, YouTube videos of girls were extremely scarce. A preliminary conclusion of this is that parkour is more dominated by male performers than what appears to be the case if one looks at some of the club websites. If this is true, one explanation could be that parts of parkour, as well as the more attention seeking way of publishing ones performances is in line with the stereotype of "hegemonic masculinity" (Connell 1995).

YouTube is a performative arena due to the vivid interplay between the users who perform aspects of community membership and identity through the interweaving of offline and offline practices, but also due to the social networking affordances allowed by the YouTube platform. Each channel on YouTube could be considered a site of engagement, as it is there that traces of social networking are collected and made visible. It is the collection and interplay of these sites, both through engaging in conversation on each individual video page, as well as subscribing and participating on a user's channel which gives meaning to the young people's performed practices within the videos. The Swedish parkour runners studied in this chapter have established practices for both their offline performances and their online interactions, and it is the reciprocity between these spaces and interactions that create a shared meaningful arena.

REFERENCES

Althusser, L. (1965). *For Marx*. London: Verso.

Apell, S. (2008). YouTube - Le Parkour Sweden Simon Apell interview. Retrieved August 26, 2009 from http://www.youtube.com/watch?v=K_C4oCDAXsg

Benkler, Y. (2007). *The Wealth of Networks: How Social Production Transforms Markets and Freedom*. New Haven, CT: Yale University Press.

Brannon, R. (1976). The male sex role: Our culture's blueprint of manhood, and what it's done for us lately. In Brannon, R., & David, D. (Eds.), *The forty-nine percent majority: The male sex role. Reading, MA.: Addison-Wesley. Buckingham, D. (2009). Video Cultures in Technology and Everyday Creativity*. New York: Palgrave Macmillan.

Burgess, J., & Green, J. (2009). *YouTube: Online Video and Participatory Culture*. New York: Polity.

Burr, V. (2003). *Social Constructionism*. London: Routledge.

Cheng, X., Dale, C., & Liu, J. (2008). Statistics and social network of YouTube videos. In Proceedings of the 16th International Workshop on Quality of Service (pp. 229–238). Enschede: IWQoS.

Connell, R. W. (1995). *Masculinities*. Cambridge, UK: Polity Press.

Glaser, B. G. (1965). The constant comparative method of qualitative analysis. *Social Problems*, , 436–445. doi:10.1525/sp.1965.12.4.03a00070

Green, J., & Jenkins, H. (2009). The Moral Economy of Web 2.0: Audience Research and Convergence Culture. In Holt, J., & Perren, A. (Eds.), *Media Industries: History, Theory, and Method*. Boston: Blackwell.

Grusin, R. (2009). In Snickars, P., & Vonderau, P. (Eds.), *YouTube at the End of New Media* (pp. 60–67). Stockholm, Sweden: National Library of Sweden.

Jenkins, H. (2006). *Convergence culture: Where old and new media collide*. New York: New York University Press.

Jenkins, H. (2007): Nine Propositions Towards a Cultural Theory of YouTube. http://henryjenkins.org. Blog post, May 28.

Jones, R. H., & Norris, S. (2005). *Discourse in Action: Introducing Mediated Discourse Analysis* (1st ed.). New York: Routledge.

Krippendorff, K. (1980). *Content analysis: An introduction to its methodology*. London: Sage.

LaBelle, B. (2006). *Background noise: perspectives on sound art*. New York: Continuum International.

Lange, P. G. (2008). Publicly private and privately public: Social networking on YouTube. *Journal of Computer-Mediated Communication*, *13*(1), 361–380. doi:10.1111/j.1083-6101.2007.00400.x

Laurendeau, J. (2003). Gender and the culture of skydiving: Misogyny, trivialization, and sexualization in a "gender-neutral" sport. Paper presented at the 2003 Annual Meeting of the American Sociological Association, Atlanta, GA.

Lave, J., & Wenger, E. (1991). *Situated Learning: Legitimate Peripheral Participation*. Cambridge, UK: Cambridge University Press.

Lefebvre, H. (1974). *The production of space*. Oxford, UK: Basil Blackwell.

Lemke, J. L. (2005). Multimedia genres and traversals. *Folia Linguistica*, *39*(1-2), 45–56. doi:10.1515/flin.2005.39.1-2.45

LeParkour Sweden. (n.d.). Retrieved from http://www.le-parkour.se/

Lincoln, Y. S., & Guba, E. G. (1985). *Naturalistic inquiry*. London: Sage.

Lindgren, S. (2009). YouTube Gunmen? Mapping participatory media discourse on school shooting videos. Paper presented at Violence and Network Society: School Shootings and Social Violence in Contemporary Public Life, Helsinki, Finland, November 6-7.

Lindgren, S., & Lelievre, M. (2009). In the Laboratory of Masculinity: Renegotiating Gender Subjectivities in MTV's Jackass. *Critical Studies in Media Communication*, *26*(5), 393–410. doi:10.1080/15295030903325313

Loret, A. (1995). *Génération glisse. Dans l'eau, l'air, la neige... la revolution des sports dans les années fun*. Paris: Autrement.

Resch, R. P. (1992). *Althusser and the renewal of Marxist social theory*. Berkeley, CA: University of California Press.

Rheingold, H. (2000). *The virtual community: Homesteading on the virtual frontier*. Cambridge, MA: MIT Press.

Rheingold, H. (2003). *Smart Mobs: The Next Social Revolution New edition*. New York: Perseus Books.

Santos, R. L., Rocha, B. P., Rezende, C. G., & Loureiro, A. A. (2007). Characterizing the YouTube video-sharing community, (Technical report). Retrieved from http://security1.win.tue.nl/~bpontes/pdf/yt.pdf.

Scollon, R. (2005). The rhythmic integration of action and discourse: work, the body and the earth. In Norris, S., & Jones, R. (Eds.), *Discourse in action: introducing mediated discourse analysis*. London: Routledge.

Sedgwick, E. K. (1985). *Between men: English literature and male homosocial desire*. New York: Columbia University Press.

Springhall, J. (1998). *Youth, popular culture and moral panic: penny gaffs to gangsta-rap. 1830-1996*. Basingstoke, UK: Macmillan.

Thomas, A. (2007). *Youth Online: Identity and Literacy in the Digital Age*. New York: Peter Lang Publishing Inc.

Van Dijk, T. A. (1991). *Racism and the press*. London: Routledge.

Vonderau, P., & Snickars, P. (2009). *The YouTube Reader*. National Library of Sweden.

Wesch, M. (2008). An anthropological introduction to YouTube – Presented to the Library of Congress, USA. Retrieved August 28, 2009 from http://www.youtube.com/watch?v=TPAO-lZ4_hU.

Wimmer, R. D., & Dominick, J. R. (2003). *Mass media research: An introduction*. Belmont, CA: Wadsworth.

YouTube(n.d.). Retrieved from http://www.youtube.com

ADDITIONAL READING

Caldwell, J. (2008). *Production Culture: Industrial Reflexivity and Critical Practice in Film and Television*. Durham, NC: Duke University Press.

Drotner, K. (1999). Dangerous Media? Panic Discourses and Dilemmas of Modernity. *Paedagogica Historica*, *35*(3), 593–619. doi:10.1080/0030923990350303

Felix, L., & Stolarz, D. (2006). *Hands-on guide to video blogging and podcasting: emerging media tools for business communication*. Amsterdam: Elsevier.

Ferraro, N. (2009). GenY Musicians the Y is for YouTube. Plus Model Magazine. Retrieved August 14, 2009 from http://www.plusmodelmag.com/General/plus-model-magazine-article-detail.asp?article-id=118239935&page=0

Flanagan, M. (2009). Critical Play: Radical Game Design. Cambridge, MA: The MIT Press.Griffith, M. (2010). Looking for you: An analysis of video blogs. First Monday 15.1. Retrieved 13 Jan 2010 from http://www.uic.edu/htbin/cgiwrap/bin/ojs/index.php/fm/article/view/2769/2430

Jenkins, H. (2006). *Fans, bloggers, and gamers: exploring participatory culture*. New York: New York University Press.

Jenkins, H. (2008). *Convergence Culture: Where Old and New Media Collide* (Revised edition). New York: New York University Press.

Jenkins, H. (2009). *Confronting the Challenges of Participatory Culture: Media Education for the 21st Century*. Cambridge, MA: The MIT Press.

Kitzmann, A. (2004). *Saved From Oblivion: Documenting The Daily From Diaries To Web Cams*. New York: Peter Lang Publishing.

Lange, P. G. (2007). Publicly private and privately public: Social networking on YouTube. *Journal of Computer-Mediated Communication, 13*(1), article 18. Retrieved August 28, 2009 from http://jcmc.indiana.edu/vol13/issue1/lange.html

Laurel, B. (1993). *Computers as Theatre New edition*. Reading, MA: Addison Wesley.

Levy, F. (2008). *15 Minutes of Fame: Becoming a Star in the YouTube Revolution*. New York: Alpha.

Lindgren, S. (2007). From Flâneur to Web Surfer: Videoblogging, Photo Sharing and Walter Benjamin @ the Web 2.0. Transformations, 2007(15), http://www.transformationsjournal.org/journal/issue_15/article_10.shtml.

Lovink, G., & Niederer, S. (2008). Video Vortex Reader: Responses to YouTube, Amsterdam: Institute of Network Cultures. Retrieved August 29, 2009 from http://networkcultures.org/wpmu/portal/publications/inc-readers/videovortex/

Lundby, K. (2008). *Digital Storytelling, Mediatized Stories: Self-representations in New Media*. New York: Peter Lang Publishing.

Montgomery, K. (2009). *Generation Digital: Politics, Commerce, and Childhood in the Age of the Internet*. Cambridge, MA: The MIT Press.

Murray, T. (2008). *Digital Baroque: New Media Art and Cinematic Folds*. Minneapolis, MN: University Of Minnesota Press.

Ricardo, F. (2008). *Cyberculture and New Media*. Rodopi.

Rodowick, D. N. (2007). *The Virtual Life of Film*. Boston: Harvard University Press.

Saco, D. (2002). *Cybering democracy: Public space and the Internet*. Minneapolis, MN: University of Minnesota Press.

Tapscott, D. (1999). *Growing up digital: the rise of the Net generation*. London: McGraw-Hill.

Tyron, C. (2007). Bringing the War Back Home: YouTube and Anti-War Street Theater. FlowTV, 6(2). Retrieved August 28, 2009 from http://flowtv.org/?p=469

Varnelis, K. (Ed.). (2008). *Networked publics*. Cambridge, MA: MIT Press.

Watkins, S. C. (2009). *The Young and the Digital: What the Migration to Social Network Sites, Games, and Anytime, Anywhere Media Means for Our Future*. Boston: Beacon Press.

KEY TERMS AND DEFINITIONS

Hegemonic Masculinity: This concept refers to a form of male identity which is based on a goal oriented model of thought typical of the hunter/leader stereotype.

Networked Publics: A gathering term for a number of socio-cultural and technological

Developments in the Field of Digitally Networked Media that have Altered the Ways in which People are Engaged: networked and mobilized.

Participatory Culture: The concept "participatory culture" refers to the lowered barriers to civic engagement and artistic expression in the new media arena. Instead of strictly divided categories of producers and consumers, we now see a number of "participants" interacting trough media according to new rules.

Performative Arena: The field of attention which is given meaning through a performative exchange between an art object and an art viewer when interwoven into a conversational exchange in which the object produces the looking/listening, and the looking/listening produced the object (LaBelle, 2006, p. 101).

Site of Engagement: the point at which mediated action happens, but also and how these points foster the construction of social identities, communities and practices.

YouTube Sublime: A term coined by Grussin (2009) describing the inability for the user to fathom the immensity and complexity of the YouTube site.

Section 3
Identity and Sexuality

Chapter 10
Young People and Cybersex in a Sexually Conservative Society:
A Case Study from Mauritius

Komalsingh Rambaree
University of Gävle, Sweden

ABSTRACT

This chapter describes the process, explains the aspects, analyses the experiences and considers the social policy implications, of cybersex among young people from the sexually conservative Mauritian society. This chapter is based on a study, in which it is found that some of young people from Mauritius are involved in cybersex. The chapter therefore argues that Internet-based technologies are further breaking down 'the traditional and moral values', which some politicians, religious leaders and parents want to preserve through social policy related to sexuality education in Mauritius. A sexual rights-based approach to policy making and interventions for a formal sex education programme in Mauritius is found more appropriate within this particular context, and therefore recommended in this chapter. Finally, this chapter concludes that an appropriate formal sex education for young people should take into account the 'Net Culture' context within which contemporary young people are growing sexually in Mauritius.

This chapter contains explicit content

INTRODUCTION

Mauritius is a small African island situated in the Indian Ocean. It has a land surface area of 1,865 square kilometres and a population of about 1.2 million. The youth population (14-25 years old) of the country is around 300,000 of which about 50% are female (MYS, 2009). It is commonly argued that the youth of today are the future of tomorrow. For economic experts, this section of the population represents the vital future human capital. For sociologists, the contemporary youth population set new patterns for social and cultural changes in a dynamic society. In traditional theories, youth is seen as distinct transitional phase of life, between the more stable categories of childhood and adulthood (Worth, 2009). In various societies, the youth population express their identities and demonstrate their sense of belonging to a particular group of young people

DOI: 10.4018/978-1-60960-209-3.ch010

through having what is commonly referred as 'youth culture' (Buchmann, 2004).

Mauritius is considered as a peaceful multicultural, politically stable and economically successful country. Since 1983, the country has maintained an average annual economic growth of around 5 per cent (Shillington, 1991; ADB/OECD, 2008). In order to further diversify its economy, Mauritius is also investing towards becoming a technology-based society (Rambaree, 2007, 2008, 2009, 2010). Mauritius has no indigenous population and the inhabitants are of European, African and Asian origins. The vast majority of inhabitants from Mauritius speak Creole (lingua-franca) and more than half of the population can speak both French and English, and more than half of them can even speak one or more Asian languages. The religions followed by the inhabitants are Hinduism (52%), Christianity (28%), Islam (17%), Buddhism (2%), and Others (1%) (CSO, 2000). Most of the inhabitants on the island are descendants of those who had been sent or brought to develop the island during the French and British colonisation periods (1710-1810 and 1810-1968 respectively). It is commonly argued that from its colonial and political history, the current Mauritian society is shaped by certain conservative values regarding sexuality (Hillcoat-Nallétamby & Dharmalingam, 2005).

Within the traditional and conservative nature of the Mauritian society, sex is still considered to be a taboo subject. Growing up as a young person in Mauritius means showing respect; importantly, by not challenging the traditional and conservative sexual norms and values. Within the traditional norms and values, sex is regarded as a private matter and no sexual activities between people before marriage is acceptable by the society. Parents guide and expect their children to follow the traditional norms and values by not bringing sexual discussions in public and not tolerating pre-marital sexual intercourse. Similarly, the community and religious leaders expect the parents and the extended family (grandparents and elder siblings) to play an important role in transferring the traditional norms and values. The success/failure of a family in Mauritius is therefore commonly judged by its ability to pass on the traditional norms and values (Schensul et al., 1994).

In Mauritius, young people are commonly categorised as either 'Bien-Elevés' (well educated/brought up) or 'Mal-Elevés' (badly educated/brought up). It is also common to find young people who try to speak about sexuality openly being labelled as 'Mal-Elevés'. Sometimes, having sex is also referred as 'faire Mal-Elevés' (misbehave) in the local jargon. Particularly, 'sexual purity' is among the basic norms and values that are shared by almost all ethnic communities in Mauritius. Within the family, more emphasis is laid on girls to remain virgin for the husband-to-be. Girls' virginity is the pride and honour of the family (Schensul et al., 1994). It is quite common in Mauritius, for families to go to the court asking for legal punishment and/or compensation from boys who have broken the promise of marriage with their female family members. Young people are given the message that they should date the one they are supposed to marry, and marry the one they have been dating (Rambaree, 2008). Sexual activities in a couple before marriage and even during the dating period are considered to be 'immoral'.

Morality related to sex, mainly the expected code of conduct and practices, is framed by the rigid moral values of certain value-based organisations such as church, schools, and voluntary ethno-socio-cultural organisations in Mauritius. Moral values are strong beliefs and opinions about what could be considered as being right or wrong. These values are therefore the guide for expected actions, directing individuals and groups to decide about what should or should not be done, and provide the basis for evaluating the conduct of others as being right or wrong (Smith, 2000). Almost all schools in Mauritius try to preach and instil moral values to the young people; and especially to girls, who are regarded as the main guardians of moral values. The Mauritian society could also be characterised as one where strong patriarchal norms and values prevail (Bunwaree, 1999). In terms of sexuality in such societies, the cultural norms demand sexual abstinence and sexual suppression before marriage, sexual fidelity, and docile and acquiescent sexual behaviour after marriage among women (Gupta and Weiss, 2005). Women in Mauritius are conditioned to abide by the dogmas of their cultural systems based on patriarchal and moral norms and values (Aumeerally, 2007). Thus, it is very common to see moral discourses especially targeted at girls in Mauritius. For example, Ajaheb-Jahangeer and Jahangeer (2004: 253) state:

"In a rather conservative society like Mauritius, it is crucial that girls are given the proper education so that they may find their way ...through life with good qualifications and strong moral values".

In a similar manner, a director from a popular secondary school gives the following statement in a local newspaper:

"One of the main characteristics of the ...college is the emphasis put on ancestral values and culture. We try to make them (the girls) understand that they can be modern women without losing their cultural heritage, if they do so they will become a pale imitation of modernity. We do not want our girls to become kites with no connections" (Hilbert, 2007: 4).

Moreover, the Mauritian education curriculum does not have a formal sex education programme. The absence of formal sex education programme is driven by the fear that such an education might incite further curiosity and encourage young people to engage in pre-marital sexual intercourse. Some of the Mauritian policy makers feel that young people need sex education because of the degradation of moral values. For instance, one of the policy-makers makes the following statement in the press:

"We decided to add sex education to the list of the new subjects (being proposed) in the light of ... the degradation of moral values" (Etienne, 2007: 5).

Many other professionals fear that the proposed sex education programme might contain moral discourses that would not attract the attention of the young people in Mauritius. For example, the Ombudsperson for Children reports that children will not listen to a moralising speech on sex and they need to be given the right information about their bodies, about relationships, the feelings involved, and the consequences a sexual relationship will have for them (Etienne, 2007: 6). Consequently, for the last 10 years the contents of the proposed sex education curriculum are still being debated within the Mauritian public discourses. The Ministry of Education and Scientific Research, the religious organisations, the Parent Teacher Associations, the private secondary school associations, and many other stakeholders have not come to a consensus on this issue. As it can be found from the above-discussion, the Mauritian society could indeed be a considered as a conservative one; however this does not mean that all contemporary young Mauritians are abiding the conservative societal values.

Hillcoat-Nalletamby and Ragobur (2005) argue that there are indications that the experience of young people in Mauritius is changing rapidly through encountering broader social networks than those offered by the immediate family. Similarly, the MFPA (Mauritius Family Planning Association) (2005) writes that the economic development has reshaped the socio-cultural environment, creating opportunities for social interaction and the development of relationships. As a result, many traditional values and norms are being eroded, giving rise to new forms of attitudes and behaviour among the young people. Several empirical studies have found that about

30% of male and 10% of female young people from Mauritius are engaged in premarital sexual intercourse (MFPA, 1993; Schensul et al., 1994; MIH, 1996; Nishimura, 2007).

The paradox is that there is still a type of moral conservatism prevailing around young people's sexuality in Mauritius, at the same time as liberal access to and use of the *Internet* is being promoted for economic development (Rambaree, 2008, 2009, 2010). More than any other medium, the Internet is a sexual medium whereby people of any age group can have access to both regulated and unregulated information, materials and services related to sex (Peter and Valkenburg, 2006). With the advent of the Internet, the social networks of young people have grown both geographically and culturally (Peter and Valkenburg, 2006; Ybarra & Mitchell, 2005; Cameron et al., 2005; Thornburg & Lin, 2002). The Internet has made it easier for the present day youth to have more possibilities for interactions and building relationships with others who are beyond their geographical and cultural boundaries. Some experts argue that the Internet is a new form of colonisation that brings about foreign influences on local people (Hall, 1999; Cline-Cole & Powell, 2004). The young people from Mauritius are therefore having more opportunities/risks to be exposed to sex and sexuality issues, activities, behaviours and practices on the Internet that come from all around the globe. However, the young Mauritians are still being denied a formal type of sexuality Information, Education and Communication (IEC) so that they can make informed decisions and cope with situations and encounters related to sex.

It could be said that the vast majority of the Mauritian young people are coping and managing the new Internet-driven development. However now and then the local media report on how the Internet is bringing a new dimension to the local young people's sexuality. The 'Shock of the New' regarding the Internet-driven development started in 2006 for the local people in Mauritius, when 14 to 16 year old students were found making and selling pornographic movies using the latest Internet technologies (L'Express, 2006a). The Internet has facilitated the connectivity between young people from all over the world. This means that young people can easily have access to attitudes, behaviours and influences coming from all around the globe (Peter and Valkenburg, 2006). An interesting quotation from a 16 year old girl in a local newspaper reads as follows:

"Nous partageons les mêmes pensées, les mêmes aspirations que n'importe quel ado francais ou anglais. Il ne faut pas croire que nous sommes moins 'avancés qu'eux. Après l'éveil à la sexualité, vers 14-15 ans, les jeunes veulent tenter l'expérience. Vouloir passer à l'acte est tout à fait normal...Mais à Maurice, nous baignons dans une culture et des principes rigides'. (We share the same thoughts, needs, and aspirations as any adolescent whether French or English. One should not believe that we are less 'advanced/ developed' than them. After the awakening of sexuality, around the age of 14-15 years old, young people want to experience sexuality. Wanting to pass to the act is just normal... But, in Mauritius, we are plunged into a culture and principles that are rigid'. (L'Express, 2006b: 6)

Thus, Mauritius provides an ideal context for studying 'Youth Culture and Net Culture'. This is the context within which this chapter is based. Within the Mauritian context, this chapter aims to explore cybersex among young people. The specific objectives of the chapter are to (a) describe the process, (b) explain the aspects, (c) analyse the experiences, and (d) consider social policy implications, of cybersex among the Mauritian young people. After this introduction, the background of the chapter is described next. This part therefore provides the reader with a guide to the conceptual, theoretical and ideological basis of the chapter.

BACKGROUND

Although young people are using the Internet for various types of sexual interactions very few studies have focused on providing a full description and the young users' experience of such online activities (Peter & Valkenburg, 2006; Ybarra & Mitchell, 2005; Cameron et al., 2005; Boies et al, 2004; Thornburg & Lin, 2002). In particular, those studies that have managed to study 'Internet and Sexuality' among young people have mostly presented it from a problematic perspective. In other words, sex on the Internet has narrowly been described as the 'problem, bad and ugly' for young people (Boies et al, 2004). If sex on the Internet is bad, why are millions of young people, from all around the world and different cultural backgrounds, interacting sexually through this particular medium (Cooper et al, 2000)? Is there an assumption that sex on the Internet is good and enjoyable only for married adults? For Subrahmanyam et al. (2004) research on the effects of young people's use of the Internet for sex is still sketchy and ambiguous. Describing the complex sexual interactions on the Internet is one possible path of future study on understanding young people's sexuality within this Internet-driven technology era (Bay-Cheng, 2001) and this suggestion is developed further towards the end of this chapter.

To start with the subject matter, it will be interesting to look at the meaning of sex and sexuality. Tiefer (as quoted in Bristow, 1997: 16) states that, "the most basic and most difficult aspect of studying sexuality is defining the subject matter". In particular, sexuality has always been an intrinsic part of human life, and it is not confined to sexual activities. Stanley (1999) argues that 'sex' defined as 'intercourse' is more often based on the assumption that there can be no variant behaviour involved beneath this visible and easily investigated aspect. Within recent social discourses, the term 'sexuality' has therefore been used to encompass a variety of meanings such as sex as biology and activity; gender as a social construct based on sex; and sexual orientation, as well as many other aspects such as eroticism and pleasure (Rival et al, 1998; Smith, 2009). A broad definition of 'sex' and 'sexuality' allows social scientists to have a better understanding of the complexities and issues surrounding the phenomena related to sex and sexuality. If it is generally agreed that sex is much more than sexual activity, what is cybersex then?

Cybersex can also be broadly defined as a collective term for all sorts of sexually related activities, entertainment, information and sexual identity exploration that are possible within cyberspace (Turkle, 1999; Döring, 2000). However, within current social science discourses cybersex has a rather narrow definition that focuses on Internet-mediated sexual activities and practices only. Cybersex is a relatively new concept within social science discourses and its conceptualisation is still in its exploratory stages; and therefore, there is still room for more debates on how cybersex should really be defined. Döring (2000) opines that the broad definition of cybersex covers so many different activities and contents that it is of practically no use for the social scientific discourse as long as individual phenomena are not differentiated from one another. By narrowing the definition of cybersex, social scientists provide a focused and detailed understanding on the subject matter through a pre-agreed delimited conceptual framework.

Ross et al (2004) refer to cybersex as the practice of engaging in sexual self-stimulation while being online with another person. For cybersex, partners meet in various online spaces such as chat rooms, community groups, gaming environments, or virtual worlds - such as 'Second Life' (Döring, 2009). Peter and Valkenburg (2006) note that the Internet offers numerous opportunities to engage in so-called cybersex, i.e. suggestive or explicit erotic messages or sexual fantasies that are exchanged with others via the Internet. During cybersex some people also exchange pictures and/or short movies of themselves and/or others found on the web to accompany text based communication (Cooper & Griffin- Shelley, 2002; Daneback et al, 2005). For Ben- Ze'ev (2004) cybersex has the basic characteristics of actual sex except for the physical contact: it has the excitement, anticipation, satisfaction, and orgasm associated with typical sexual activities. Some researchers (Cooper & Griffin-Shelley, 2002; Daneback et al., 2005; Attwood, 2009) present cybersex as a subcategory of online sexual activities (OSA) and define it as when two or more people engage with each other in a sexual manner through Internet technologies, for the purposes of sexual pleasure that may or may not include masturbation. This is the conceptualisation of cybersex that is used in this chapter.

Just like it is important to have a conceptual framework for cybersex, a theoretical framework is also required for structuring research and debate within social science discourses. But, social researchers do not have to feel coerced to choose the already existing theories to frame their studies. Researchers have the flexibility to inductively approach their studies/subjects and come up with their own explanation, understanding and interpretation that are grounded on empirical evidence collected from the field. This is what is commonly known as grounded theory approach in doing social research. In fact, grounded theory is found to be an appropriate approach where limited studies exist in the specific domain of the research and/or where the researcher/s do/es not want to be influenced by preconceived theoretical ideas (Hunter et al., 2005). This is the approach that is adopted in the writing of this chapter.

The chapter is therefore grounded on empirical data that were collected through three different qualitative data collection techniques. First, 137 narrative interviews were carried out through giving some open-ended questions and vignettes to the research participants. Each research participants had their questions and vignettes in a personal floppy-disk in which they had to type and save their responses. Given that sex is a taboo subject in Mauritius, such technique of data collection from the research participants was found to be most appropriate. Research participants had opportunity to provide anonymous responses and privacy to express themselves freely without the discomfort or fear of others watching what they were writing.

Second, 8 Focus Group Discussions (FGDs) were carried out with the volunteers from those who participated in the narrative interviews. The FGDs were mainly used to enhance the validity through checking whether the participants would agree, in a collective setting, with the findings that had been gathered from individuals through the narrative method. In the FGDs, it could be said that not all of the participants had experiences related to cybersex, but almost all of them had some knowledge and experience regarding sexual interactions within the Internet based chat settings.

Finally, 53 rapid Internet-mediated semi-structured interviews (through Internet Chat) with young people from the Mauritius was carried out to probe certain specific issues related to the research subject, that is cybersex among the Mauritian young people. The 53 young Mauritians were those who volunteered as informant to provide information about their own experience on cybersex. The recruitment was done in privacy through individual contacts and call for volunteers. The volunteers were asked to join a social network link by creating an anonymous identity by a given date. There were 72 volunteers on the specifically created 'research social network', but only 53 of the 72 were interviewed. The others were not available. It is worth pointing out that 53 interviews were felt more than enough for obtaining theoretical fit (that is, data were felt adequate enough for carrying out theoretical discussion). These three mixed-techniques of data collection have therefore taken care of ensuring about some the challenges and difficulties that researchers sometime encounter in carrying out empirical studies related to sex in conservative societies.

The grounded theory data analysis was carried out with the Atlas-ti 5.2 (Computer Aided Qualitative Data Analysis Software). All the data collected through the narrative interviews, the digitally recorded FGDs and the chat logs for semi-structured interviews were plugged into Altas-ti 5.2. The grounded theory analysis was based on Glaser's (1978) non-linear method of theory generation through data analysis. It is worth pointing out that this research started in 2006 (some further studies are still being undertaken) and had the ethical approval from University of Manchester and University of Mauritius, and informed consent of all participants and clearance for data collection were obtained from various authorities in Mauritius (Ministry of Education, Ministry of Technology, Ministry of Youth and Sports, parents). The process of getting the ethical clearance took time, as several presentations were made to the various stakeholder groups such as parents, representatives from Ministries, board of ethics etc.

MAURITIAN YOUNG PEOPLE'S CYBERSEX

The Process

Mauritian young people interact within the same global cyberspace as millions of people from all over the world and therefore are exposed to the risk/chance to encounter the globally connected online sexual space. Like young people from all corners of the planet, Mauritian young people discover the Internet services, functions, modes and capabilities for online sexual interactions either by themselves or through the help of others (mostly friends). For example, some young people who use Internet-based social networks encounter sexual overtures for experimenting in new ways of having Internet mediated sexual interactions such as cybersex (Wolak et al., 2004; Mitchell et al., 2001). For instance, one of the research participants reports that:

"Young people may get involved in cybersex with or without their consent as some have been forced to ... (the force here is rather expressed as being coerced by others while exploring the Internet)... They surf on the internet and choose what they want to do as they are curious to discover the unknown" (Girl, 22 years old)

Young people learn about cybersex through their peers and also from the media like TV, newspapers and the Internet itself. It could also be said that the discovery about cybersex among some young Mauritians is driven by their curiosity (Rambaree, 2007 & 2009). It could be said that the majority of Mauritian young people are curious to discover this new adventure of cybersex. This eagerness to know about sexuality is a basic facet of human nature (Turkle, 1999; Yates, 1978). Sexuality is known to exist as a powerful inner force right from infancy and childhood (Yates, 1978; Runeborg, 2004). When sexual maturation starts especially during early adolescence, one becomes more eager to know about sex (Rambaree, 2007, 2008 & 2009). Young people are curious about sex and the Internet is a readily available source for dealing with their sexual curiosities (Walsh and Wolak, 2005; Thornburgh and Lin, 2002). In fact, the Internet presents a plethora of sexual services that fascinate the young people. The fascination that young people have with sexuality and sexual communication is therefore manifested on the Internet in a multitude of ways and cybersex is one of them (Cooper et al., 2000).

It also happens that sometimes, through online discussions, experimentation, and exposures, some young people find themselves intentionally/unintentionally involved in cybersex (O'Connell et al, 2004; Lybarra et al., 2004; Boies et al; 2004; Cameron et al., 2005). This starting phase is an interesting part of the cybersex that needs to be highlighted. In this study it is found that very rarely cybersex starts directly between two people interacting on the Internet for the first time. The first interaction between two people usually starts by an introduction and then gradually move towards what could be termed as 'cyber-flirting'. Flirting is commonly defined as the initial actions one takes to convey a message of sexual interest or attraction (Whitty and Carr, 2003). Within the cyberspace, 'cyber-flirting' could be considered as an important phase of cybersex. For Ben-Ze'ev (2004: 149) cyber-flirting is "a type of verbal dance in which the boundaries of sexuality are not clearly drawn". In fact, it could be said that 'cyber-flirting' is the stepping stone towards cybersex. Through cyber-flirting young people also get the sense of the move towards cybersex. Within the process of 'cyber-flirt' trust is built and the intention towards the move to sex is informally negotiated. In a similar manner, Longmore (1998) writes that the Internet provides an important medium for cybersex, through which people do go beyond flirting. Therefore, some people might have intention for cyber-flirting but get unintentionally involved in cyber-sex.

The Aspect

The aspect of cybersex among Mauritian young people is usually clandestine. Some Mauritian young people do not want others on the 'actual space' (referring to outside the cyberspace) to know about what they do as 'sex' on the 'cyberspace'. Cybersex, just like 'actual space sex', is considered as a private matter that remains between the parties involved in the act. The vast majority of the study sample who are involved in cybersex do not use webcam and prefer to keep anonymity through nick names. Most of the Mauritian young people believe that it is 'disgusting' and 'immoral' to talk about their private sexual activities with others. But, from the collected data, the great majority of Mauritian young people believe that most of the young people having Internet facilities at home (more than 80% do have) have been involved in cybersex at least once. Gathered evidence from Internet-based social network users also confirms that cybersex is a common practice among a small minority of young people from Mauritius (Rambaree, 2005). Usually, it is at late night or during the absence of the parents/siblings that a small minority of the Mauritian young people secretly go on the Internet for social interaction and cyber-flirt, which sometimes end up with cybersex. For instance, consider the below-given quotation from a research participants.

'At night when my parents go to sleep I go on the internet secretly searching for someone to have some excitement...' (Girl, 19 years old)

The aspect of cybersex among the Mauritian young people could also be looked at from a gender perspective. A very small minority of boys have been reporting that they discuss or relate their cybersex activities and experiences with their male peers as a way to be 'macho' (masculine). A very small minority of Mauritian young boys also report that they have been having cybersex in a group of friends together at their friends' places, but this was mostly texting sexual messages to girls who they come across online as a way of having fun among boys. Most of the time, the Mauritian girls try to avoid talking about cybersex (even if they are involved it in), because girls who talk about sex are easily stereotyped as being 'easy' and 'cheap'. For example one girl from this study comments.

"Girls who get involved in cybersex are stigmatised in Mauritius and viewed as those without social status..." (Girl, 21 years old)

As mentioned earlier, a small number of Mauritian young girls report that they get 'forcefully driven' by their friends (peer pressure) and get carried away by curiosity towards 'cybersex'. Girls, who have cybersex, also report that they choose someone 'unknown' as they are concerned about their image in society. According to some of the young boys, there are a small minority of Mauritian girls who are involved in cybersex and, while being online, enjoy it, but cannot assert themselves outside the cyberspace in relation to sexual encounters. Some of boys (very small number) report that they would like to meet the people with whom they have cybersex, but that is difficult for them. For example some of the typical gathered quotations from the Mauritian young people in this study are:

"They (girls) are influenced by their friends. New technology forces them in a way to get involved in cybersex" (Girl, 22 years old)

"I have had cybersex with maybe more than 100 Mauritian girls, but in real life I cannot get a single girl to talk about sex. When I start talking about sex with some of my female friends at the University they call me pervert, they are 'Sociables' (respectable persons in society)" (Boy, 24 years old)

In addition, some studies have reported that cybersex liberates women from the patriarchal sexual norms and allow them to be more assertive than offline (Döring, 2000). Smahel and Subrahmanyam (2007) report that their findings run counter to the stereotype that males are more interested in sex and more actively seek partners for this purpose. Another study has found that in cybersex, men identify themselves in a wide variety of ways; tender and tough, dominant and submissive, playful and provocative, high minded and dirty minded (Attwood, 2009). Some researchers report that, women seem to have a stronger preference for cybersex than men (Cooper et al., 1999; Döring, 2009). However in the case of Mauritius, women report that cybersex is a man's world.

Another aspect of cybersex - among the young people within the Mauritian context - is that it is often considered as being 'not real'. For some Mauritian young people, going for cybersex is a game where people play within the 'virtual' space and therefore they do not consider it as a 'real' sexual activity. Moreover, given that the partners do not have physical contact with each other, a small proportion of the young people take cybersex as being 'not real'. Some young people also argue that they do not see cybersex as 'real' because they do not know who is at the other end, whether a boy or a girl, old or young etc. The anonymity and the absence of physical (and sometimes visual) contact between the partners therefore make cybersex as 'not real' for some of young people in Mauritius.

In fact, the Internet has a reality of its own, which is normally termed the 'virtual reality'. Ben Ze'ev 2004:2) states: "The cyberspace is Virtual in the sense that imagination is intrinsic to that space". In actual space, the Internet is mainly regarded as a technological tool, but in cyberspace the Internet is a social environment with its own virtual realities. According to Riva (2005), modern technological advancements are transforming the way people interact on the Internet among themselves and with objects around them by gradually shifting the focus away from the computer as such, towards the user, creating a new kind of reality. In addition, sexual activities cannot be narrowed to those involving physical contacts between the partners only. Several researchers argue that cybersex should not be classified as a deficient substitute for 'actual sex', but should instead be understood as a specific form of sexual expression that can play a legitimate role in the sexual and relational life of its participants (Carvalheira & Gomes, 2003; Döring, 2000 & 2009; Ross et al., 2004). Therefore although cybersex is 'Virtual' does not mean that it is completely 'not real'.

The Experience

A small number Mauritian young people – in this study - describe their experiences of cybersex as being positive. Those who report positive experiences perceive cybersex as being exciting, pleasurable, relaxing and entertaining. Cybersex also brings a feeling of achievement and maturity to some Mauritian young people. Being able to perform certain sexual tasks and activities such as cybersex, give some young people the sense of sexual maturity, growth and development (Rambaree, 2007, 2009). For those who have done it, cybersex is something that young people should do at least once in their youth and therefore it is what could be referred as being among the 'developmental tasks' (Coleman and Hendry, 1999; Kuttler and La Greca, 2004). For more than half of young Mauritians interviewed in this study, cybersex is also the only way to express themselves sexually, as sex outside the cyberspace represents health risks as well as negative societal sanctions. Thus, it could be said that cybersex has made it possible for some Mauritian young people to deal with their sexual drives and enjoy sex which otherwise would not have been possible, as they would not dare to take the risks of negative societal sanctions.

Sexual drives are part and parcel of human sexuality and involve sexual fantasies, sexual desires, sexual maturity, sexual stimulation, and sexual capacity. During youth (especially during adolescence), sexual maturation is accompanied by increased sexual drives and therefore an increasing interest in sex (Weinstein & Rosen, 1991; Subrahmanyam et al., 2004). As a consequence, many young people begin to interact sexually and start searching for sexual stimulation through the Internet (Rice, 2001, Subrahmanyam et al., 2004; Peter & Valkenburg, 2006). Cybersex is therefore one of many ways young people can enjoy their sexualities. For instance, some gathered quotations from young Mauritians are as follow:

"they (young people from Mauritius) are curious to discover the unknown and meet other people to relax who share the same interests such as sex where they can express themselves fully..." (Boy, 23)

"Youngsters are easily influenced by what arouse their pleasure. It is not easy to control... Technologies are available and make it possible for us to explore about sexuality" (Girl, 21)

Even in their 20s, young Mauritians are still considered as 'children' who need to be controlled by parents and protected from the bad influences that are related to sex, such as having pre-marital sexual intercourse. Yates (1978) argues that, this is more a threat to parental self-esteem by evoking fears of loss of control or moral disintegration. In most societies young people are therefore expected to control their sexual activities. But, the norms and control regarding sexuality are more rigid within sexually conservative societies. In a sexually conservative society like Mauritius, sex among unmarried young people (even above the age of 18 years), especially girls, is labelled as a 'bad thing'.

Indeed, cybersex can be 'bad' for young people. Cybersex has some highly problematic aspects such as those involving children, especially those below the age of 16. In Mauritius, having any type of sexual contact (physical and non-physical) with minors (below the age of 16), is a criminal offence. This is based on the reasoning that individuals below the age of consent, which in many countries is around the age of 15, do not know what they are doing and are vulnerable to abuse and traumatic experiences. In addition, cybersex involving pornographic materials are also considered as inappropriate. Some researchers argue that pornography could generally be defined as those materials that combine sex and/or the exposure of genitals with abuse or degradation in a manner that appear to endorse, condone, or encourage such behaviour (Russell, 1998). Pornography is usually negatively loaded, and often associated with the socially unacceptable, the deviant, the censured, or the shameful (Traeen et al, 2006; Traeen et al., 2004; Beggan and Allison, 2003). Morgan (1980) opines that pornography is the theory and rape the practice (Morgan, 1980). Moreover, several researchers (particularly feminists) argue that pornography degrades, dehumanises, and debases people, especially women (Russell, 1998; Zillmann and Bryant, 1989; Fisher, 2001; Gossett and Byrne, 2002).

Moreover, any form of abuse, such online sexual harassment and cyber-stalking, should also not be tolerated. As mentioned earlier, a minority of young people using the Internet found themselves 'forced' to cybersex. Some young people start engaging with others online and gradually the conversation moves towards sex. When someone has not agreed to get involved in cybersex and he/she is coerced into sexual communications as a form of 'cyber-stalking' (online harassment), this could be considered as 'cyber-rape' (also referred as virtual rape). For instance, a female research participant reports that, it was difficult for her to complete her assignment for a distance learning module, because one of her classmates - who was assigned to work with her - was constantly forwarding sexual comments and materials instead of discussions on the academic tasks. Young people can therefore be sexually molested and raped online in situations in which they are not aiming for cybersex as well as in situations in which they engage in cybersex with an illusionary feeling of safety (Döring, 2000). Most of the time, it was the young girls who complained about being forced by boys into cybersex. In a couple of cases, some young boys have also been reporting that they have been harassed by other boys in gay cybersex (Refer to the below given quotation, for instance).

"I don't like chatting with gays, but they send messages that they want to go out with me...they say that they like my voice ...they do something when I talk with them" (Boy, 14 years old)

In several cases, young people from Mauritius have described their cybersex experience as being disgusting, confusing, traumatic, and something that they feel ashamed of, guilt for, and fear about. In the majority of cases, it was the girls who found their cybersex disgusting. Their experiences have been termed disgusting and shameful mainly because they have been either exposed to some pornographic materials or have been asked to do certain things that they are not prepared for and cannot think of doing. A small number of young people also feel ashamed of and guilt for using their study times 'unnecessarily' on activities like cybersex. A very small number of young people also report the fear of being found out by others about their clandestine activities. Those who have been having cybersex with complete anonymity also report about having the fear of their identity being discovered or in some cases disclosed to others. For a very small number of young people, the first experience of cybersex has been a fearful one, as they felt vulnerable and uncertain about the consequences. A small number of young people report their cybersex experience as being confusing, as they do not know whether they have been doing the right thing or not. Some of the young people also described their cybersex experience as something sinful, things they should not have done. In general, it could be said that there are some young people in Mauritius (maybe about quarter from this study) who feel confused about sexuality as they are just exposed to it without being prepared for how to deal with it. Some of the quotations gathered from those young people who have been reporting about their cybersex as a bad experience are mentioned below.

"It was disgusting and horrifying the things I was asked to do". (Girl, 22 years old)

"I feel really ashamed of what I did...I had the fear for a week whether someone could discover who I am..." (Boy, 24 years old)

"I did it because it just happened...what a shame I felt...I still worry about whether I did the right thing" (Girl, 21 years old)

SOLUTIONS AND RECOMMENDATIONS

The Internet is more or less integrated with everyday life in which sexuality is a part for most people (Daneback et al, 2006; Mustanski, 2001). Inevitably, young people from diverse socio-cultural backgrounds are exposed to a variety of perspectives on sexuality from the globally interconnected net-world in which they have virtually more power to choose sex information, communication and education than what they have in their actual world context. The Internet is facilitating exposure to sex and sexuality issues beyond the 'controlled' socio-cultural boundaries. Within this context, young people in Mauritius need support and guidance to grow sexually in a healthy manner. More attention should be paid to supporting the broad population. Given that the young people are the defining users of the Internet, they should be prepared for coping positively with Internet through new forms and methods of formal sexuality education (Döring, 2009). In the absence of formal sex education in schools and lack of communication with responsible adults on sexuality, Mauritian young people can be confused and might have difficulties in dealing with sexuality issues that are dominantly present on the Internet in various forms.

In particular, one of the major failures of the absence of a formal sex education programme is that young people do not learn about sexuality as a social part of human life. Through the informal sex education in schools what young people in Mauritius are learning is 'Biology' and/or 'Moral' driven discourses related to sexuality. What young people are exposed to on the Internet is completely different and therefore they need to be prepared and supported to face it. In fact, the moral approach is problematic where sex is still considered very much taboo and communities are often in denial about youth sexual activities (Francis, in Press). It is therefore important for Mauritian policy makers, religious leaders and parents to realise that as part of human nature it is normal for young people to have sexual drives and sexual activities form part of human growth and development. It is argued that a "just say no to sex' strategy does not work with young people (American Psychological Association, 2002).

One of the possible solutions to deal with sexuality issues in Mauritius would be to take a sexual rights-based approach for promoting healthy growth and development of young people. Sexual rights are grounded within existing international human rights treaties, covenants and conventions (Maticka-Tyndale & Smiley, 2008). The sexual rights-based approach is based on the principle that young people have the right to grow safely into adulthood, and to gradually become sexual human beings according to their level of maturity and therefore formal IEC for informed decision making and their protection from sexual coercion, incest, rape and trafficking should be provided as part of their basic rights (Hendriks, 1995; Sundby, 2006). Young people's need for sex education is evidenced by their involvement and interest in sexual activity and the often involuntary context within which they feel abused and confused by their inadequate levels of knowledge of how to protect themselves and others (Singh et al, 2005). Within such a context, young people have rights to IEC on sex for protection and for making informed decisions. As McDonald (2004:1) puts it:

"When sexual rights and reproductive rights are placed squarely in a human rights framework,

it is impossible for governments to deny their responsibility to provide comprehensive sexual and reproductive health information, education and services".

FUTURE RESEARCH DIRECTIONS

Gross et al. (2002) argue that social research is needed to further examine the relationship between rapidly evolving uses of the technology and the sexual development of young people. However, it is important that researchers do not take a narrow view of young people's sexuality. Most studies on adolescent sexuality, especially from developing country contexts, have focused on problem behaviour, abuse, outcomes and risks, such as rape, assault, pregnancy and STIs (Sexually Transmitted Infections), resulting in a narrow and one-dimensional view regarding the sexual lives of young people (O'Sullivan, 2006; Ussher and Mooney-Somers, 2000; Tolman and Szalacha, 1999). Sex is pleasurable, enjoyable and full of emotion for young people. According to some experts, future research therefore needs to explore young people's sexuality in the era of technology from a broader perspective. For instance, further studies should also include the love, emotion, desire, pleasure, and eroticism aspects of sexuality that are essential parts of young people's sexual development (Stevenson, 2002).

In particular, empirical evidence on young people's sexuality from low/middle income countries is under-represented in academic discourses (Rambaree, 2007). Parker (2004) opines that in an increasingly globalised era, most research activities have continued to be conducted in economically developed countries. Scientific publications concerning sexual Internet use rarely address inter- and cross-cultural issues and do not cover sexually conservative societies, for example, Islamic countries (Döring, 2009). As a result, the complexity of issues related to young peoples' sexuality has remained largely distorted and under-explored (Parker, 2004). It is therefore important to encourage study on young people's sexuality from a variety of different socio-economic and cultural contexts.

Particularly, both inductive and deductive studies from a variety of geographical, social, economic and cultural contexts should be encouraged. For instance, Döring (2009) opines that quantitative data representative of national populations at large is lacking for numerous aspects of Internet sexuality and there is also a marked deficit of qualitative studies concerning emotional, cognitive and relational dimensions of online sexual activities. Such studies are vital in theorising young people's sexuality within the Internet era. Contemporary youth is much different from the 1900s youth, and therefore new theories on young people's sexuality have become essential. Time has come to rethink theories on new forms of sexuality from the perspective of a social difference paradigm, which would consider the contributions of culture and society to sexual development; as well as the expanding role played by social meaning systems and social practices in defining features of sexual cultures including the 'Net Culture' (Herdt, 1999; Parker, 2004; Rambaree, 2008).

Daneback et al (2005) opine that cybersex is a growing phenomenon with a significant impact on participants but very little research has been done on this subject to date. According to Döring (2009), the implications of Web 2.0 on Internet sexuality, especially in the fields of online pornography and sexual education are under-researched. Such a study can therefore bring a better understanding on where, how and why young people use the Internet for cybersex. This chapter presents an exploration of cybersex among young people from a sexually conservative country context; however several layers of such empirical evidence are needed for social scientists to re-theorise young people's sexuality, especially with the emergence of the 'Net Culture'. In fact, within the field of sexuality there are not enough theories to structure debate and provide basis of proving and disapproving ideas and hypotheses (Baumeister & Tice, 2001). At the same time that theories are being constructed, it is also important that social scientists look critically at the emerging theories. Emerging theories should be challenged with new empirical evidence, thereby allowing for further theoretical solidification and development.

CONCLUSION

This chapter brings understanding on the process, aspects and experiences of cybersex within a transitional country which is trying to become a cyber-island and at the same time still wants to preserve its conservative ways of dealing with young people's sexuality. In particular, the 'Net Culture' is in some ways altering the 'Youth Culture' regarding sexuality in Mauritius. Young Mauritians are exposed to sex on the Internet and therefore they need to be supported and guided so that they can grow and develop healthily. Within this context, adopting a sexual rights-based approach could help those engaged in policy making and politicians who want to make formal sex education a reality in Mauritius.

REFERENCES

AfDB/OECD (African Development Bank/ Organisation for Economic Co-operation and Development). (2008) '*Mauritius*'. Retrieved December 10 2008 from www.oecd.org/dataoecd/13/7/40578285.pdf

Ajaheb-Jahangeer, S., & Jahangeer, A. C. (2004). School Culture in a Private Secondary Institution in Mauritius. *International Education Journal*, *5*(2), 247–254.

American Psychological Association. (2002). *Developing Adolescents: A Reference for Professionals*. Washington, DC: American Psychological Association.

Attwood, F. (2009). Deepthroatfucker' & `Discerning Adonis': Men & cybersex. *International Journal of Cultural Studies*, *12*(3), 279–294. doi:10.1177/1367877908101573

Aumeerally, N. L. (2007). The Ambivalence of Postcolonial Mauritius. *International Journal of Cultural Policy*, *11*(3), 307–323. doi:10.1080/10286630420003l2543

Baumeister, R. F., & Tice, D. M. (2001). *The Social Dimension of Sex*. Boston: Allyn & Bacon.

Bay-Cheng, L. Y. (2001). SexEd.com: Values and norms in Web-based sexuality education. *Journal of Sex Research*, *38*(3), 241–251. doi:10.1080/00224490109552093

Beggan, J. K., & Allison, S. T. (2003). Reflexivity in the pornographic films of Candida Royalle. *Sexualities*, *6*(3-4), 301–324. doi:10.1177/136346070363003

Ben-Ze'ev, A. (2004). *Love Online*. Cambridge, UK: Cambridge University Press. doi:10.1017/CBO9780511489785

Boies, S. C., Knudson, G., & Young, J. (2004). The Internet, Sex, and Youths: Implications for Sexual Development. *Sexual Addiction & Compulsivity*, *11*, 343–363. doi:10.1080/10720160490902630

Bristow, J. (1997). *Sexuality*. London: Routledge.

Buchmann, M. (2004) Sociology of Youth Culture. *International Encyclopedia of the Social & Behavioral Sciences*, 1: *16660-16664*

Bunwaree, S. (1999). Gender inequality: the Mauritian experience. In Heward, C., & Bunwaree, S. (Eds.), *Gender, Education and Development: Beyond Access to Empowerment*. London: Zed Books.

Cameron, K. A., Salazar, L. F., Bernhardt, J. M., Burgess-Whitman, N., Wingood, G. M., & DiClemente, R. J. (2005). Adolescents' experience with sex on the web: results from online focus groups. *Journal of Adolescence*, *28*(4), 535–540. doi:10.1016/j.adolescence.2004.10.006

Carvalheira, A., & Gomes, F. A. (2003). Cybersex in portuguese chatrooms: A study of sexual behaviors related to online sex. *Journal of Sex & Marital Therapy*, *29*(5), 345–360. doi:10.1080/00926230390224729

Cline-Cole, R., & Powell, M. (2004). ICTs, 'Virtual Colonisation' and Political Economy. *Review of African Political Economy*, *31*(99), 5–9. doi:10.1080/0305624042000258388

Coleman, J. C., & Hendry, L. B. (1999) *The Nature of Adolescence, (3rd Ed)*. London: Routledge.

Cooper, A., & Griffin-Shelley, E. (2002). The internet: The next sexual revolution. In Cooper, A. (Ed.), *Sex & the internet: A guidebook for clinicians*. New York: Brunner-Routledge.

Cooper, A., Mcloughlin, I. P., & Campbell, K. M. (2000). 'Sexuality in Cyberspace: Update for the 21st Century'. *Cyberpsychology & Behavior*, *3*(4), 521–536. doi:10.1089/109493100420142

Cooper, A., Scherer, C. R., Boies, S. C., & Gordon, B. L. (1999). Sexuality on the internet: rom sexual exploration to pathological expression. *Professional Psychology, Research and Practice*, *30*(2), 154–164. doi:10.1037/0735-7028.30.2.154

CSO (Central Statistics Office). (2000). *Population Census 2000*. Retrieved December 23, 2009, from www.gov.mu/portal/sites/ncb/cso/report/.../census5/index.htm

Daneback, K., Cooper, A., & Månsson, S.-A. (2005). An internet study of cybersex participants. *Archives of Sexual Behavior*, *34*(3), 321–328. doi:10.1007/s10508-005-3120-z

Daneback, K., Ross, M. W., & Månsson, S.-A. (2006). Characteristics and behaviors of sexual compulsives who use the internet for sexual purposes. *Sexual Addiction & Compulsivity*, *13*(1), 53–67. doi:10.1080/10720160500529276

Döring, M. D. (2009). The Internet's impact on sexuality: A critical review of 15 years of research. *Computers in Human Behavior*, *25*, 1089–1101. doi:10.1016/j.chb.2009.04.003

Döring, N. (2000). Feminist Views of Cybersex: Victimisation, Liberation, & Empowerment. *CyberPsychology & Behaviour*, *3*(5), 863–884. doi:10.1089/10949310050191845

Etienne, P. (2007). *'Sex Education is Becoming Urgent for Self-Protection'*. Retrieved May 05, 2007 from www.lexpress.mu/display_article_sup.php?news

Fisher, W. A. (2001). *'Internet Pornography: A Social Psychological Perspective on Internet Sexuality'*. Retrieved December 12 2009 from http://www.findarticles.com

Francis, DA. (in press). Sexuality Education in South Africa: Three Essential Questions. *International Journal of Educational Development*, in Press, Corrected Proof, Available online 31 December 2009

Gross, E. F., Juvonen, J., & Gable, S. L. (2002). Internet use and well-being in adolescence'. *The Journal of Social Issues*, *58*(11), 75–90. doi:10.1111/1540-4560.00249

Gupta, G. R., & Weiss, E. (2005). Women's lives and sex: Implications for AIDS prevention. *Culture, Medicine and Psychiatry*, *17*(4), 399–412. doi:10.1007/BF01379307

Hall, M. (1999). Virtual Colonization. *Journal of Material Culture*, *4*(1), 39–55. doi:10.1177/135918359900400103

Hendriks, A.The right to health promotion and protection of women's rights to sexual & reproductive health under international law. The economic covenant and the women's convention. *The American University Law Review*, *4*, 1123–1134.

Herdt, G. (1999). Clinical ethnography and sexual study. *Annual Review of Sex Research*, *10*, 100–119.

Hilbert, P. (2007). '*Academic Success with Special Attention to Moral Values'*, Retrieved May 05, 2007 from www.lexpress.mu/display_article_sup.php?news

Hillcoat-Nallétamby, S., & Dharmalingam, A. (2005). *The influence of historical cultural identity in shaping contemporary reproductive behaviour in Mauritius*. Paper presented at the International Union for the Scientific Study of Population, XXV International Population Conference, Tours, France, July 18-23, 2005.

Hillcoat-Nalletamby, S., & Ragobur, S. (2005). 'The Need for Information on Family Planning Among Young, Unmarried Women in Mauritius'. *Journal of Social Development in Africa*, *20*(2), 39–63.

Hunter, K., Hari, S., Egbu, C., & Kelly, J. (2005). 'Grounded Theory: Its Diversification & Application Through two Examples from Research Studies on Knowledge & Value Management'. *The Electronic Journal of Business Research Methodology*, *3*(1), 57–68.

Kuttler, A. F., & La Greca, A. M. (2004). 'Linkages among adolescent girls' romantic relationships, best friendships, and peer networks'. *Journal of Adolescence*, *27*, 395–414. doi:10.1016/j.adolescence.2004.05.002

L'Express. (2006a). '*More schoolgirl porn "shows" under scrutiny*.' Retrieved March 03 2006 from http://www.lexpress.mu

L'Express. (2006b). '*Opinions De Jeunes: Réaction au "Show": "Nous ne sommes pas des enfants modele'*. Retrieved March 03 2006 from http://www.lexpress.mu

Longmore, M. A. (1998). Symbolic Interactionism and the Study of Sexuality - The Use of Theory in Research and Scholarship on Sexuality. *Journal of Sex Research*, *35*(1), 44–58. doi:10.1080/00224499809551916

Lybarra, M., Leaf, P. J., & Diener-West, M. (2004). Sex Differences in Youth-Reported Depressive Symptomatology & Unwanted Internet Sexual Solicitation. *Journal of Medical Internet Research*, *6*(1), 10–22.

Maticka-Tyndale, E., & Smylie, L. (2008). Sexual Rights: Striking a Balance. *International Journal of Sexual Health*, *20*(1&2), 7–24. doi:10.1080/19317610802156996

McDonald, K. (2004). At the UN: Mainstreaming Sexual Rights. *Sexual Reproductive Health & Rights*, *1*(1), 1–4.

MFPA (Mauritius Family Planning Association). (1993). *Research Report on Young Women, Work & AIDS-related Risk Behaviour in Mauritius*. Port Louis: Mauritius Family Planning Association.

MFPA (Mauritius Family Planning Association). (2005) *'Mauritius Reproductive health Education for Women Workers'.* Retrieved December 21, 2009 from http://www.icomp.org.my/South-South/S-S-Catalogue.htm

MIH (Mauritius Institute of Health). (1996). *Research Report on National Survey on Youth Profile*. Pamplemousses, Mauritis: MIH

Mitchell, K. J., Finkelhor, D., & Wolak, J. (2001). Risk Factors for & Impact of Online Sexual Solicitation of Youth. *Journal of the American Medical Association*, *285*, 3011–3014. doi:10.1001/jama.285.23.3011

Morgan, R. (1980). Theory and Practice: Pornography and Rape. In Lederer, L. (Ed.), *Take Back the Night: Women on Pornography* (pp. 134–140). New York: William Morrow.

Mustanski, B. S. (2001). Getting wired: Exploiting the Internet for the collection of valid sexuality data. *Journal of Sex Research*, *38*, 292–302. doi:10.1080/00224490109552100

MYS (Ministry of Youth & Sports). (2009). National [Port Louis: Government of Mauritius]. *Youth & Policy*, 2010–2014.

Nishimura, Y. H., Ono-Kihara, M., Mohith, J. C., Ngmansun, R., Homma, T., & Diclemente, R. J. (2007). Sexual behaviors and their correlates among young people in Mauritius: a cross-sectional study. *BMC International Health and Human Rights*. Published online 2007 October 5. .doi:10.1186/1472-698X-7-8

O'Connell, R., Price, J., & Barrow, C. (2004) '*Cyber Staking, Abusive Cyber Sex, & Online Grooming: A Programme of Education for Teenagers'*. Retrieved November 15 2009 http://www.uclan.ac.uk/host/cru/docs/NewCyberStalking.pdf

O'Sullivan, L. F. (2006). The Sexual Lives of Early Adolescent Girls. *Journal of Sex Research*, *43*(1), 6–2.

Parker, R. (2004). Introduction to Sexuality and Social Change: Toward an Integration of Sexuality Research, Advocacy, and Social Policy in the Twenty-First Century. *Sexuality Research & Social Policy: Journal of NSRC*, *1*(1), 7–14. doi:10.1525/srsp.2004.1.1.7

Peter, J., & Valkenburg, P. M. (2006). Adolescents' Exposure to Sexually Explicit Material on the Internet'. *Communication Research*, *33*(2), 178–204. doi:10.1177/0093650205285369

Rambaree, K. (2005) `*The Ecology of Sexuality in a Mauritian Internet Chat Room (MICR): An Internet Mediated Research (IMR)*. Paper presented for the IFRD Conference in Mauritius, January 2005

Rambaree, K. (2007).*The Ecology of Internet & Early Adolescent Sexuality in a Technology-driven Mauritian Society.* PhD Thesis submitted to The University of Manchester, UK

Rambaree, K. (2008). Internet-Mediated Dating/Romance of Mauritian Early Adolescents: A Grounded Theory Analysis. *International Journal of Emerging Technologies & Society*, *6*(1), 34–59.

Rambaree, K. (2009). 'Internet, Sexuality & Development: Putting Early Adolescents First'. *The International Journal of Environmental, Cultural. Economic & Social Sustainability*, *5*(2), 105–119.

Rambaree, K. (2010) "Children and the Janus-faced Internet: Social Policy Implications for Mauritius as a Developing Country Case Study'. In I, Berson & M. Berson (Eds) *High-Tech Tots: Childhood in a Digital World*. Greenwich, CT: Information Age Publishing

Rice, F. P. (2001). *Human development*. Upper Saddle River, NJ: Prentice Hall.

Riva, G. (2005). The Psychology of Ambient Intelligence: Activity, Situation and Presence. In Riva, G., Vatalaro, F., Davide, F., & Alcañiz, M. (Eds.), *Ambient Intelligence* (pp. 17–33). Amsterdam: ISO Press.

Rival, L., Slater, D., & Miller, D. (1998). Sex & Sociality: Comparative Ethnographies of Sexual Objectification. *Theory, Culture & Society*, *15*(3-4), 295–321. doi:10.1177/0263276498015003015

Ross, MW, Rosser, BRS, & Stanton, J (2004). Beliefs about cybersex & Internet-mediated sex of Latino men who have Internet sex with men: relationships with sexual practices in cybersex & in real life. *AIDS CARE*, 16 (8), 1002_1011

Runeborg, A. (2004). *Sexuality – a super force: Young People, Sexuality & Rights in the era of HIV/AIDS*. Stockholm: Sida, Department for Democracy and Social Development, Health Division.

Russell, D. (1998). *Dangerous Relationships: Pornography, misogyny, and rape*. Thousand Oaks, CA: Sage.

Schensul, S., Oodit, G., Schensul, J., Seebuluk, S. U., Bhowan, J., & Aukhojee, P. (1994). *Young Women, Work, & AIDS-Related Risk Behavior in Mauritius*. Washington, DC: International Center for Research on Women.

Shillington, K. (1991). *Jugnauth: The Prime Minister of Mauritius*. London: McMillan.

Singh, S., Bankole, A., & Woog, V. (2005). Evaluating the need for sex education in developing countries: sexual behaviour, knowledge of preventing sexually transmitted infections/HIV and unplanned pregnancy. *Sex Education, 5*(4), 307–331. doi:10.1080/14681810500278089

Smahel, D., & Subrahmanyam, K. (2007). Any Girls Want to Chat Press 911: Partner Selection in Monitored and Unmonitored Teen Chat Rooms. *Cyberpsychology & Behavior, 10*(3), 346–353. doi:10.1089/cpb.2006.9945

Smith, C. (2009). Pleasure & Distance: Exploring Sexual Cultures in the Classroom. *Sexualities, 12*(5), 568–585. doi:10.1177/1363460709340368

Smith, D. S. (2000). *Moral Geographies: Ethics in a World of Difference*. Edinburgh, UK: Edinburgh University Press.

Stanley, L. (1999). The Cultural Bias of Sex Surveys. In Nye, R. A. (Ed.), *Sexuality*. Oxford, UK: Oxford University Press.

Stevenson, M. R. (2002). 'Conceptualizing Diversity in Sexuality Research' in M.W. Wiederman & B.E. Whitley, Jr. (Eds) *H& book for Conducting Research on Human Sexuality*, (pp. 455-479). Mahwah, N.J: Lawrence Erlbaum Associates

Subrahmanyam, K., Greenfield, P. M., & Tynes, B. (2004). Constructing sexuality & identity in an online teen chat room. *Applied Developmental Psychology, 25*, 651–666. doi:10.1016/j.appdev.2004.09.007

Sundby, J. (2006). Young people's sexual and reproductive health rights. *Best Practice & Research. Clinical Obstetrics & Gynaecology, 20*(3), 355–368. doi:10.1016/j.bpobgyn.2005.12.004

Thornburgh, D., & Lin, H. S. (Eds.). (2002). *Youth, pornography, and the Internet*. Washington, DC: National Academic Press.

Tolman, D., & Szalacha, L. (1999). Dimensions of Desire: Bridging Qualitative & Quantitative Methods in a Study of Female Adolescent Sexuality. *Psychology of Women Quarterly, 3*, 7–39. doi:10.1111/j.1471-6402.1999.tb00338.x

Traeen, B., Nilsen, T. S., & Stigum, H. (2006). Use of pornography in traditional media and on the internet in Norway'. *Journal of Sex Research, 43*(3), 245–255. doi:10.1080/00224490609552323

Traeen, B., Spitznogle, K., & Beverfjord, A. (2004). Attitudes and use of pornography in the Norwegian population 2002. *Journal of Sex Research, 41*(2), 193–200. doi:10.1080/00224490409552227

Turkle, S. (1999). Identity in the Age of the Internet. In Mackay, H., & O'Sullivan, T. (Eds.), *The Media Reader: Continuity & Transformation* (pp. 287–305). London: Sage.

Ussher, J. M., & Mooney-Somers, J. (2000). Negotiating Desire and Sexual Subjectivity: Narratives of Young Lesbian Avengers. *Sexualities, 3*(2), 183–200. doi:10.1177/136346000003002005

Walsh, W. A., & Wolak, J. (2005). Nonforcible Internet-related sex crimes with adolescent victims: Prosecution issues and outcomes. *Child Maltreatment, 10*, 260–271. doi:10.1177/1077559505276505

Weinstein, E., & Rosen, E. (1991). The development of adolescent sexual intimacy: Implications for counseling. *Adolescence*, *26*, 331–340.

Whitty, M. T., & Carr, A. N. (2003). Cyberspace as Potential Space: Considering the Web as a Playground to Cyber-Flirt. *Human Relations*, *56*(7), 861–891. doi:10.1177/00187267030567005

Wolak, J., Finkelhor, D, & Mitchell, K.J. (2004). Internet-initiated Sex Crimes against Minors: Implications for Prevention Based on Findings from a National Study. *Journal of Adolescent Health*, 35(5), 424.e11– 424.e20.

Worth, N. (2009). Understanding youth transition as 'Becoming': Identity, time and futurity. *Geoforum*, *40*, 1050–1060. doi:10.1016/j.geoforum.2009.07.007

Yates, A. (1978). *Sex without Shame: Encouraging the Child's Healthy Sexual Development*. New York: William Morrow & Company.

Ybarra, M. L., & Mitchell, K. J. (2005). Exposure to Internet Pornography among Children and Adolescents: A National Survey'. *Cyberpsychology & Behavior*, *8*(5), 473–486. doi:10.1089/cpb.2005.8.473

Zillmann, D., & Bryant, J. (1989). *Pornography: Research Advances and Policy Considerations*. Hillsdale, NJ: Erlbaum.

KEY TERMS AND DEFINITIONS

Sexually Conservative Society: A society that resists change regarding sexuality and tries to promote moral and traditional ways of dealing with sexuality issues.

Young People: Those who are of the age between 14-25 years old.

Sex: More than just the biological organ and penetrative intercourse, includes identity, eroticism, orientation and pleasure etc.

Cyberspace: The resourceful space people can have access to on the internet for undertaking various sorts of activities such as for entertainment, communication, sex etc.

Cybersex: Internet-mediated sexual activities that involve sexual communications, textual and/or through webcam.

Grounded Theory: As an approach it is an inductive way of doing scientific research where the reporting is grounded on the empirical evidence collected from the field. It is also a way of analysing qualitative data.

Sexual Rights: Part and parcel of human rights that deal with sexuality related issues such as access, services, protection and respect of dignity and so on.

Chapter 11
Youth, Sexuality and the Internet:
Young People's Use of the Internet to Learn about Sexuality

Kristian Daneback
University of Gothenburg, Sweden

Cecilia Löfberg
Stockholm University, Sweden

ABSTRACT

Prior research on the use of the internet for sexual purposes has primarily focused on its negative and problematic aspects, such as compulsivity and addiction. Thus, little is known about any possible benefits. The purpose of the current chapter is to focus on how young people aged 12-24 use the internet as a source of knowledge about sexuality. The results rely on qualitative and quantitative data collected in Sweden at various points in time between 2002 and 2009. Young people seek information about various things in relation to sexuality. The primary reason is to gain knowledge about bodily functions and sexual performance. The qualitative data also indicated gender differences in how sexual issues were communicated online. Some young people sought sex information to become sexually aroused while others viewed online pornography to gain knowledge about sexuality, suggesting a possible link between knowledge and sexual arousal. The internet encompasses specific characteristics making it easier to communicate about sexuality in ways sometimes impossible offline. The fact that one can type while being physically distant to others appears to be a particular advantage with using the internet as a source for knowledge about sexuality compared to other ways of communication.

This chapter contains explicit content

INTRODUCTION

The purpose of this chapter is to discuss how the internet can constitute an arena of knowledge and learning about sexuality for young people in today's society. In many western cultures and societies sexuality is surrounded by shame and guilt and often consigned to the private areas of life. Nevertheless, sexuality is one part of the reflexive self which is being negotiated in a never-ending project (Giddens, 1992). Gagnon

DOI: 10.4018/978-1-60960-209-3.ch011

and Simon (1973) claim that the sexual learning process from early age is often non-verbal or negative in the sense that sexual behaviours are either ignored or deemed inappropriate. Sexuality or knowledge about sexual intercourse, for example, is not included in the cultural norms embedded in modern discourses of childhood and therefore neglected (Renold, 2005). According to Gagnon and Simon (1973), most of our sexual knowledge is gained from peers when we are young. As we grow older, we make the association between sexuality and guilt which makes us keeping sexual matters secret from, for example, parents and friends.

The consequence is that sexual arousal and practices are restricted to the individual and kept as secrets, away from others sometimes including one's partner(s). At the same time, and as Foucault (1976) observed some decades ago, sexuality is discussed more openly today than ever before; it is constantly present in public settings in various media and defined as the individual's responsibility. This reveals two parallel tendencies: privatization by individualization on the one hand and collectivization by increasing public exposure on the other. One can assume that this becomes a complex area to handle for the individual in general and for young people in particular since sexuality is a topic in progress for this group. Young people and the complexity of the topic combined with the restrictions for social performances in offline environments and the alternative conditions for interaction that the internet provides make an interesting point of departure for further investigation about how this topic is handled by young people in online environments. This point of departure becomes even more interesting when considering the technological shift from the early days when the World Wide Web was a means to communicate contents to an almost endless number of users to Web 2.0 platforms where the users create the content themselves in a more interactive version of the web.

Relatively few studies focus on how young people seek information or talk about love and sexuality online and when this is the case they primarily focus on the negative or problematic aspects associated with this usage. Concerns arise in adult society when young people spend time interacting or gathering information about sexuality. One example of this is a study of youth magazines in the UK (Tincknell et al., 2003). This study shows that there is a "societal concern" arising about some magazines that provide too much sexual information to young girls. One can easily come to the conclusion that young people's use of the internet for developing sexual skills is a burning issue and therefore strongly related to negative aspects. So far users have been seen as passive victims of problematic online content (Döring, 2009). This may be particularly true when referring to the younger segment of users. However, research suggests that most young people manage to navigate in the sexual landscape while being both reflective and critical (Månsson & Löfgren-Mårtenson, 2007). Thus, young users should be perceived as active internet users who decide what content they want to find and what content they want to avoid online. Furthermore, young users may be perceived to encompass a competence which allows them to evaluate and reflect over the sexually related content they find or avoid on the internet.

Although the number of published articles about love and sexuality on the internet has grown exponentially since 1993, a closer look reveals a fragmented field of research. A comprehensive literature review of the scientific journal papers published between 1993 and 2010 concludes that a lions share of the studies have focused on the negative – or problematic – consequences of the use of the internet for sexually related activities (Döring, 2009). Another literature review suggests that it is possible to discern three problematic areas that have caught the researchers' attention since the mid 1990s. One is the medical, or epidemiological, area where the major focus has been on the internet in relation to HIV and sexually transmitted infections (primarily in the subgroup of men who have sex with men). Another area is more psychologically oriented and focuses primarily on internet addiction and compulsivity raising questions whether or not the use of the internet for sexual purposes will lead to addiction or compulsive behaviour? The third area concerns moral aspects such as infidelity and pornography use and the effect on partners and children (Daneback, 2006). Perhaps a legal area could be added including online sexual solicitation, prostitution, and child pornography.

Among the studies conducted so far, only a few focus on how the internet is used to find information and increase knowledge of sexual matters (Döring, 2009). Tentative results from a recent Norwegian study suggest that 9 out of 10 women and practically all men (99%) aged 18-30 to use the internet for sexual purposes. Compared to earlier studies, using the internet to find information about sexual matters, to view pornography and to buy sexual products seem to have increased among young people. A recent Swedish study showed that seeking information about sexual matters on the internet was the most common online activity for young women (18-24 years). For men in the same age group, viewing pornography and flirting with others ranked higher than information seeking, but at the same time they sought information to almost the same extent as women did (Daneback & Månsson, 2009).

Some studies have focused on evaluating the quality of the information about sexuality found on various web sites while others have focused on educational interventions, for example how the internet could be used in sex education and sex therapy. The common denominator is the fact that they are created by experts. Less focus has been put on how internet users create the content themselves on so called Web 2.0 platforms (Döring, 2009). This could perhaps explain the persistent view on the users as passive receivers rather than active users creating the content through interaction.

In this chapter we will discuss how the use of the internet to talk about and seek information about sexual matters can be understood as a source of knowledge for different kinds of sexual questions that young people may have.

BACKGROUND

Young People and Sexuality

Young people start acting out sexual identities in the way they socialize with each other around the age of ten (Renold, 2005). Of course this also includes thoughts about the sexual body and what this body can/should look like and how it can be used. Earlier, these thoughts were often kept to oneself because of the difficulties one had as a child or adolescent to enter the "world of sexuality" controlled by the adult world. Barriers that used to separate childhood from adulthood and reproduce the cultural representation of childhood and children as innocent, pure and natural are now being overcome through media (Buckingham, 2000; Prout, 2008). Today, talking about the sexual body online and the knowledge that comes out of it is a source for information about the sexual body available to most young people in western societies. Discussion fora on the internet on the topic sexuality are used by children and youths to gain knowledge of sex and relationships (Löfberg, 2008). Heterosexuality as a norm and as an everyday practice is a "key theme" when children are making sense of themselves as girls or boys (Renold, 2005). Renold's (2005) ethnographic study of children's (age 10-11) gender and sexual relations in primary school clarifies the importance of the heterosexual institution for becoming a "proper" girl or boy. Being able to attract a person of the opposite sex becomes a signifier on how well one is succeeding with the task of being a proper girl or boy. When it comes to sex and sexuality these are phenomena that are not included in the concept of childhood. Renold states that "sexual innocence then is something that adults wish upon children, not a natural feature of childhood itself" (Renold, 2005, p. 22). Media and societal notions about sexuality influence the cultural notions that develop in young people's creations of meaning on a day-to-day basis. In adolescence the interest in sexuality and the sexual drive increases (Subrahmanyam et al., 2004). Earlier research shows that the primary source for information about sex is peer-communication followed by the media (Gagnon & Simon, 1973; Subrahmanyam et al., 2004). At the same time, for many adolescents sexuality is a controversial topic; a topic where confidentiality is of particular concern (Suzuki & Calzo, 2004).

Young people, the internet, and sexuality brought together in one constitute a topic of highest interest for media and have become a major concern in the adult world. From being "a forbidden area of the adult world" sexuality is now accessible to children and young people through digital media (Buckingham, 2000). How these changed conditions are being used by young people is a question that is crucial in this chapter.

Young People, the Internet and Sexuality

Researchers in the field of virtual communication argue that conditions for communication online offer possibilities for people to explore their identities and to express ideas that are not easily expressed in offline contexts (see for example boyd, 2008; Hine, 2000; Holloway & Valentine, 2003; Jaquemot, 2002; Jones, 1998; Livingstone & Bober, 2005; Stern, 2008). On the internet, young people can create a sphere of intimacy where the things that are being expressed will not reach the local peer group and be a cause for slander (Holloway & Valentine, 2003). Susannah Stern (2008) means that the possibility to present oneself, in the shape of an identity or a self-image, that cannot be presented elsewhere, is embedded in online activity.

Social interaction online implies acts without physical bodies. This can be experienced as a feeling of safety that encourages an intimacy that would not be expressed in face-to-face interaction (Holloway & Valentine, 2003; Mann & Stewart, 2000; Sharf, 1999). Researchers like Holloway and Valentine (2003) mean that social relationships sometimes benefit from the absence of the body and state that:

Bodies can get in the way of social relationships because of the meaning that are read off from them, or the judgments which are made about particular physical characteristics such as age attractiveness and gender (Holloway & Valentine, 2003, p. 133)

Young people use the internet as part of their everyday lives; they do not log on to the internet for a specific reason or to engage in a specific activity, but this is a part of their everyday usage. Daneback (2006) suggests that young people easily and willingly adopt new techniques and arenas whereas older users prefer to stick with techniques and arenas they are used to and that satisfy their needs. Another distinction between the groups concerns with whom they interact online. While young people often interact with people they also know offline, older users interact with people they have first got to know on the internet (however, this may change with the increasing popularity of social networks among adults). Furthermore, young people spend many hours online engaging in several activities simultaneously, for example, using the internet to do their homework while surfing the web and interacting with their peers. The older users, on the other hand, use the internet in their spare time and log on for specific reasons to, for example, visit dating sites (Daneback, 2006). Because the internet generation often knows with whom they interact, they make up identities less often (although slight modifications and "white lies" occur) while older users more often engage in role playing games where they create their identities (Ibid.). How the internet is integrated in the users' everyday lives, thus, influences how they use the internet for love and sexual purposes. This has to be taken into consideration when researching love and sexuality on the internet.

The internet recreates the social context and redefines with whom it is possible to share sexual matters. The sexual material accessible online; information, erotic texts, or images, is found within a secure distance from one's parents, friends, or children and, therefore, the internet constitutes a legitimate source of sexual material. Reinforced by anonymity, sexual matters are experienced as even further away from the peer group education described by Gagnon and Simon (1973). The consequence is that we do not have to keep our sexual secrets to ourselves but at the same time let them remain secrets for the offline surroundings. We can keep the secrets and reveal them simultaneously without experiencing any negative sanctions or consequences. This means that it is possible to engage in sexual activities that would otherwise be restricted by norms and instead change them through exploration and practice as we, according to Gagnon and Simon (Ibid.), did as adolescents before the internet.

Summing up

We began this chapter with discussing sexuality as something private, often kept from parents, friends, and sometimes also from partners. We also discussed how this becomes especially clear in relation to young people. Sexuality in relation to children and young people is a burning issue because of the strong discursive connection to adulthood that the topic has. Even so, young people in today's society gain knowledge about sexuality through digital media. Earlier studies show that it is principally the problematic consequences of internet use in relation to sexual behaviour that, so far, has caught the researchers' interest.

In this chapter we will take another turn and discuss *how the use of the internet to talk about and seek information in these matters could be understood a source of knowledge.* We will explore this by providing empirical examples from four studies conducted in the 2000s. They comprise both qualitative and quantitative data collected in Sweden at various points in time (2002, 2003-2004, and 2009). In the following section focus will be on 12-15 year-olds before we move on to young adults, 18-24 years old.

LOOKING FOR SEXUAL INFORMATION AND ADVICE ONLINE

The data material for this section is collected from a discussion group on a website for young people in 2004[1]. The reason for choosing these particular discussion groups was that they included longer and more elaborative conversations. This allowed for an analysis of the social interaction, and the topics could also be seen as "difficult to talk about" in the everyday life offline. The methodological approach was inspired by ethnographic methodology. To understand the social and cultural norms that were embedded in the communication on the discussion board the discussions were followed on a daily basis for seven months. The stated age of each participant is the age reported upon registration on the website. For a more detailed discussion about the method see Löfberg (2008).

What do young people ask questions about? Below are some examples of the questions that were stated on a discussion forum with the topic sexuality on a social network site designed for young people.

Girl 13 years: *I know I will get a lot of answers like for example "you're too young for this and so on...." But who says that I'm going to use the information now. It's just good to know. Is it harmful with anal sex or not?*

Boy 13 years: *My dick is so small… is that abnormal? It's only 14 centimeters and I'm 13 years….*

Girl 13 years: *How should you do it so it turns out the right way when you shave down there… so it won't blead??? It always does ☹*

Boy 12 years: *Does anyone have a good way to wank??????*

Questions are postulated out of curiosity; they want to gain knowledge about sex. These questions show that online contexts are used by young people to seek knowledge about the sexual body. The questions concern both expected sexual performance (e.g., how to have anal sex, how to masturbate) and the sexual look (e.g., how to shave genitals, size of the sexual organ). The way young people use this forum to gain information and knowledge varies. Below we will discuss this and illustrate it with two empirical examples of how young people use social network sites to gain information and knowledge about sexual matters. The examples have been chosen due to their high representativity regarding the topics chosen for discussion among the younger age groups.

Discussion groups are concurrently a space for trying out personal thoughts and ideas about one's own sexual activity as well as a space for gaining information and knowledge from other participants. Online it is possible to ask questions and to get information without anyone knowing who you are or how you will use the information. This can be specifically important when the topic is sexual information and advice. Let us look at the first example:

Girl 14 years: *The first time one has sex… that is maybe the first time one sees each other's bodies in the nude. Can the guy get "turned off" by seeing the body of the girl?*

Young woman 24 years: *I think that it should be more to it to make that happen. You want to have sex partly because your partner turns you on, and less clothes doesn't change that in the first place*

Girl 14 years: *Okay, because I'm a little bit worried about that because me and my boyfriend has talked about having sex and I'm afraid that this will happen when he sees me, sort of.*

Young woman 24 years: *I wouldn't worry if I were you. (Hopefully) he's with you because he likes you and wants to have sex with you because you turn him on. But wait with sex as long as you feel unsure. You don't need to have sex the first time you see each other naked. Take it slow so you feel safe and comfortable in the situation. You will benefit from that in the long run. Having sex at the same time as you're worrying about your body isn't fun for either of you.*

Girl 14 years: *Nope, you're right… But I feel ready for it and we've been together for six months now… and talked very much about it. And we were actually close to having it once but he backed out because two of our friends were there, not in the same room, but he didn't want to do it because they were there. And that one could understand, sort of, but I don't know what to do..*

Young woman 24 years: *If you don't know what to do it's a clear sign that you should wait. If it's right you will feel it, and then it will just happen. Be sure that you have condoms at home when it happens so you can protect yourselves.*

Girl 14 years: *Mm but it's so strange cause I want to!! Every time I'm with him I hope for it to happen!!! But then I get that feeling that he will* *think of my body as ugly or something so he won't like to do it or something...*

Young woman 24 years: *I know it's easy to say, but try to let go of those thoughts. Your boyfriend is with you, thus you ought to be attractive in his eyes. As I said try to be naked with each other without having sex, soon you will notice if your thoughts are confirmed or not. My personal guess is that you will have nothing to worry about.*

Girl 14 years: *Okay, I get it..Mm feels good talking to you...you understand ...!:P hehe*

I will try that because I want to do it with him!

Young man 24 years: *I just have to say that I agree with (Young woman, 24 years) completely in everything she says. If your boyfriend is with you it's cause he likes you. And if you have your clothes on or not hardly makes any difference. And most of it he knows already, I mean you can see rather much of what a body looks like by watching it with clothes on? Breasts for example, you can see rather precisely what size they are without taking the clothes of, right? So I guess he likes them already? And if you're worried about how you look down there I can assure you, as a guy, that most guys find all pussies nice and lovely, regardless of size and shape of the lips and so on. The same goes for the breasts. Try to be proud of what you look like and try to enjoy instead of having thoughts about what he will think of your body. If you're satisfied with your body you will relax more easily and it will be much easier for you to get real wet and besides, a girl who you clearly can see enjoys her body is MUCH more sexy to look at than one who's trying to hide it.*

As we can see in the example above the young girl looks for knowledge about her sexual debut, her first sexual intercourse. In seeking this information she reveals her anxiety about the grade of attractiveness of her body. This can be understood as feelings and thoughts that are not so easy to reveal in other settings. Due to the anonymity provided by the internet, the girl can elaborate her private thoughts about sex in a collective context designed for questions about sexual matters without including her partner or other members of her local environment. The uncertainty mirrored in the girl's questions also shows how different aspects of sexuality, such as shame, guilt and desire, co-exist in the individual. Ambivalent feelings can be a hard task to handle, online this ambivalence can be (and is) expressed in an outspoken way.

On this discussion board, girls who initiate communication about sexuality are answered by older participants most of the time. Moreover, the older participants also put the girls in specific positions (Löfberg, 2008). This means that the information they get introduce them to positions created by adults about adult sexual activity. This can be understood as empowering where a girl's curiosity about her sexual acts and how to perform sexually can be expressed and answered to respectfully by adults in a way that probably could not take place in settings where participants confront each other face-to-face. On the other hand the boys' questions were answered more equally by both peers and adults. Interestingly, the mediation of information can be done beyond the regular limits of social norms and structures that, in offline settings, make sexuality a secret.

In the next example a 14-year-old boy postulates a question about sexual practice. Like the girl in the example above he expresses some kind of uncertainty when he talks about the sexual activity he is interested in. But compared to the girl's question this question is formulated from a different angle and the questions are also answered differently. While the notion the girl expresses and is confronted with can be seen as a mix between insecurity and curiosity, the boy below states a more concrete and upfront question.

Boy A 14 years: *Can anyone tell me how to lick my girlfriend in a good way because my girlfriend doesn't seem to get turned on so I feel as a failure ....anyone who can help me with this? /Please*

Boy B 14 years: *Use your tongue over clitoris! that usually helps*

Girl 16 years: *Not right away!! I don't like that anyway... it is nicer if he goes softly and get more aggressive after a while...If you expose clitoris and start licking frenetically right away it could turn out as a very unpleasant feeling due to my experince....*

Boy B 14 years: *Okay due to my experience it usually works fine*

Girl 16 years: *Yeah it is nice... but not at once... you have to be really turned on first...due to my experience...*

Boy B 14 years: *Yeahyeah...*

Boy 15 years: *It depends on what she likes... but I mean... if you want to succeed at once... see to it that she is positioned like you should go doggy style and then you lick her that way... from behind that is... I mean then you can even fuck her with your tongue much easier than in the regular position... anyway it is a good position to succeed... and see to it that you're taking it easy with the tongue... be real careful and tease her ;0) It almost always works...*

The 14-year-old boy who initiates the discussion asks a question about a specific sexual technique and the participants in the group engaged in this conversation are mostly his own age. Here it is peer-communication going on where rather detailed advice is given. In this way, discussion fora can be seen as contexts where children and young people have access to social knowledge and spaces where to reflect upon ideas, norms and values about sexuality and gender identity created both in their own peer culture and in adult society. In this example the knowledge produced revolves around concrete techniques that can be used.

As we can see in these examples different ways of expressing and discussing sexual matters appear. For example, for the girl asking about the first time a couple is to undress in front of each other the internet can be seen a space where doubts and anxieties in relation to sexual activity can be expressed and reflected upon. Support and care can be mediated from other participants. Most of the support is provided by participants from an older age group. Recalling the example where the boy asks the question about oral sex, we can see how the internet gives space for more practical, technically oriented advice, where participants of similar age elaborate together upon sexual practices and their benefits and losses. These different knowledge pools are in some extent gendered where girls are more disposed to discuss their anxieties and search for support and boys to a greater extent turn to communication for practical matters (Löfberg, 2008).

The girl in the example above clearly states her ambiguity to a sexual situation and asks for help. The boy wants to know how he should act sexually with his girlfriend to please her. The question is: could this be done in a public offline setting? Probably not. Online it is possible to express feelings and notions in public settings to a greater extent than in public settings offline. This means that the knowledge one gains – by support, advice or alternative actions - is based on the possibility to ask questions that are rarely asked in other public settings. Moreover, what is learned emanates from participants of different age and of both genders who have a mutual interest in discussing sexual issues. It is also possible for all participants to contact each other at any time.

Young people consider sexual activity with partners as an option to create meaning, a possibility to develop towards adulthood and to continue their constructions of gender (Löfberg, 2008; Renold, 2005; Tincknell et al., 2003). The meaning they identify in sexual activity must be weighed against other, already established, notions about sexuality in relation to gender construction. Their ways of trying to incorporate sexuality in their lives come across difficulties due to the cultural tension sexuality as a topic in itself holds, but also in the relation it carries in intersectionality with the social group "young people". For young people the internet is an arena for creating and exploring cultural meanings that society holds about sexuality and to express them on an arena where privacy can be obtained in a public room. Young people communicate ideas, norms and values about sexuality and how to perform and handle this phenomenon that exists in society. What is suggested, too, is that the ways of seeking information and knowledge about sexual activity differs to some extent between girls and boys in younger age groups (12-15 years in this case). Girls are sometimes more prone to discuss their personal shortcomings or anxieties while boys are more interested in practical matters.

FROM ADOLESCENCE TO ADULTHOOD

Using the internet to look for information about sexual matters is a significant activity for young people who are leaving their adolescent years and are about to enter adulthood as young adults. The following section is primarily based on three data materials, collected in 2002, 2003-2004, and 2009 respectively and comprises interview data and survey data (For more detailed descriptions of the methodological procedures and characteristics of data see Cooper, Månsson, Daneback, Tikkanen, & Ross, 2003; Daneback, 2006; Daneback & Månsson, 2009).

In a survey study conducted in 2002, 22 percent men and 35 percent women aged 18-24 claimed to use the internet to look for information about sexuality and to seek support in sexual issues. When asked if the internet had influenced their knowledge about sexuality, 6 out of 10 reported an increase in knowledge while 4 out of 10 reported no change (Månsson, Daneback, Tikkanen, & Löfgren Mårtenson, 2003). Notably, almost no one reported the internet to have had a negative impact on their knowledge about sexuality, their sexuality, or their sexual relationship (those who indicated that they had partners). However, this and other studies of the same scale conducted around the millennium provided little qualitative insight in what kind of sexual information and sexual knowledge was sought and its consequences for the users.

A qualitative interview study, conducted in 2003 and 2004, showed that seeking information and knowledge on the internet could refer to a variety of topics relating to sexuality. The following is an excerpt from an interview with an 18 year old male informant:

Young man 18 years: *Regarding sex information I don't remember the name of the site...I was introduced to this site and from there I clicked on links to find what I was looking for...*

Interviewer: *Ok, do you remember what kind of information you were looking for?*

Young man 18 years: *Guys' development into manhood and sexual orientation.*

***Interviewer:** Did you get better information on the web compared to from school, friends, or parents?*

***Young man 18 years:** I got better info about guys' development at the youth clinic and concerning sexual orientation I got better info through homosexual friends.*

***Interviewer:** What's your opinion about the [sex] information on the web in general?*

***Young man 18 years:** It's mostly about the general thing... everyone is developing differently and so on... you can't find the precise answers you are looking for... and concerning sexual orientation they said that it wasn't abnormal and that there are many organizations that support this issue... I also found many "coming out of the closet"-stories... generally I think the info was good for those who want to know some things, but most things you can't find... so it's probably best to search IRL.*

As is shown, this informant was looking for developmental information relating to biological questions, but also about sexual orientation, questions about sexual identity. However, it is quite clear from the example that the internet is not the sole way of searching for information. This informant combined searching the internet with visits to the youth clinic to find information about homosexuality. He stresses that there are many "coming out stories" and information that homosexuality is not abnormal; that is information that may help and empower those who have not come out yet. This is supported by prior research claiming that the internet may help homo- and bisexuals in their "coming out" process, either by talking to other homosexuals or by creating open homo- or bisexual identities online (Ross & Kauth, 2002). Interestingly, in the excerpt above, the informant says that he obtained the best information offline; that there is adequate information about certain things to be found online, but a lot of information one looks for is lacking. The question is if this could be related to the time the interview was conducted or to this informant in particular. As we stated in the beginning of this chapter, research about online sexual information is largely absent and maybe this in turn is because information about sexual matters has been lacking too. It is also known that many young people, especially in rural communities, neither have access to non-heterosexual networks nor the means to access them (Ibid.).

Besides looking for information about sexual development and sexual orientation, young people are also interested in sexual techniques and sexual performance. As we suggested above for the young people 12-15 years, how these questions are postulated on the internet can vary in relation to the gender position of the one who asks the question. Below is an excerpt from an interview with an 18 year old female informant who says that it is easier for her to talk about sexuality when she is online:

***Interviewer:** Can you give an example of what you talk about regarding sex and sexuality?*

***Young woman 18 years:** You talk about what you like sexually and what kind of experiences you have and so on.*

***Interviewer:** Do you talk about this with guys or girls on the web chat?*

***Young woman 18 years:** Only guys.*

***Interviewer:** And you talk about it with girls in real life?*

Young woman 18 years: *Yes, exactly.*

Interviewer: *Is it different talking about it with a guy rather than a girl?*

Young woman 18 years: *Yes, guys are more curious…and when on the internet you don't really know who the other one is and that makes it less hard to tell or ask.*

Interviewer: *Which is an advantage?*

Young woman 18 years: *Yes it can be.*

Interviewer: *Are there differences regarding what you talk about too?*

Young woman 18 years: *I don't talk about details with my IRL friends, but that's what guys on the internet want to hear all about.*

Interviewer: *And you're happy to tell them about details?*

Young woman 18 years: *Not really. I usually say that it's private.*

Interviewer: *Ok, can you still feel that it's fun/relieving talking to guys about sexuality?*

Young woman 18 years: *Yes, you get tips about what guys like…and get advice and so on.*

Interviewer: *So you may learn things? Can you give me an example of what you've learned?*

Young woman 18 years: *How guys prefer to have oral sex.*

Interviewer: *How to do it so to speak?*

Young woman 18 years: *I guess you can say that.*

This example shows that it is easier to talk about specific sexual issues on the internet, that it is possible to be more detailed, perhaps even more sexually explicit, in questions and answers than in offline contexts. But it also shows that it is easier to cross gender boundaries. Gagnon and Simon (1973) mean that the sexual education is gendered, that young females' sexual education is separated from young males' and that it differs in content. Here this young woman can ask questions to "guys" about sexual techniques, in this case how to perform oral sex. And this without getting stigmatized due to the breaking of the traditional sexual scripts that regulate female sexuality, branding a "too" sexually interested young woman a "slut" or a "whore" available to any man. In fact, for this young woman, these discussions can be labelled learning experiences that may enhance present or future relationships. Especially if we believe in Gagnon and Simon's notion that heterosexual sexuality is surrounded by silence.

In general, but not always, it seems like sex information and education on professional web sites on the internet primarily consist of socio-biological, formal, explanations of sexuality while the sexual performance, the action so to speak, is taught by peers. In a way, this division mirrors our contemporary sexuality education, where biology is taught by schools and the sexual performance is taught by non-professionals. Gagnon and Simon (1973) mean that sex education occurs until people reach into their 20s. After that, sexuality is discussed to a significantly lesser degree and sometimes even silenced. In addition, Gagnon and Simon mean that sex education differs greatly between genders, where female sexuality – compared to male sexuality - is repressed, related to love and romance, or not talked about at all. Even though female sexuality is discussed in schools nowadays, a recent Swedish study confirms the gender differences observed by Gagnon and Simon; gender differences in today's sex education remain despite the opposite intentions (Bäckman, 2003). However, non-professional knowledge of sexuality is shared in same-sex groups between peers and through practical experience. The findings in this and other studies on internet sexuality show that this may have changed now. With the internet, people of all ages, both men and women, search for information on sexuality and talk about sex with other interested people. Further, the internet has made it possible to talk about sexuality with others in ways that have been impossible before. For young men (18-24), learning how to have sex was the primary reason why they used the internet to seek information and support in sexual issues. In addition, this was also among the top reasons for young females in the same study (Daneback & Månsson, 2009).

Some informants incorporate pornography as a source of knowledge in their relationships, either to become aroused or to find inspiration for expanding their sex life. In the example below, a young man (18 years old) says that by viewing pornography he has become more curious and interested in sex:

Interviewer: *You said you view porn on the internet, why?*

Young man 18 years: *I like sex and it's nice to explore new things.*

Interviewer: *I see, did you learn something from it or have you become interested in something that you haven't known or liked before?*

Young man 18 years: *I guess I have learned something and maybe become more interested in the opposite sex...I've become curious as you say...*

Interviewer: *Of what possibilities sex has to offer you?*

Young man 18 years: *Yes, exactly.*

Interviewer: *Is there something you have "liked" on the internet that you have tried in real life? Something you had never tried before?*

Young man 18 years: *I can't really remember... maybe trying bondage.*

In a study conducted in 2002 it was found that 74 percent men and 24 percent women aged18-24 used the internet to view pornography. In comparison, among those aged 50-65, 72 percent men and 4 percent women claimed to view pornography (Månsson, Daneback, Tikkanen, & Löfgren Mårtenson, 2003). In 2009, 88 percent men and 35 percent women aged 18-24 reported to use the internet to view pornography indicating an increase for this online sexual activity for both females and males (Daneback & Månsson, 2009). Perhaps we should not consider sexual knowledge to revolve only around biology and identity, but also around sexual practices, sexual excitement and arousal. And perhaps, for some, the latter could mirror the interest for viewing pornography on the internet along with the perceived anonymity online.

Some of the young people (18-24) who use the internet to seek information and advice or support claimed to do it to become sexually aroused; in comparison more young men than young women (Daneback & Månsson, 2009). It is unclear if they become aroused in the process of looking for information or when interacting with others or if their intent is to become sexually aroused and seeking information about sex is a legitimate way of becoming aroused. It could perhaps be erotic chatting with the purposes of arousal, masturbation and orgasm, the activity that has sometimes been referred to as cybersex (e.g., Daneback, Cooper & Månsson, 2003). As Foucault (1976) noted, detailed sexual talk was commonplace 400 years ago. In history, fantasies, desires and impure thoughts were told behind the screens in confession booths. By talking in this way, the details were filled with arousing and exiting qualities. Besides the internet constituting a place making it easier to talk about sexuality in detailed ways, the above example also shows how this makes it possible to make it a learning experience, how to increase one's knowledge in a specific area of sexuality. It is clear that for the girl in the example above, approaching a boy and ask how to have oral sex would be impossible offline.

What we have focused on so far has been a few activities and some reasons to engage in these. However, many young people also turn to the internet to seek information and advice because they are curious. Others claim that they do not have anyone they can talk with about these issues. Daneback and Månsson (2009) found that more than one out of ten among young men (18-24) claimed this. This means that not only does the internet create a context where it is possible to talk in ways not possible offline, but it also provides some young people with someone they can talk about sexuality with.

Summary and discussion

By making the spoken word silent (typed; textualized) and changing physical proximity for virtual proximity two of the main obstacles for talking about sexual matters are removed. By extending the possible group of people to talk to, from family and peer groups to virtually anyone on the internet, even more obstacles are removed. The spoken word and familiar people are usually what makes it sexuality difficult to talk about, at least when it comes to one's private thoughts and experiences. Perhaps we can compare the internet to the confession booths of yore but without having to pronounce those words we may have terrible difficulties pronouncing in the presence of others.

Reinforced by anonymity children and young people, on for example discussion boards, raise their questions in a mix of curiosity and insecurity. Instead of guessing, young people can ask (diverse) questions about what it is like being gay, how to use a dildo and where to buy it, how to have oral or anal sex, about their bodily functions, and how to shave their genitals. They can try their thoughts and experiences on others and they can reflect upon them together. As the society is getting more sexualized and the places to discuss this are obscure, non-existent, or restricted (i.e. do not leave room for these, sometimes new, questions), the internet provides arenas where questions can be asked, answered, and reflected upon. And this without risking personal negative sanctions; without being called a whore if one expresses a highly sexual interest, without being beaten up due to one's sexual orientation.

GAINING INFORMATION AND KNOWLEDGE ABOUT SEXUALITY ON THE INTERNET

As has been shown in this chapter, young people seek information about their sexual bodies online. Through media in general and digital media specifically children and young people of late modern society are getting access to sexuality in a way that differs from that of earlier generations. Due to this change older discourses of childhood and youth are blurred (Buckingham, 2000; Prout, 2008). Sexuality becomes a topic which can be explored at a greater extent through the internet. In the same time, as has been mentioned before, young people's search for knowledge about sexuality awakens concerns in the adult society (Tincknell et al., 2003). This clash between the current agency of young people and institutionalised discourses about sexuality - telling who can perform sexual activity, who can construct a sexual body and what preferences of sexuality should be expressed - manifests itself in general ideas about sexuality online as entirely a problem.

As we have discussed, sexual education for young people turns out to be gendered in offline settings (Gagnon & Simon, 1973; Bäckman, 2003). Tolman and Diamond (2001) also mean that there is an invisible cultural pressure on girls to act in line with the limited sexuality that cultural norms of society offer them. Sexuality has not been a field where girls have been culturally prioritized (Ibid.). For example young girls do not want to discuss lust and sexuality in educational situations in school settings (Epstein & Johnson, 1998). Girls in this context experience that they have to protect their reputation to avoid slander (Ibid). This is also manifested online. Cultural norms are not absent. The results from the discussion board study mentioned above suggest that girls, to a greater extent than boys, discuss their personal shortcomings or anxieties whereas boys ask questions about practical matters. This can be understood as a signifier for gendered cultural norms in society. However, the internet blurs these gendered boarders and makes way for alternative expressions about sexuality and sexual behaviour beyond cultural norms for what to say and to whom. People of all ages, both men and women, search for information on sexuality and talk about sex with other interested people. In their study, Daneback and Månsson (2009) found that young females who sought sexual information on the internet primarily wanted to gain knowledge about their bodies. However, for almost half of the young men in the sample this was the reason they used the internet too. Both Löfberg (2008) and Daneback and Månsson (2009) show that young people in different age groups use internet to get information and learn about their bodies, how they work sexually or how one can sexually construct the body. This is a source of knowledge for young people, which also ought to be acknowledged for the positive aspects it can have.

Besides discussing bodily functions and sexual identities online, it is also possible to share detailed descriptions of sexual techniques. The willingness, regardless of gender and age, to learn how to have sex and the way it is expressed online makes us believe that these are questions that are ignored or silenced, for various reasons, in offline settings such as in schools, at home, and among peers. But it can also be perceived in the light of sexual lust, sexual excitement and arousal – ways to enhance sexual encounters and relationships. This is a connection that seems to have been little researched so far. And perhaps the increasing interest for both young men and women to view pornography on the internet can be interpreted in line with this – as a way to become excited and inspired to explore one's sexuality, alone and with partners.

The way to express oneself online changes because of changed conditions for communication (boyd, 2008; Hine, 2000; Holloway & Valentine, 2003; Jaquemot, 2002; Jones, 1998; Livingstone & Bober, 2005; Stern, 2008). When young people seek information and knowledge about sexuality online we can see two overall aspects that influence the learning conditions.

The possibility to express. The terms for what to express change online, in comparison to offline settings. Young people claim that they can talk about things in a different, often more authentic, way online (Holloway & Valentine, 2003; Stern, 2008; boyd, 2008). In relation to sexuality this condition becomes central for the internet as a source for learning experiences. Social norms that restrict young people's agency offline are not structured in the same way online. This means that they can express sexual matters that they have on their minds. By expressing their thoughts and questions in public online settings they create an arena for learning that earlier was limited to peers and thoughts of one's own.

The possibility to extend knowledge. What is learned online comes from participants with a variation in geographical location, age and gender. This means that gaining information and knowledge about sexuality online is a process performed in a cultural environment separated from the local environment, local norms and local structures about how to understand sexuality. Turning to the internet means that one can contact other individuals with a mutual interest in discussing these questions and who are possible to get in contact with at any time.

What is specific for the internet, but not touched upon in this chapter, is that the interactions between participants in, for example, discussion fora are available to non-participants or lurkers. However, we know little about why some users choose to participate and others do not. We also do not know what non-participants make of the discussions they follow and the consequences they might have for them. This should be considered in future research.

CONCLUDING REMARKS

As shown in this chapter the internet is a place where young people can get support in understanding their own sexual bodies. Sexuality is a concern even for younger age groups in society (Castells, 1997; Daneback & Månsson, 2009; Giddens, 1991; Löfberg, 2008; Renold, 2005). The internet contributes with a space where this sexual body can be reflected upon and where thoughts and feelings of insecurity about sexual matters can be revealed in a way that in many aspects differs from offline settings. This possibility needs to be considered in our understanding of young people in media society of today.

As we can see there are many aspects of how to use the internet for seeking sexual information and knowledge, but also the content of the information and knowledge searched for. Young people are interested in how their bodies work, but also in sexual techniques. Sexuality and surrounding discourses are changing with the landscape. Research has shown that the abundance of sexual sites on the internet can be handled and reflected upon by young people, that they have ways and strategies to navigate in the sexual landscape online. They are active users, not passive victims of the new technology. And maybe the internet is a way for young people to succeed in learning to express and reflect upon their sexuality in ways older generations have failed. And maybe this can influence the way sexual matters are discussed offline, promoting sexual experiences and sexual health. Social arenas online are used by young people for these purposes. This needs to be acknowledged in a reflexive way where the children's and young people's perspectives are taken into consideration. This activity challenges well established discourses about children's and young people's agency in the "adult world". And maybe it will make room for alternative understandings of how to construct gender and how to be young. For this to happen, we believe that more research focusing on the positive aspects of internet sexuality is needed.

REFERENCES

Bäckman, M. (2003). *Kön och känsla: samlevnadsundervisning och ungdomars tankar om sexualitet* [Sex and emotion: Education in sexual matters and young peoples thoughts about sexuality]. Stockholm, Sweden: Makadam. boyd, d. (2008). Why youth love social network sites: The role of networked publics in teenage social life. In Buckingham, D. (Ed.) *Youth, identity, and digital media*, pp. 119-142. The John D. and Catherine T. MacArthur Foundation Series on Digital Media and Learning. Cambridge, MA: The MIT Press.

Buckingham, D. (2000). *After the death of childhood. Growing up in the age of electronic media.* Cambridge, UK: Polity Press.

Castells, M. (1997). *The power of identity. Volum 2, information age*. Oxford, UK: Blackwell.

Daneback, K. (2006). *Love and sexuality on the internet* [Doctoral dissertation]. Göteborg, Sweden: Göteborg University.

Daneback, K., Cooper, A., & Månsson, S.-A. (2005). An internet study of cybersex participants. *Archives of Sexual Behavior, 34*, 321–328. doi:10.1007/s10508-005-3120-z

Daneback, K., & Månsson, S-A. (2009). Kärlek och sexualitet på internet år 2009

Döring, N. M. (2009). The internet's impact on sexuality: A critical review of 15 years of research. *Computers in Human Behavior, 25*, 1089–1101. doi:10.1016/j.chb.2009.04.003

En enkätundersökning bland män och kvinnor 18-24 år [Love and sexuality on the internet 2009: A survey among men and women 18-24]. I Ungdomsstyrelsen, *Se mig. Unga om sex och internet [See me. Young people on sex and the internet],* pp 182-237. Ungdomsstyrelsen 2009:9.

Foucault, M. (1976). The history of sexuality: *Vol. I. The will to know*. London: Penguin.

Giddens, A. (1992). *The transformation of intimacy: sexuality, love and eroticism in modern societies*. Stanford, CA: University.

Herring, S. C. (2008). Questioning the generational divide: Technological exoticism and adult constructions of online youth identity. In Buckingham, D. (Ed.) *Youth, identity, and digital media*, pp 71-92. The John D. and Catherine T. MacArthur Foundation Series on Digital Media and Learning. Cambridge, MA: The MIT Press.

Hine, C. (2000). *Virtual Ethnography*. London: Sage.

Holloway, S. L., & Valentine, G. (2003). *Cyberkids, children in the information age*. London: Routledge.

Jacquemot, N. (2002). *Inkognito. Kärlek, relationer & möten på Internet* [Incognito. Love, relations & encounters on the internet]. Stockholm, Sweden: Bokförlaget DN.

Livingstone, S., & Bober, M. (2005). *UK Children Go Online. Final report of key project findings.* E.S.R.C, Economic & Social Research Council, @ Society. http://www.children-go-online.net/ (retreived 071121).

Löfberg, C. (2008). *Möjligheternas arena? Barns och ungas samtal om tjejer, killar och sexualitet på en virtuell arena.* [Arena of Possibilities? Children's and Young People's conversations about Girls, Boys, Emotions and Sexuality on a Virtual arena] [Doctoral dissertation] Stockholm, Sweden: Pedagogiska institutionen, Stockholms universitet.

Månsson, S.-A., Daneback, K., Tikkanen, R., & Löfgren-Mårtenson, L. (2003). *Kärlek och sex på internet* [Love and sex on the internet]. Göteborg University and Malmö University.

Månsson, S.-A., & Löfgren Mårtenson, L. (2007). Let's talk about porn: On youth, gender and pornography in Sweden. In Knudsen, S. V., Löfgren-Mårtenson, L., & Månsson, S.-A. (Eds.), *Generation P? Youth, gender and pornography* (pp. 241–258). Copenhagen: Danish School of Education Press.

Prout, A. (2008). Culture-nature and the construction of childhood. In Drotner, K., & Livingstone, S. M. (Eds.), *The international handbook of children, media and culture*. London: Sage.

Renold, E. (2005). *Girls, boys and junior sexualities. Exploring children's gender and sexual relations in the primary school*. London: Routledge Falmer.

Ross, M. W., & Kauth, M. R. (2002). Men who have sex with men and the internet: Emerging clinical issues and their management. In Al Cooper (Ed.) *Sex and the internet; A guidebook for clinicians,* pp 47-69. New York: Brunner-Routledge.

Stern, S. (2008). Producing Sites, Exploring Identities: Youth Online Authorship. *Youth, Identity, and Digital Media*. In Buckingham, D. (Ed.), *The John D. and Catherine T. MacArthur Foundation Series on Digital Media and Learning* (pp. 95–118). Cambridge, MA: The MIT Press.

Subrahmanyan, K., Greenfield, P. M., & Tynes, B. (2004). Constructing sexuality and identity in an online teen chat room. *Applied Developmental Psychology, 25*, 651–666. doi:10.1016/j.appdev.2004.09.007

Suzuki, L. K., & Calzo, J. P. (2004). The search for advice in cyberspace: An examination of online teen bulletin boards about health and sexuality. *Applied Developmental Psychology, 25*, 685–698. doi:10.1016/j.appdev.2004.09.002

Tincknell, E., Chambers, D., van Loon, J., & Hudson, N. (2003). Begging for it: "New femininities", social agency, and moral discourse in contemporary teenage and men's magazines. *Feminist Media Studies, 3*, 47–63. doi:10.1080/14680770303796

Tolman, D. L., & Diamond, L. M. (2001). Desegrating sexuality research: Cultural and biological perspectives on gender and desire. *Annual Review of Sex Research, 12*, 33–74.

ENDNOTE

1 The examples comprise questions frequently asked on a public discussion board for questions about sex in the Swedish web community Lunarstorm.se. The community's ambition is to develop an affinity between young people (12-20) but it is open for everybody. During the period for data collection, February to September 2004, Lunarstorm.se had 320.000 visitors daily and around 1.1 million members. 171 .567 young people, aged 12-17, visited the site on a daily basis. On average every visit was approximately 20 minutes and every member spent more or less 40 minutes there on a daily basis. The questions illuminate areas that are of interest for children between 12-15 years. From these empirical data examples will be discussed of what and how young people approach sexuality online. In exemplifying from data the discussions are translated from Swedish to English. The discussions are not shown in its entirety. We have chosen to show the contributions that are of specific interest in relation to the purpose of this chapter.

Chapter 12
Adolescents and Online Dating Attitudes

Olugbenga David Ojo
National Open University of Nigeria, Nigeria

ABSTRACT

The introduction of Internet technology worldwide has brought out different behaviors in human beings; old and young. One of the vehicles for this is the online dating social network; a means by which people meet other people with whom they enact social relationships through the internet. This chapter considers the attitudes of adolescents with African background in relation to online dating. The findings show that there is no significant difference in the attitudes of both male and female adolescents in relation to online dating and also that the perception of parents in relation to adolescents getting involved in online dating activities is in tandem with the African belief that adolescents are too young to get tangled in any aspect of sexual activities, online dating inclusive. It also reveals that both male and female adolescents have coping strategies for dealing with online dating activities. Solutions and recommendations to resolve the emanated issues are also highlighted.

INTRODUCTION

Adolescents' sexuality behavior worldwide is seriously undergoing a transformation from what it used to be in the past taking into the cognizance their social adventure in the area of computer technology. This new phase of development that could be described as a technology-based society entered into by virtually all the countries of the world. Due to the need for development and global compliance, government in each society is positively and actively engaged in encouraging computer based services; computer ownership/access and Internet use in order not to be left behind in the global development among the comity of the countries of the world. This move is virtually turning every society into Cyber societies. Consequently with these efforts, young adolescents are very much fascinated and attracted by new technologies and hence, computers and

DOI: 10.4018/978-1-60960-209-3.ch012

computer-based technologies have become the new craze for adolescents. It could be said that a very large percentage of adolescents have access to the Internet world wide. This might clearly confirm the fact that adolescents use the Internet generally more than any other age category, with e-mailing and chatting being the most popular of all the Internet activities.

Although the formal school system may not allow adolescents to use the Internet for anything other than academic activities, these adolescents still have access to its use whenever they need it because the services are readily available in their homes or at the cybercafés that are available in towns in their societies. It is during these access periods that many adolescents express their sexual behaviors on the Internet as they chat with different categories of friends and/or as they view pornographic sites.

A lot of concern has been expressed about the delinquent behavior of adolescents in connection with sexual matters and various analysis of adolescents' sexuality behavior in Africa have observed a transformation in their sexual activities. With rapid urbanization, technological development and breakdown of moral values, sexual permissiveness has increasingly become a characteristic of the social life of adolescents.

In African setting, adolescents are not permitted to be involved in any sexual activities. Discussing sexual related subjects is in fact a taboo and anyone caught is seriously punished by parents and elders. Although there are traditional structures that govern marriage and sexual behaviors as well as the family and community support networks that ease the transition to adulthood. During this period, an adolescent has little opportunity for full sexual expression until his late twenties (Bledsoe and Cohen, 1993). In Ghana, like other sub-Saharan African countries, Barbara et al, (1999) and Grindal, (1982) remarked that adolescents' sexual experience was not encouraged and that the elders were concerned that the very young participate in sexual activities. In the past, before the advent of western education, young people were given informal education to prepare them for life (Fafunwa, 1974) The traditional system of education in Nigeria saw to it that young men and women were taught to acquire a healthy attitude towards sexual activities as part of their preparation for adult life. The belief in the African setting is that sexual activities' including dating is a prerogative of adults.

However, the disorganization of primary societies, the resort to urban life, the move away from traditional roots and loss of cultural values and the advent of technological development through the use of telephones, computer and the Internet, has paved the way for adolescents like other members of the society to acquire and display different kinds of attitudes and behaviors. The Internet has broken cultural barriers by providing information, communication and opportunities to all and sundry, an opportunity tapped into by adolescents to carry out their dating activities. Thus, it could be inferred that formal sanctioning of dating or other forms of sexual activities by adults is leading adolescents to pursue such relationships albeit in secret due to opportunities provided through the Internet (as no adolescent would inform an elder that he/she is involved in online dating).

Adolescence is the period between childhood and adulthood. However, some cultures may not recognize this term because there is no distinct period to identify what transpires between childhood and adulthood though teens all over the world display the similar characteristics. It is the period from the onset of puberty wherein several physical and emotional changes occur in both male and female children and occurs within the ascribed teenage age range of their growth and development. It significantly ends at the age of nineteen. During this period, there are noticeable changes in height and shape of their bodies and also, signs of pubescence. Significantly, adolescence period is not a period for only changes in physical and emotional growth and developments; it is a time when the adolescents adjust to several resultant effects of growth and development that occur in them. It is a period to adjust to unfamiliar and a strange body, when they react to events and issues basically only on the basis of their own understanding without recurring to anybody, the time they develop interest in the members of opposite sex, the period of confusion and despair, and also the period they need to contend with new unfolding intellectual power and prowess. It is for these reasons that Hall (1904) tagged the period as a period of storm and stress. Hall sees the adolescence period as a time that is characterized by youthful belief and idealism, and also filled with concerns about sexual development all of which can be turbulent. Many other psychologists too regard the period as very stressful.

As stressful, complex and complicated the adolescence period is in all ramifications, so also is the attitude of the adolescents. This is the time for physical development, cognitive development which translate to moral reasoning, the time for social and personality development which nurture the formation of personal identity. It is all these developments coupled with the influence of the home and peer groups that form the basis for the kinds of attitudes that is being observed in adolescents. These developments can bring about role confusion when they are not properly handled and channeled. It is on the basis of this that identity is formed and invariably, would inform the kind of attitude an adolescent would exhibit in the society. During adolescence, more than any other influence, peer group factor which can be regarded as social influence has a lot of impact on the beliefs and attitudes of adolescents.

It is for the above reiterated facts that various attitudes are formed among youths who participate in the varieties of technological activities like Internet dating common in the World Wide Web within the societies and cultures of the world. For healthy development, there is nothing wrong for adolescents to have goal oriented social relationships. This opinion is corroborated by Prezza, Guiseppina & Dinelli (2004); Baurnesister & Leary, (1995) who refer to social relationship by adolescents as a fundamental need and a crucial activity for healthy development. Furman & Wehner (1993) also believes that early dating experiences are central in the social life and emotional experience and development of adolescents while adolescents' early dating experiences are sources of development of adolescent's self identity and capacity for intimacy because of the role it plays in the personality make up of the adolescents (Montgomery and Sorell, 1998). The onus therefore, is that such adolescents' social relationships should be such that are done with good intentions and not with ulterior motives which may be detrimental to the positive growth and development of the adolescent concerned.

As technological development and innovations are opening up avenues for adolescents to venture into lots of activities through the Internet; dating inclusive, a research outlets for studying and understanding adolescents' attitudes and beliefs in various activities are also being opened. Hardey (2004) observes that Internet allows for a new means of meeting people and establishing relationships. The observation is based on the belief that meeting people through the Internet would continue for a long time. It is also the belief of Ben-Ze'ev (2004) that understanding of dating and amorous relationships is been altered from what it used to be by the Internet. According to him, Internet dating allows people to interact with one another over a distance within a very short time and at very less expensive cost than it used to be and also with an opportunity for continuity of the relationship.

Purposefully, this chapter looks into the attitudes of African adolescents in relation to online dating with a view to discussing their characteristics in order to expose and reveal the hidden intentions. This will enable the young innocent adolescents to be aware of eventualities involved in such online dating. It will be a source of comparative study and an avenue to promote good attitudinal values among young adolescents. It is a source of literature for study on young adolescents' attitudes regarding relationship development. It helps reveal the Internet attitudes of young adolescents' of African backgrounds to online dating. Lastly, it helps to promote knowledge and attitudinal change in relation to heterosexual friendship and dating among the youth in general.

The overall objectives of this chapter include ascertaining the characteristics of young male and female adolescents in relation to online dating; determining if there is relationship between the attitudes of the male and female adolescents in relation to adolescents' online dating; examining the coping strategy of the male and female adolescents when handling online dating between the opposite sex; and relate the views of the male and female adolescents on the perception of their parents considering the attitudes of adolescents online dating.

BACKGROUND

Heterosexual friendship among the youth generally has been an accepted occurrence universally and it has been the media through which adolescents express their sexuality. Sexual activity among adolescents is clearly increasing, as well as the rapidity of occurrences of adolescents' sexuality activities and the degree of magnitude of these activities worldwide. Adolescence period is a time of storm and stress. Part of their attitudes is capacity for complex and highly abstract reasoning. They misunderstand life's problems because they place themselves at the centre of every question to the exclusion of others due to their egocentrism belief. It is the time they want to establish a strong sense of self; self identity in order for them to be able to cope with many changes that surround the adolescent years. These changes include those that occur through physical, emotional and moral developments.

Since the advent of Internet via technological advancement worldwide, adolescents have demonstrated their intrinsic abilities to demonstrate the fact that the computer age is actually theirs. Comparing what they perform attitudinally on different available websites with that of adults is the beginning of an attempt to reduce the value attached to human dignity. This is not to exonerate the adults from some of these attitudes but the fact remains that adults who involve themselves in those various activities are more mature and are old enough to face consequences of the negative effects of cybernetic activities.

Due to modernization and rapid technological developments, adolescents are spending more time with peers, social groups or alone than being with families. Most people join social groups because there are many benefits for being part of such groups. Adolescent dating as a trend in social networking will continue to increase and also be inextricably linked with one another as long as there is continuity in Internet usage. Adolescents can get in touch with one another through chat rooms, Facebook or online dating websites, and feelings can blossom without physical face to face contact. Any of them can meet, greet and chat for hours with strangers from Nigeria or Sweden without leaving their rooms in South-Africa or Haiti, via personal advertisement or matchmaking websites such as Match.com.

This development has already made teens online dating to become increasingly common. However, by replacing personal contact with online dating, teens miss opportunities to build personal relationships. As teens become romantically involved via online networks, they are potentially putting themselves and their families in danger. As with any online exchange, there is no guarantee that the person on the other end is being honest and sincere. Predators can approach unsuspecting teens, especially teen girls, with little effort. No matter how cautious the teenagers are, social networking relationships can be dangerous. Not only are the teens at risk from sexual predators, they could meet someone who is abusive, wants to take advantage of them or someone who simply is not right for them.

As inundated above, Internet dating can be difficult enough for teens. Dating online does not offer the benefit of facial expression or body language to gauge a person's interest or honesty. It is safer for teens to experience the fun and also learn the opportunities of real world dating before they try it online. There is no doubt that dangers of Internet dating are real. The active adolescent participant is just frolicking on the Internet browsing from site to site and all over sudden meets a person by chance who he or she calls a friend. The person is so interesting and soon he/she becomes addicted to the person. He/She start to feel like dying and raring to sit in front of the computer to chat based on the emotional dictates of the feelings they have for one another. The assumption is the belief that both of them have so much in common that either of them actually feels they have finally met the right person. The young man or woman is so excited about the online partner that he or she eagerly anticipates the online chatting on the Internet. He or she feels the day is blue anytime his/her Internet partner was not available for chatting. None of them ever sit down to reflect on the danger of Internet dating whereas it could become the beginning of insanity if it turns the other way due to the degree of emotion that would have been expended.

It is quite easy to manipulate a young mind. After everything has fallen into place, this predator arranges for a physical meeting in an exotic or just a common place with pretence to blow the partner's mind. It is only luck, ability to discern and wisdom that can get the adolescent out of the entanglement once she/he has met with the date; it will be difficult to redeem him/her. Everything has been arranged prior to this meeting and this person promises you heaven right here on this poor earth. The Internet date might propose a trip or an overseas vacation. This might turn out not to be a vacation as the adolescent might be used to make another pornographic video, be introduced into drugs, be turned into a sex slave or much more. All these are likely features of online dating after a relationship has fully developed between the partners.

In other to rejuvenate the positive attitude that promotes high sense of value with the intent of developing and promoting healthy sense of self esteem among adolescents, this chapter therefore looks into the adolescents attitude about online dating with a view to discuss the vices that are embedded therein.

CHARACTERISTICS AND ATTITUDES OF ADOLESCENTS IN RELATION TO ONLINE DATING

It is common occurrence for dating partners to meet one another through some distinct context such as school, place of work, malls and stores, friends, place of worship, local market and parties. The Internet is another avenue added to this context for dating partners to meet. The Internet now amongst other prominent functions fosters social relationship among different levels of ages and at times, participants crisscross age wise. All levels of people that surf the Internet meet people they establish relationship with and which may end in marriage. This is attested to by Walther and Burgoon (1992) when they observed that the Internet opens up another channel for people to meet with people for the first time and allows initiation of meaningful and satisfying conversations, builds a stable, long-term relationship similar to face-to-face interactions. This is suffice to say that lots of people use Internet to meet with persons and to also maintain social relationships. This set of people includes adolescents. Like every other age group that engaged in online dating, the adolescents who are seriously involved have some characteristics that distinguish them and make them easily recognizable anytime they are in an Internet centre or cybercafé.

As observed and noted, the adolescents involved in the activity, daily visit Internet centres or cybercafés at regular intervals since they have to keep to the agreed times and appointments of their online partners. This is due to the fact that they need to establish a personality of somebody who is upright and responsible, and also, the need to continually share and exchange views on different issues and personal information.. This issue of regular interval visits of the Internet was equally part of the findings of Levine (2000) when she referred to adult and Internet dating in her work on Virtual attraction. She found that in the virtual world, people who develop relationships tend to be online at regular intervals. She also reported that such regular internet visits leads to increase in social interaction and in the level of intimacy. This is a resultant effect of regular adolescent disclosures about self during their virtual interactions.

Another notable characteristic and attitude of adolescents who engage in online dating activities is that they show an attachment for the use of the Internet. They do not spare any leisure time since they have to search for their intending online partner and this might take a while because they always have a specification they are looking for. They search the net for profiles of opposite sex based on their choices of descriptive characteristics such as age, race, colour, and religion. Secondly, their attachment to the Internet may be due to the fact that online dating, like face-to –face dating, is embedded in physical expression of emotional feelings that waters the continued growth of intimacy. Ben-Ze'ev (2004) refers to the phenomenon as detached attachment which means a relationship that is characterized by physical distance and emotional closeness.

Other attitudes and characteristics displayed by adolescent online daters like their adult counterparts is that they tend to use hidden identities and use irregular e-mail addresses when they are online. They communicate with their supposed online partners with the hidden identities and do not reveal their real identities until they are sure that they would want to proceed with the partner. The reason for such behavior may be connected to their need to feel safe and secure before the disclosure is made so as not to get involved with the same person twice and also because they do not want to be seen on the Internet by another old date with whom they have crossed each other's paths. After disclosure of true identities and exchange of authentic e-mail addresses, a common feature in their online exchange of communication is the use of emoticons to express emotional feelings they have to each other. These are non-verbal cue signs that make their emotional expressions to be covert to a third party. This feature marks the communication of adolescents involved in online dating, while the affair lasts. Cerpas (2002) note that romantic gestures convey a lot of emotions while Albright and Conran (2003) attest to the fact that emoticon in online dating is similar to face to face dating relationships in that it reflects partners' emotions such as regular smiles to a huge grin.

COPING STRATEGY OF ADOLESCENTS INVOLVED IN ONLINE DATING

Individuals engaged in dating relationships, be they face to face or online, contend with the same emotional feelings. Dating relationship can be rewarding, fulfilling, and memorable looking at various experiences that the act affords the players. The period it may last can also be extremely trying and difficult considering emotional tensions that may culminate into disappointments and frustrations that can be generated at any point in time when there is breach of trust or expectations regarding what makes a good relationship. This is what could be regarded as the norms that should guide the relationship. To survive the tensions that arise from online dating relationships, adolescents who have been involved in dating have their coping strategies that help them face the aftermath of tensions generated by their affairs with their online partners. These coping strategies are made up of each participant's personality traits and character through which strength to withstand and deal with stressful situations is drawn.

Just like other adolescents on other continents, African adolescents who are involved in online dating cope with whatever the outcomes and experiences of their unfulfilled expectations standard from online dating tensions using strategies that are applicable to their personalities. As pointed out earlier, it is mainly through peer groups that the adolescents learn or discuss issues that are related to dating relationships and/or any other sexual activities. Therefore, whatever happens in a dating relationship is usually informed by these interactions without recourse to older persons especially since such activities are regarded as a taboo. The onus is on the involved adolescents to address the issues since there are no existing measures for coping efforts used specifically in dating relationships. It is unclear if partners in dating relationships use different coping strategies other than those relied on by partners in marital relationships. There can only be speculations on the types of coping efforts people rely on when dealing with unfulfilled expectations in dating relationships.

In order to survive the stress that emanates from their dating activities, strategies adopted by the adolescents to cope with the stress include deliberate termination of the relationship, wait and see attitude by staying in unsatisfactory relationships while waiting for the partner to take the lead on what action he or she prefers, ignoring the issue of contention or confronting the partner, and lastly to seek advice from peers. In retrospect, these coping strategies are what other people who are involved in dating relationships rely on to resolve and cope with what eventually happens in their dating relationships when there are issues that border on unfulfilled expectations.

When an online dating relationship is positively on course, the only inference that could be drawn is that the expectations are being met or exceeded by the parties involved. In that wise, for its continuity the participants would have to maintain the status quo and make sure that what is sustaining the relationship is always provided. This is the only coping strategy required for sustaining any dating relationship be it online or otherwise. It is only then that the partners can continue to conceive the idea of spending time together either on the Internet or by planning to meet face to face, exchanging views on goals and values and on any other ideas they deem important to their relationship. It is for the avoidance of the opposite of this type of outcome that Baucom, Epstein, Rankin & Burnett (1996) suggest that romantic relationships can be particularly distressing when relationship standards are not fulfilled. Sprecher and Metts (1999) therefore opined that dating partners should enter into romantic relationships with standards, or pre-existing beliefs, about what makes a good relationship as this will guide participants' attitude and thereby avoid the tension that could arise if standard is not fulfilled.

PERCEPTIONS OF PARENTS ABOUT ONLINE DATING BY ADOLESCENTS

The percentage of parents that are computer literates in Africa is very low compared with the rate of those in the developed world (Wikipedia encyclopedia). This definitely has its impact on the number of parents that are aware of adolescents' online dating. In spite of this, the parental perception of adolescents' online dating activities just like the majority of those who are enlightened and are computer literate is not different from what it is when compared with face to face dating. Their belief is that adolescents are not mature enough to be involved in any sexual exploits although there are some parents who are indifferent. African adolescents' sexual attitude and behavior is highly influenced by their family; most importantly, by the mother. The mother more than the father except in a few cases, is always showing concern for the adolescents' sexual attitude. They monitor the adolescents to see that they comply with the parental rules on dating and other aspects of sexual attitudes. Unfortunately, most adolescents do not agree with their parents' rules, many show only a moderate level of compliance. With the advent of technology which also has opened up the avenue for internet dating, many adolescents are involved in the online dating without the knowledge of their parents. Adolescents would claim to be working on the computer under the guise of attending to their school assignments or the act of perfecting their computing skills. They would log on to the Internet and navigate their way to the dating site or chatting room. Where personal computers are not available, they would visit the cybercafé under the same guise.

It is the general belief among the parents that current adolescent's sexuality behavior in Africa has defied all culturally accorded respect given to it in the past. The assumption is that adolescents do not have respect for culture and are not ready to learn about life from elders. Inferences from a focus-group discussion on adolescents' sexuality behavior by parents (Ojo & Fasubaa 2005) depict what the perception of parents is about all aspects of adolescents' sexuality; online dating inclusive. In like manner, parents see attitudes of adolescents regarding online dating as blatant indiscipline. It is the belief of the parents that disciplinary measures of high standard should be adopted at schools and home to check the menace. The belief is that the dating attitude is out of place for their age, while many also use it to defraud innocent partners of their emotional feelings and property which they collect as gifts having developed a romantic relationship with faceless individuals through virtual medium of communication. Also, parents remarked that it is the foreign culture and way of life that is affecting the adolescents as well as even some older people who are morally corrupt. They attribute the scenario of adolescent online dating to the power of civilization and modernization through technological advancement. They claim that the youth generally feel that they know better than their parents so they do comply with parental instructions as a result of their exposure via the Internet.

ISSUES, CONTROVERSIES AND PROBLEMS ASSOCIATED WITH ADOLESCENTS AND ONLINE DATING ATTITUDES

The main focus of this study are the problems embedded in online dating activities engaged in by adolescents. Based on the premise of maturity, adolescents are viewed to be too young to get involved in sexual activities until they are mature for such emotional venture. What is involved in dating or any other sexual activities are so much that adolescents would not be able to cope with the attendant effect the repercussions might have. As observed earlier, adolescence is the period and time for the consolidation of self concept and self esteem that started from the childhood period. The time for identity formation that is very stressful during which they respond to myriad of physical, cognitive and social changes. The issues therefore lie in the fact that it is difficult for adolescents to adjust well to all the changes, happenings and situations that make the adolescence period to be the time of storm and stress because of their immaturity and lack of experience in the ways of life. Based on this premise, it would be difficult for adolescents to cope with the emotions, knowledge and intricacies that are involved in online dating considering the fact that adults who are involved gets their fingers burnt in face to face dating irrespective of the number of years the partners would have been in the relationship. Secondly is the problem of how will each party on the either sides of the divide cope with the fraud that has been introduced to the online dating activities. The party who wants to defraud the partner plays on the emotions of the other partner due to the strong friendship and relationship that has been established by requesting for financial assistance (or other sorts of gifts) in the name of a relation such as mother, father or sister being critically ill in the hospital while he/she required huge sum of money to pay for the hospital bill since he or she is the only surviving son or daughter whereas nobody is ill. He/She collects money for personal use and enjoyment at the detriment of the partner who is romantically committed and has feelings in a relationship that is non-existent. In the same way, it can be difficult for adolescent to survive the trauma of being defrauded emotionally via online dating activities. It is a known fact that it is easier to manipulate a young mind for the purpose of taking advantage of the young person.

Adolescents who fall prey to emotional fraudsters find it difficult to recover due to the stage of their development because they are inexperienced and find such things difficult to fathom. And if the adolescent's life is rectified through rehabilitation; counseling in most cases, it may never be the same. There would have been encumbrances on his or her path.

As good and useful as technology advancement is for development, modernization and growth of human endeavor locally and globally also it has negative influences on the society. Technology has removed borders and rendered time irrelevant. The dangers of adolescents' online dating are evident because they believe in their partners even though it may be lies. Online dating does not enable adolescents who possess little experience in matters of the heart to observe the facial expression or body language of the partner on the other side of the digital divide to confirm the genuineness of the intention. All these issues confirm the fact that the dangers of Internet dating are real for adolescents' and mature persons alike. Many of the adolescents just like their peer groups whose life experience is minimal are breaking the boundaries of culture and tradition through online dating via technology to look for idealistic love as an outlet to find their life partners; a mirage for their age and the level. As a result of development, modernization and civilization, the impact of technology from this view might be seen as liberating due to the positive impact it has brought to the society locally and globally in many areas of life. However, with regards to upholding the societal values, culture and tradition especially in the African context the practice of online dating can be said to open the door to deception, misrepresentation and sometimes downright criminal activities.

METHODOLOGY

This study is essentially a product of survey research. The study collected the views, beliefs and opinions of adolescents and parents through a five-point Likert type questionnaire administered on the adolescents and a structured oral interview of parents who has children and wards of adolescent age. This is complemented by secondary data materials. The sample for the study is made up of 86 adolescents comprising of thirty-two (32) girls and fifty-four (54) boys while interviews were conducted on thirty (30) parents who were purposefully selected. The ages of the adolescents ranges between 14 and 19. Seventy-one (71) of the sampled adolescents are in the University and Polytechnic while the remaining few (11) are in the senior classes of secondary school. The Questionnaire used named 'Adolescents and Online Dating Attitude Questionnaire- AAODAQ' has a two week interval test-retest reliability coefficient of 0.70. The items on the questionnaire were also subjected to scrutiny of experts in the areas of Computer Science and Psychology in order to establish the clarity and appropriateness of the inventory items. After questionnaire administration, the data gathered from interpretation after coding were analyzed using t-test statistics.

DATA ANALYSIS

The data collected for analysis were from the thirty-two girls (32) and fifty-four boys (54) who completed the administered questionnaire. The characteristics of adolescents that go to the internet for online dating were not in any way different from those observed globally. The female and the male adolescents satisfy the characteristics of those involved in online dating. These adolescents claimed that they visit the Internet regularly at appointed times. They also claimed that in some days, it is possible for them to log on to the internet as much as five times in a day. A few of them agreed to the fact that they visit cybercafés on their way home from school since they do not have computers at home. Forty-seven percent (47%) of the adolescents do not have Computers or Internet connection in their homes. Thirty-three percent (33%) of the adolescents which is one third of the total population sampled claimed that they stay on the internet as late as three o'clock in the morning in order to chat with their partners. This confirms the fact that Nigerian adolescents engage in online dating have partners across the globe if the time zone is observed. Some observed that the cybercafés they use do overnight browsing and you have to register in good time during the day; mostly before twelve noon or else one would be shut out since there are lots of people who want to browse over night. The data analysis show data, it is evidently show that there is no significant difference in the attitude and characteristics of both male and female adolescents who are involved in online dating ($t=1.78, df=31, p>0.05$). Another interesting finding is that some adolescents chat with as many as three to five partners at the same time using different chatting names in order to conceal their identities. This portends the fact that from abinitio, there is tendency for the relationship not to have any true disclosures. Although, this is common amongst the males, a few of the girls confirmed they chat with as many as two to three partners too at the same time. One wonders how these young girls and boys cope with assumed intricacies that will be emotionally involved in dealing with two or more different people who are not located in the same place at the same time, and each with different emotional involvement coupled with the kind of different discussions each of the partners would have had the last time they met on the Internet. Their activities suggest that they are motivated by gifts from the relationships and it is the same reason that sustains the relationships. The adolescents also mentioned that they sometimes complement the online chatting with phone calls. According to them therefore, it is a question of establishing adequate closeness and deep intimacy before emotional sentiment is whipped up for demand for gifts.

The adolescents in this study like every other persons show reliance on the same coping strategies to resolve issues and problems that might arise between themselves and their online partners. They believe that they should allow the partner to dictate pace of relationship and make their move based on the observed trend of the attitude of the partners. The proviso is that if the relationship fails suddenly for whatever reason, that is, if the partner's attitude is unlike what it was before or becoming unpredictable since expected standard was not met, they will move on to other relationship. Some of the female adolescents' opinion is that one should just let the relationship stay without further emotional commitment. They are ready to still chat with their online partners but will not indicate eagerness to initiate online chats again. Going by the expressed feelings of the male and female adolescents observed in this study, it can be said that the status of the relationship will be best determined by who handles one another better. However, the male adolescents do not show any emotional attachment which suggests that they are just in the online relationships for selfish reasons.

On what their views about what their parent's perception of their attitudes are in relation to online dating, all the adolescents revealed the fact that they go online dating without the knowledge or consent of their parents. They claim that they got involved through the influence of their classmates and friends and due to the advantage of their knowledge of computing; exposure to computing technology. They confirmed that their parents would never be in support of them chatting with opposite sex when they should be reading or studying. Thirty-seven percent (37%) of them actually remarked that their parents are not computer literate and could not have known that they are involved in online dating even if it is done in their presence. They however affirm that their parents must not know that they are chatting with members of the opposite sex. They all agreed that they are too young for dating activities and that at their age; the opposite sex cannot visit them at home, though they go to parties and do all sorts of things outside of their homes and without the knowledge of their parents. The comments of these adolescents show that their parents will not be positively disposed to their involvement in online dating activities. There is no significant difference in the views of both male and female adolescents regarding their parental perception of adolescents' involvement in online dating ($t=1.15$, $df=31$, $p>0.05$). They are all aware of the fact that they are not yet mature for dating venture as they do not possess the experience required considering the nature of their culture and tradition where issues of sexuality is concerned. It is important to observe that those adolescents whose parents even have the knowledge of computing at whatever level do not claim anything different from others whose parents are computing illiterates. Thus it can be inferred that the adolescents in general agreed to the fact that they are too young and inexperienced in the way of sexual responsibilities and therefore, in the African settings, the parents cannot be in support of adolescents' online dating. This is not to deny the fact that there are parents who do not hold tenaciously to the rules and standard dictated by the African culture and customs regarding adolescents' sexual activities.

SOLUTIONS AND RECOMMENDATIONS

The onus for resolving the problems that are identified the online dating in relation to adolescents lies with the society. The society is made up of the people hence it is the people who are to guide and watch themselves with a view to behave according to the social norms of the society. The social norms enable people to behave in a standard pattern of behavior that is considered normal in any society. Scientific technology which online dating came about is a good omen for the whole world and it is for the purpose of development. It was made for the human use in order for life to be easy; it is the human angle to the use that has caused various problems that are being encountered in different human endeavors. Technology is nothing other than tool for human use. It is the human beings that always decide to use it in whatever way he or she feels. If the society allows the societal norms guide their attitudes and action in their daily affairs, the world would be better. Critically, in the African settings, adolescents are regarded as too young to be involved in sexual activities; dating inclusive. It is assumed that they do not have the maturity and experience to cope with what it entails. By getting involved in it either through the Internet or face to face is to go against the ethics of human living and since such things are regarded as taboo, there are always repercussions. The attitude of any adolescent taking part in online dating activities can be said to be unAfrican. They always hammer on teaching of culture and values continuously in virtually all African settings. In the case of online dating, the repercussions might be a situation whereby an adolescents who is involved get damaged emotionally through broken heart and/ or bruised ego, falls into the hands of kidnappers, Pedophiles, traffickers of child pornography and homosexuals, get pregnant prematurely when she agreed to meet her date for rendezvous, might lead to procurement of abortion and its complications if there arise any problem, loose out in school since he or she has been preoccupied by the thought of the partner instead of his or her studies or get swindled financially by the partner who does not have any emotional feelings but focus for what he or she could get materially from the relationships. With all sense of modesty, nothing is wrong with dating provided it is platonic wherein it is for the purpose of assessing each other for the possibility of becoming potential mates and not otherwise. Such date is devoid of all what the online dating with sexual activities connotation involves.

Finally, the solution to the problem lies with the parents, the schools and the adolescents themselves considering the importance of the roles each would perform in relation to afore discussion. It will therefore be recommended that Parents should be there to guide the adolescents in all their ways. They should see that there is a two-way communication between them and the adolescents, show support when it is necessary and be very open to them such that adolescents will be very free to discuss with them any issues that might be bothering him or her without fear. Parents should see that the adolescents imbibe religious and moral values which are always in tandem with societal norms. The school on its side should see to the administration of the expected discipline and teaching of those subjects or courses that will take care of enlightenment of the adolescents. This might include courses such as Family Life Education and other sociological courses that will help teach culture and values in the school system. As part of the discipline, school computers should be secured against dating sites and other sites that are not educative and there should be rules and regulations guiding navigations. The portion that falls on the lap of adolescents is the fact that they should learn to live responsibly and heed the teachings of their parents and elders.

FUTURE RESEARCH DIRECTIONS

Youth as they are known are very vibrant and this global period is actually theirs. The emerging trend in every part of the globe shows that none of the adolescents are ready to be left behind no matter the continent his or her village belongs. The rate at which they blend with technological advancement is surprising and amazing. This points to the fact that development no matter how minute gets to every nook and cranny of the world an idea which is highly encouraging because it shows that development and integration of human mind is progressing in every part of the globe. For a shift in paradigm, further researches relating to the youth can be organized in the area of adolescents' attitude to family life education because the rate at which adolescents are getting involved in sexual activities can only be imagined.

CONCLUSION

In this chapter, the characteristics and attitudes of African adolescents in relation to online dating are discussed. Their means of coping with the outcome of online dating is also discussed. The chapter also reveals the position and belief of African parents regarding adolescents' online dating activities which they believe adolescents are delving to too early since adolescence period is the time for adolescents to grow, develop and learn how to cope with later life eventualities. Inference is also drawn to the position of the African culture and tradition regarding adolescents' online dating behavior; an aspect of sexual activities.

Conclusively, it shows that African parents will not support an adolescent getting involved in any aspect of sexual activities including online dating and the fact that the adolescents participate in the online dating game without the knowledge of their parents and the fact that the majority were actually participating in the game not because they expected the relationships to blossom since that can only happen by chance but because their counterparts all over the world are into it since it is part of skills of being computer and internet literate and they want to be part of it. Also, they are involved because of both financial and material gain that can come out of it.

REFERENCES

Albright, J. M., & Conran, T. (2003). Desire, Love and Betrayal: Constructing and Deconstructing Intimacy Online. *Journal of Systemic Therapies*, *22*(3), 42–53. doi:10.1521/jsyt.22.3.42.23352

Alzae, H. (1978). Sexual Behaviour of Columbia Female University Students. *Archives of Sexual Behavior*, *7*(1), 43–53. doi:10.1007/BF01541897

Barbara, S. M., Daniel, B., Wesley, H. C., & Binka, P. (1999). The Changing Nature of Adolescence in the Kassena-Nankana District of Northern Ghana. *Studies in Family Planning*, *30*(2), 95–111. doi:10.1111/j.1728-4465.1999.00095.x

Baucom, H. D., Epstein, N., Rankin, L. A., & Burnett, C. K. (1996). Assessing Relationship Standards: The Inventory of Specific Relationship Standards. *Journal of Family Psychology*, *10*(1), 72–88. doi:10.1037/0893-3200.10.1.72

Baumeister, R. F., & Leary, M. R. (1995). The Need to Belong: Desire for interpersonal Attachment as a Fundamental Human Motivation. *Psychological Bulletin*, *117*, 497–529. doi:10.1037/0033-2909.117.3.497

Ben-Ze'ev, A. (2004). *Love Online*. Cambridge, MA: Cambridge University Press. doi:10.1017/CBO9780511489785

Bledsoe, C. H., & Cohen, B. (1993). *Social Dynamics of Adolescents Fertility in Sub-Sahara Africa.* Washington, D.C: National Academy Press.

Cerpas, N. (2002). Variation in the Display and Experience of Love between College Latino and Non-Latino Heterosexuals Romantic Couples. *The Berkley McNair Research Journal*, *10*, 173–189.

Diepold, J., & Young, R. D. (1979). Empirical Studies of Adolescents Sexual Behaviour: A Critical Review. *Adolescence*, *16*(53), 45–64.

Fafunwa, A. B. (1974). *History of Education in Nigeria.* London: George Allen and union Ltd.

Furman, W., & Wehner, E. A. (1993). Romantic Views: Toward a Theory of Adolescent Romantic Relationships. In Montemayor, R., Adams, G. R., & Gullota, G. P. (Eds.), *Advances in Adolescent Development* (pp. 168–195). Thousand Oaks, CA: Sage Publications.

Grindal, B. T. (1982). *Growing Up in Two Worlds: Education and Transition Among the Sisala of Northern Nigeria.* New York: Irvington Publisher.

Hall, G. S. (1904). *Adolescence*. New York: Appleton.

Hardey, M. (2004). Mediated Relationships: Authenticity and the Possibility of Romance. *Information Communication and Society*, *7*(2), 207–222. doi:10.1080/1369118042000232657

Levine, D. (2000). Virtual Attraction: What Rocks Your Boat. *CyberPsychology & Behaviour*, *3*(4), 565–573. doi:10.1089/109493100420179

Montgomery, M. J., & Sorell, G. T. (1998). Love and Dating Experience in Early and Middle Adolescence: Grade and Gender Comparisons. *Journal of Adolescence*, *21*, 677–689. doi:10.1006/jado.1998.0188

Ojo, D. O., & Fasubaa, O. B. (2005). Adolescent Sexuality and family Life Education in South Western Nigeria: Responses from focus Group Discussion. *Journal of the Social Sciences*, *10*(2), 111–118.

Owuamanam, D. O. (1982). Sexual Activity of School-going Adolescents in Nigeria. *Adolescence, 17*(65), 81–87.

Prezza, M., Guiseppina, M., & Dinelli, S. (2004). Loneliness and New Technologies in a Group of Roman Adolescents. *Computers in Human Behavior, 20*, 691–709. doi:10.1016/j.chb.2003.10.008

Soyinka, F. (1979). Sexual Behaviour among University Students' in Nigeria. *Archives of Sexual Behavior, 8*(1), 15–26. doi:10.1007/BF01541209

Sprecher, S., & Metts, S. (1999). Romantic Beliefs: Their influence on relationships and patterns of change over time. *Journal of Social and Personal Relationships, 16*, 834–851. doi:10.1177/0265407599166009

Walther, J. B., & Burgoon, J. K. (1992). Relational communication in Omputermediated interaction. *Human Communication Research, 19*, 50–88. doi:10.1111/j.1468-2958.1992.tb00295.x

Wikipedia: *The free encyclopedia*. (2004, July 22). FL: Wikimedia Foundation, Inc. Retrieved August 10, 2004, from http://www.wikipedia.org

ADDITIONAL READING

Baker, A. (2002). What makes an online relationship successful? Clues from couples who met in cyberspace. *Cyberpsychology & Behavior, 5*(4), 363–375. doi:10.1089/109493102760275617

Biddlecom, .,A., Awusabo-Asare & Bankole, A. *(2009*). Role of Parents in adolescents sexual Activity And Contraceptive Us in Four African Countries. *International Perspectives on Sexual and reproductive Health, 35(12).*

Bolig, R., Stein, P. J., & McKenry, P. C. (1984). The Self-Advertisement Approach to Dating: Male-Female Differences. *Family Relations, 33*, 587–592. doi:10.2307/583839

Buss, D. M., & Barnes, M. (1986). Preferences in Human Mate Selection. *Journal of Personality and Social Psychology, 50*(3), 559–570. doi:10.1037/0022-3514.50.3.559

Ciairano, S., Bonino, S., Kliewer, W., Miceli, R., & Jackson, S. (2006). Dating, Sexual Activity, and Well-Being in Italian adolescents. *Journal of Clinical Child and Adolescent Psychology, 35*(2), 275–282. doi:10.1207/s15374424jccp3502_11

Ellison, N., Heino, R., & Gibbs, J. (2006). Managing Impressions Online: Self-Presentation Processes in the Online Dating Environment. *Journal of Computer-Mediated Communication, 11*(2). doi:10.1111/j.1083-6101.2006.00020.x

Esere, O. E. (2008). Effect of Sex education programme on at-risk sexual behaviour of school-going adolescents in Ilorin, Nigeria. *African Health Sciences, 8*(2), 120–125.

Fiore, A. T., Taylor, S. T., Zhong, X., Mendelsohn, G. A., & Cheshire, C. *(1899). Who's Right and Who Writes: People, Profiles, Contacts, and Replies in Online Dating.* In proceedings of Hawaii International Conference on System sciences, *43.*

Kiesler, S., Siegel, J., & McGuire, T. W. (1984). Social psychological aspects of computer-mediated communication. *The American Psychologist, 39*, 1123–1134. doi:10.1037/0003-066X.39.10.1123

Kiesler, S., Zubrow, D., Moses, A. M., & Geller, V. (1985). Affection in computer-mediated communication. *Human-Computer Interaction, 1*, 77–104. doi:10.1207/s15327051hci0101_3

Koniak-Griffin, D., Lesser, J., Uman, G., & Nyamathi, A. (2003). Teen pregnancy, motherhood, and unprotected sexual activity. *Research in Nursing & Health, 26*, 4–19. doi:10.1002/nur.10062

Madden, M., & Lenhart, M. (2006). *Online Dating*. Pew Internet & American Life Project.

Makinwa-Adebusoye. (1991). Sexual Activity of Adolescents in Youth and Reproductive Health in Africa. *An Annotated Bibliography on Adolescents Reproductive Health UNFPA, Nigeria.*

Miller, K. S., Kotchick, B. A., Dorsey, S., Forehand, R., & Ham, A. Y. (1998). Family communication about sex: what are parents saying and are their adolescent listening? *Family Planning Perspectives*, *30*, 218–222. doi:10.2307/2991607

Mturi, A.J. (2003). Parents' Attitudes to Adolescent Sexual Behaviour in Lesotho. *African Journal of Reproductive Health / La Revue Africaine de la Santé Reproductive*, 7(2), 25-33.

Muuss, R. (1996). *Theories of adolescence* (6th ed.). New York: McGraw-Hill.

Schoofield, M. (1967). *He Sexual Behaviour of Youngpeople*. Boston: Little,Brown and Company.

Shibazaki, K., & Brennan, K. A. (1998). When birds of different feathers flock together: A preliminary comparison of intraethnic and inter-ethnic dating relationships. *Journal of Social and Personal Relationships*, *15*, 248–256. doi:10.1177/0265407598152007

Tang, S. M., & Zuo, J. P. (2000). Dating attitudes and behaviours of American and Chinese college students. *The Social Science Journal*, *37*, 67–78. doi:10.1016/S0362-3319(99)00066-X

Wekerle, C., & Wolfe, D. A. (1999). Dating Violence in Mid-Adolescence: Theory, Significance, and Emerging Prevention Initiatives. *Child Psychology Review*, *19*, 435–456.

Wells, K. R. (2006). *Teenage Sexuality*. New York: Thomson Gale.

Wight, D., Williamson, L., & Henderson, M. (2006). Parental Influences on young people's sexual behaviour: A longitudinal analysis. *Journal of Adolescence*, *29*, 473–494. doi:10.1016/j.adolescence.2005.08.007

Xie, B. (2007). Using the Internet for offline relationship formation. *Social Science Computer Review*, *25*(3), 396–404. doi:10.1177/0894439307297622

Yum, Y., & Hara, K. (2005). Computer-Mediated Relationship Development: A Cross-Cultural Comparison. *Journal of Computer-Mediated Communication*, *11*(1). doi:10.1111/j.1083-6101.2006.tb00307.x

Zimmer-Gembeck, M. J., Siebenbruner, J., & Collins, W. A. (2001). Diverse aspects of dating: Associations with psychosocial functioning from early to middle adolescence. *Journal of Adolescence*, *24*, 313–336. doi:10.1006/jado.2001.0410

KEY TERMS AND DEFINITIONS

Adolescents: Somebody who has reached puberty but is not yet an adult

Adolescence: the period from puberty to adulthood in human beings. It is the stages that leads to maturity in the development of human being

Coping Strategy: A carefully devised plan of action to cope successfully with a difficult problem or situation

Culture: the beliefs, customs, practices, and social behavior of a particular nation or people

Online Dating: The activity of making contact and communicate with other people over the Internet, usually with the objective of developing a personal romantic or sexual relationship

Family Life Education: is the educational effort to strengthen individual and family life through a family perspective with the intention of enriching and improving the quality of individual and family life

Attitude: an opinion or general feeling about something that guides personal reaction to the thing

Traditional: something that is based on a long-established action or pattern of behavior in a community or group of people, often one that has been handed down from generation to generation.

Chapter 13
The Role of Internet Newsgroups in the Coming-Out Process of Gay Male Youth:
An Israeli Case Study[1]

Avi Marciano
University of Haifa, Israel

ABSTRACT

The study examines internet newsgroups as a potential mitigating tool in the complex coming-out process of gay male youth. Employing a qualitative discourse analysis of the newsgroup's messages, the chapter focuses on an Israeli newsgroup that appeals to GLBT (gay, lesbian, bisexual, transgender) youth and operates within the most popular UGC (user-generated content) portal in Israel. The findings indicate that the researched newsgroup functions as a social arena that offers its participants an embracing milieu, where for the first time in their lives they are free of moral judgment of their sexuality. Through four distinct yet interrelated ways, the newsgroup helps its participants to cope with one of the most significant milestones in a gay person's life – the coming-out process: (1) refuting prevalent stereotypes of homosexuality; (2) facilitating the acceptance of one's sexual orientation; (3) prompting its disclosure; and (4) creating social relations within and outside the virtual environment.

INTRODUCTION

The 1960's and 1970's were charged with significance for the lesbigay struggle. Two decades after the publication of Alfred Kinsey's startling findings (Kinsey, [1948] 1998) regarding the prevalence of homosexuality, three pivotal events occurred: The Stonewall riots in 1969,[1] the removal of homosexuality from the DSM (The American Psychiatric Association's Manual of Mental Disorders) in 1973, and the publication of Foucault's ([1976] 1978) renowned book "The History of Sexuality: The Will to Knowledge" three years later. In tandem with these events scholars abandoned the pathological focus in favor of social and cultural explanations. Consequently, a rich body of research that examined the interrelationship between media and homosexuality took shape.

Alongside these processes, from the 1970's onward the Israeli GLBT community achieved

DOI: 10.4018/978-1-60960-209-3.ch013

impressive developments in societal, cultural and legal spheres, with the founding of the Society for the Protection of Personal Rights (SPPR). Among the societal achievements were the nationwide deployment of the society's branches, the formation of various community-based organizations, and the increasing popularity of the Gay Pride Parade (Kama, 2005; Moriel, 2000). In addition, the community experienced a cultural upsurge, reflected in publication of several GLBT journals, the emergence of queer movies, and a significant improvement in media attitudes toward homosexuality (Kama, 2005; Padva, 2005). In the legal-judicial field too, numerous achievements constituted a "gay legal revolution", as Harel (2000) put it.[2]

Starting in the 1990's, the unique characteristics of the internet attracted substantial attention in media research. However, in spite of the growing body of research that focuses on the internet and homosexuality, there is a notable absence of research assessing the internet's role in the lives of gay youth and more particularly its role in the coming-out process, entitled a "rite of passage" (Bridgewater, 1997) due to its importance for the gay individual. Research linking the internet and gays tends to focus on random sexual relationships, Aids, and other elements (Grov et al., 2007) that formerly played a central role in reducing the homosexual individual (who later became a "gay") to his sexuality.

This investigation seeks to contribute another layer to "the science of oppression", a term coined by Monique Wittig ([1981] 1993) in relation to feminist and lesbian insights based on personal oppressed experience, in order to shed fresh and critical light on the homophobic reality. Many gay male youth undergo an onerous life experience as a result of continuing homophobia. The innovative nature of the internet, historically a new medium, opens up new possibilities. Examining the interaction between these possibilities and the peculiar life experience of gay youth is, therefore, of primary importance. This chapter examines newsgroups as a potential mitigating tool in the complex coming-out process, by tracking the impact of gay youth's participation in those newsgroups, the quality of this impact, and the ways it's achieved. However, the chapter focuses on a specific newsgroup, and therefore there is no intention of generalization. This research is placed in a broader context in which the internet is examined as an empowering tool for various marginalized minority groups.

BACKROUND

Coming-Out of the Closet during Adolescence: Meanings and Implications

The coming-out process consists of three chief stages during which the individual recognizes his sexual orientation, adopts an appropriate identity, and discloses it to others. This process, as depicted by informants, is a linear progression of self discovery whereby the heterosexual identity – enforced and artificial – is gradually replaced by a substantive and genuine gay one (Rust, 2003).

During the 1970's and 1980's several models depicting the gay identity formation were introduced (Cass, 1979; Coleman, 1982; McDonald, 1982). By and large, these models are composed of gradational stages through which the individual faces dilemmas and obstacles, while the last stage is the desired one. Although each model underscores different facets, all of them indicate an initial stage in which the individual identifies himself as heterosexual and sequential stages in which he acknowledges his uniqueness, tries to explain it, strives toward contact with peers, and finally accepts his new identity. While early awareness of homosexual attraction may begin as early as the age of nine (Herdt & Boxer, 1993), acknowledging the orientation and exposing it occur at 15 and 17 respectively. These ages are under a steady decline (Grove et al., 2006).

Coming-out during adolescence is even more complex. As a formative and sensitive period, adolescence poses harsh challenges (Erikson, 1968) that appear to be even harsher for gay adolescents (Garnets & Kimmel, 2003). The fundamental challenge is the heterosexual socialization gay youth go through. Unlike members of other minority groups who have easy access to socialization agents similar to themselves, for gay youth such agents are neither visible nor accessible (Kama, 2002). Consequently, they grow with a default heterosexual identity. During the coming-out process this identity is replaced by a new, extrinsic and stereotyped one, in a process of resocialization. This process is portrayed in the literature as a loss that situates the individual in a state of psychological vulnerability, requiring support and relief (Rust, 2003).

In addition to that fundamental and ongoing challenge, gay male youth come up against quotidian impediments that burden their daily routine and hinder the acquisition of positive identity. Paradoxically, it is the schools – pivotal socialization agents – that fail to function properly. This failure finds expression in the nurturing of a homophobic and oppressive atmosphere (Smith, 2007); in demeaning attitudes of teachers (Meixner, 2006; Mudrey & Medina-Adams, 2006); and in the failure to provide relevant and accurate information on homosexuality (Uribe & Harbeck, 1992). A research report of The Israeli Gay Youth Organization supports this state of affairs and provides data regarding the Israeli education system: Out of 390 GLBT respondents, 31% complained that most students had uttered homophobic comments, 55% reported that teachers had ignored these comments and 23% complained that teachers themselves had made such comments. As to information resources in schools: 45% reported to have access to GLBT internet sites, 15% reported that books and other resources existed in the school library and 50% noted that at least one staff member in the school was "open to the issue" (Pizmony-Levy et al., 2008). The direct implication of the shortfall of information is the inability to refute homophobic prejudices. More importantly, the lack of relevant, accurate information in the first stage of Cass' model (1979) may impede the progression to the next stage and therefore hinder or prevent the consolidation of an affirmative gay identity.

Two additional factors compound the difficulties that gay youth encounter. The first and most significant is frequent abuses, including verbal insults, threats of physical violence, and chasing or following – in ratios of 80%, 44%, and 31% respectively (Pilkington & D'Augelli, 1995). Here too the research report of The Israeli Gay Youth Organization reveals that 20% of the respondents suffered from physical assaults and 36% felt insecure, both as a result of their sexual orientation (Pizmony-Levy et al., 2008). The second factor is parental reaction, which tends to be severe and uncompromising (Mallon, 1998; Thompson, 2001). This reality explains the social isolation that many gay youngsters suffer (Martin & Hetrick, 1988). In this reality, coping with sexual orientation can often end in failure (Beaty, 1999; McDonald, 1982), which is reflected among others in suicide statistics that are three times higher than those of heterosexual youth (Gibson, 1994). The importance of this state of affairs intensifies in view of the positive impact of coming-out on the emotional (Lasala, 2000) and physical (Larson & Chastain, 1990) state of the gay individual. Considering these difficulties, and the life circumstances of marginal groups in general, many researches stressed the actual contribution of the internet as an empowering tool for these groups (Barak & sadovsky, 2008; Bowker & Tuffin, 2007; Mehra et al., 2004; Radin, 2006).

The Newsgroup as a Social Arena

"Virtual communities are social aggregations that emerge [...] when enough people carry on [...] public discussions long enough, with sufficient human feelings, to form webs of personal relationships in cyber-space" (Rheingold, 2000: xx). Virtual communities are based among others on newsgroups, defined as a specific forum style method of communication (Long & Baecker, 1997). Newsgroups belong to the category of computer-mediated communication (Marcoccia, 2004) and are actually a hybrid of interpersonal and mass communication (Baym, 1998). They constitute a multi-participant arena that is not dependent on time (asynchronous) or place (Granit & Nathan, 2000), and characterized by the public nature of its messages (Baym, 1998). Whereas the asynchronicity disrupts its inner dynamics and makes the discussion structure more complex (Marcoccia, 2004), the technical interface allows progressive interactions that counteract this complexity (Donath et al., 1999).

Granit and Nathan (2000) list five types of communities. Two of them are relevant for this study: A supportive community and a social community. The goal of the first is to improve its members' condition, while its principal advantages are the common denominator that unites them and the interactions based on it. The virtual form of the supportive community benefits its members with accessibility in time, accessibility in place and anonymity. The goal of the second type is to create social relations while the benefit of its virtual form lies in the willingness to accept members who suffer from social inferiority, and to help them to create relations as equal peers.

On-line communities and newsgroups in particular provide their members with a sense of belonging (Wellman, 2001) and help them to form and maintain social ties (Boase et al., 2006). These insights were demonstrated in two different empirical studies, where 66% of respondents reported that they felt a sense of belonging to the group (Roberts, 1998) and 61% reported that they formed a personal relationship with someone they "met" in the newsgroup (Parks & Floyd, 1996). Some scholars even credited newsgroups as propitious agents of a social and civil revival (Connolly, 2001).

As to adolescents, the newsgroup functions as an unprecedented arena of information.[3] This merit becomes more prominent in light of the substantial difficulties of youth to obtain information on the web (Dresang, 1999), as well as the unique information needs of GLBT adolescents and particularly the blocks to the provision of information they face (Fikar & Keith, 2004). On-line searches reduce the difficulties of traditional search methods, such as problems of access and embarrassment, whereas the heterogeneity that typifies the newsgroups – where members of different ages and experiences gather – may help the neophytes with guidance. Another merit of the internet at large relates to its role as an arena for mental-aid seekers (Gould, Munfakh, Lubell, Kleinman & Parker, 2002) and to the assuasive potential it may consequently have.

The researched newsgroup is a lesbigay medium. Lesbigay media, as depicted by Kama (2007), are produced by and for the members of the lesbigay community and are able to compensate for the deficiency in the mainstream media. Lesbigay media hold several objectives: Forming a shared consciousness; fostering empowerment and communal consolidation that provides a sense of belonging; validating the self, and remedying alienation. It also serves as a socializing agent that fosters the development of positive gay identity, being a symbolic sphere where gay adolescents feel free of judgment. These merits were stated in research studies that had examined the internet's role in the lives of minority groups, including sexual minorities (Mehra et al., 2004), and gay male youth in particular (Nir, 1998). The first study examined the importance of computer-mediated communication for adult GLBT, and stressed its role as a social support system as well as its contribution to the development of positive "queer" identity, and to the establishment of political awareness. The second study focused on Israeli gay teenagers' involvement patterns in newsgroups. However, both studies refrained from relating to the coming-out process which, as stated, constitutes a major milestone that merits a thorough study.

The above literature portrays the newsgroups as social arenas. Considering the unique life experience of gay male youth, and especially the social consequences of the coming-out process, these arenas seem to be an ideal sphere to embrace this process.

The significance of the internet and newsgroups in particular intensifies due to new statistics that reveal their popularity among youth: 93% of the American youth surf the internet (Lenhart et al., 2010) and over half of them regularly participate in on-line groups, whose number constantly increases (Joyce & Kraut, 2006). Similar data characterize the Israeli case: 92% of the Israeli Jewish youth surf the internet, whereas "only" 34% participate in newsgroups (Rafaeli et al., 2010). The general internet penetration-rate, it should be mentioned, is higher in Israel than in the United States (Internet World Stats, 2009). In light of these statistics, today's youth can be best described as "digital natives" (Palfrey & Gasser, 2008).

METHOD

The research is aimed at illuminating the role of newsgroups for gay male youth in coping with the coming-out process and is based on a qualitative discourse analysis of the messages posted on the newsgroup "Young Pride". The newsgroup, launched in 2002, applies to GLBT youth and operates within the most popular user-generated content portal in Israel.

Since the portal's management decline to furnish details regarding the newsgroup (out of confidentiality), the only available information is the one accessible on the "About" page, where the newsgroup is depicted as a "home for GLBT youth", open to various discussions with emphasis on pride.

The number of active participants is unavailable as well since posting messages is not conditional on subscribing. Nevertheless, during the research the newsgroup included 66 members aged 14 to 20 who had chosen to subscribe and maintain a visible profile, and many more unsubscribed members.

At the beginning of the research, which extended over five months in 2008, the newsgroup's archive contained 9,000 messages (on 600 pages, with 15 messages per page). Of these, 250 messages were sampled by non-probability purposive sampling: Out of the 600 pages, 100 were sampled (in equal increments of six pages), while two or three messages pertaining to the coming-out process or related aspects were sampled from each page. Sampling 250 messages out of 9000 inevitably restricts my interpretation; yet the fact that the messages' topics tend to recur allowed me to classify them into eight thematic categories.[4] The recurrence of the discussion topics, and thus the ability to categorize them, made it possible to study the newsgroup's prevailing mindset and to identify the most pertinent messages.

A discourse analysis of the messages as a sociocultural text has been conducted, assuming that within different social contexts these messages embody concealed meanings (van Dijk, 1990). Those meanings were interpreted on the basis of insights gained from the theoretical section (of the research) in compliance with the inner dynamics of the newsgroup, shaped predominantly by interpersonal relationships. Examining these dynamics through the messages, I assumed that they may reflect reality but also – and maybe primarily – serve as a significant constructing factor of it.

This examination adopts Foucault's ([1969] 1972) insight regarding the nexus between power-relations and discourse, and Butler's (1993) assumption as to the constructive quality of discourse and its ability to shape our perceptions of the normal as opposed to the aberrant.

Discourse Analysis is an umbrella term used to characterize different approaches to the study of textual components and discursive practices

(Tracy, 1995). The messages examined in the research were analyzed by Tracy's (ibid) Action-Implicative Discourse Analysis (AIDA),[5] which focuses on the examination of communicative practices, and the problems and contradictions elicited thereby. Moreover, it aims at revealing the meanings behind them. The text was examined by addressing the interrelationship between the explicit meaning of the text and the broader context in which it appears, in order to indicate how the individual's world, values and tenets are reflected in the text. During the research I avoided any intervention in the newsgroup's inner dynamics in order to grasp its natural atmosphere, unaffected by external manipulations.

DISCUSSION

Participation in the newsgroup helps its participants cope with the coming-out process in four main ways: Refutation of stereotypes; coming to terms with their sexual orientation; encouraging them to reveal it; and creating social ties. The majority of the posts describe feelings of relief, evoked by the participation itself. Among such posts are explicit declarations of the participant's wish to come out of the closet:

(1) It encourages me: the very fact that I posted here brought me **huge relief**, knowing I've got someone to share all my dilemmas and problems with ☺, and since I started posting here (just a few days ago), I feel an even greater need to **shout out loud** and tell everyone.[6]

(2) I think I'm close to being happy: writing here has been so helpful…I feel I'm about to become really **free**! I'm happy to say that I think my close friends will accept it really well… huge thanks to all of you!

As noted, in these messages the writers connect the sense of relief with their desire to come out. According to Rust (2003), coming-out of the closet is indeed easier for gay youth who feel they can connect socially with other gays. That is, the very fact of posting on the newsgroup and creating contact with similar members encourages coming-out.

The Newsgroup's Contribution to Refutation of Stereotypes

A major part of the messages reflect stereotypical perceptions shaped by a biased world of images through which the newsgroup's members see the adult community. Researchers (Beentjes et al., 2001) assert that adolescents devote much of their time to media which inevitably become a chief source that informs their world of images. The importance of the media for young gays intensifies due to the lack of similar agents of socialization, a lack that leads them to rely more strongly on gay images presented in the mass media. These images, to put it mildly, are far from being ideal and were accompanied over a long period by symbolic annihilation and negative stereotypes (Kama, 2002) in various media: Film (Dyer, 2002), television (Croteau & Hoynes, 200) and the press (Alwood, 1996).

(3) Dissatisfied with myself: […] don't get me wrong, I haven't got a problem with lesbians and gays but their way of life looks frightening. I don't want to go to drak [drag] queen parties and live in an **isolated** community, that's **totally frightening** for me […].

(4) A quickie or a serious relationship?: And now there's something that really **worries me**. I'm 17 and I'm so looking for a serious relationship with someone. I feel it's something I really lack in life. […] I know that **life alone is tough** […] I'm afraid that the stigma is true and most gays are looking for quickies. **I don't want to remain alone for the rest of my life** […].

The finding that gay and lesbian youth suffer from a lack of actual role-models (Vincke & van-Heering, 2004) explains the stereotypes discernible in these messages. In the absence of appropriate models, young people rely on partial representations that picture gays as lonely people who waste their time at parties.

Perceiving the stereotypes as true is liable to cause the young people to reject their gay identity, in an attempt to escape such a lifestyle. Beyond that, however, social tagging entails significant psychological implications: Meyer (2003) argues that gays, like any tagged minority group, suffer from chronic tension as a result of the tagging they receive. Society's reaction to deviance causes the tagged individual to develop defensive behavior, with attendant mental symptoms. In practice, it finds expression in self-hatred, in over-shyness, in obsession deriving from the stigmas, and even in rebellion. Goffman (1963) holds that among the implications of the perceived stigma, the individual feels he is "less deserving" and develops mistrust in-, and a sense of alienation from the dominant culture. These feelings coincide with the low self-esteem that characterizes young GLBTs (Garnets et al., 2003), and are evident in messages dealing with love. The vast majority of such messages relate to the lack of love, to perceptions of love as the root of all evil, and as the source of the general depression expressed there. It is hard to assess whether this attitude toward love is the cause or the outcome of low self-esteem, but it can be assumed that the second option is more probable because gay youths are exposed to numerous risk factors such as negative stereotypes and heterosexist attitudes that may foster poor self-esteem.

The discourse on the newsgroup reflects an agreed-on division between two groups: Confused "newbies" who are unfamiliar with lesbigay matters, and senior, self-aware surfers who have gained their position after having "gone through the mill" of experiences typifying the early phases of the coming-out process. The main advantage of this division lies in the tendency of the senior members to refute prevailing stereotypes, as a part of a conscious attempt to ease feelings of distress. In other words, the juxtaposition of the two groups enables the experienced members to provide useful instructions, as reflected in many posts, such as posts nos. 8, 10 and 18 below.

In contrast with the negative influence of stereotypes, positive gay images are likely to help prevent denial of the identity by reducing internalized homophobia (Rust, 2003). Indeed, some of the messages contain links that provide positive role-models, such as GLBT Israeli artists. For example, in a reply to a message claiming that gay couplehood doesn't exist, surfers provided links to articles relating the story of a well-known gay Israeli couple – Dr. Amit Kama and Prof. Uzi Even – who contributed substantially to advancing the Israeli gay community. Gay Israeli respondents, it should be noted, state real longing for normative images of gays in the media (Kama, 2002).

The Newsgroup's Contribution to Reconciliation with Sexual Orientation

Discourse on the newsgroup reveals that some of the youths are not reconciled with their conscious sexual orientation. Unlike the confusion entailed in the sexual orientation, which may occur because they are still in the early stages and due to immaturity, the failure to reconcile with the *conscious* tendency should be examined in terms of internalized homophobia. It has been claimed (Garnets et al., 2003) that this is one of the outcomes of the heterosexist ideology prevailing in society. It goes without saying that internalized homophobia is a delaying factor that one must overcome in order to complete the coming-out process. Many of the messages posted on the newsgroup indeed contain expressions of self-homophobia (examples 5 and 6 below).

(5) **I hate myself:** hate those thoughts, that attraction, that I'm the way I am [...] you don't think perhaps it's true what they say and if I do everything possible, I'll get over it? [...] I'm simply angry at myself [...] I was always different, I've had enough of it!!

(6) "I can't accept it: [...] I don't want to be gay, or a gay in denial. Every day I battle my tendencies, and it's **destroying me from the inside**. What do I have to do to become straight? [i.e., heterosexual]. I don't want to accept myself. Maybe witchcraft would help me, or hormone treatment [...] because I can't live this **split** [...] but I have to, I mustn't give in, otherwise I'll be lost, it's so stupid that God created me a boy and I'm not attracted to girls, what a huge joke, just to see if people can cope with their lives? [...] Is there anyone in the whole world who managed to **change his sexual tendency**, or is it in your genes for ever?

These messages fit the situation described in the first stage of Cass' model of identity formation (Cass, 1979). At this stage, the individual wonders whether his behavior fits the description of homosexual behavior. The thoughts undermine his heterosexual identity and, to resolve the contradiction, he either seeks relevant information or reacts with denial. Denying one's sexual orientation is common on the newsgroup's messages, and it mostly finds expression in posts where the writer wants to learn something from the newsgroup members, but claims that he's straight. Responding to a message where the writer maintained he was straight, but wanted to know how other participants' fathers reacted when they heard about their sons' sexual tendencies – another participant wrote:

(7) You're curious, as a straight guy? Why do you want to hear what our fathers think about it? Do you want to share something with us, maybe? Regards.

According to Rust (2003), such a response, despite its lack of sensitivity, can render an unthinkable situation conceivable. It validates the individual and his denied tendency, helping him to imagine an alternative reality in which he accepts himself. The individual's understanding of his power to bear and cope with that reality may well advance him along the coming-out stages by reducing denial reactions. As noted, in contrast with the option of denial, the healthy option involves attempts to obtain information about the community. The internet in general, and particularly the messages posted on the newsgroup, facilitate the individual to obtain information relevant to him, and thus backs-up the transition to the second stage. In many cases, the information needed to come out of the closet is contained in the talkbacks to messages describing lack of acceptance:

(8) Coming-out of the closet toward yourself: society conditions us to think that the only normative lifestyle is heterosexual. Everyone raised in this world assumes that he's straight [...]. That's the precise reason that coming-out of the closet is such a big deal - it shatters some kind of assumption. Regrettably, the situation you're in is the rule for everyone who isn't straight. A gay who doesn't go through this stage [...] is the exception to the rule [...] gradually you'll start to realize there's nothing wrong about your sexual orientation [...] just think of yourself as a caterpillar that eventually becomes a butterfly.

This talkback was apparently written by a young man somewhat older than the others, and it projects several ideas to the addressee: First of all, there is an attempt to inform the youth about his condition, to locate him on a linear development axis, and also to make it clear that despite the deviance (due to heteronormative definitions) – it's OK. Additionally, the writer tries to show what the future holds ("you'll start to realize") and to sow hope about it ("a caterpillar that eventually becomes a butterfly"). In other words, this response tries to compensate for the instability that, according to Savin-Williams (2005), characterizes many young gays. Similar to that message, hinting that linear development does exist, other messages detail the coming-out stages, and those details are likely to be beneficial since they help identify which stage the youth is at, and its typical problems (Rust, 2003).

The most common reason for rejecting the gay option and for the uncompromising desire to be "straight", as elicited from the messages, is unwillingness to give up the family framework which is perceived possible only as part of a heterosexual lifestyle. Rust (ibid) asserts that during the process of mourning the lost heterosexual identity – a normal and necessary stage in the reconciliation process – the individual forces himself to abandon certain expectations, such as the expectation for a family, while in fact he only should adapt his expectations to the new circumstances. Some of the messages on the newsgroup supply information as to the feasible coexistence of a family framework and a gay way of life:

(9) Just wanted to ask your opinion: [...] I really want to be straight. [...] I know you'll say it's ok to be gay or bi [...] but I want to be straight. I always **dreamed** about having a wife and kids, and I could travel the country to meet my parents and hers. [...] but I still want a regular family [...].

A response:

(10) Want to be straight: [...] I have to say that's one of the stages I went through myself [...] I thought about not having a wife, a home, two kids and a white picket-fence, like I always dreamed. It seems that some dreams are never fulfilled [...] what will happen if in a few years time you fall in love with the man of your dreams? Will you go back to that childhood dream when you were 16 and tell him 'Listen, it's over between us, because I'm looking for a woman?' I'm positive it won't happen. Take your time, and think about it [...].

Other responses offered more concrete information:

(11) If the reason is children, you can take it off the list, because lesbigay parenting [a link to Gay-Lesbian Parenting newsgroup is provided] is starting to be run-of-the-mill, including court cases and adopted kids [...]. If you're sure about yourself, it's a pity to keep on 'trying' to be straight, because it's impossible, **it simply takes your strength away** [...].

(12) And now it's official: gays and lesbians can be the heirs of their partners [a link to an article is provided] – a major step in the struggle for rights.

Unlike response 10 which attempts to "soften" the young man's feelings by expressing identification and pointing to his error, the two other responses provide URLS to internet sites where concrete information is available. The very awareness of the existence of a newsgroup for gay and lesbian parenting may settle the erroneous discrepancy between the familial framework and the gay way of life. As noted, this awareness may encourage self-reconciliation and assists coping with the coming-out process.

In addition to this kind of information, many messages provide information relating to historical, psychological and legal matters relevant to the GLBT community. For example, responses to questions like "what is a bisexual?", "can gays get married?" or "how does sexual identity take shape?". Other messages explain how the Gay Pride Parade was created, or invite newsgroup members to a meeting that will discuss similarities and differences between the gay and feminist struggles. According to Shiloh (2007), as part of the consolidation of identity, youth need information about lesbigay culture, the community's history, the local community's activity, safe sex and so on. Sharing information, particularly during meetings, he continues, may serve as a catalyst for advancing personal development processes and helps boost self-confidence. It is a vital condition for disclosing one's sexual orientation.

The Newsgroup's Contribution to Disclosing Sexual Orientation

The various messages analyzed so far contribute in diverse ways, mostly indirect, to completing the coming-out process. Alongside these messages are more than a few that deal directly with coming-out and its ramifications. Most of them address the connection between coming-out and the family's reaction, which young gays consider an influential and significant factor. Studies indicate that concerns about parental reactions and about damaging the relationship with them after disclosing one's sexual orientation is the central issue that worries teens who attend support-groups (D'Augelli et al., 2002). The newsgroup provides a suitable platform for bringing up troubling questions and dilemmas regarding these issues:

(13) What did you feel when you came out of the closet?! Did you come out to your parents / friends / family?! How did they react?! Did you lose any friends?! Did you come to terms with it right away, or did it take time?! Write here everything that happened!"

(14) I want to come out of the closet: how do I do it, do I tell all my friends and my parents? In a way they will accept?

Answers to these sort of questions are immensely important because the questioners receive reliable answers from people like them, who went through the same experience. This is all the more important when it concerns young gays who suffer during adolescence from a central deficit – the lack of socialization agents who can provide information and advice on related matters. Moreover, messages like these are often answered with practical recommendations and salient information likely to help in the coming-out process:

(15) Warmly recommend: I want to recommend the book "Mom, I have something to tell you", which I think most of you know. It's about gays and lesbians, how they live with it, coming-out of the closet, and how their parents reacted [...]. It also has explanations and answers to questions we all ask, and always wanted to get answers [about]. I'm sure you'll feel better about yourselves after you read the book, which will **fill you with hope** about the next stage [...].

Other messages include declarations about the intention of coming-out, and are generally followed by encouraging, motivating replies. They tend to support the declarations and to urge the writer to take the next step, offering backing and advice on how to come out, as well as noting that another reality is possible – to those participants who view coming-out as an impossible option:

(16) Coming-out: ...me?: Guys, it sounds a bit unexpected but **my head's starting to change**... and I'm actually starting to think about coming-out...[...] yes...I'm aware of the problems ahead but they'll be there anyway. I can already predict their hate for me...the mortification... not allowing me to meet boys ... but I won't give in [...] I'll fight to the end...I'll struggle ... I'll rebel – everything, until they start accepting me the way I am...**it can be done**...[...].

(17) Fateful decision: OK, that's it... I've finally decided! I'm going to tell my sister in a day or two, and then my mother [...] afterwards I'll tell my father and older brother... wish

me luck!!! And a hug won't do any harm, either.

In response:

(18) First of all, **way to go for taking the step!** I hope you're really ready for it, and your parents and friends too. There are lots of ways to do it, but first of all I think you should read an article about coming-out, and then decide if it's really right for you. Then **we'll** move to ways of telling them and how they'll accept it. Look at the Tehila site [gives the URL and a link to a specific article]. [...] And afterwards if you like, I'll gladly help you and answer questions. Anyway – lots of luck!

As discernible in Message 16, this is a participant who changed his mind about the feasibility of coming-out, although he understands that problems may result from taking the step. Coming-out of the closet after taking part in the newsgroup (even passive participation) is done *despite* the exposure to the tough implications it may have. While participating in the newsgroup, the young people read about coming-out experiences that include stringent reactions from society and from the family too. Therefore, the choice to come out of the closet *nevertheless* is necessarily a considered, calculated step that requires self-acceptance. Its significance lies not only in accepting one's sexual orientation but also in the willingness to deal with the results of disclosing it, even if they are harsh. Participating in the newsgroup, as the messages show, helps to overcome some of the barriers and speeds up the disclosure of orientation through support and help, practical hints or suggestions, and providing an attentive ear for the young men's dilemmas.

The Newsgroup's Contribution to Creating Social Ties

Taking part in the newsgroup plays a central role in creating a consolidated sub-community (of newsgroup members) and in linking it up with the broader GLBT community. The contribution is reflected at two levels: At the first conscious level, it seems that the newsgroup members feel part of the community whose members share similar circumstances. At the second physical level the newsgroup serves as a platform for creating social ties with members of the community, while simultaneously offering official and unofficial support.

The messages on the newsgroup provide gay youth with three types of information that helps them join a new social community: Information about communal social groups, about places of entertainment for the community, and about professional support organizations. A supportive social community composed of people with similar identity provides its members with a sense of shared destiny and according to Rust (2003), contributes to coping with the coming-out process, which is inevitably accompanied by the loss of social relationships. Examining the community using Cass' model (1979) makes that contribution clear: Unlike the first stage in the model, where the individual confronts lack of self-acceptance, in the second stage – although he is beginning to accept his queer orientation and consider its implications – he still presents himself as a heterosexual. At the same time, at this stage the individual is likely to allow trusted friends to give him support, and he tries to disclose his identity gradually, attempting to test initial reactions before the "great revelation". At the third stage – the Identity Tolerance stage – the individual is more strongly committed to his new identity, and seeks a GLBT community; the next stage includes strengthening ties with its members. Familiarity with the GLBT community and access to it are therefore almost vital conditions for progressing along the various stages of the coming-out process. Participating in a newsgroup lets the members learn about the very existence of the community, get to know its members, gradually become part of it, and enjoy its potential advantages in coping with the inherent difficulties of the coming-out process. A great number of messages that describe distress and isolation are answered with a recommendation to join a community-based social group:

(19) It's tough in this glass closet: Okay. So I'm gay in the closet and of course everybody guessed it because of my nickname […] I feel really alone […] I'm very frustrated at school, and ashamed to come out in the classroom […]. I tried to meet other gays in chat rooms but they've only got sex on their mind. **Please help me!!!**

In response:

(20) Go somewhere like the Open House: there's one in Jerusalem and one in Tel Aviv, if you come to Jerusalem I'll introduce you to friends... How about it? […]. Every city has its own community, **and you're not the only gay where you live!!!** Go out, get to know gays, go to places where gays get together, but the most recommended thing is meeting groups of gays and lesbians, who talk, enjoy themselves, and make contacts… and you get to meet lots of new people. I hope I've helped you.

An answer like this, which could be perceived as a rather amateur, trivial recommendation, can in fact contribute immensely to consolidating sexual identity and helping the individual with the coming-out process. According to Yalom (1995), lesbigay youth who were in social contact with other lesbigay youth, have a similar health condition to that characterizing heterosexual youth. This finding is notably important in view of the findings presented previously that GLBT youth suffer from low self-esteem and problematic psychological state. Yalom's findings (ibid) also coincide with Cass' (1979) claim that the quality of social ties is a significant factor in the process of coming to terms with one's sexual orientation.

Social groups of lesbigay youth, as the cited reply recommends, help youth to share their feelings with others and to receive validation for their feelings from other young people who have undergone similar experiences. Such groups also help to develop social skills and to get accustomed to a resocialisation process that matches the new circumstances (Shiloh, 2007). The importance of these groups increases in places where the lack of a GLBT community intensifies the difficulty (Rochlin, 1994). Other people post to the newsgroup express a conscious need for involvement in the community:

(21) I assume I won't change schools at this stage […] but what I can do is get to know people from activities outside school. I need to meet new people because the primitive homophobes at my school are driving me crazy!!! I want to know if there are normal people in the world! Can you recommend me a volunteering place […] in the central Israel?

Alongside information about social meetings and places of entertainment, there is great demand for information about professional support and help organizations. Obtaining help from sources that take a positive approach to sexual orientation is likely to be strongly significant in creating a positive gay identity (Klein, 1993). Professional advisers can ask questions that help the individual to imagine himself as gay. Questions such as "how will people around you react to your coming-out, and how will you react to their reaction?" allow the individual to assess more realistically if the coming-out process will be a "worthwhile step". Moreover, questions of this kind can reduce the social alienation that typifies closeted gays, because they allow the individual to imagine himself creating social ties that he perceives as impossible (Rust, 2003).

Participation in the newsgroup therefore enables users obtain information about social groups, places of entertainment, and support groups. The social relationships created within the newsgroup join forces to create a communal framework that facilitates the coping with the coming-out, due to its physical existence but no less because the young boys become aware of its embracing nature. At this stage, when the social isolation is replaced by social ties and feelings of belonging, expressions of pride are likely to surface.

Pride and Subversion

Some of the discussions on the newsgroup transcend the everyday preoccupation with sexual orientation. As such, the newsgroup is also a social arena that encourages debate on general matters that can divert members' attention from the daily difficulties entailed in their sexual orientation. Thanks to this role of the newsgroup, and taking into account the psychological situation of many young gays, as described previously, the newsgroup is important not only as an arena encouraging coming-out, but also as an "island of sanity" that lets them take a break from their complex life circumstances – an ever-present part of their adolescence. Another kind of messages, less common but still noticeable, are those messages expressing a discourse of pride as well as subversion of heterosexual norms:

(22) *Ucht'ch-ism.*[7] A way of life or an incurable illness? [...] I was always more sensitive and had no tendencies for the usual male brutality, but recently I took on the burden of being an *ucht'cha.* Why? **Just because!** [...] if I would want to keep my sexual attraction secret I could talk and act about it differently, more concealed, but that's not my goal [...] You ask yourselves - so why be an *ucht'cha*?! Why?! And I say – **I feel like it.** It's the figure **I choose** to represent myself with [...] If I really want to do something, **I immerse myself in the stereotype.** [...] And so, my dear freaks [...] **standing before you is an *ucht'cha*** [...].

Nicknaming himself as a sissy attests to the acceptance of his identity and even pride in its queerness. Moreover, the content of the message and particularly the way it's phrased, correspond with the central idea of the political lesbigay struggle, during which gays and lesbians adopted the term "queer" precisely because it is homophobic slander. Similar to adopting the term "queer", a young man proudly defining himself as a sissy challenges the "compulsory heterosexuality", a term coined by Rich (1993 [1986]) in the lesbian context. At a deeper level, this message can be identified with Butler's (1990) notion of gender as a performance. She argues that gender is not an essentialist-intrinsic component but merely an expression of physical gestures, subjected to a strict social policing that catalogues them into two genders: Woman and man. A deviant gendered appearance – by adopting gestures of the other gender – enables the individual to undermine the naturalness of the concept and disclose its inherent depressive nature. The message's author **chose** "to appear" as a sissy, a figure characterized by those deviant physical gestures that Butler discusses. In doing so, he rejects the policing norms and refuses to accept the gendered identity he was designated for. In other words, the writer's choice deconstructs the enforced trinity (biology, gendered identity, gendered appearance) and thus subverts its naturalness. In another message, where the writer relates that he came out of the closet during a class, and was applauded by his classmates, he says:

(23) [...] and I know I'm proud and did what I felt, and no one can take my pride away! I wish you all happiness, love and pride!

Although in this case it is not easy to determine conclusively if participation in the newsgroup is what inspired the feelings of pride, or whether it is a case of feelings that took shape outside the newsgroup, in both cases the newsgroup's contribution is discernible. Even if the second option is correct, it is an expression of affirmative feelings on a newsgroup whose members gathered due to shared traits, principally those of deviance, alienation, and rejection by family and society - and hence its importance. For participants who perceive coming-out as an unreasonable option, expressions of pride may sow hope for the future, being proof of absolute self-acceptance.

CONCLUSION

The Young Pride newsgroup is an empowering arena that is capable of easing the coping of young gays with the coming-out process. Its contribution is achieved through four chief ways: Refuting stereotypes; reconciliation with one's sexual orientation; encouraging young people to reveal their orientation; and helping them build social ties.

A discourse analysis of the newsgroup's messages and a study of the inner dynamics prevailing there allow charting an imaginary line. At one end of the line is the closeted individual, confused and lacks social ties with the GLBT community, and at the other is one who has come to terms with his sexual orientation and is seeking to integrate into the community. Between these two poles the four mentioned processes unfold, in parallel with the process of consolidating a gay identity. The newsgroup allows its members to "find their place" along that line, in accordance with their progression (as they perceive it) and to choose the contents most relevant to them. As such, it becomes a social arena that can help a broad spectrum of young people, at whichever stage they are (attesting to this, a 16-year old who presented himself as transgender, a situation considered extreme on this newsgroup, also received appropriate help and support).

The newsgroup's content is embedded in three levels. At the first level – the personal – the individual experiences inner processes, which lead him to reconcile to his orientation as a vital but insufficient condition for coming-out. The second level - the familial – includes processes that help the individual understand his interaction with the family, which he perceives as a significant factor in the process. At the third level – the social – the individual comprehends the relationship between himself as a gay and his social setting. The importance of the three levels (myself, family, society) is that they serve as criteria by which young men assess their progress in the coming-out process, and thus their position as to the gay matter.

In addition, an overall view of the newsgroup's dynamics reveals two contributions at the macro level. They are ostensibly contradictory, but in fact complementary. Above all, the newsgroup is an "island of sanity", precisely because of its ability to detach its members from the homophobic, heterosexual environment – where the young men are subject to constant, inflexible judgment. Moreover, for young gays further along the coming-out process the newsgroup functions as a bridge that connects them with the external GLBT community. In other words, these are two gradual stages: First, the newsgroup is a protective sanctuary allowing them to consolidate their identity as gays, while receiving support from similar people, and only afterward to come out into "the real world".

Nir (1998), who examined gay youths' involvement patterns in newsgroups, regards her respondents as an imagined community (Anderson, 1983), and she distinguishes between the space where it exists and the tangible world of its members. Although the newsgroup initially constitutes an isolated arena, social relations are subsequently formed outside the newsgroup, where the imagined community becomes a tangible one. This is how the newsgroup compensates for the lack of a central need – a healthy socialization process – and socializes its members into the gay world, where for the first time in their lives, their sexual orientation is not an indication.

The potential contribution of newsgroups for young gays coping with the coming-out process supports research studies pointing out the positive influences of the internet on different minority groups, such as cancer patients (Radin, 2006), low-income families (Mehra et al., 2004), the hearing-impaired (Barak & Sadovsky, 2008), and the physically handicapped (Bowker & Tuffin, 2007).

In the light of these findings and the varying patterns of coming-out of the closet among populations with varying socio-demographic characteristics, such as ethnicity, race, and gender (Grov et al., 2006), there is room for future quantitative research that examines the correlation between internet access, internet usage patterns and scope, and surfers' coming-out experiences. The percentage of internet use is higher among the young, educated, and affluent (Slevin, 2000). A potential correlation between internet-use patterns of gays belonging to different cultural groups, and the coming-out experience (if it occurs, at which age, and so on) might be complex, but certainly interesting and important.

ACKNOWEDGMENT

* I want to express my heartfelt gratitude and appreciation to Dr. Einat Lachover for her supporting and enriching guidance. I also want to thank Dr. Rivka Ribak, Dr. Michele Rosenthal, and Mr. Elad Hamo for their contributions and support.

REFERENCES

Alwood, E. (1996). *Straight news: Gays, lesbians and the news media*. New York: Columbia University Press.

Anderson, B. (1983). *Imagined communities: Reflections on the origin and spread of nationalism*. London: Verso.

Barak, A., & Sadovsky, Y. (2008). Internet use and personal empowerment of hearing impaired adolescents. *Computers in Human Behavior*, *24*(5), 1802–1815. doi:10.1016/j.chb.2008.02.007

Baym, N. K. (1998). The emergence of on-line community. In Jones, S. G. (Ed.), *Cybersociety 2.0: Revisiting computer-mediated communication* (pp. 35–68). Thousand Oaks, CA: Sage.

Beaty, L. A. (1999). Identity development of homosexual youth and parental and familial influences of the coming out process. *Adolescence*, *34*(135), 597–601.

Beentjes, J. W. J., Koolstra, C. M., Marseille, N., & van der Voort, T. H. A. (2001). Children's use of different media: For how long and why? In Livingstone, S., & Bovill, M. (Eds.), *Children and their changing media environment: A European comparative study* (pp. 85–112). Mahwah, NJ: Lawrence Erlbaum.

Boase, J., Horrigan, J. B., Wellman, B., & Rainie, L. (2006). *The strength of internet ties*. Pew Internet and American Life Project. Retrieved March 10, 2010 from http://www.pewinternet.org/~/media//Files/Reports/2006/PIP_Internet_ties.pdf.pdf

Bowker, N. I., & Tuffin, K. (2007). Understanding positive subjectivities made possible online for disabled people. *New Zealand Journal of Psychology*, *36*(2), 63–71.

Bridgewater, D. (1997). Effective coming out: Self-disclosure strategies to reduce sexual identity bias. In Sears, J. T., & Williams, W. L. (Eds.), *Overcoming heterosexism and homophobia: Strategies that work* (pp. 65–75). New York: Columbia University Press.

Butler, J. (1990). *Gender trouble: Feminism and the subversion of identity*. New York: Routledge.

Butler, J. (1993). Critically queer. *GLQ: A Journal of Lesbian and Gay Studies, 1*(1), 17-32.

Cass, V. C. (1979). Homosexual identity formation: A theoretical model. *Journal of Homosexuality, 4*(3), 219–235. doi:10.1300/J082v04n03_01

Coleman, E. (1982). Developmental stages of the coming-out process. *Journal of Homosexuality, 7*(2/3), 31–43. doi:10.1300/J082v07n02_06

Connolly, R. (2001). The rise and persistence of the technological community ideal. In Werry, C., & Mowbrey, M. (Eds.), *Online community: Commerce, community action and the virtual university* (pp. 317–364). Upper Saddle River, NJ: Prentice Hall.

Croteau, D., & Hoynes, W. (2000). *Media society: Industries, images, audiences*. Thousand Oaks, CA: Pine Forge.

D'Augelli, A. R., Pilkington, N. W., & Hershberger, S. L. (2002). Incidence and mental health impact of sexual orientation victimization of lesbian, gay, and bisexual youths in high school. *School Psychology Quarterly, 17*(2), 148–167. doi:10.1521/scpq.17.2.148.20854

Donath, J., Karahalios, K., & Viegas, F. (1999). Visualizing conversation. *Journal of Computer-Mediated Communication, 4*(4). Retrieved March 10, 2010 from http://jcmc.indiana.edu/vol4/issue4/donath.html

Dresang, E. T. (1999). More research needed: Informal information-seeking behavior of youth on the internet. *Journal of the American Society for Information Science American Society for Information Science, 50*(12), 1123–1124. doi:10.1002/(SICI)1097-4571(1999)50:12<1123::AID-ASI14>3.0.CO;2-F

Dyer, R. (2002). *The matter of images: Essays on representation*. London: Routledge.

Erikson, E. H. (1968). *Identity, youth, and crisis*. New York: W. W. Norton.

Fikar, C. R., & Keith, L. (2004). Information needs of gay, lesbian, bisexual, and transgender health care professionals: Results of an internet survey. *Journal of the Medical Library Association, 92*(1), 56–65.

Foucault, M. (1972). *The archeology of knowledge and the discourse on language*. New York: Pantheon. (Original work published 1969)

Foucault, M. (1978). *The history of sexuality 1: An introduction*. New York: Random House. (Original work published 1976)

Garnets, L. D., Herek, G. M., & Levy, B. (2003). Violence and victimization of lesbians and gay men: Mental health consequences. In Garnets, L. D., & Kimmel, D. C. (Eds.), *Psychological perspectives on lesbian, gay and bisexual experiences* (pp. 188–206). New York: Columbia University Press.

Garnets, L. D., & Kimmel, D. C. (2003). *Psychological perspectives on lesbian, gay and bisexual experiences*. New York: Columbia University Press.

Gibson, P. (1994). Gay male and lesbian youth suicide. In Remafedi, G. (Ed.), *Death by denial* (pp. 15–64). London: Alyson Publication.

Goffman, E. (1963). *Stigma: Notes of the management of spoiled identity*. Upper Saddle River, NJ: Prentice-Hall.

Gould, M. S., Munfakh, J. L. H., Lubell, K., Kleinman, M., & Parker, S. (2002). Seeking help from the internet during adolescence. *Journal of the American Academy of Child and Adolescent Psychiatry, 41*(10), 1182–1189. doi:10.1097/00004583-200210000-00007

Granit, E., & Nathan, L. (2000). Virtual communities: A new social structure? [Hebrew]. *Megamot, 40*(2), 298–315.

Grov, C., Bimbi, D. S., Nanin, J. E., & Parsons, J. T. (2006). Race, ethnicity, gender and generational factors associated with the coming-out process among gay, lesbian, and bisexual individuals. *Journal of Sex Research*, *43*(2), 115–121. doi:10.1080/00224490609552306

Grov, C., Debusk, J. A., Bimbi, D. S., Golub, S. A., Nanin, J. E., & Parsons, J. T. (2007). Barebacking, the internet, and harm reduction: An intercept survey with gay and bisexual men in Los Angeles and New York City. *AIDS and Behavior*, *11*(4), 527–536. doi:10.1007/s10461-007-9234-7

Harel, A. (2000). The rise and fall of the Israeli gay legal revolution. *Columbia Human Rights Law Review*, *31*(2), 443–471.

Herdt, G., & Boxer, A. M. (1993). *Children of horizons: How gay and lesbian teens are leading a new way out of the closet*. Boston: Beacon.

Internet World Stats. (2009). *List of countries classified by internet penetration rates*. Retrieved March 10, 2010 from http://www.internetworldstats.com/list4.htm

Joyce, E., & Kraut, R. E. (2006). Predicting continued participation in newsgroups. *Journal of Computer-Mediated Communication*, *11*(3), 723–747. doi:10.1111/j.1083-6101.2006.00033.x

Kama, A. (2002). The quest for inclusion: Jewish-Israeli gay men's perceptions of gays in the media. *Feminist Media Studies*, *2*(2), 195–212. doi:10.1080/14680770220150863

Kama, A. (2005). GLBT issues in Israel. In Sears, J. T. (Ed.), *Youth, education, and sexualities: An international encyclopedia* (pp. 448–453). Westport, CT: Greenwood.

Kama, A. (2007). Israeli gay men's consumption of lesbigay media. In Barnhurst, K. G. (Ed.), *Media Q, Media\Queered: Visibility and its discontents* (pp. 125–142). New York: Peter Lang.

Kinsey, A. C. (1998). *Sexual behavior in the human male*. Indiana: Indiana University Press. (Original work published 1948)

Klein, F. (1993). *The bisexual option*. New York: Harrington Park.

Larson, D. G., & Chastain, R. L. (1990). Self-concealment: Conceptualization, measurement, and health implications. *Journal of Social and Clinical Psychology*, *9*(4), 439–455.

Lasala, M. C. (2000). Lesbians, gay men and their parents: Family therapy for the coming-out crisis. *Family Process*, *39*(1), 67–81. doi:10.1111/j.1545-5300.2000.39108.x

Lenhart, A., Purcell, K., Smith, A., & Zickuhr, K. (2010). *Social media and mobile internet use among teen and young adults*. Retrieved March 10, 2010 from http://pewinternet.org/~/media//Files/Reports/2010/PIP_Social_Media_and_Young_Adults_Report.pdf

Long, B., & Baecker, R. (1997). A taxonomy of internet communications tools. Retrieved April 1, 2008 from http://www.dgp.toronto.edu/people/byron/webnet/Taxonomy.html

Mallon, G. P. (1998). Knowledge for practice with gay and lesbian persons. In Mallon, G. P. (Ed.), *Foundations of social work practice with lesbian and gay persons* (pp. 1–30). New York: Haworth Press.

Marcoccia, M. (2004). On-line polylogues: Conversation structure and participation framework in internet newsgroups. *Journal of Pragmatics*, *36*(1), 115–145. doi:10.1016/S0378-2166(03)00038-9

Martin, A. D., & Hetrick, E. S. (1988). The stigmatization of the gay and lesbian adolescent. In Ross, M. W. (Ed.), *Psychopathology & psychotherapy in homosexuality* (pp. 163–184). New York: Haworth Press.

McDonald, J. G. (1982). Individual differences in the coming-out process for gay men: Implications for theoretical models. *Journal of Homosexuality*, *8*(1), 47–60. doi:10.1300/J082v08n01_05

Mehra, B., Merkel, C., & Bishop, A. P. (2004). The internet for empowerment of minority and marginalized users. *New Media & Society*, *6*(6), 781–802. doi:10.1177/146144804047513

Meixner, E. (2006). Teacher agency and access to LGBTQ young adult literature. *Radical Teacher*, *76*, 13–19.

Meyer, I. H. (2003). Minority stress and mental health in gay men. In Garnets, L. D., & Kimmel, D. C. (Eds.), *Psychological perspectives on lesbian, gay and bisexual experiences* (pp. 699–731). New York: Columbia University Press.

Moriel, L. (2000). Israel and Palestine. In Haggerty, G. E. (Ed.), *Gay histories and cultures: An encyclopedia* (pp. 481–484). New York: Garland Publishing.

Mudrey, R., & Medina-Adams, A. (2006). Attitudes, perceptions, and knowledge of pre service teachers regarding the educational isolation of sexual minority youth. *Journal of Homosexuality*, *51*(4), 63–90. doi:10.1300/J082v51n04_04

Nir, L. (1998, July). *A site of their own: Gay teenagers' involvement patterns in IRC and newsgroups*. Paper presented at the international communication conference, Jerusalem, Israel.

Padva, G. (2005). Israel filmmaking. In Gerstner, D. (Ed.), *Routledge international encyclopedia of queer culture: Gay, lesbian, bisexual and transsexual contemporary cultures* (pp. 312–313). New York, London: Routledge.

Palfrey, J., & Gasser, U. (2008). *Born digital: Understanding the first generation of digital natives. New-York*. Basic books.

Park, M. R., & Floyd, K. (1996). Making friends in cyberspace. *Journal of Computer-Mediated Communication*, *1*(4), 80–97.

Pilkington, N. W., & D'Augelli, A. R. (1995). Victimization of lesbian, gay and bisexual youth in community settings. *Journal of Community Psychology*, *23*(1), 34–56. doi:10.1002/1520-6629(199501)23:1<34::AID-JCOP2290230105>3.0.CO;2-N

Pizmony-Levy, O., Kama, A., Shilo, G., & Lavee, S. (2008). Do my teachers care I'm gay?: Israeli lesbigay school students' experiences at their schools. *Journal of Gay and Lesbian Youth*, *5*(2), 33–61.

Radin, P. (2006). "To me it's my life": Medical communication, trust, and activism cyberspace. *Social Science & Medicine*, *62*(3), 591–601. doi:10.1016/j.socscimed.2005.06.022

Rafaeli, S., Ariel, Y., & Katsman, M. (2010). *Adolescents online: Patterns of Internet usage and online consumption*. Ministry of Industry, Trade and Labor: Research and Economy Administration. Retrieved March 10, 2010 from http://www.moit.gov.il/NR/rdonlyres/6F6300A3-336D-4A8D-8D44-9213065DF8F7/0/9784.pdf.

Rheingold, H. (2000). *The virtual community: Homesteading on the electronic frontier*. Cambridge, MA: MIT Press.

Rich, A. (1993). Compulsory heterosexuality and the lesbian existence. In Abelove, H., Barale, M. A., & Halperin, D. M. (Eds.), *The lesbian and gay studies reader* (pp. 227–254). New York, London: Routledge. (Original work published 1986)

Roberts, T. L. (1998). *Are newsgroups virtual communities*? Proceedings of the Annual ACM SIGCHI Conference on Human Factors in Computing Systems (CHI 1998). Pp. 360-367. New York: ACM Press.

Rochlin, M. (1994). Sexual orientation of the therapist and therapeutic effectiveness with gay clients. In Gonsiorek, J. (Ed.), *A guide to psychotherapy with gay and lesbian clients* (pp. 21–29). New York: Harrington Park Press.

Rust, P. C. (2003). Finding sexual identity and community: Therapeutic implications and cultural assumptions in scientific models of coming out. In Garnets, L. D., & Kimmel, D. C. (Eds.), *Psychological perspectives on lesbian, gay and bisexual experiences* (pp. 227–269). New York: Columbia University Press.

Savin-Williams, R. C. (2005). *The new gay teenager*. Boston: Harvard University Press.

Shilo, G. (2007). *Pink life: Gay, lesbian, bisexual and transgender youth*. TA, Israel: Resling. (Hebrew).

Slevin, J. (2000). *The internet and society*. Cambridge, UK: Polity Press.

Smith, J. (2007). ''Ye've got to 'ave balls to play this game sir!": Boys, peers and fears: The negative influence of school-based "cultural accomplices" in constricting hegemonic masculinities. *Gender and Education, 19*(2), 179–198. doi:10.1080/09540250601165995

Thompson, M. (2001). Coming out inside. In Berzon, B. (Ed.), *Positively gay: New approaches to gay and lesbian life* (pp. 32–38). Berkley, CA: Celestial Arts.

Tracy, K. (1995). Action-implicative discourse analysis. *Journal of Language and Social Psychology, 14*(1-2), 195–215. doi:10.1177/0261927X95141011

Uribe, V., & Harbeck, K. M. (1992). Addressing the needs of lesbian, gay and bisexual youth: The origins of Project 10 and school-based intervention. In Harbeck, K. M. (Ed.), *Coming out of the classroom closet: Gay and lesbian students, teachers and curricula* (pp. 9–29). New York: Haworth.

van Dijk, T. A. (1990). Discourse analysis in the 1990's'. *Text, 10*(1-2), 133–156. doi:10.1515/text.1.1990.10.1-2.133

Vincke, J., & van-Heering, K. (2004). Summer holiday camps for gay and lesbian young adults: An evaluation of their impact on social support and mental well-being. *Journal of Homosexuality, 47*(2), 33–46. doi:10.1300/J082v47n02_02

Wellman, B. (2001). Physical place and cyber place: The rise of personalized networking. *International Journal of Urban and Regional Research, 25*(2), 227–252. doi:10.1111/1468-2427.00309

Witting, M. (1992). *The straight mind and other essays*. Boston: Beacon Press.

Yalom, I. D. (1995). *The theory and practice of group psychotherapy*. New York: Basic Books.

KEY TERM AND DEFINITIONS

Computer-Mediated Communication: The full range of communicative acts that occur through the use of networked computers. Research in the field of CMC usually deals with the social implications of networked technologies.

Heterosexism: An ideological array of attitudes, acts or institutions that nurture the belief in the superiority of heterosexuality and discriminate none-heterosexual people on that basis.

Homophobia: Antagonism, fear or hatred toward those who love and sexually desire others of the same sex. Homophobic attitudes derive from prejudice and find expression in discrimination, harassment or acts of violence.

Newsgroup: One kind of a networked discussion group. Newsgroup is a multi-participants forum mostly focused on a specific topic of interest. Unlike some other kinds of discussion groups, the newsgroup is asynchronous and usually public.

Stereotype: A common public belief directed toward a social group or "representatives" individuals. The stereotype is a simplified conception based on a prior assumption in order to reduce complexity, and it may result in generalization and prejudice.

Virtual Community: A network of individuals who maintain social ties in cyber-space. The virtual community is not place-bound and therefore it potentially transcends traditional boundaries such as geographical, racial, political etc.

ENDNOTES

1. The Stonewall Inn was a New York bar for gays that, like similar venues, suffered from frequent police raids accompanied by violence, humiliation and arrests. On the evening of 27 June 1969, the raiders encountered powerful resistance from the guests in the bar, which ended after three days of riots (Alwood, 1996).
2. On 1 August 2009, a gunman opened fire at a gay youth center in Tel Aviv and killed two people. The incident served as a rude awakening to the local community that has been experiencing a very gay-friendly atmosphere.
3. Surfing the internet, it should be noted, is fraught with dangers. This chapter focuses on the ways by which the participation in newsgroups shapes the coming-out process, and therefore it does not elaborate on these dangers. However, this is not to deny the concrete negative aspects of surfing.
4. These are the subjects of the posts on the newsgroup: requests for information on a specific subject; love matters; expressions of confusion and disappointment about the writers' sexual orientation; a direct reference to the newsgroup's centrality in the young people's lives; the coming-out experience; expressions of depression alongside expressions of pride; announcements about newsgroup meetings; and discussions of stereotypes.
5. Under this approach, the discourse-analysis process is principally ethnographic, and requires the researcher to be familiar with the examined environment. In this context, it should be noted that my personal experience with the coming-out process naturally structures the worldview that runs through the textual analysis.
6. Message structure: a chronological number added for convenience and for common language (in parentheses); the message title as written by the surfers (unless no title was written); a colon, followed by the message's content. The emphases as well as all the text in square parentheses are mine. Three dots in square parentheses [...] reflect deleted text.
7. From the word "*Ucht'cha*". It is commonly used in Israel's GLBT community as a derisive nickname for feminine flamboyant gays (analogous to "sissy"), and is part of a whole language that has developed in Israel's GLBT community. Israeli gays tend to talk about themselves using the feminine (Hebrew is a gender-biased language). *Ucht'cha* is a corrupted form of the word *Uchti* - Arabic for "my sister".

Section 4
Contemporary Challenges

Chapter 14
The Competent Youth's Exposure of Teachers at YouTube.se

Marcus Samuelsson
Linköping University, Sweden

ABSTRACT

Swedish children are generally well informed, from preschool all the way through the entire school system, about the meaning of the Convention on the Rights of the Child, CRC (United Nations, 1989), and other similar ideas about their position in relation to adults in school, adults at home and adults in the society. As a result of that experience, almost 30.000 hours in school as an institution, it is possible to argue that a lot of Swedish children, through discussions are well trained to act according to the idea of the competent child. But how do adults, teachers, researchers and journalists react when competent youth use their knowledge to expose what they have experienced in school on websites like YouTube? In order to answer such a question this chapter examines two movies of exposed teachers on YouTube.se. These movies, one with an angry teacher and one with a jocular teacher, are subsequently used in order to argue that exposures like these of teachers are reasonable to understand as being made by the competent youth.

INTRODUCTION

Schools and the events going on during ordinary lessons have in one sense, at least in the western society where school buildings exist, been closed from direct insight until the Internet exploded during the 1990s. Even so youth, parents, teachers, headmasters, authors as well as news media have been aware of some of the doubtful things that happen in certain classrooms, as part of certain teachers' ways of managing their work behind closed classroom doors. Many teachers and headmasters previously used to look the other way. They were so to speak, able to turn their back at children's, youth's and in some cases parent's stories about what was going on at the lessons in school. This phenomenon has earlier been described as an example of teachers' ethical

DOI: 10.4018/978-1-60960-209-3.ch014

dilemmas, called the *collegial paradox* (Colnerud, 1997; 2002). The collegial paradox was shown in such a way that teachers constructed silent alliances with each other when doubtful rumours or accusations, mostly from offended students, where exposed. By doing that, they took sides for their colleagues against the students, even if they knew that there was no smoke without fire. But, teachers and headmasters who recently used to turn a blind eye when other teachers' behaviours towards children or youth opposed for example *Convention on the Rights of the Child, CRC* (United Nations, 1989), *Lärares Yrkesetik, LY* (Lärarförbundet & Lärarnas Riksförbund, 2002) or a legislation such as *The Child Protection Law, TCPL* from 1 of April 2006 are nowadays forced to react one way or another, as a consequence of witnesses of teachers' violations that are to be found in movies at different websites. The growth of Internet that started 20 years ago has, together with the development of mobile phones, especially the new techniques for sound recording and filming, challenged the until now more or less closed school system. One of the latest, but possibly not the last, ways of obtaining insight in great parts of youth institutionalized school life (Sigsgaard, 1992; Samuelsson, 2008) was the introduction of websites as YouTube.com in February 2005 and later on YouTube.se in October 2008.

The amount of movies exposing teachers' classroom leadership during 2008-2009 where described as a major problem for the teacher profession

One out of twenty teachers has during the last twelve month period sometime been harassed in relation to his or her occupation by being registered with a mobile phone. (Lärarnas Riksförbund, 2008, p. 5)

The same evaluation reported that as many as eleven out of hundred answered that they knew another teacher who during the last twelve month period had been harassed on the Internet. According to such experiences it is quite easy to understand that teachers feel as if their hands are tied in such a way that they were, or became, afraid to interfere and as a consequence of that perhaps where filmed and later on exposed on websites like YouTube. Almost the same conclusions were drawn by Honawar (2007) from interviews with teachers in USA. The general council for the American Federation of Teachers said that fear of being taped could change how teachers interact with students and is therefore "disturbing to the educational process" (Honawar, 2007, p. 2). Another consequence, according to the general council was that the whole environment could be affected.

Now, concern is growing among teacher advocates that the proliferation of such videos is causing stress among teachers and some students, and could have a chilling effect on classroom discussions. (Honowar, 2007, p. 2)

Such an experience can affect the classroom situation as well as teachers' ways of managing the classroom. Woolfolk Hoy and Weinstein (2006) suggest that students' and teachers' beliefs in good classroom management can be seen as opposite conceptions. For students good classroom management was built up around a

Fair and reasonable system of classroom rules and procedures that protect and respect students. Teachers are expected to care for the students, their learning and their personal lives, before the students will respect and cooperate with the teachers. (Woolfolk Hoy & Weinstein, 2006, p. 209)

Students didn't mind differential treatment and they wanted, as long as there were no classism, favoritism, racism or sexism, teachers to maintain order without punishment while teachers seemed to believe that students needed to earn their concern, interest, respect and relationship as well as their caring.

Choices and autonomy support come with successful self-regulation and not before. And some teachers believe that being "mean" may be necessary, in the beginning at least, to establish authority – don't smile until Christmas. (Woolfolk Hoy & Weinstein, 2006, p. 209)

As Sheets and Gay (1996) stated, it is possible to describe these contrasting views as a downward spiral of mistrust where students only start cooperating first when the teacher has earned it and the teacher withholds their caring until the students have earned it.

In this chapter I aim, with this knowledge in mind, to problematize how the content of movies of teachers at websites such as YouTube.se could be understood. The starting point of such a problematization is an interpretation (Kvale 1997). Using an interpretation one could argue that the exposure could be understood as an act of the competent youth. Such an interpretation could perhaps be understood as provocative since it challenges a great part of the news media and teacher unions' discourse about the phenomena. Regardless of that I start by describing what kind of movies of teachers are to be found at YouTube.se. Next, I provide a description of the content in the movies with different sorts of teachers. This is followed by a discussion where the result is discussed in the light of the idea of the competent youth. The chapter is finally concluded with some implications where I argue for some lessons that teachers and people working in school, as well as adults could learn from the interpretations made in this chapter.

ADOPTING NEW LOGICS AND BEHAVING PROVOCATIVELY

In order to understand the content of this chapter it is important to also bear in mind that YouTube was first launched as a way of exposing the creators of different kinds of movies. Exposure on websites in general and YouTube specifically is, as the logo says *Broadcast Yourself*. It's about the directors showing themselves through the movies they have uncovered on the website. On the basis of this it is reasonable to understand that youth as *digital natives* (Prensky, 2001) or *internet natives* (Dunkels, 2005) would not exclude classrooms, schools or teachers as phenomena interesting enough to record and expose on YouTube. A second function of websites like YouTube has to do with the different sorts of movies that are found. Movies on YouTube could be categorized according to two sorts of logics: *entertainment* or *enlightenment.*

An ordinary search at YouTube showed many examples of all *entertainment logics*. Through further and more sensible searches examples of *enlightenment logics* were also found. A recently exposed example of the first *enlightenment logic*, namely *give publicity to unfairness* was possible to take part of in November 2009 when Russian policeman Alexsej Dymovskij revealed a video where he spoke of experienced corruption among the Russian police force and demanded that Russia's Prime Minister Vladimir Putin would do something. In less than a week the movie was seen by 700.000 visitors on YouTube. Examples of the second *enlightenment logic*, i.e. *illustrations of role models* are movies from television productions like Britain's Got Talent, where Paul Potts and later on Susan Boyle were exposed. Through examples like these youth are, over and over again, taught how effective websites like YouTube are

Table 1. Exposing logics for YouTube movies

Logic Category
Entertainment Disadvantage of others
Irony
Self-exposing
Enlightenment Give publicity to unfairness
Illustrations of role models
Protecting oneself

in attending to problems as well as showing good examples. A well-known example of the third *enlightenment logic - protecting oneself* are the snapshots that were taken by the prison guard Charles Graner at Abu Ghraib prison showing him and other female soldiers such as Lynndie England offending Iraqi prisoners (Gourevitch & Morris, 2008s) i.e. that youth through websites like YouTube are given a new arena to use in order to affect and establish changes in the bigger perspective all over the world and in the smaller perspective in the classroom and at school.

Teachers' experiences of websites like YouTube are described by Honawar (2007) who stated that English-based websites exposed angry teachers as well as enthusiastic teachers. Honowar continues and reports from several interviews with teachers from different parts of USA who find it unpleasant to see small parts, sequences of their up to 45 minutes long lessons grow out of context and later on be presented for a worldwide public. As a way to handle such a phenomenon one of the interviewed teachers said that he, once in a while, checked the web in order to find out if he had been exposed in any kind of movies. Brown (2007) found out that Canadian teachers unanimously had voted for creating a national policy on how cell-phones were to be used in schools and how students who hurt someone's feelings, students or teachers, should be punished. The background for this was that

Cyberbullying in all its forms, from Facebook insults to YouTube ambushes, has skyrocketed in recent months to become the number one non-academic problem facing classroom today, says CTA [Canadian Teachers' Federation] Winston Carter. (Brown, 2007, July 14).

Honawar (2007) meant that there was a tendency in USA for that as well

School districts tend to ignore videos that are simply embarrassing to a teacher, but do act when they find that the taping is a threat to the school or teacher or is disruptive to learning. (Honawar, 2007, p. 4)

In 2006, The Professional Association of Teachers in England went so far that as demanding that YouTube should be forbidden since the responsible company encouraged net bullying (Honowar, 2007). Another aspect was found in Canada where four students were expelled from school after spreading a video of a teacher scolding a student. The other way round was described in an article by Pickett and Thomas (2006, p. 41) describing how a teacher in USA was dismissed as a consequence of saying to the students that "you are a bunch of [replaced word]".

Research like the above seem to focus on the problems with new technique such as cell-phones and websites while others (Barack, 2007) report about several educational debaters who highlights the possibilities rather than the threats of new technology. One of them depicts about cell-phones: "Eventually they'll be like pencils. No different" (Barack, 2007, p. 1). The same former teacher continues by saying

Schools need to reconsider their perspective on cell phones. They're potentially a powerful tool for collaborative learning. (Barack, 2007, p. 2)

Similar arguments were expressed by the chairman of the Unicef UK and the educational think tank Futurelab who warned that

Technology-savvy children were switching off and becoming 'emotional truants' because schools were not relevant in a digital age. (Milne, 2008, p. 1)

A similar description is found in Shuler (2009) who describes how digital and mobile technologies are changing young people's knowledge and ideas about learning outside the classroom while they experience that "none of that expertise or engagement currently has currency inside the classroom" (Shuler, 2009, p. 21).

THE COMPETENT CHILD AS PART OF THE YOUTH CULTURE

Twenty years ago the *Convention on the Rights of the Child, CRC* (United Nations, 1989), was ratified. The convention was later on followed by expanded focus on different sorts of children's perspective. As a part of such a focus, a broader perspective on research and thinking developed the ideas of *the competent child*, a contemporary representation where the child is seen as reflexive, autonomous and robust (Alanen, 1992; Prout & James, 1997; Brembeck, Johansson & Kampman, 2004) The conception of *the competent child* has since *CRC* was ratified been developed in both child culture research as well as in educational research. Brembeck, Johansson & Kampmann (2004, p. 7) argue that "the idea of the competent child is a fundamental part of the modernity project in most Nordic countries, relating to modern institutions". Even so, children and youth, being seen as immature are often thought of as different from those who are mature, like the difference between those who educate and bring up and those who are educated and brought up (Brembeck, Johansson & Kampman, 2004).

On the contrary, the fast changes in the lives of today´s children and in the notions of children as competent have caused a certain amount of turbulence. (Brembeck, Johansson & Kampman, 2004, pp. 19-20)

In relation to facts like these, Brembeck, Johansson and Kampman, (2004) mean that today's child who, in her or his own words or acts, is supposed to be able to express "needs" in order to be an autonomous and self-regulating child must be addressed and treated in other ways than children previously were.

Another aspect of the contemporary competent child is discussed by Kryger (2004) who reminds us not to forget some of the elements of the term lifelong learning. According to this idea the learners are supposed to design their own routes, where learning is seen as a never ending process, "at one's own risk and one's own responsibility" (Kryger, 2004, p. 155). In this learning the teacher becomes more of a facilitator, a grown up helper, than an ordinary teacher while the flexible citizen is created. As a consequence of that children are more or less seen as entrepreneurs, shaping their own learning pathways.

In relation to such an approach it is important to remember that a lot of Swedish children go to pre-, primary and secondary schools where placards such as *Jäntelagen* (Palm, 2000), an antonym to *Jantelagen* by Sandemose (1968), are nailed to the walls or on bulletin boards:

You shall believe that you are something

1. You shall believe that you are something.

2. You shall believe that you are as good as others and that everyone else are as good as you.

3. You shall believe that you are as wise as others, and sometimes wiser.

4. You shall know that you are as good as others. Knowing that you did your best gives you an opportunity to appreciate those who are better.

5. Sometimes you know more than others.

6. You are not more than others but you are outstanding like others.

7. You are good enough.

8. Laugh at yourself and your world, it releases you.

9. You shall believe that people care about you.

10. You shall believe that you can teach others quite a lot and that you can learn from others.

11. Why? Because you are someone, one who is needed.

Placards like these work beneficiary to help a lot of Swedish children from early ages to learn about their rights as well as learning about parents' and teachers' restrictions for behaving one way or the other towards them.

Besides these parts of ordinary school life and teaching, younger and younger children are given mobile phones (Frietas & Buckenmeyer, 2009) not seldom so that their parents are able to reach them whenever or for whatever reason they need to, but also, especially for youth, in order to protect themselves, among other things through *ICE: In Case of Emergency* (Medierådet, 2007) numbers on their mobile phones. Parents and several nongovernmental organizations such as for example (Friends, 2009) and recently reported research such as (Livingstone and Haddon, 2009) put great effort in teaching youth about accurate risk calculations in daily life as well as on the Internet in order to prepare them for the contemporary society.

The youth of today can therefore be described as on the one hand well aware of their rights, throughout information and discussions (Elvstrand, 2009), but on the other hand greatly depending on adults in home and at school. Even so most youth are in good shape to recognize and put up a fight towards unfairness and last but not least through websites like YouTube given effective tools to share their experience with others.

WHAT KIND OF TEACHERS ARE EXPOSED ON YOUTUBE

One aim for this chapter is to describe the variations of movies supposed to be found on YouTube.se that expose Swedish teachers. For such a reason I conducted several hours of watching movies at YouTube.se connected to the Swedish word for teacher, *lärare*. The purpose was, irrespective of when the searches were conducted, to search for an understanding of the movies. It was also to become familiar with the context and the logics of YouTube. My approach was inspired by *Ethnographic content analysis*Altheide (1987). He argues that "Ethnographic content analysis is used to document and understand the communication of meaning, as well as to verify theoretical relationships" (Altheide, 1987, p. 68). He also states that the researcher during the process continuously tries to change and develop the themes growing out of the analysis.

ECA consists of reflexive movement between concept development, sampling, data collection, data coding, data analysis, and interpretation. The aim is to be systematic and analytic, but not rigid. Although categories and "variables" initially guide the study, others are allowed and expected to emerge throughout the study. Thus, ECA is embedded in constant discovery and constant comparison of relevant situations, settings, styles, images, meanings and nuances. (Altheide, 1987, p. 68)

An immediate result of the search was that there seemed to exist fewer movies with teachers than news media had reported. At the same time I realized how difficult it was to get a grip of the content of the teacher movies. It seemed that YouTube.se was built up around some sort of viewer logic, which wasn't always compatible with my research logic. As a result of my first search, December 2008, I found 900 movies. A more recent search, December 2009, got the result of about 2270 movies. During one year the amount of movies was multiplied one and a half times when I used the word teacher. If this is to be seen as an ordinary, big or small change is beyond this chapter's possibility to answer. What can be noted besides the fact of multiplication is that movies with teachers are as old as the website YouTube.com, or in other words, youth embraced the possibilities of using these channels to highlight teachers' ways of behaving while leading the work in classrooms from the beginning. As a result of my searches I have so far found two sorts of movies with teachers; movies showing *angry teachers* or movies showing *jocular teachers*. That does not mean that I have watched all the 2270 movies connected with the Swedish word for teacher, lärare. Angry teachers often scolded at the students and made threats against them irrespective of what the student did in the movies. The studied movies with angry teachers showed students in what appeared to be secondary schools or high schools. Jocular teachers sang, danced or laughed with the students in movies that seamed to be from middle school, secondary school as well as high schools. In the same way as in the movies with the angry teachers the jocular teachers seemed to play around irrespective of what the students did at the moment. An important result so far is that it has revealed that a third possible category of teachers' movies is not yet found. That is movies showing *ordinary teachers*, the reasonably most common teacher that acts in front of students who more or less does their usual schoolwork.

From the description I continued with the *ethnographic content analysis* by looking at one movie showing an *angry teacher* and one movie showing a *jocular teacher*. I scouted for movies that had lots of verbal and nonverbal interaction in order to find out about differences and similarities between movies with an *angry teacher* and those with a *jocular teacher*. The movies *Anders flipp* and *Lennart goes wild* were selected based on those criteria.

The movie showing the angry teacher, *Anders flipp,* starts with the camera directed at a male person, through the movie association assumed to be a teacher. The teacher is standing in between a whiteboard and some kind of desk where some books and papers are lying. The teacher is commonly dressed, wearing a light blue shirt and a sweater hanging over his shoulders. Over his head, from the ceiling on top off the room hangs something that looks like a world map and a white screen. Boys and girls mumbling are heard when the male teacher with a forced facial look over his face after two seconds says:

*2 **Teacher Says:** To hell with you, quite simply, I get my damn pay anyway!*

*6 **Teacher Says:** Silence, while several students says something inaudible*

*8 **Teacher Says:** Be quiet when I am speaking to you*

*10 **Teacher Says:** Don't you understand this is serious, that there is always someone who learns something when we are going through something in a lesson, instead of just sitting and laughing at me, and talking a lot of crap*

*21 **Girl Says:** Anders, but are we not allowed to (inaudible)*

*24 **Teacher Interrupts the girl and says:** Yes, but you must be quiet anyways*

27 ***Boy Says:*** *Okay, (inaudible) Girl Says: I am not saying that we should (inaudible)*

29 ***Teacher Interrupts and beats his clenched right hand on his desk while he screams:*** *THERE MUST BE SILENCE*

31 TeacherRaises both his fists up over his head before they sink down to the desk

33 ***Students Talk:*** *A girl says something (inaudible) when a boy says: Lets go, let's go out anyway*

35 Teacher Wipes perspiration from under his nose with a finger. Stops for a moment before he shouts out loud: DO YOU NOT

UNDERSTAND WHAT I AM SAYING?

41 Teacher Stands and swings his arms along the side of his body

43 ***Teacher Watches the students, pointing with his left index finger and says with a loud voice:*** *It´s not a damn, it's no damn kindergarten*

46 Teacher Takes a book with both his hands and throws it at the desk

47 Teacher Stands still and puts his hands into his pockets

49 ***Teacher Raises his hands at the side of his head and says:*** *You just babble without stop, dju, dju, dju, dju*

51 Teacher Flutters his fingers at the side of his mouth

54 ***Teacher Says:*** *Words just draining out from you, completely unrestrained, åp, åp, åp, åp while he at the same time gestures with his hands beside his face*

1.02 Students Laughs

1.03 Teacher Wipes himself on the forehead with his right hand

1.04 Students Laughs

1.08 ***Teacher Says:*** *All the time (inaudible)*

In this excerpt from the movie *Anders flipp* we quite directly observe an angry teacher, later on addressed as Anders by the secondary school or high school students. The teacher is standing behind his desk and looking out over what seems to be a whole class setting. The camera is during the movie set in a specific place, directed at the teacher. Even so the teacher as well as several students are heard during the movie. The interaction between the teacher and the students as a group starts after two seconds. The direction, teacher - group, teacher- girl or boy are then possible to take part of through out the entire excerpt. While doing this it is important to remember that we know nothing about was has being going on before the teacher's reaction in the movie and we do not know anything about what is to follow after the movie ends. What we can observe is that the teacher's tension is reduced during the interaction from the way his voice shift changes during the interaction starting with an expression of anger and a threat of withdrawal of love. This is followed by a demand for the students to be quiet. When the demand doesn't make the students listen, the teacher instead appeals to them to show respect for those who wish to learn. The teacher's appeal can be understood as questioned by a girl. Her question is interrupted by the teacher who clenches his hand in the desk and screams in a high voice. The teacher's scream does not change the students talk and is therefore followed by another scream and a second threat of withdrawal of love. The climax is then to come when the teacher throws a book at his desk. This action seems to change the interaction in the classroom. The teacher's anger decreases while the tension in his body seems to ease off so he manages, in some kind of ordinary way, to perform his work in the classroom. We can also notice the girl, the same at both times according to voice recognition, who tries to calm the teacher while a boy at one time gives response to what the teacher said. Throughout the movie the teacher is the one leading the interaction and only once or twice he actually listens to the students. The questions he during the movie directs to the students can therefore be understood as rhetoric. To summarize, this movie shows a disciplinary teacher scolding his students who, according to what is said by the teacher, act irresponsible during the lesson by not taking him or his teaching seriously. Another way of understanding the content of this movie is to state that it shows some students that make their teacher loose control so he starts to act irresponsible by swearing and scolding them.

The movie showing a jocular teacher, *Lennart goes wild,* starts with the camera directed at a male teacher who sits behind a table, on which lays several books, papers, two pens and two boxes of some sort in front of a white board. When the movie starts there are laughs heard in the classroom. Beside the teacher there is a bookshelf where some ringbinders and some box files are seen. The teacher is wearing a pair of dark shoes, a pair of dark trousers and a black, brown and white striped rugby sweater. He looks relaxed and smiles during the entire movie. Laughs are heard from male secondary school or high school students in the background, when a boy, suddenly says, directed to the teacher:

*2 **Boy Says:** One cannot be sure of that (inaudible)*

*3 **Teacher Says:** No, it is not certain one can see (inaudible)*

*4 **Boy Says:** Is there, is there no fly you can hunt then Lennart?*

*5 **Teacher Looks around and says:** No I can't see anyone*

8 Teacher Points at a pupil in the other part of the room and says: You had, you had problem with one recently

*9 **Boy Says:** Hm, yes, exactly, precisely*

*11 **Teacher Says:** well then, lets take up a book*

12 Students Laughs loud

***13 Teacher Takes up a book over his head, turns from one side of desk to the other and smashes the book down, presses it against the bench and at the same time says:** aaaaahhh ... poooff with a smile on his face*

13 Teacher Smiles at the student with the mobile

13 Students Laughs loudly

18 Teacher Puts the book, opened, aside

19 ***Boy Says:*** *You are on YouTube now, Lennart*

19 Students Laughs loudly

20 ***Teacher Says:*** *Yes, that´s nice*

21 ***Boy Says:*** *(inaudible)*

22 ***Teacher Says:*** *Ok let´s move on*

In this movie the interaction between the jocular teacher called Lennart and a group of students is quite immediately showed. We then follow how the interaction between the students and the teacher changes second by second. The interaction is introduced by one boy and answered by the teacher. The teacher then answer the second question from the same boy, close to the camera, and address a question to another boy at the opposite part of the classroom. The teacher replies to the answer from that boy and starts to chase the imaginary fly in front of his desk were he sits tight looking relaxed mostly watching the students who are heard quite clearly, through the entire movie. In this movie, as the one above with the angry teacher, we know nothing about what has occurred before the reaction we have seen in the movie and we know nothing about what will follow after the movie has stopped. We therefore know nothing of the reactions from the students as a response to the teacher's positive answer about being exposed at YouTube.se or what follows after he says "Ok let´s move on". To summarize, this movie shows interaction between a teacher and some students who seems to be equal parts. The students involve the teacher to play a role in their YouTube movie and the teacher plays his part so well that he can easily bring the students back to work when the imaginative fly is crushed.

THE CONTEMPORARY COMPETENT YOUTH'S EXPOSURES OF TEACHERS ON YOUTUBE.SE

In this chapter I have described an interpretation (Kvale, 1997) of movies that expose angry teachers as well as jocular teachers on YouTube.se. The results confirm Honowar's (2007) findings. In addition to other earlier research, I also described that a third sort of teacher, the *ordinary teacher*, perhaps nineteen out of twenty teachers according to an evaluation from (LR, 2008) weren't able to be found exposed on YouTube.se. These new findings, not earlier described at least according to research I have read, support my interpretation that *the competent youth* has less to gain by exposing all sorts of teachers. That is probably because it would be seen by fewer viewers, lead to fewer comments and more importantly, that there is less to learn from such a movie even if those teachers truly follows *CRC* and acts according to *LY*. This, in turn, means that this chapter argues that *the competent youth* upload movies from school on websites such as YouTube.se for particular reasons. I interpret that such a reason could be understood according to the *enlightenment logics: Give publicity to unfairness, Illustrations of role models* or *Protecting oneself.* Such a conclusion about *the competent youth's* way to handle new technology is also useful for teachers to use in order to keep students engaged in schoolwork (Barack, 2007; Brown, 2007; Milne, 2008).

Conclusively it was quite easy to find movies that exposed *angry teachers* or *jocular teachers* on the website YouTube.se. As a consequence, it could be understood that the contents of those movies offered viewers something for their lifelong learning (Kryger, 2004). Such an interpretation would challenge argumentations made by news media and teacher unions who argue that exposing teachers is to be understood as some of the *entertainment logics* that I have presented in this chapter. It is reasonable to argue that exposure of *angry teachers* while they are acting against *CRC*, or *LY* follows as a consequence of the ideas youth have been trained through the entire school system. They act so to speak according to the idea of *the competent child* just as we expect them to do. Such an act is also similar to other examples of whistle blowing (Hedin, Månson & Tikkanen, 2008) as well as a challenge towards the collegial paradox (Colnerud, 2002). The interpretation of such an exposure could be understood as if *the competent youth* has much knowledge to gain from the movies of *angry teachers* crossing some important agreements about how pupils and youth should be treated in school. Youth's use of contemporary technologies, mobile phones and computers in order to confront injustice and protecting themselves starting with things happening in their own classrooms at their own school should be taken seriously according to the downward spiral of mistrust that Sheats & Gay (2006) described.

Youth's competent ways of acting is not just about revealing unfairness such as those seen in the movies with the *angry teachers,* it is also about offering alternative ways to act through the examples of the *jocular teachers*. In this way *the competent youth's* acts could be understood as if they have recognized that power and threats are less useful in order to change people's way of acting while reinforcements such as encouragement and rewards are more useful in order to change people´s way of behavior according to well-established psychological and educational research (Samuelsson, 2008).

Without having any exact numbers this chapter assumes that far more youths than adults take part in movies at websites like YouTube. From such an expectation one could also assume that the movies are most likely made to inform other youths. These youths are through the movies urged to be aware of how they are treated by teachers or other adults in or outside school. They are also urged to take part in how the classroom is managed (Woolfolk Hoy & Weinstein, 2006).

One could perhaps think that the above discussed viewpoints of *the competent youth* is frequently described among earlier research connected to the phenomenon that this chapter describe, but as far as I have found that is not the case. Firstly because there seems to be less research done about youth and YouTube connected to the ideas of *the competent child* and secondly because most of the research that I have found connect youth movies on websites with cyber bullying and mobile phones and thereby becomes quite normative while this chapter tries to be descriptive while offering a qualified support for discussion among teachers and adults as well as youths themselves.

FUTURE RESEARCH DIRECTIONS

Although this chapter is based upon an interpretation (Kvale, 1997) of watched movies exposed at YouTube.se it would be interesting to interview youths that have exposed teachers on YouTube. I would like to ask them about the reason for their exposure and if they have got any reactions of their behavior as well as of their movies. It would also be interesting to find out more about what kind of experience they have got out of the exposure. Furthermore, what are the possible differences between those who have exposed *angry teachers* and those who have exposed *jocular teachers?* Likewise, it would valuable to meet teachers or other adults working in schools that have been exposed on YouTube.se in order to hear their stories; what did they feel and what happened after they got aware about the movies that exposed their work. If there are differences between those who have been exposed as *angry teachers* and those who have been exposed as *jocular teachers,* what are the specifics of their respective experiences? And how can schools or the society as a whole, work proactive in order to handle these new sorts of phenomena? It would also be interesting to analyze the comments made about some of the movies with angry teachers as well as with jocular teachers. Finally it would be of great interest to find out how both students as well as teachers reacted to my interpretation that exposure of *angry teachers* as well as of *jocular teachers* is to be understood as an act of the competent youth.

REFERENCES

Alanen, L. (1992). *Modern Childhood? Exploring the 'Child Question' in sociology*. Jyväskylä: University of Jyväskylä.

Altheide, D. (1987). Ethnographic Content Analysis. *Qualitative Sociology, 10*(1), 65–77. doi:10.1007/BF00988269

Barack, L. (2007). Schools Mute Cellphones: Teachers nationwide hear one ringy-dingy too many. *School Library Journal, 10*(1), 1–2.

Brembeck, H., Johansson, B., & Kampman, J. (2004). *Beyond the competent child: Exploring contemporary childhoods in the Nordic welfare societies*. Fredriksberg, Denmark: Roskilde University Press.

Brown, L. (2007). Cellphone policy needed, teachers says. *The Star, July 14*, 1-3

Colnerud, G. (1997). I de MÖRKASTE vrårna av skolans värld. *Pedagogiska Magasinet, 1*(4), 61–65.

Colnerud, G. (2002). Den kollegiala paradoxen. *Pedagogiska Magasinet, 4*(4), 24–30.

Dunkels, E. (2005). Nätkulturer - vad gör barn och unga på Internet? *Tidskrift för lärarutbildning och forskning, 1-2*, 41-49.

Elvstrand, H. (2009). *Delaktighet i skolans vardagsarbete*. Linköping, Sweden: Linköping universitet.

Friends. (2009). *Koll på nätet: En bok om att skapa samtal mellan barn, unga och vuxna om nätet*. Stockholm, Sweden: Friends.

Frietas, D., & Buckenmeyer, J. (2009). Cell Phones in American High Schools: 21st Century Connections. *FSC News, January 19*, 1-3.

Gourevitch, P., & Morris, E. (2008). *STANDRAD OPERATING PROCEDURE: A WAR STORY*. London: Pan Macmillian.

Hedin, U.-C. Månson. S-A., & Tikkanen, R. (2008). *När man måste säga ifrån: Om kritik och whistleblowing i offentliga organisationer.* Stockholm, Sweden: Natur och Kultur.

Honowar, V. (2007). Cellphones in Classrooms Land Teachers on Online Video Sites. *Education Week, 27*(11), 1–12.

Kryger, N. (2004). Childhood and "New Learning" in a Nordic Context. In H. Bremback, B. Johansson, & J. Kampman (Eds.), *Beyond the competent child: Exploring contemporary childhoods in the Nordic welfare societies.* (pp. 153-176). Fredriksberg, Denmark: Roskilde University Press.

Kvale, S. (1997). *Den kvalitativa forskningsintervjun*. Lund, Sweden: Studentlitteratur.

Lärarförbundet., & Lärarnas Riksförbund. (2002). *Lärares Yrkesetik.* Stockholm, Sweden: Lärarförbundet, Lärarnas Riksförbund.

Livingstone, S., & Haddon, L. (2009). *EU Kids Online: Final report*. London: LSE.

Medierådet. (2007). *Tips till vuxna om barn/unga och mobiltelefoner*. Stockholm, Sweden: Regeringskansliet.

Milne, J. (2008). What have we got to be scared of? *Times Educational Supplement*, *25*(January), 1–5.

Palm, G. (2000). *Den svenska högtidsboken*. Stockholm, Sweden: En bok för alla.

Pickett, A. D., & Thomas, C. (2006). Turn OFF That Phone. *The American School Board Journal*, (April): 40–45.

Prensky, M. (2001). Digital Natives, Digital Immigrants. *On the Horizon,* 9 (5). New York: NCB University Press

Prout, A., & James, A. (1997). A new paradigm for the sociology of childhood? Provenance, promise and problems. In James, A., & Prout, A. (Eds.), *Constructing and reconstructing childhood* (pp. 7–17). London: Falmer press.

Samuelsson, M. (2008). Att förhålla sig till institutionalisering: Utmanande för såväl lärare som elever. *LOCUS, tidskrift för forskning om barn och ungdomar, 3-4(4),* 83-99.

Sandemose, A. (1968). *En flykting korsar sitt spår*. Stockholm, Sweden: Forum.

Sheets, R. H., & Gay, G. (1996). Student Perceptions of Disciplinary Conflict in Ethnically Diverse Classrooms. *NASSP Bulletin*, *80*(580), 84–95. doi:10.1177/019263659608058011

Shuler, C. (2009). *Pockets of potential: Using Mobile Technologies to Promote Children's Learning*. New York: The Joan Ganz Cooney Center at Sesame Workshop.

Sigsgaard, E. (1992). Forholdet mellem menneskesyn, paradigme, forforståelse og data i humanistisk forskning. *Nordisk Psykologi*, *44*(4), 9–19.

United Nations. (1989). *Convention on the Rights of the Child, CRC*. New York: United Nations.

Woolfolk Hoy, A., & Weinstein, C. S. (2006). Student and Teacher Perspectives on Classroom Management. In Evertson, C. M., & Weinstein, C. S. (Eds.), *Handbook of Classroom Management: Research, Practice and Contemporary Issues* (pp. 181–219). Mahwah, NJ: Lawrence Erlbaum Associates.

Chapter 15
Moving from Cyber-Bullying to Cyber-Kindness:
What do Students, Educators and Parents Say?

Wanda Cassidy
Simon Fraser University, Canada

Karen Brown
Simon Fraser University, Canada

Margaret Jackson
Simon Fraser University, Canada

ABSTRACT

The purpose of this chapter is to explore cyber-bullying from three different, but interrelated, perspectives: students, educators and parents. The authors also explore the opposite spectrum of online behaviour - that of "cyber-kindness" - and whether positive, supportive or caring online exchanges are occurring among youth, and how educators, parents and policy-makers can work collaboratively to foster a kinder online world rather than simply acting to curtail cyber-bullying. These proactive efforts tackle the deeper causes of why cyber-bullying occurs, provide students with tools for positive communication, open the door for discussion about longer term solutions, and get at the heart of the larger purposes of education – to foster a respectful and responsible citizenry and to further a more caring and compassionate society. In the course of this discussion, they highlight the findings from two studies they conducted in British Columbia, Canada, one on cyber-bullying and a later study, which addressed both cyber-bullying and cyber-kindness.

INTRODUCTION

The proliferation of electronic media in recent years has allowed children and adolescents to take schoolyard bullying to an entirely new level – into the realm of cyber-bullying. Cyber-bullying in the school context involves using emails, websites, text messaging, camera phones, blogs, YouTube, Facebook and other forms of social networking technology to spread hurtful, nasty, derogatory,

DOI: 10.4018/978-1-60960-209-3.ch015

vulgar or untrue messages to or about other students, teachers or acquaintances. Cyber-bullying typically threatens the reputation, wellbeing, security and/or safety of the targeted victim. Unlike face-to-face bullying, which typically happens at a given location, in a particular moment in time, and by a perpetrator that is known or seen by the victim, messages posted online can be spread globally, can exist in perpetuity, and the cyber-bully can hide his/her identity by using an avatar. Cyber-bullying can impact the victim in serious ways: low self-esteem; inability to concentrate on schoolwork; anger; anxiety; depression; and even suicide (Brown, Jackson & Cassidy, 2006; Gradinger, Strohmeier & Spiel, 2009; Willard, 2006; Ybarra and Mitchell, 2008).

Unfortunately, cyber-bullying has expanded into a global phenomenon (Kowalski et al, 2008), while educators, parents, and policymakers struggle to develop effective solutions (Campbell, 2005; Shariff, 2005; Belsey, 2006; Brown et al. 2006; Willard, 2006). While most of the attention in the current literature has been on trying to understand cyber-bullying, the cyber-bully and the impact on the victim (Beale & Hall, 2007; Brown et al. 2006; Li, 2007; Patchin & Hinduja, 2006; Shariff, 2008; Smith et al. 2008; Willard, 2007; Worthen, 2007; Ybarra, Diener-West and Leaf, 2007), little or no attention has been given to examining whether technology is also being used by youth to communicate thoughtful, kind and caring messages (Cassidy, Brown & Jackson, 2010; Cassidy, Jackson & Brown, 2009; Jackson, Cassidy & Brown, 2009a; Jackson, Cassidy & Brown, 2009b). Further, the focus in schools primarily has been on punishing the cyber-bully and implementing anti-cyber-bullying programs, rather than on developing holistic solutions that cultivate more respectful on-line exchanges and build a more caring school culture.

While it is important to understand why young people cyber-bully, the extent of cyber-bullying, and the characteristics of perpetrators and victims, it is equally as important to investigate whether cyberspace is also being used in positive ways. Do children and adolescents use technology in ways that communicate care and kindness to one another and foster peers' self-esteem? Can young people's negative exchanges be re-directed in more positive ways that assist parents, educators, school counselors and other professionals in diminishing the potential risks associated with hurtful cyber-bullying victimization?

Green and Hannon (2007) suggest that digital technologies offer a "third space" (p. 60) between official and informal contexts, where young people can "create portfolios of digital media, engage in peer teaching and develop their confidence and voice" (Sharples, Graber, Harrison & Logan, 2009: 72). Ybarra et al. (2007) agree that online access can afford positive initiatives for youth, providing important information on a range of questions and concerns. Certainly children and adolescents are using the Internet as a positive and interactive venue for creating blogs (journaling), social dialoguing, sharing of ideas, and engaging in scholastic endeavours or innovative pursuits (Dowell, Burgess and Cavanaugh, 2009). As a result, the Internet offers potential for positive, helpful and caring communications between youth - or what we have termed, "cyber-kindness."

This chapter examines the spectrum of students' on-line exchanges, from cyber-bullying to cyber-kindness, highlighting information from two research studies we conducted in British Columbia, Canada. In our first study (2005-2007), we examined 11-15 year old (Grades 6 to 9) students' perceptions and experiences with cyber-bullying and also interviewed fifteen teachers, vice-principals and principals. In our second study (2007-2009), we expanded our research to examine cyber-kindness among the same age/grade group of students, although at different schools, and also extended our research of school personnel to include school counselors and youth workers as well as teachers, vice-principals and principals (n=17). In the second study, we also surveyed parents, since cyber-bullying most often occurs using the home computer, and because parents play a critical role in preventing cyber-bullying and fostering cyber-kindness.[1] It is also important to solicit information from parents, as most parents underestimate the extent of cyber-bullying happening in their children's lives as well as the amount of time their children spend online each day (Bhat, 2008; Cassidy et al. 2010; Jackson et al. 2009a&b; Chou, Yu, Chen and Wu, 2009). While some researchers have found that close parental supervision is unrelated to youth engaging in risky online behaviour (Liau, Khoo & Ang, 2008), others find that lack of parental supervision may lead to increased chat room use (Sun et al. 2005). Further, parents tend to envision the potential risks of stranger online contact but often neglect to discuss with their children the dangers that can occur from friends and classmates (Bhat, 2008).

In the chapter we posit that school personnel need to work collaboratively with parents and with youth to develop policies and practices that proactively foster kindness, caring and respect towards one another rather than merely focusing on curtailing or punishing negative behaviour, such as cyber-bullying. This chapter, therefore, is situated within and contributes to two literatures: cyber-bullying, which to date primarily has focused on students and teachers and only recently on parents, and not at all on cyber-kindness (Brown et al. 2006; Cheever & Carrier, 2008; Cassidy et al. 2009; Cassidy et al. 2010; Jackson et al. 2009a&b; Kowalski, Limber & Agatston, 2008; Liau et al. 2008; Rosen, Cheever & Carrier, 2008; Shariff, 2006); and the ethic of care literature, which stresses modeling, practice, dialogue and confirmation as a way to foster positive behaviour and which has not been sufficiently addressed in the cyber-world (Cassidy & Bates, 2005; Gregory, 2000; Noddings, 2002, 2005; Owen & Ennis, 2005; Rauner, 2000).

RESEARCH METHODS

Because our Study #1 is reported elsewhere in the literature (Cassidy et al. 2009; Jackson et al. 2009a; Jackson et al. 2009b), including the methods used to collect the data, we will not describe this study here, other than to note that it involved students in Grades six to nine (ages 11-15) (n=365) completing a fourteen-page questionnaire (192 variables), which included: closed-ended questions on cyber-bullying and open-ended questions asking respondents to provide examples of cyber-bullying; which students were most likely to be bullied online; solutions to the cyber-bullying problem; and other information on cyber-bullying they felt relevant. This study also included hour-long semi-structured interviews with fifteen teachers, vice-principals and principals. When discussing this study, we will cite published papers.

Study #2 on cyber-bullying and cyber-kindness was recently completed and the results have yet to be published (see Cassidy et al. 2010). In this study, 339 students in Grades six to nine (ages 11-15) were canvassed from a large metropolitan region of British Columbia, Canada. The researchers designed the student survey to accommodate youth of varying language abilities, meaning the wording used was relatively simple and the font was stylized for easy reading. Twenty background questions were asked of students; for example, their age, grade level, gender, ethnicity, home and first language, success with school, the number of computers in their home, the location of these computers, how often they were online, when they were online, and so on. The closed-ended questions (single-response questions, categorical-response items, rating scales), as well as the open-ended questions about cyber-bullying in the students' survey, inquired about the frequency and methods used in cyber-bullying and being cyber-bullied, the situation in which bullying took place, their reactions to being cyber-bullied, whom they talked to after they were cyber-bullied, information regarding cyber-bullying teachers and principals, and possible solutions to cyber-bullying.

Students were also asked a number of closed-ended and open-ended questions related to "cyber-kindness". This section included baseline statements to determine if respondents knew the differences between kind, thoughtful and caring messages and bullying types of comments, assessed the prevalence of, and methods used, in cyber-kindness, determined the situations in which cyber-kindness took place, asked for other examples of sent and received kind, thoughtful and caring messages, and canvassed students' opinions regarding what the school, students and parents could do to encourage kind and caring online behaviours.

In addition, in-depth, semi-structured interviews were conducted with 17 educators from these schools; respondents included two principals, four vice-principals, five teachers, four counselors and two youth workers. The 45-60 minute taped interviews included closed-category questions using a five-point *Likert* response scale ranging from 1 (extremely concerned) to 5 (not concerned at all) on the issue of cyberbullying, and another scale ranging from 1 (very familiar) to 5 (not familiar at all) on familiarity with technology (email, cellular phones/text messaging, Facebook, MSN/chat rooms, YouTube and Blogs). Another closed-category question using a five-point Likert response scale from 1 (not important at all) to 5 (extremely important) asked respondents how important is in the life of the school, to (a) prevent cyber-bullying and to (b) encourage cyber-kindness. Several open-ended questions probed their experiences and suggestions regarding cyber-bullying and cyber-kindness, giving voice to their perspectives (Barron, 2000; Cook-Sather, 2002; Palys, 2003). Interview transcripts were reviewed and re-reviewed for emergent themes (and sub-themes), using a grounded theory approach and an open coding method (Miles & Huberman, 1994; McMillan & Schumacher, 1997).

Three hundred and fifteen parents of students at the participating schools were surveyed. The parents' survey included single-response questions, categorical-response items, rating scales and open-ended questions. The first section of the questionnaire assessed their knowledge of technology, concerns about cyber-bullying, supervision of their child's computer use, hours their child spends online, and knowledge of their child's cyber-bullying victimization and/or perpetration. The second section, comprised of open-ended questions, canvassed parents' knowledge of cyber-kindness and how educators and parents might help cultivate more considerate, respectful and caring online interactions among youth. Similar to the qualitative analysis undertaken with the educators' interviews, the themes which surfaced in the open-ended responses from parents, were determined through a process of review and re-review according to the frequency of responses and the strength of response (Miles & Huberman, 1994; McMillan & Schumacher, 1997).

WHAT STUDENTS, EDUCATORS AND PARENTS SAY ABOUT CYBER-BULLYING

Students

Where does Cyber-Bullying occur?

Since youth are very selective when engaging in popular online forums, determining their trends and behaviours and the media they choose for cyber-bullying, is important for policy-makers. In our first study (Cassidy et al. 2009), we discovered that the most common vehicle for cyber-bullying was chat rooms (53% of participants chose this option), while 37% said through emails or MSN, and only 7% said text messages. These results are similar to what Patchin and Hinduja (2006) found during the same time period; 384 respondents under age 18 reported that cyber-bullying most often occurred in chat rooms, but was less prevalent when texting or emailing. In our second study of 339 youth from the same age group, social networking sites like Facebook and MySpace had gained popularity among youth, such that this vehicle was the choice venue for cyber-bullying (52% of respondents), followed by email and MSN at 32%, and text messaging at 2.5%. Chat rooms were only recorded at 12%. This trend may change, of course, as new technology emerges and youth engage in new online agencies. This also underscores the importance of up-to-date research on this topic.

A disquieting conundrum, among educators and parents, is whether cyber-bullying starts at home or at school and the connection between the two domains. The findings from our first study (Cassidy et al. 2009) showed that 64% of youth claim that cyber-bullying is most likely to start at school and then continue at home, meaning that an incident at school precipitated the negative on-line exchange on the home computer. In this regard, findings from our second study are consistent with the first study, with 65% of respondents reporting that cyber-bullying starts at school. Schools typically place tighter controls and monitoring on computer use than do parents, although these controls are often sporadic and inconsistent, as demonstrated by the Valcke, Schellens, Van Keer and Gerarts (2007) study, which showed that approximately 50% of students found the controls to be sporadic, with youth in lower grades (Grade Three) admitting to significantly fewer computer controls than students in the higher grades (Grades Five and Six). If the research continues to demonstrate that cyber-bullying on the home computer is a reaction to incidents that happened at school, this raises the issue of whether victims of schoolyard face-to-face bullying may become empowered on the home computer and become cyber-bullies. Conversely, it is also possible that students who are victims of face-to-face bullying at school will continue to be victimized in online ways through the home computer (Brown, et al., 2006) - a form of "double jeopardy" as described by Raskauskas and Stotz (2007).

Cyber-Bullying Victims

Online bullying can cause victims to experience a myriad of psychosocial effects ranging from depression, low self-esteem, anger, school absenteeism, poor grades, anxiety, and a tendency towards suicidal thoughts or suicide (Brown et al. 2006; Willard, 2006; Ybarra and Mitchell, 2008). The negative impact of the written word or a posted picture or video in cyberspace can be far-reaching and long-term, especially if the perpetrators are peers or someone they know (Ybarra and Mitchell, 2008). Victims often repeatedly re-visit the postings, causing continual re-victimization (Brown et al. 2006). Given the negative psychosocial effects of online victimization, educators and parents must be vigilant in reviewing possible academic and personal anomalies that may occur after an incident. School attendance and performance may suddenly drop or physical/mental ailments may occur (Bhat, 2008). Rigby (2005) noted that children who are cyber-bullied may be more prone to diminished levels of mental health. Bhat (2008) confirmed that victims tend to internalize feelings of anxiety, loneliness and depression, which can result in disengagement from school and peer relationships. One counselor, who was interviewed our second study, expressed concern about an adolescent male who was the victim of a lunch-hour school prank, conceived and videotaped by classmates and posted on YouTube. This student had totally disengaged from school activities and classmates after the incident, especially during the lunch hour. Ybarra and Mitchell (2008) reported a higher likelihood of youth who have been victimized online to bring a weapon (e.g. gun) to school.

A school youth worker from our second study, who teaches sex education and drug education, said in her interview that she finds Facebook scary, and gave the example of a fifteen-year-old youth attending a party, drinking alcohol or taking drugs, and then participating in sexual activity. The next day the sexual encounter is posted on Facebook because some reveler at the party photographed the event. In her view, this type of unfortunate exposure could ruin the adolescent's personal life as well as impact future academic opportunities (such as scholarships or university entrance), or, in the longer term, employment prospects. Dowell et al. (2009) confirmed that university personnel and potential employers have begun to review postings on the Internet (especially Facebook) for possible (mis)behaviours that might impact future employment or admissions of applicants.

Cyber-Bullying as Relational Aggression Among Youth

Ybarra and Mitchell (2008) posit that youth who are cyber-bullied are often targeted by someone they know. In our first study, (Jackson et al. 2009b), we examined the cyber-bullying acquaintanceship issue through a theoretical relational aggression lens. That is, we sought to determine whether cyber-bullying could be considered a form of relational aggression among youth. Relational aggression is defined by Crick et al. (1999) as "behaviours that harm others through damage (or threat of damage) to relationships or feelings or acceptance, friendship, or group inclusion" (p. 177). Examples in the literature include spreading rumors with the intent to harm others or to socially exclude them through covert gossip or rumor (Crick and Grotpeter, 1995). The technology of the Internet provides a perfect conduit to achieve these ends, and there are numerous examples cited in the cyber-bullying literature (Juvonen & Gross, 2008; Slonje & Smith, 2008).

Gender is an important variable to consider when examining cyber-bullying as relational aggression, given the observed preferences of girls to engage in covert gossip and rumor. Boys engage more in face-to-face bullying but girls employ covert tactics that affect social acceptance and friendships (Campbell, 2005). In analyzing the relationship between cyber-bullying and relational aggression (Jackson et al. 2009b), we found that slightly more girls (29%) than boys (21%) admit to engaging in cyber-bullying practices. This is consistent with the notion that girls will engage more often than boys in an activity, which can be defined as relational aggression. Further, almost 33 percent of the girls in our study reported that they have witnessed someone cyber-bullying online or on a cellular phone, as opposed to only 22 percent of the boys.

Living in fear or contemplating suicide after cyber-bullying are serious problems that educators, parents and clinicians must be prepared to handle. In our first study, (Cassidy et al. 2009), we found that 2.2 percent of boys and 1.6 percent of girls admit to being afraid as a result of cyber-bullying messages they had received, and another 4 percent confess to having suicidal thoughts (n=365). In the same study, approximately 40 percent of the respondents reported being a victim of cyber-bullying. Given the seriousness of the symptoms, medical personnel and school counselors/psychologists need to play a role in educating parents on how to reduce their children's risks for online victimization and to monitor their psychosocial health and online behaviour (Genuis and Genuis, 2005; Ybarra and Mitchell, 2008). Genuis and Genuis (2005) point out that health-care workers and clinicians must be prepared to become involved in the education of youth and their parents, because of their unique position that affords them the opportunity to interact with both groups. Similarly, Dowell et al. (2009) assert that school personnel need to be educated about cyberspace demeanors in order to better assess young people who may be at risk so they can intervene or refer youth to those who can help. In this vein, medical and psychologists needs to work collaboratively with schools in fostering better mentally and emotionally healthy young people.

Who are the Cyber-Bullies?

If cyber-bullying is to be effectively addressed, more needs to be known about who participates in this type of behaviour and whether there are some common characteristics among youth who cyber-bully. Ybarra and Mitchell (2004) recognized three cogent psychosocial antecedents: substance abuse issues; conventional bullying victimization; and delinquency. These researchers found that cyber-bullies are more likely to drink alcohol, smoke tobacco and engage in fights, and display significant academic under-achievement. Insubstantial caregiver-child relationships may also be a factor, with the same researchers demonstrating that 44 percent of cyber-bullies claim low emotional attachment to caregivers. They also suggested that frequent Internet use may lead to increased online abuse. Certainly the digital age is here to stay and most youth spend a significant amount of time online, primarily on home computers.

Although Ybarra and Mitchell (2004, 2008) found that these characteristics were present in those who cyber-bullied, in both our studies we found that this behaviour was far too prevalent in the population we surveyed to be only restricted to those with emotional, attachment, academic or psychosocial problems. We found that between one-quarter (first study) and one-third (second study) of students in Grades six to nine reported participating in cyber-bullying, and this behaviour crossed socio-economic status, culture, gender and academic success. We did not, however, specifically collect data relating to attachment, psychosocial problems, delinquency, or substance abuse issues, but it is unlikely that these characteristics would be evident in all or even most of the youth we researched.

Bhat (2008) explained that young people may possess impulsive tendencies that make them inclined to entertain cyber actions in haste, without realizing the full ramifications of such behaviour. Thus, impulsivity in adolescence may underscore the lack of understanding or consideration for victims of their actions - a potential lack of insight into the full thrust of what they have done and the harm that befalls their victims. In our interviews with school personnel, they said that on a regular basis they were having to deal with students who had posted something online in an impetuous moment, only to be faced with the realization that the victim was devastated not only by the action itself but also by the magnitude of the audience that views it, be it strangers, friends or peers. The educators said that they reminded their students that their actions had short and long-term consequences; however, this admonition typically fell on deaf ears, and instances of cyber-bullying continued.

We found that age is also an important factor with cyber-bullying. In both studies, our findings indicate that cyber-bullying behaviour typically escalates between ages 13 and 14 and then diminishes as youth grow older. In our first study, between one-quarter and one-third of students aged 12 to 14 reported bullying others online versus those students aged eleven (17%) and fifteen (19%). The findings from our second study are consistent – 9 percent (aged 12), 34 percent (aged 13) and 35 percent (aged 14) occasionally cyber-bullied, with 2 percent (aged 13) and 7 percent (aged 14) often engaging in such behaviour. Younger and older students outside of this age range engaged less often.

Reporting Practices: To Whom do Youth Turn for Help?

If cyber-bullying occurs, to whom do the victims go to report the incident – parents, teachers, counselors, friends, police, or no one? Determining the answer to this question helps guide policymakers, educators and parents and sketches the possible isolation that many youth feel when targeted online. If youth are not confiding in adults or other professionals, then schools and parents maybe unaware of the extent of cyber-bullying, or downplay its impact, and victims may suffer in silence or confide only in peers who are not in a position to help with the long term effects of victimization. It also raises the important issue of bystander reporting, and how schools can work with peer groups in encouraging them to support their friends by informing adults about possible cyberspace infractions.

Studies show that youth are more likely to confide in friends rather than parents or educators (Patchin & Hinduja, 2006; Bhat, 2008). Our research (Cassidy et al. 2009) corroborated this, with 74% saying that would tell their friends, 57 percent would tell their parents, and only 47 percent would tell school officials. Almost no one would tell the police. We also found that students are more likely to report the cyber-bullying to school officials if they witnessed it than if they experienced it themselves. In our second study, approximately 27 percent of students would report cyber-bullying to school officials if they were victims as opposed to 40 percent who would report to a teacher, counselor or school administrator if they witnessed cyber-bullying taking place.

For those respondents who would not tell school personnel, 30 percent fear retribution from the cyber-bully (Jackson et al. 2009a). This robust response contravenes much of the current literature, which posits that youth are reluctant to report incidents to adults primarily out of fear that parents will limit or remove online computer time (Brown et al. 2006). Consonant with this finding is Bhat's (2008) tragic recollection of Alex Teka, a young New Zealand girl, who was the victim of cyber-bullying, who took her own life at the age of twelve because the cyber-bullying escalated once her mother complained to school authorities. Such findings impact the type of policies and practices educators need to develop in school to protect the victim from further retaliation if incidents are revealed to adults. Ybarra and Mitchell (2008) emphasize the need for funding initiatives such as online youth outreach programs and online mental health services which may assist youth who have been cyber-bullied.

Our study (Cassidy et al. 2009: 392) revealed other reasons why youth do not report cyber-bullying incidents to school personnel: the students see it as their problem not the responsibility of adults (29%); school personnel could not prohibit the bullying in any case (27%); friends could get into trouble if the cyber-bullying is revealed (26%); caregivers would restrict Internet access (24%), and other students would label them as "informers" (20%). As well, age is a paramount factor in reporting practices. Younger students are more likely to report incidents to school officials than those youth 14 years and older. Not surprisingly, as children enter puberty, confiding in adults (parents, school officials) is sometimes seen as the least viable option; they would rather turn to peers for support during their teenage years. Further, Williams and Guerra (2007) say that students tend to perceive adults as untrustworthy and schools as unsupportive. Unfortunately, in many cases, students may opt to remain silent, internally wrestling with the consequences of cyberspace victimization (Juvonen and Gross, 2008; Slonje and Smith, 2008).

Although there is a distinct gap in technological knowledge between youth and adults, youth need to be encouraged to discuss cyberspace problems with authorities, and adults need to reach out to youth to work collaboratively with them. Youth may fear that adults, coined digital immigrants by Prensky (2001), may not understand their digital culture and the philosophy of online life, so divulging confidential information may be hindered by this additional concern.

Students' Opinions about Cyber-Bullying

It is also vital to ascertain young people's opinions about cyberspace bullying in order to determine where they are coming from and any possible misconceptions they might have. Using a four-point *Likert-type* scale, we (Cassidy et al.

2009) asked our 365 respondents to rate certain statements, ranging from "strongly agree" to "strongly disagree". Responses to some questions were revealing. For example, almost 47 percent ("strongly agree", "agree") that freedom of expression is a right and online speech is borderless; this is despite Canadian legislation and the Canadian *Charter of Rights and Freedoms* that clearly defines limitations to freedom of expression (Shariff, 2006; Shariff and Gouin, 2005; Shariff and Johnny, 2007). Similar legislation exists in other democracies. Also of concern is that almost 50 percent ("strongly agree", "agree") that cyber-bullying is a normal part of the online world, and 32 percent ("strongly agree", "agree") that online bullying is less injurious than conventional face-to-face bullying, as cyber-bullying is merely "words in cyberspace" (Cassidy et al. 2009: 397). Also revealing is that almost three-quarters of the respondents indicate that cyber-bullying is more of a problem than prior years, suggesting that policy and practice initiatives undertaken by school districts and parents to counteract this growing problem are not working effectively to alleviate cyberspace transgressions. In this vein, more than 60 percent of respondents suggest that the solutions lie with young people who are far more familiar with technology than adults, again underscoring the tension between digital immigrants and digital natives (Brown et al. 2006; Prensky, 2001). Such admissions give credence to the view that policymakers, school officials and parents should collaborate with youth on finding workable solutions.

Students' Solutions to Cyber-Bullying

When we asked our 11-15 year old participants (Cassidy et al. 2009) to choose the three best solutions for curtailing cyber-bullying behaviour from a list of ten options, the three most commonly selected solutions were (in the following ranked order):

1. Set up anonymous phone-in lines;
2. Develop programs to teach students about cyber-bullying and its effects;
3. Work on creating positive self-esteem in students.

In our second study, we asked a similar question. The results were as follows:

1. Develop programs to teach students about cyber-bullying and its effects (47%);
2. Set up anonymous phone-in lines (42%);
3. Focus on developing cyber-kindness instead of trying to stop cyber-bullying (40%);
4. Work on creating positive self-esteem in students (38%).

It is interesting that none of the most commonly selected choices had to do with punishment; rather students wanted to tell the authorities in confidence about what was happening, wanted curriculum on the topic, and saw the need for students to develop a better sense of self.

Educators: Their Views on Cyber-Bullying and Cyber-Kindness

In this digital age, it is important to probe educators' knowledge and experiences of cyber-bullying and their views on ways in which schools might promote, through the formal and informal curriculum, kinder, more caring and respectful online behaviours among youth. In our second study, we interviewed 17 educators to ascertain what is being done in their schools to address cyber-bullying and to foster "cyber-kindness". The results of these interviews are reported here.

Concerns about Cyber-Bullying

We used a recoded five-point *Likert-style* scale (1=not concerned at all; 5=extremely concerned), to determine the educators' level of concern about cyber-bullying. Surprisingly (given the level of cyber-bullying among youth), seven of out the seventeen educators (or about 42%) admitted that they were "not concerned at all" about the issue (M=3.6, SD=.9). However, when asked how important it was to prevent cyber-bullying compared to other competing priorities in the life of the school, almost 85 percent of these educators deemed it "extremely important", on a five-point scale between "not important at all" to "extremely important" (M=4.3, SD=.9). They stressed the need for teachers to learn new technology and proposed that the primary method for preventing cyber-bullying is education. It also was acknowledged that parents needed to be more involved in their children's online activities, particularly as many parents are not very knowledgeable about various forms of technology and, as Attewell, Suazo-Garcia and Battle (2003) point out, parents generally do not supervise their children's computer use. In our study, we found that almost 40 percent of youth have their computers in their bedrooms (Cassidy et al. 2010), thus preventing any real monitoring of behaviour.

Examples of Cyber-Bullying Provided by Educators

Each of the educators interviewed in our study was able to relay at least one story of a cyber-bullying incident that occurred among students at their school. One counselor confirmed that all of the bullying incidents at his school that year were cyber-oriented, not face-to-face, and that all involved girls – gossiping, backstabbing, excluding and criticizing. Lenhart, Madden, Macgill and Smith (2007) posited that girls are the primary bloggers, dominating the teen blogosphere, whereas boys are chiefly involved in video-sharing websites such as YouTube where they upload more so than girls who use these sites. Certainly the creation of blogs and Web pages are increasingly becoming more common among youth, as indicated in Chou et al.'s (2009) Taiwanese study; these sites could be used in positive ways instead of a vehicle for cyber-bullying. The examples of cyber-bullying among students reported by educators in our study showed both genders using Facebook to cyber-bully, with girls more likely to use words to communicate hurt and boys more likely to use visual media like pictures. However, as noted above, most of the cyber-bullying activity at their schools was happening under the radar of educators; only very few incidents were brought to their attention.

Here are three examples of cyber-bullying reported by educators:

1. A group of boys scattered spaghetti spray on another student, called the student names, videotaped the incident using a video camera and then posted it on Facebook. Once the video was uploaded to Facebook, other students made derogatory comments about the victim.
2. Educators at one secondary school were shocked when six or seven cyber-bullying incidents occurred within a month or so after school administrators conducted a school-wide seminar on cyber-bullying and the consequences of such behaviour. Administrators came to the conclusion that students were testing the school because one student on Facebook (who was bullying another student) admitted the idea originated from the school presentation and he wanted to see if the school would actually follow through with its reprimand.
3. A video was posted on YouTube showing two students engaged in a physical fight off school property. The video also showed student bystanders watching the fight and cheering them on, some of whom were students in the school.

A number of educators in this study expressed concern that when parents were notified about their child's cyber-bullying adventures, several were defensive and showed surprise that their child was involved, even when it was clear that the negative behaviour occurred on the home computer. One vice-principal discussed the strong link between the home and school life vis-à-vis the Internet, and how cyber-bullying that starts at home negatively impacts the school milieu, and vice versa.

Teachers and Principals Being Cyber-Bullied

Several educators on our study expressed alarm at the growing propensity for teachers and principals to be the target of cyber-bullies. Indeed, in the student survey part of our study, 12 percent of the sample (n=334) admitted to personally participating in posting a mean, nasty, rude or vulgar message about a school official online. In addition to making reference to the "Rate Your Teacher" sites used by students, teachers provided examples such as the following:

1. Over the holiday season, a teacher was targeted on Facebook by a student for no apparent reason. It involved rude comments, swearing and name-calling. The teacher had set up a school Facebook account for a school challenge and the student created a fake Facebook account targeting the teacher.
2. Individual students created a group on Facebook to attack a teacher. This type of behaviour then spread to different groups attacking other teachers on Facebook.
3. Students hid their cellular phones under their school desks and made fun of a teacher (hair, clothes, etc.) by secretly texting their friends. The teacher was oblivious to these interactions.

Indeed, media stories abound about educators being targeted online through Facebook, harrassed through polling sites or being blasted by pictures posted on YouTube (Froese-Germain, 2008). In a recent 2008 national poll conducted by the Canadian Teachers Federation, one in five teachers reported being victims of cyber-bullying (ATA News, 2009). This appears to be a growing phenomenon.

School Policies on Cyber-Bullying

Educators from the schools in both our studies confirmed that there was no specific cyber-bullying policy in place, and any problems that surfaced fell under the auspices of the face-to-face bullying policy, which tend to be district wide and described in the school students' handbook. One principal hoped that any policy concerning cyber-bullying would not be too specific so as to limit the options the principal could choose, and which could be tailored to the specific context.

Although it is important for adults to understand what youth are saying online, they also need to frame this understanding within youth culture and their styles of communication. Communication among youth is very different than that of the older generation. Wright and Lawson (2004) espouse the benefits of using social networking sites and other online tools as pedagogy in the classroom, as this is the form of communication youth use and this knowledge can be re-directed in positive ways in the classroom. In a Flemish study, however, 78 principals from primary schools were asked to describe the way technology was used in their classrooms. All but one principal listed rules, restrictions and monitoring placed on the use of computers to provide safe Internet use. Only one principal used technology as curriculum, incorporating a project that collaborated with teachers and students to focus on safe Internet use (Valcke et al. 2007).

Further, parents are often left out of the loop when it comes to developing policy involving social communication tools and Internet use. This restricts the ways in which they might help their children develop more positive ways to communicate online. In the end, parents and teachers, who are typically cyberspace foreigners, must strengthen their knowledge base when it comes to online forums and technology, as children and adolescents are fast becoming the experts in this area. In doing so, parents and teachers will be able to better understand the attitudes and philosophies that guide young people's Internet use (Chou et al. 2009). As it stands now, children in households are the computer experts, which Stahl and Fritz (2002) suggest, "disrupts the guiding role of parents" (p. 9). Ybarra and Mitchell (2008) suggest that schools re-focus their efforts from restricting online access in schools to employing mental health interventions for susceptible youth and cyberspace safety education that teaches youth positive and caring online communications.

Parents: Views, Roles, Supervision

Parents are key to preventing cyber-bullying and to addressing it when their children are victims or perpetrators, yet very little research has been done on parents' perceptions, understandings, or experiences. This is why we surveyed parents (n=315) in our second study, and linked these findings with what students were saying in their survey and what educators were saying in their interviews. We uncovered five important findings.

First, parents seriously underestimate the amount of time their children spend on line each day. From our parents' questionnaire (see Table 1), just under one-quarter of parents believe their child spends less than one hour online (*n*=303). Only 6 percent believe their child is online five hours or more. Alternatively, as shown in Table 2, the significance of variation between parents' estimation and actual usage by their child is evident, as 11 percent of students admit spending more than five hours a day online. Approximately 52 percent of parents estimate their children spend one to two hours per day online, which is fairly constant with the students' self-reports of 48 percent. The same can be said of Internet usage of three to four hours in which 15.5 percent of parents versus 16.8 of students self-report such involvement. However, at this point an inverse pattern begins to emerge (3 hours or more) in which the parents' estimation and student self-reports reverse, with far more parents underestimating the actual hours their child spends online. As the literature reveals, many parents may miscalculate the amount of time their child spends on the Internet, or are simply unaware of their child's computer usage (Liau et al. 2008).

Bjørnstad and Ellingsen (2004) also found the same result as did Liau et al. (2008). In some cases, a youth may spend hours online after the parent(s) go to bed, particularly in some immigrant communities where he or she is communicating with friends or family from a different time zone,

Table 1. (Parents) (recode) How many hours per day (on average) do you think your child spends online?

		Frequency	Percent	Valid Percent	Cumulative Percent
Valid	1-2 hours	163	51.7	53.8	53.8
	3-4 hours	49	15.6	16.2	70.0
	5-6 hours	13	4.1	4.3	74.3
	More than 6 hours	6	1.9	2.0	76.2
	Less than 1 hour	67	21.3	22.1	98.3
	None	5	1.6	1.7	100.0
	Total	303	96.2	100.0	
Missing	System	12	3.8		
Total		315	100.0		

Table 2. (Children) How often are you online?

		Frequency	Percent	Valid Percent	Cumulative Percent
Valid Total	not every day	107	31.6	31.6	31.6
	usually 1-2 hours per day	138	40.7	40.7	72.3
	usually 3-4 hours per day	57	16.8	16.8	89.1
	usually 5-6 hours per day	19	5.6	5.6	94.7
	usually more than 6 hours per day	18	5.3	5.3	100.0
		339	100.0	100.0	

or where the youth is involved in blogs, chat rooms, or global games. This miscalculation of computer use time is more serious when parents are not aware of what their children are doing on the computer. In our study, approximately 40 percent of parents either do not or only minimally supervise their child's time on the computer. This corroborates Lenhart and Madden's (2007) study, where they found that although some parents do invoke computer rules governing their child's Internet use, others feel disinclined to do so because they feel their child is safe or mature enough to handle specific circumstances. Even in cases where there is some degree of supervision or parameters around computer use in the home, this does not mean that youth adhere to the rules. It may be that rules are ignored or circumvented by youth. In any case, the supervisory techniques identified by parents in our study were broad in scope – from walking back and forth occasionally or walking into the room at regular intervals, installing filters/trackers, or checking the online history. A few parents used more invasive strategies, like prohibiting social networking and chat sites, removing Internet access from the home, or strictly monitoring computer time.

Second, parents' familiarity with technology fluctuated widely. Most parents (77%) were very familiar with older technology such as email and cellular phones but were ambiguous about weblogs (69%) and Facebook (65%). Overall, we found no significant relationship between the parents' degree of technological knowledge and their level of concern about cyber-bullying, but correlation analysis was significantly related ONLY with parents' knowledge of older technology (emails) ($p=0.025$; $p<.05$). Accordingly, there was no link between those parents who were more familiar with new technology such as weblogs and Facebook and those who expressed more concern about cyber-bullying and the potential dangers of online communications. As set out in Figure 1, approximately 20 percent of parents indicated that they were "not too concerned" about cyber-bullying while 44 percent were "quite concerned" ($n=276$, $M=3.5$, $SD=1.19$).

Third, we found no significant differences in the level of supervision and the level of concern about cyber-bullying ($x^2=8.145$, $p>.05$); the extent of supervision had no correlation with the level of concern parents had about cyber-bullying.

Fourth, we noted a disparity between what parents said about the extent to which their children were involved in cyber-bullying, and what the youth on the surveys reported. Approximately 10 percent of parents reported their child had been involved in some form of cyber-bullying, while in the student survey, 35 percent admitted to cyber-bullying and 32 percent of students reported being victimized. Seemingly parents are unsuspecting of their children's cyber-bullying experiences, or their children have kept them out of the knowledge loop.

Lastly, when we asked parents about cyber-bullying prevention, five themes emerged: tighten controls and restrict behaviour; focus on preven-

Figure 1.

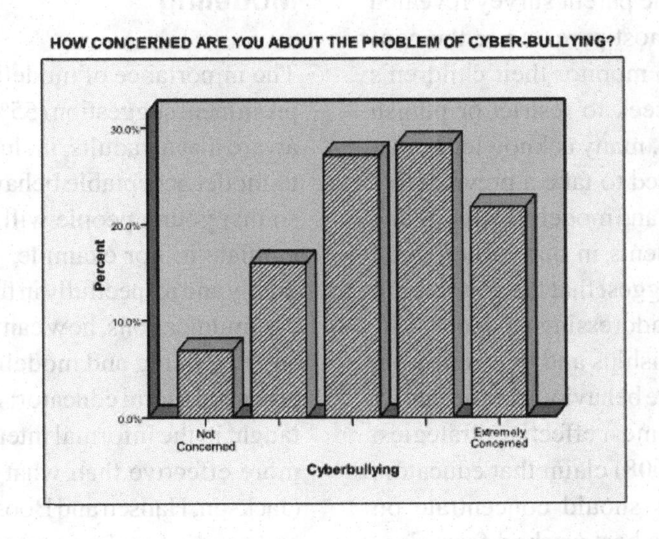

tion, education and positive communication; unfamiliarity with cyber-bullying and possible solutions; holding the school responsible; and cyber-bullying is inevitable. Approximately 41 percent of respondents (*n*=108) said that the best way to prevent cyber-bullying was to impose strict controls and punish if breached. However, 36 percent (*n*=95) offered a more holistic, contextual approach to solving cyber-bullying problems (discussed in the next section as this connects to the discussion on cyber-kindness). Ten parents reported that they were unfamiliar with the term "cyber-bullying" and its consequences, and therefore had no suggestions for solutions. Three parents felt the onus should be on schools to take the lead in thwarting cyber-bullying, while another three cynical parents admitted that cyber-bullying was inevitable and prevention was impossible.

CYBER-KINDNESS: CARING AND RESPECTFUL ONLINE RELATIONSHIPS

Electronic arenas are not all negative and can be a haven for positive discourse where youth seek a safe, nurturing environment for behaviours that promote social responsibility and encourage caring and respectful interactions. For example, social networking sites are primarily used by young people to keep in contact with their friends and engage in conversation and social planning. A Pew Internet study, for example, revealed that 91 percent of teens who interact on social networking sites do so to maintain contact with frequent friends (Lenhart et al. 2007). Overall, online engagement can create and further friendships (Kowalski and Limber, 2007). Some of the school-based interventions, however, are insufficient in teaching youth the necessary skills and behaviour when online, with many school programs focusing on restrictive methods rather than instructional strategies and collaboration with students to develop skills and knowledge (Valcke et al. 2007). When we asked educators in our second study (five-point *Likert-style* scale, 1=not important at all, 5=extremely important) about how important it was to encourage cyber-kindness in relation to the various competing priorities in the school, 65 percent (*n*=17, SD=1.03, M=4) claimed that this issue is an important one that needs to be developed and practiced in the classroom and in personal communications. Only one youth worker in the school did not feel that teaching about cyber-kindness was important at all.

Our findings from the parent survey revealed similar results. While most parents use the control method as a way to monitor their children's online behaviour, and seek to restrict or punish those who cross the line, many acknowledge that schools and families need to take a preventative approach and encourage and model positive online behaviour. Indeed, students in our earlier study (Cassidy et al. 2009), suggest that the solutions to cyber-bullying lie with addressing students' self-esteem and peer relationships and that education and modeling appropriate behaviours in the family and in the school are the most effective strategies. Ybarra and Mitchell (2008) claim that educators and other professionals should concentrate on youths' behaviour as the best method for reducing online victimization rather than governing specific online sites.

Bhat (2008) recognizes the importance of parents teaching their children appropriate online etiquette and working with them to distinguish the difference between appropriate and inappropriate online behaviour. Lee (2009) makes the same observation, and discusses the impact of the Internet on "social interaction, social relationships, and social capital" (p. 510). Approximately 71 percent of parents (*n*=224) in our second study suggested ways to cultivate a more considerate, kind, respectful and caring online environment, although fifteen parents were mystified about the concept of cyber-kindness – its authenticity and usefulness. The following themes emerged from the analysis of their suggestions, offered in response to an open-ended question: *If cyber-kindness is the opposite of cyber-bullying, how might educators and parents help cultivate more considerate, kind, respectful and caring online interactions between youth today?*" - the importance of adults' modeling the right behaviour; the importance of the home environment; the need for classroom instruction and rewards in this area; and the importance of providing online examples to show what cyber-kindness might look like. Each of these suggestions is discussed in turn.

Modeling

The importance of modeling was by far the most prominent suggestion (55%, *n*=124). Parents were aware that all adults, including educators, needed to model acceptable behaviour in their own lives so that young people will see this behaviour and emulate it. For example, if adults are not acting kindly and respectfully in their own personal online communications, how can youth be expected to do so? Practicing and modeling the right values is a powerful moral educator; in fact, typically what is taught in the informal interactions of life are often more effective than what is said or admonished (Jackson, Hansen and Boostrom, 1995). Modeling and practice are key components for fostering the ethic of care, as articulated by philosopher and educator, Nel Noddings (2002, 2005) and others (Cassidy and Bates, 2005; Gregory, 2000; Rauner, 2000). Parents in our study understood that they needed to "practice what they preached", so their words were not empty admonitions. One parent stressed the need to make kindness and caring among youth the "cool" thing to do, so others would follow in this trendsetting behaviour. Another parent explained that if kindness and respect were prevalent in the real world, then perhaps such behaviour would spill over into the virtual world.

Indeed, the cyberspace realm offers much potential for positive initiatives. For instance, just over 50 percent of the four million (and growing) blogs are created by youth between the ages of 13 and 19, and this is an arena for youth and adults alike to share stories, voice their opinions, and record journals (Chou et al. 2009). Tynes (2007) explained that youth, especially those who may not have siblings or parents to help them with homework, often ask online friends for help with homework or advice about particular classes. Facebook, in particular, can be an invaluable resource of information and consultation.

Home Environment

In our study, parents emphasized that learning begins at home, and that parents had to assume more responsibility to teach and model respect for others. As one parent said: "it all starts in the home. If children do not see kindness and respect from siblings, friends and parents, the war is nearly lost." Tynes (2007) argued for parents to discuss the positive aspects of the Internet, rather than focus on the negative aspects. Problems that are inadequately addressed at home may predispose youth to vulnerable online activities. Unfortunately, as we discussed earlier, parents often lack the skills and knowledge about online technology; this fact was acknowledged, however, by only two parents in our study who stated the need to become more technologically savvy in order to help their children. This lack of knowledge, however, should not stop parents from discussing and modeling respectful behaviours, so youth can recognize and maintain such actions in their online world.

Classroom Instruction and Resources

Parents presented the idea of schools developing homework assignments based on random acts of cyber-kindness, in collaboration with parents and peer groups at the school and at home. Other parents suggested that educators should provide examples of positive language that youth could practice online with a friend or a senior citizen such as a grandparent. In a similar vein, Bhat (2008) stressed the need for school counselors to take an active role in offering training to students and parents about "netiquette" standards of behaviour since many youth may not understand what are acceptable online expressions. Other parents suggested in-class activities such as role-playing, a cyber communication course and basic instructions on the ethics of Internet use and respect.

The website Net-Detectives (www.net-detectives.org) uses a role-play model that could be adapted for cyber-bullying challenges and as a way for students to identify with different perspectives using the role-play strategy (Wishart, Oades and Morris, 2007). An Internet safety advice website, Kidsmart (www.kidsmart.org.uk), is helpful for schools, parents and students and provides lesson plans and slides for parents. Bhat (2008) outlines a number of online programs and resources that educators might want to use or adapt for their classroom: for example,

i-SAFE (http://www.i-safe.org), Bullying, No Way! (http://www.bullyingnoway.com.au/), SAMHSA (http://www.samhsa.gov/index.aspx), Steps to Respect: A Bullying Prevention Program (http://www.cfchildren.org/programs/str/overview/). Media Awareness Network (http://www.media-awareness.ca/english/index.cfm), has some specific resources on cyber-bullying for classroom use.

Contests and Rewards/ Online Examples

Some parents in our study thought that if students regularly saw examples of positive messages online they might be more likely to do the same. For example, each week the school could install a new screensaver on the school computers showing examples of positive online exchanges submitted by students in the school. Others thought rewards or contests might foster positive initiatives – rewarding students with I-tune dollars or celebrate kindness through an award system at school. Overall parents expressed their concern about how cyber-bullying seemed part of the wider culture of how youth related to one another, and because of this it required a deep and sustained effort on the parts of parents and educators.

TEACHERS: OPINIONS ABOUT CYBER-KINDNESS

In our second study, the 17 educators in their interviews were asked to give examples of times when students used technology to communicate positive, supportive, respective and kind messages to each other or to school personnel. They were also queried on ways in which they might incorporate cyber-kindness into the curriculum. When asked on a five-point *Likert-style scale* how important it is to teach cyber-kindness in relation to competing priorities, 65 percent (n=11, SD=1.03, M=4), stated that cyber-kindness is "important" or "very important."

Educators, though, had few examples of cyber-kindness to share, and even fewer suggestions for fostering cyber-kindness. One counselor speculated that the majority of the interactions between youth online are positive. Another teacher offered an example of a student who initiated an act of kindness towards a classmate, by hooking up a camera in class to MSN and connecting with the student who was ill at home so he could follow the lesson. This same teacher involved his students in two charities, which operate online and provide resources and support to the needy. A vice-principal mentioned that one teacher was using a Wiki site for students to post positive annotations about their class discussions. The administrators at this school also used Wiki for teachers to share curriculum ideas and readings. This school also used Facebook to raise $75,000 for juvenile diabetes.

Another counselor who had worked in the school system for 29 years viewed cyber-kindness as being practiced in the school environment when students helped each other with homework online or initiated correspondence with students who were new to the school. Several educators mentioned kind or considerate thank you emails or examples of students bolstering others' self-esteem by saying flattering comments about their appearance.

ENCOURAGING CYBER-KINDNESS

Suggestions from educators and parents for ways to encourage a kinder online world are in line with the current literature: educating teachers about social networking sites (Worthen, 2007); engaging parents and teachers in collaborative solutions (Beale and Hall, 2007); designing effective curriculum (Sharples et al. 2009); modeling appropriate values and behaviours in the school and home (Noddings, 2002, 2005); and building trusting relationships with youth so that open and respectful dialogue can occur (Cassidy & Bates, 2005; Palmer, 1998). Our study of cyber-bullying and cyber-kindness, however, raises questions for further research.

For example, because technology is changing so rapidly and young people are on the cutting edge of the new media, what will be the impact of this increasing gap in knowledge between the adult generation and youth? Should young people be given the responsibility for setting up their own ways of monitoring online behaviour and also play a role in developing school curriculum for addressing the deeper issues as to why they and their peers cyber-bully? If cyber-bullying is so widespread among the youth culture, then researchers need to investigate why this is so. Is it because, as our study suggests, that youth see negative online exchanges as just a regular part of the online world and something to be tolerated? If so, then how can this perception be addressed? Further investigation also needs to be undertaken with respect to gender differences in cyber-bullying (Jackson et al. 2009b), and since there are differences in how girls verses boys cyber-bully perhaps there needs to be differential approaches according to gender for curtailing cyber-bullying. Further, additional studies need to be undertaken to assess whether there are differences in perceptions and experiences depending on culture, or social-economic status or other characteristics. Also, the new field of how to foster cyber-kindness is an area full of opportunity for further research.

CONCLUSION AND IMPLICATIONS

It is obvious that there is a generational gap between educators and youth regarding familiarity with technology. This restricts the ways in which teachers might creatively address issues of cyber-bullying and work towards cultivating a kinder online world. Educators' priorities seem more geared to curtailing cyber-bullying than to building a positive school culture of cyber-kindness. It is apparent that few educators can provide examples of cyber-kindness and even fewer can suggest ways of fostering a more positive online world with their students. It is also apparent from the literature that schools do not provide sufficient opportunity for discussing cyber-bullying among stakeholders and for engaging youth in collaborative efforts to develop policies and practices for countering this growing problem. Nor are parents and other related professionals (medical staff, psychologists, counselors, police) invited to participate in the dialogue and to work with schools to develop effective measures and to deal with the aftermath of victimization (Tynes, 2007).

It is also evident that much of the cyber-bullying activity is happening under the radar of school staff and parents. Educators who participated in our two studies were only able to provide a few examples of cyber-bullying incidents that happened in their schools, yet up to 32 percent of students in these schools had been cyber-bullied in the last year and up to 36 percent had participated in cyber-bullying activities towards others. Also, 12 percent admitted to posting mean, nasty, rude or vulgar messages about a school official online, another aspect of cyber-bullying that is worrisome. Similarly parents were generally unaware of the extent of cyber-bullying being experienced or perpetrated by their children. Only thirty-three parents (approximately 10%) were aware that their child had been involved in various types of cyber-bullying activities, either as a victim or perpetrator, indicating a glaring discrepancy between what the parents know about and what is actually happening in their child's life. Parents, then, are generally uninformed about the extent of cyber-bullying being experiencing by their children. It is apparent that concerted and attentive dialogue about cyber-bullying involving all the major players in the lives of youth should occur and that stakeholders should work collaboratively to form long-term solutions that encourage kindness and caring in the online world.

REFERENCES

Attewell, P., Suazo-Garcia, B., & Battle, J. (2003). Computers and Young Children: Social Benefit or Social Problem? *Social Forces*, *82*(1), 277–296. doi:10.1353/sof.2003.0075

Barron, C. (2000). *Giving Youth a Voice: A Basis for Rethinking Adolescent Violence*. Halifax, Nova Scotia: Fernwood Publishing.

Beale, A., & Hall, K. (2007). Cyberbullying: What school administrators (and parents) can do. *Clearing House (Menasha, Wis.)*, *81*(1), 8–12. doi:10.3200/TCHS.81.1.8-12

Bhat, C. (2008). Cyber Bullying: Overview and Strategies for School Counsellors Guidance Officers and All School Personnel. *Australian Journal of Guidance & Counselling*, *18*(1), 53–66. doi:10.1375/ajgc.18.1.53

Bjørnstad and Ellingsen. (2004). *Onliners: A report about youth and the Internet*. Norwegian Board of Film Classification.

Brown, K., Jackson, M., & Cassidy, W. (2006). Cyber-bullying: Developing policy to direct responses that are equitable and effective in addressing this special form of bullying. [), http://umanitoba.ca/publications/cjeap/]. *Canadian Journal of Educational Administration and Policy*, *57*,

Campbell, M. (2005). Cyberbullying: An older problem in a new guise? *Australian Journal of Guidance & Counselling*, *15*(1), 68–76. doi:10.1375/ajgc.15.1.68

Cassidy, W., & Bates, A. (2005). "Drop-outs" and "push-outs": finding hope at a school that actualizes the ethic of care. *American Journal of Education*, *112*(1), 66–102. doi:10.1086/444524

Cassidy, W., Brown, K., & Jackson, M. (2010). *Redirecting students from cyber-bullying to cyber-kindness.* Paper presented at the 2010 Hawaii International Conference on Education, Honolulu, Hawaii, January 7 to 10, 2010.

Cassidy, W., Jackson, M., & Brown, K. (2009). Sticks and stones can break my bones, but how can pixels hurt me? Students' experiences with cyber-bullying. *School Psychology International*, *30*(4), 383–402. doi:10.1177/0143034309106948

Chou, C., Yu, S., Chen, C., & Wu, H. (2009). Tool, Toy, Telephone, Territory, or Treasure of Information: Elementary school students' attitudes toward the Internet. *Computers & Education*, *53*(2), 308–316. doi:10.1016/j.compedu.2009.02.003

Cook-Sather, A. (2002). Authorizing students' perspectives: Toward trust, dialogue, change in Education. *Educational Researcher*, *24*(June-July), 12–17.

Crick, N., & Grotpeter, J. (1995). Relational aggression, gender, and social-psychological adjustment. *Child Development*, *66*, 710–722. doi:10.2307/1131945

Crick, N., Werner, N., Casas, J., O'Brien, K., Nelson, D., Grotpeter, J., & Markon, K. (1999). Childhood aggression and gender: A new look at an old problem. In Bernstein, D. (Ed.), *Nebraska symposium on motivation* (pp. 75–141). Lincoln, NE: University of Nebraska Press.

Dowell, E., Burgess, A., & Cavanaugh, D. (2009). Clustering of Internet Risk Behaviors in a Middle School Student Population. *The Journal of School Health*, *79*(11), 547–553. doi:10.1111/j.1746-1561.2009.00447.x

Froese-Germain, B. (2008). Bullying in the Digital Age: Using Technology to Harass Students and Teachers. *Our Schools. Our Selves*, *17*(4), 45.

Genuis, S., & Genuis, S. (2005). Implications for Cyberspace Communication: A Role for Physicians. *Southern Medical Journal*, *98*(4), 451–455. doi:10.1097/01.SMJ.0000152885.90154.89

Gradinger, P., Strohmeier, D., & Spiel, C. (2009). Traditional Bullying and Cyberbullying: Identification of Risk Groups for Adjustment Problems. *The Journal of Psychology*, *217*(4), 205–213.

Green, H., & Hannon, C. (2007). *TheirSpace: Education for a digital generation*. Demos, London. Retrieved July 6, 2009 from: http://www.demos.co.uk/publications/theirspace

Gregory, M. (2000). Care as a goal of democratic education. *Journal of Moral Education*, *29*(4), 445–461. doi:10.1080/713679392

Jackson, M., Cassidy, W., & Brown, K. (2009a). Out of the mouths of babes: Students "voice" their opinions on cyber-bullying. *Long Island Education Review*, *8*(2), 24–30.

Jackson, M., Cassidy, W., & Brown, K. (2009b). *"You were born ugly and youl die ugly too": Cyber-bullying as relational aggression. In Education (Special Issue on Technology and Social Media (Part 1),* 15(2), December 2009. http://www.ineducation.ca/article/you-were-born-ugly-and-youl-die-ugly-too-cyber bullying-relational-aggression

Jackson, P. W., Hansen, D., & Boomstrom, R. (1993). *The moral life of schools*. San Francisco: Jossey-Bass Publishers.

Juvonen, J., & Gross, E. F. (2008). Extending the school grounds? Bullying experiences in cyberspace. *The Journal of School Health*, *78*(9), 496–505. doi:10.1111/j.1746-1561.2008.00335.x

Kowalski, R., & Limber, S. (2007). Electronic Bullying Among Middle School Students. *The Journal of Adolescent Health*, *41*(6), S22–S30. doi:10.1016/j.jadohealth.2007.08.017

Kowalski, R. M., Limber, S. E., & Agatston, P. W. (2008). *Cyber bullying: Bullying in the digital age*. Malden, MA: Blackwell Publishers. doi:10.1002/9780470694176

Lee, S. (2009). Online Communication and Adolescent Social Ties: Who benefits more from Internet use? *Journal of Computer-Mediated Communication*, *14*(3), 509–531. doi:10.1111/j.1083-6101.2009.01451.x

Lenhart, A., & Madden, M. (2007). Teens, Privacy & Online Social Networks: How teens manage their online identities and personal information in the age of MySpace. *Pew Internet & American Life Project*. Retrieved from: http://www.pewinternet.org

Lenhart, A., Madden, M., & Macgill, A. (2007). Teens and Social Media: The use of social media gains a greater foothold in teen life as they embrace the conversational nature of interactive online media. *Pew Internet & American Life Project*. Retrieved from: http://www.pewinternet.org

Li, Q. (2007). New bottle but old wine: A research of cyberbullying in schools. *Computers in Human Behavior*, *23*(4), 1777–1791. doi:10.1016/j.chb.2005.10.005

Liau, A., Khoo, A., & Ang, P. (2008). Parental awareness and monitoring of adolescent Internet use. *Current Psychology (New Brunswick, N.J.)*, *27*(4), 217–233. doi:10.1007/s12144-008-9038-6

ATA News (2009). *RCMP and CTF join forces to fight cyberbullying*, *43*(10), 6.

Noddings, N. (2002). *Educating moral people: A caring alternative to character education*. New York: Teachers College Press.

Noddings, N. (2005). *The challenge to care in schools: An alternative approach to education*. New York: Teachers College Press.

Owen, L., & Ennis, C. (2005). The ethic of care in teaching: An overview of supportive literature. *Quest*, *57*(4), 392–425.

Palmer, P. J. (1998). *The courage to teach: Exploring the inner landscape of a teacher's Life*. New York: Jossey-Bass.

Palys, T. (2003). *Research Decisions: Quantitative and Qualitative Perspectives* (3rd ed.). New York: Thomson Nelson.

Patchin, J. W., & Hinduja, S. (2006). Bullies move beyond the school yard: A preliminary look at cyber bullying. *Youth Violence and Juvenile Justice*, *4*(2), 148–169. doi:10.1177/1541204006286288

Prensky, M. (2001). Digital natives, digital immigrants. *Horizon*, *9*(5), 1–6. doi:10.1108/10748120110424816

Raskauskas, J., & Stoltz, A. D. (2007). Involvement in traditional and electronic bullying among adolescents. *Developmental Psychology*, *43*(3), 564–575. doi:10.1037/0012-1649.43.3.564

Rauner, D. (2000). *They still pick me up when I fall: The role of caring in youth development and community life*. New York: Columbia University Press.

Rigby, K. (2005). What children tell us about bullying in schools. *Children Australia Journal of Guidance and Counselling*, *15*(2), 195–208. doi:10.1375/ajgc.15.2.195

Rosen, L., Cheever, N., & Carrier, L. (2008). The association of parenting style and child age with parental limit setting and adolescent MySpace behavior. *Journal of Applied Developmental Psychology*, *29*(6), 459–471. doi:10.1016/j.appdev.2008.07.005

Shariff, S. (2006). Cyber-Dilemmas: Balancing Free Expression and Learning in a Virtual School Environment. *International Journal of Learning*, *12*(4), 269–278.

Shariff, S. (2008). *Cyber-bullying: Issues and solutions for the school, the classroom and the home*. New York: Routledge.

Shariff, S., & Gouin, R. (2005). *Cyber-Dilemmas: Gendered Hierarchies, Free Expression and Cyber-Safety in Schools*.paper presented at Oxford Internet Institute, Oxford University, U.K. International Conference on Cyber-Safety. Paper available at: www.oii.ox.ac.uk/cybersafety

Shariff, S., & Johnny, L. (2007). Cyber-Libel and Cyber-bullying: Can Schools Protect Student Reputations and Free Expression in Virtual Environments? *Education Law Journal*, *16*(3), 307.

Sharples, M., Graber, R., Harrison, C., & Logan, K. (2009). E-safety and Web 2.0 for children aged 11-16. *Journal of Computer Assisted Learning*, *25*(1), 70–84. doi:10.1111/j.1365-2729.2008.00304.x

Slonje, R., & Smith, P. (2008). Cyberbullying: Another main type of bullying? *Scandinavian Journal of Psychology*, *49*(2), 147–154. doi:10.1111/j.1467-9450.2007.00611.x

Smith, P., Mahdavi, J., Carvalho, M., Fisher, S., Russell, S., & Tippett, N. (2008). Cyberbullying: its nature and impact on secondary school pupils. *Journal of Child Psychology and Psychiatry, and Allied Disciplines*, *49*(4), 376–385. doi:10.1111/j.1469-7610.2007.01846.x

Stahl, C., & Fritz, N. (2002). Internet safety: adolescents' self-report. *The Journal of Adolescent Health*, *31*(1), 7–10. doi:10.1016/S1054-139X(02)00369-5

Sun, P., Unger, J., Palmer, P., Gallaher, P., Chou, P., & Baezconde-Garbanati, L. (2005). Internet accessibility and usage among urban adolescents in southern California: Implications for web-based health research. *Cyberpsychology & Behavior*, *8*(5), 441–453. doi:10.1089/cpb.2005.8.441

Tynes, B. (2007). Internet safety gone wild? Sacrificing the educational and psychosocial benefits of online social environments. *Journal of Adolescent Research*, *22*(6), 575–584. doi:10.1177/0743558407303979

Valcke, M., Schellens, T., Van Keer, H., & Gerarts, M. (2007). Primary school children's safe and unsafe use of the Internet at home and at school: An exploratory study. *Computers in Human Behavior*, *23*(6), 2838–2850. doi:10.1016/j.chb.2006.05.008

Willard, N. (2006). Flame Retardant. *School Library Journal*, *52*(4), 55–56.

Willard, N. (2007). The authority and responsibility of school officials in responding to cyberbullying. *The Journal of Adolescent Health*, *41*(6), 64–65. doi:10.1016/j.jadohealth.2007.08.013

Williams, K., & Guerra, N. (2007). Prevalence and Predictors of Internet Bullying. *The Journal of Adolescent Health*, *41*(6), S14–S21. doi:10.1016/j.jadohealth.2007.08.018

Wishart, J., Oades, C., & Morris, M. (2007). Using online role play to teach internet safety awareness. *Computers & Education*, *48*(3), 460–473. doi:10.1016/j.compedu.2005.03.003

Worthen, M. (2007). Education policy implications from the expert panel on electronic media and youth violence. *The Journal of Adolescent Health*, *41*(6), 61–62. doi:10.1016/j.jadohealth.2007.09.009

Wright, E. R., & Lawson, A. H. (2004). *Computer-mediated Communication and Student Learning in Large Introductory Sociology Courses.* Paper presented at the Annual Meeting of the American Sociological Association, Hilton San Francisco & Renaissance Parc 55 Hotel, San Francisco, CA. Retrieved from: http://www.allacademic.com/meta/p108968_index.html

Ybarra, M., Diener-West, M., & Leaf, P. (2007). Examining the overlap in Internet harassment and school bullying: Implications for school intervention. *The Journal of Adolescent Health*, *41*(6), 42–50. doi:10.1016/j.jadohealth.2007.09.004

Ybarra, M., & Mitchell, K. (2004). Youth engaging in online harassment: associations with caregiver-child relationships, Internet use and personal characteristics. *Journal of Adolescence*, *27*(3), 319–336. doi:10.1016/j.adolescence.2004.03.007

Ybarra, M., & Mitchell, K. (2008). How Risky are Social Networking Sites? A comparison of places online where youth sexual solicitation and harassment occurs. *Pediatrics*, *121*(2), 350–357. doi:10.1542/peds.2007-0693

ENDNOTE

[1] Parents and other primary caregivers such as grandparents, legal guardians, etc. will be referred to here collectively as "parents".

Chapter 16
Electronic Aggression among Adolescents:
An Old House with a New Facade (or Even a Number of Houses)

Jacek Pyżalski
Wyższa Szkoła Pedagogiczna w Łodzi, Poland & Nofer Institute of Occupational Medicine, Poland

ABSTRACT

The chapter is focused on the problem of electronic aggression (conducted via the Internet or mobile phones) in the context of young people as potential victims and perpetrators of such aggression. The text addresses two main issues: the potential novelty of electronic aggression and its potential distinctive features and the diversity of electronic aggression acts (with a proposal of typology). The first aspect is analyzed through the new model – ABACUS that could be used to compare electronic and traditional aggression. The chapter presents also a typology of electronic aggression based on the victim's identity and his/her relationship with a young person who is a perpetrator. The presented theories and discussions are illustrated with new data from two Polish projects on students and teachers experiences with electronic aggression.

INTRODUCTION

The fact that new communication technologies (mostly Internet and mobile phones) can be used as tools to conduct hostile acts is obvious for both the scientific community and the general public. The cases where the electronic aggression attacks led to disastrous effects particularly among young people e.g. suicidal attempts are causing much alarm in the popular media. Those emotional transmissions often reinforce an oversimplified and exaggerated picture of the phenomenon of aggression conducted via ICT (electronic aggression), as well as its causes and effects. This attitude sometimes affects scientists as well with a kind of "moral panic" about the phenomenon. Not neglecting the potential negative consequences of electronic aggression, I stand on the position that in order to obtain a real picture of this phenomenon, two main areas should be scientifically explored and discussed. First, we should find out to what extent electronic aggression is qualita-

DOI: 10.4018/978-1-60960-209-3.ch016

tively new, comparing to traditional aggression (where the new technologies are not involved) in terms of influencing factors, psychological and social mechanisms and potential consequences. The answer is interesting for scientists, but is also of importance for practitioners involved in prevention and intervention in electronic forms of aggression. Still the aggression where the new media are used as tools is an important topic, since in the modern society media "emerged as a social institution, assuming many of the functions formerly served by traditional social institutions such as the church, school, government and family" (Silverblatt, 2005, p. 35.). This may be reinforced by the extensive usage of new media by young people and their significant role in development of young people (Living and Learning with New Media, 2008). Taking this into account, the media used to conduct aggression may cause a serious danger particularly for the contemporary young generation.

The second important issue concerns the diversity of electronic aggression acts that have one thing in common – a usage of new communication technologies as a tool to conduct hostile acts. In reality, it is vital to develop a typology that will enable us to make some order in electronic aggression variety. Some kind of typology should be developed since electronic aggression acts differ substantially when the psychological and social mechanisms involved as well as their consequences are considered.

In this chapter we will start from the overview of terms and definitions concerning the aggressive acts conducted through new communication technologies. Here we will also look closer at the features that various authors typically attribute to electronic aggression. Although electronic aggression is not exclusive for children and adolescents most of the chapter is based on research in this age group. That does not limit the validity of majority of presented interpretations to electronic aggression in other age groups.

Afterwards, a new theoretical model useful for the analysis of distinctive features of electronic aggression will be presented (ABACUS model). The proposed model underlines that a particular act of electronic aggression may be more or less distinctive according to how many "new" features are present. It also shows that all features referred to sometimes as typical for electronic aggression may be present while acts of traditional aggression are conducted. Despite this some mechanisms lying behind those electronic aggression features are novel in the context of electronic aggression due to specific qualities of computer mediated communication (CMC) and online interactions in a broader sense; this back up a position that electronic aggression is a qualitatively different phenomenon comparing to traditional aggression.

Then the need for typology of electronic aggression will be underlined and the proposal of such typology based on the victim's identity and his/her relationship with a perpetrator will be presented and discussed in terms of their potential harmful effects.

The discussion will be illustrated by partial data from two research projects described briefly at the end of the chapter[1].

ELECTRONIC AGGRESSION: DEFINITIONAL ISSUES

Electronic aggression is a general term that defines all hostile acts conducted with a help of new communication technologies or new media (often referred as the Internet and mobile phones) (David-Ferdon, Herz, 2007). Technically speaking, what distinguishes electronic aggression from its traditional version is a tool used by perpetrators to conduct harmful acts. Lists of such behaviors included in various definitions vary sometimes significantly from one study to another. To complicate issues more, till now many researchers use different terminology (e.g. electronic harassment, cyberbullying, etc.) without distinguishing the terms used, particularly when they operationalize the different concepts in measuring tools. Usually, regardless of the terminology, the definition consists of longer of shorter lists of hostile behaviors that can be conducted via the Internet or mobile phones. For example the long definitional list presented by Aricak, et al. (2008) includes lying, hiding the user identity, introducing oneself as someone else, threatening, teasing, insulting, defamation, intimidation, rumor, displaying others' pictures without their consent. But those authors include also different types of computer criminal acts as hacking or sending infected software to other users. Such a list is only one example as other researchers present different lists which add other activities and skip others.

In one of their studies, Patchin & Hinduja (2006) use the term "online bullying" and define it as "bothering someone online, teasing in a mean way, calling someone hurtful names, intentionally leaving persons out of things, threatening someone and saying unwanted sexually related things to someone". Some other authors include also such activities as spamming (sending unwanted electronic messages) or impersonation (stealing someone's electronic identity, e.g. through password theft). It is worth emphasising that some of the listed behaviors are distinguished based on the descriptive technical aspects of behavior (what the perpetrator actually does – e.g. publishing pictures without consent). At the same time, the others are distinguished based on perpetrators' and/or victims' feelings/interpretations (e.g. insulting, teasing in the mean way, doing those acts purposefully).

In reality, broadly defined and operationalized electronic aggression (also when the other terms are used) covers a wide range of behaviors that differ substantially in terms of victim identity or the technical methods of conducting aggression. For example, electronic aggression may be targeted against a victim known in a real world, at a person known only in the cyberspace or a stranger (e.g. a celebrity person) or a group of people (e.g. racial/sexual minority). That issue was taken into account by some researchers. Wolak, et al. (2007) explored the aspect of identity of a harasser from a victim's perspective and proved that harassment by a known peer and an online only contact are different particularly when potentially hurtful effects are discussed. Another aspect is connected to the fact that some electronic aggression acts are targeted directly at a victim (e.g. someone sends vulgar e-mails, MNS messages) while the other harm a victim indirectly (e.g. someone posts unwanted visual material about a victim on the Internet or spreads rumors).

All those problems with definitions and operationalizations are particularly true in terms of cyberbullying that should not be simply defined and operationalized as the aggression conducted by electronic means, but the aggression with certain characteristics described below. In this case the definition goes beyond the "technical" description of the act but moves rather to the question "how" the act is conducted also in terms of psychological and social mechanisms present.

The roots of the term 'cyberbullying' should be placed in the classical research of Olweus (1978; 1993), who clearly differentiates school bullying from the general school aggression. Based on his definition most authors agree that bullying should be defined as aggression with a number of specific features. Those distinctive aspects are: the negative intentions of perpetrators, the repetition of hostile acts and the imbalance of power between perpetrator(s) and victim(s), so that the latter have difficulty in defending themselves from aggressors. Most researchers agree on those three distinctions[2] (Monks, et al., 2009). Additionally, hostile acts are to be found in traditional bullying carried out by the people from familiar social groups – a class, a school, or neighborhood or at least the group of people that are frequently met in a spatial world (e.g. prison, children's home, etc.) (Stassen-Berger, 2007; Griffin & Gross, 2004; Monks, et al., 2009). One should note that to define a series of acts as bullying, all the above listed features must appear simultaneously. To sum up, "not all aggression is bullying, but bullying is always aggression" (Stassen-Berger, 2007, p. 194). Unfortunately, some authors use the term cyberbullying without clearly mentioning those characteristics, associated with bullying – they are more focused on electronic tools to conduct aggression. In some other cases they do involve those characteristics in the definition but they do not operationalize them in the research project. It seems that in many cases the term cyberbullying is used in too broad a sense, defining many aggressive acts even when the vital bullying characteristics are not present (Wolak, et al., 2007). Another issue is that bullying characteristics are sometimes problematic and not easy to operationalize when describing the cyber form of bullying (Agatson, et al., 2008; Slonje&Smith, 2008) – that important issue will be analyzed later in the chapter.

To sum up, what we need at this stage of research on electronic aggression is an awareness concerning the variety of electronic aggression types and clear definitions of them – as well as the measurement tools that take this diversity into account. This is inevitable in order to achieve relevant data that can be used in practice to prevent electronic aggression conducted and experienced by young people and its consequences.

Electronic Aggression: New or Old Phenomenon?

Most of the authors of the above-cited definitions state that if we use new media (the Internet/mobile phones) to conduct aggression/bullying or harassment we will produce electronic forms of those phenomena, respectively electronic aggression, cyberbullying and electronic harassment (Williams&Guerra, 2007). Additionally, when certain characteristic should be present in traditional forms (e.g. repetition in bullying), they are expected to occur in electronic forms as well. However, is the new tool enough to speak about the novel phenomenon? If yes, the question is why such an issue is raised with the Internet and mobile phones (referred also as new media) and had not been put forth so strong in the past, e.g. in case of traditional phone that may be used as a tool to conduct hostile acts as well? Some researchers define "new" features characteristic for electronic aggression but at the same time they underline that such particularities of online communication technology are present only in some electronic aggression cases (Juvonen&Gross, 2008). Thus, it is worth exploring them to find out whether there are any characteristics we can perceive as ambiguous attributes of the electronic form of aggression. Last but not least, what are the potential consequences of the presence of those potential new features at least in some electronic aggression cases? Below I will analyze three characteristics that are perceived as attributes of electronic aggression, namely anonymity, unintentionality and continuity.

Anonimity

Anonymity is often perceived as a basic characteristic of all communication that takes place in online environment (McKenna&Bargh, 2007; Wang et al., 3009), though its continual and common presence is far from the true. In many cases people who communicate via the Internet know each other from spatial world or have exchanged so much information during online interactions that full anonymity is clearly not existing in their relationships (Subrahmanyam,, et al., 2008).

When focusing on electronic aggression, anonymity seems to be the main justification and facilitator of hostile acts particularly when the Internet is in use (Juvonen&Gross, 2008). But what actually is meant by this anonymity? There are some vital aspects that should be taken into consideration. Part of them are connected to a potential perpetrator while the others lay on a victim's side. Starting from a perpetrator, we should discuss a mechanism of deindividuation (McKenna&Bargh, 2007). This mechanism connected to anonymity means that an individual experiences problems to control his/her own behavior and tends to react immediately according to external cues without making rational decisions. It is also connected to the tendency of ignoring how others assess the behavior of an individual. All those states are also associated with presence of a large number of people and feeling unity with them. (McKenna&Bargh, 2007).

Anonymity often means a lack or a great reduction of nonverbal cues in CMC – in cases when it takes textual form. In traditional communication, those cues play an important role in both relationships formation and its later maintenance. They are also vital for proper recognizing emotional states of those involved in communication act. Due to this lack of nonverbal cues may lead to situations where potential perpetrators will be unaware of actual consequences of their behavior online on the other people. They may also easily engage in the communication that they assess to be a hoax or a joke, while the other side will "read" them as aggressive acts – this will be further described below. On the Internet is also easier to reshape and construct someone's identities that are different from those in offline world (McKenna&Bargh, 2007). This is particularly important when an individual communicates with the others not known from offline environments. This may mean that those young people who do not engage in traditional aggression would engage in its electronic form in online environment.

On the side of someone attacked through new communication technologies anonymity of a perpetrator may reinforce victimization. When a perpetrator is hidden, even a trivial case may be assessed as a very serious issue. Also the possibilities to defend are restricted as there is no one who can be approached.

Unintentionality

Most scholars perceive negative intention as the feature that distinguishes aggression from other behaviors (Hasset&White, 1989; Wang et al., 2008). Although it is reasonable not to call aggression the acts that were not intended to harm another person that are a few things to consider in case of electronic form of aggression. First of all, a lot of electronic aggression perpetrators claim not to be aware of the fact that they online behavior and communications could have seriously harm anybody. In some cases we may assess those statements are mere excuses but taking into account the features of CMC (e.g. reduced nonverbal cues) it may be also their true experience. At the same time those unintentional acts may have really negative impact on the victims. Moreover, when we take into account the dimension of intentional harm on the side of a perpetrator and the dimension of being harmed on the side of a victim we can analyze four situations visualized in the Table 1.

The most obvious case of electronic aggression is presented in Situation 1 (Table 2.). There a perpetrator intended to harm a victim and they succeed with it. Such situations were described by some adolescents - respondents in the qualitative part of one of my studies (Pyżalski, 2009) – they claimed that their intention was to harm someone and then they were describing the subsequent consequences proving they have reached their aim (for example a victim's breakdown or refusal to go to school).

However, a lot of perpetrators engaged in behaviors that were "aimless" or that they had some other goals far removed from harming their victim (Situation 3). Still, they may claim that

Table 1. Perceived and intended harm in electronic aggression from the perspective of a victim and a perpetrator

		Victim's subjective feeling of being harmed	
		Yes	**No**
Perpetrator's intention to harm the victim	Yes	Situation 1	Situation 2
	No	Situation 3	Situation 4

Table 2. Involvement of Polish adolescents as perpetrators of different electronic aggression types in the previous 12 months.

Target of electronic aggression	% of respondents that attacked the target in the previous year	Type of electronic aggression
People known only from the Internet	42,5	Cyberbullying ?
Young people known from school/class or the place of living who are not close friends of a respondent	39	Cyberbullying ?
Close friends	26,8	Cyberbullying ?
Unknown people, totally randomly chosen	24,2	Random electronic aggression
Former girlfriend/boyfriend	16,9	Cyberbullying ?
Not individuals but groups of people, e.g. fans of the particular football team or a musician.	15,8	Electronic aggression against groups (bias cyberbullying)
Popular people, e.g. singers, actors, etc.	11,1	Electronic aggression against celebrities
Other people like homeless/alcoholics	10,8	Electronic aggression against the vulnerable/ Cyberbullying?
Teachers	9	Cyberbullying ?
Known adults	8,9	Cyberbullying ?

In all cases indicated by (?) the classification of electronic aggression act as cyberbullying is not obvious.

they were very surprised when some real problems and consequences for the victim occurred.

Among those reasons some respondents indicated that electronic aggression acts brought them respect among their peers, linked to high competences in using new communication technologies. The good example was one of the perpetrators who used to break into accounts of others and then using their e-mail addresses to send out fake messages or to break into their accounts in online games to steal their "identities". When asked about the advantages of such behavior he said that everyone living in neighborhood knew him as "hacker" and that gave him a high social position. One of the respondents a teacher from a secondary school describes the similar experience: *"The student has put on his blog vulgar comments about some of his teachers. He used really vulgar and insulting language. When asked about the reasons for his behavior he said he wanted to show off in front of peers. (...) A teacher (victim) was really touched and her self-esteem decreased."*. This finding is in line with study of Vandebosch, & Van Cleemput (2008). Some of the perpetrators of cyberbullying were claiming that through the acts they wanted to show their computer skills.

Additionally, involvement in the acts that are not intentionally harmful but end badly for someone may be particularly facilitated by electronic communication features. For example, perpetrators do not experience the emotional reactions of their victims that could have act as emotional meter that serves to temper (…) behaviors (Kowalski, Robin&Agatson, 2008, p. 65).

There were also many situations described by a victim (Situation 2) where someone attacked them electronically by way of chat, but they were not harmed at all and simply ignored or blocked the perpetrator. Such situations were often claimed when only a single attack occurred and the perpetrator was thought to be a stranger. The victims simply defined such perpetrators as "stupid" and were also saying that such situations were numerous but it was hard to recollect details as they are not important.

The most difficult to interpret is situation 4, where neither intention to harm nor the harm itself is present. How then do we know that what has taken place is electronic aggression? The only perspective we can adopt here that of the social norms of an external observer. Such norms are sometimes not accepted by the participants involved. For example, the respondent was involved for a long time in exchanging vulgar insulting chat messages with his peer. However, both involved were presenting the belief that there is nothing wrong with such behavior and what they do is a normal way to talk *"in the chat"*. Of course, it is disputable whether such a situation should be labeled aggression – still this is an interesting point for discussion – as those involved even have been exosed to extremely vulgar language.

Continuity

Some authors state that electronic forms of aggression may be a novel phenomenon since the impact of aggressive act is ongoing (Slonje&Smith, 2008; Walrave&Heirman, 2009). Victims of cyberbullying have no place to escape from electronic aggression acts as materials placed on the Internet are persistent, replicable and easy to find by the interested users (Boyd, 2007). That means that a young victim is never safe – there is no place and time when hostile acts are not present. This is analyzed as completely different to traditional aggression, where the hostile acts could have been conducted only when a victim and a perpetrator were situated in the same place at the same time (Walrave&Heirman, 2009). The same is sometimes extended even to those electronic aggression acts where there are no materials published on the Net. The so-called "always on" generation uses new communication technologies extensively as a tool to socialize with peers (Living and Learning with New Media, 2008). That means that even a young person that is targeted directly, through, say, unwanted messages may perceive it as ongoing as he/she is always possible to reach through the Internet or a mobile phone. Additionally, decision about giving up the usage of communication technologies would often mean social exclusion from a peer group (Kowalski, et al., 2008).

The ABACUS Theory of Electronic Aggression

While we analyze the features of electronic aggression (e.g. anonymity, unintentionality, continuity) we should put forth the most important question: Is there, except the tool used to conduct hostile acts (namely new communication technologies), any distinctive feature there is present in electronic aggression acts and is not present in traditional aggression at the same time? The answer to this question seems to be negative. In order to explain this I designed the simple theoretical Abacus model. Its aim is to compare different electronic aggression acts as well as electronic and traditional aggression as separate phenomena. The model uses the picture of abacus to explain the differences between electronic aggression and traditional aggression (Figure 1).

In the model there are horizontal rods serving as symbols for criteria that are used to analyze the features that are often referred as typical for electronic aggression. We have also the left and right frames that respectively illustrate electronic and traditional aggression. Then there are beads that can be moved to the traditional or electronic frame according to the situation of the particular act of aggression. The model clearly shows that all the characteristics may be present (or absent) in both kinds of aggression.

To sum up the presented model allows "dynamically" analyze the aggression acts on the framework that uses criteria described as typical for electronic aggression. Below I use the Abacus model to discuss three key features described above: anonymity, unitentionality and continuity.

Anonymity is present only in some electronic aggression acts while it is completely absent in others – so according to the ABACUS model the

Figure 1. The ABACUS model of electronic aggression

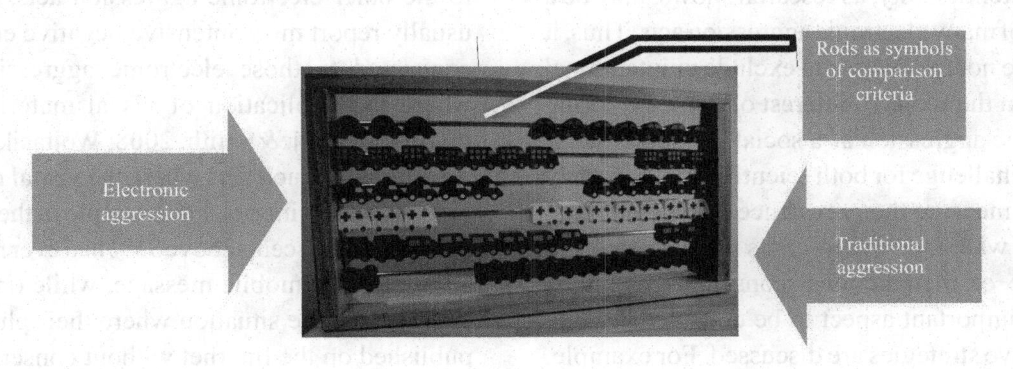

beads only sometimes can be moved in the direction of electronic aggression. For example, in a big study conducted by me on the representative sample of Polish adolescents (N=2143), only about one third of respondents admitted to having sent something unpleasant to the other person and purposefully hiding his/her identity, although a lot more admitted perpetration of electronic aggression. The findings of the other authors also show that anonymity is only potentially present in electronic aggression acts. Juvonen and Gross (2008), in the study of American adolescents discovered that two thirds of cyberbullying victims knew their perpetrators and a half of them revealed that it was someone they know from a school. In the research of Hinduja and Patchin 80% of the victims knew the perpetrators' identity. Thus, anonymity cannot be treated as the distinctive feature of electronic aggression in general sense. Additionally, some of traditional aggression acts where young people are involved may also be anonymous. For example, someone may spread rumors or scrap insulting words about a victim in the school toilet. According to the ABACUS model, in this case the beads on the "anonymity" rod will be moved towards traditional aggression frame. Taking this into account one can ask why anonymity is so much discussed as the hallmark of electronic aggression? The answer seems to lie in mechanisms connected to anonymity and described in the first part f this chapter (e.g. deindividuation). To sum up, anonymity should be considered when electronic aggression is discussed, as it brings with it a lot of significant psychological mechanisms (and potential consequences) on the side of a victim and a perpetrator.

The same refers to unintentionality, which is present only in some electronic aggression acts. In the research on a big sample of Polish adolescents more than 37% revealed that they have send something via the Internet that was intended to be a joke, but actually harmed someone seriously. But at the same time more than 25% revealed that during the previous year they were harassing intentionally some peers via a mobile phone or the Internet for a longer period of time. That means that we can witness traditional intentional acts while considering electronic aggression. Unintentional acts that led to disastrous consequences may also be present in in a number of traditional aggression cases where a perpetrator is not able to predict consequences of his/her actions. Again then uninentionality cannot be treated as the distinctive feature of electronic aggression. However, in case of electronic aggression unintentionality may be facilitated by some features of CMC (e.g. lack of nonverbal cues in textual communication). Those features may cause problems on the side of a perpetrator to read early signs of negative emotions in a victim and make him/her to continue unconsciously harmful attacks.

Unintentionality, as research shows, may be a feature of many electronic aggression acts. Thus, it would be not reasonable to exclude unintentional acts from the scope of interest of those exploring electronic aggression as a social phenomenon. It is a big challenge for both scientific methodology (how to measure the prevalence of unintentional acts? – where in many cases perpetrators are unaware of their actions' consequences). It is also an important aspect to be considered while preventive strategies are discussed. For example, how can we prevent someone from being victim to unintentional aggressive acts?

The third feature – continuity, is similarly to anonymity and unintentionality, is also present only in some electronic aggression acts. It is worth underlying that this feature is present mainly in case of those electronic aggression acts where a perpetrator publishes online some materials about a victim. As Boyd (2007) states, electronic publication makes the material persistent, replicable and searchable by what she calls *invisible audience*. Due to this, hostile attacks are not dependant on the place and time where they occur (Walrave&Heirman, 2009) and make a potential victim more vulnerable. This seems to be a difference when compared to the majority of traditional aggression attacks. However, continuity is not completely absent in traditional aggression. The situation where the student is attacked publicly in a way that many students also those not involved are aware of the attack also causes a kind of continuity. This is reinforced by a victim's awareness that his/her situation is known to many people. Of course, here we should take into account the scale – in traditional aggression the group of people is restricted to those from the same institution or neighborhood but in electronic aggression this group may be and sometimes is extended to a significantly great number of other people that watch the materials on the Internet. It is also worth underlying that potential harmful effect of those electronic aggression acts where the material is published is higher comparing to the other electronic aggression acts. Victims usually report more intensive negative emotions connected to those electronic aggression acts where the publication of visual materials was involved (Slonje&Smith, 2008; Wojtasik, 2009). Additionally, such acts where the visual material is used are in minority. For example in the sample of Polish adolescents above 35% had ever received an unpleasant mobile message, while only 13% experienced the situation where their photo was published on the Internet without consent.

In conclusion – The ABACUS model clearly visualizes that dimensions used to describe electronic aggression, namely anonymity, unintentionality and continuity are only potential features of this kind of aggression and may be also present in a traditional version of that phenomenon. Based on this we can analyze such electronic aggression acts that are anonymous, unintentional and continuous making electronic aggression really different from typical traditional aggression where those features are absent. In such cases the situation will involve many additional mechanisms connected to the specifity of computer mediated communication and human behavior in the cyberspace.

Electronic Aggression: Not All Cats are White

Electronic aggression is too complicated a phenomenon to be analyzed without a typology that allows us to recognize the variety of its different types. Depending on who is attacked and what is the actual behavior of a perpetrator, we can have diverse electronic aggression types that vary in terms of potential mechanisms and consequences involved. For example swearing once at unknown user in the chat room is completely different to regular publishing on the Internet offensive materials about a classmate. Those differences are influenced by the actual relationships between perpetrator and victim and the perceived importance of a perpetrator to a victim as well as the seriousness of the electronic aggression act itself.

Based on this five different electronic aggression types are recognized: cyberbullying, electronic aggression against the vulnerable, electronic aggression against random victims, electronic aggression against celebrities and electronic aggression against groups.

Cyberbullying

As stated above, cyberbullying is an electronic form of bullying – so it should be described and analyzed through the lens of bullying theory (Due at al., 2005; Monks et al., 2008). In the case of young people, cyberbullying is a particular kind of peer aggression.

The first problem with cyberbullying is connected to the fact that traditionally both victims and perpetrators belong to the same social groups in a spatial worlds, e.g. classroom, family or at least have to live and/or meet in the same institutions, e.g. prison, children's home, etc.(Monks, et al., 2009). Of course, when cyberbullying is an extension of traditional bullying – just using "electronic" means it is very easy to claim the situation is similar. In my research on the sample of Polish adolescents (N=2143) 33% were the victims of bullying or cyberbullying in the past year. 61,3% of the victims suffered only from traditional bullying; 12% from cyberbullying and 26,7 from both forms. For perpetration, the respective figures were: 48,6%; 11,2% and 40,2%. This means that in schools, both phenomena of bullying and cyberbullying are connected, but the relation is not perfect. Indeed, it is very seldom that someone involved in cyberbullying without being a perpetrator or a victim of traditional bullying at the same time. Some authors claim that the term cyberbullying should be restricted only to the situations where online aggression is connected to peer aggression and school relationships (Wolak, et al, 2007). This seems to be a valuable attempt to be more precise with the terminology concerning electronic aggression. But maybe that attitude is too restrictive and does not take into account that a lot of Internet users gather online only groups where the close interpersonal links and relationships develop (McKenna, 2008). If we acknowledge that interpersonal aggression in such groups can also take a form of cyberbullying. To make things more complicated, cyberbullying should have all three explicitly defined characteristics as described above, namely the imbalance of power, negative intentions on the side of perpetrators, and repetition. Those features, relatively easy to operationalize, are obvious in the spatial world. In cyberspace, those features may have different mechanism or even become questionable due to specific characteristics of CMC (Dooley, et al., 2009). It may be the case with the imbalance of power that can be caused by mechanism that, as stated above, may be potentially present in CMC, e.g. anonymity. Sometimes those features are even used purposefully. One of the respondents who encouraged his peers to send threatening mobile messages to his friend from many unknown numbers openly expressed such awareness:

"the guy is in real trouble.... he doesn't know who is sending this – doesn't know what can happen – it's better when he's uncertain what can happen..."

Sometimes the imbalance of power attributed to some specific features of electronic communication is also expressed by victims. One of the 15 year-old girls described a situation where she had received a series of anonymous cell phone messages criticizing her harshly. When she was asked about her subsequent emotions she confessed to having cried nearly all night after the incident. Following this, she was asked about the worst aspect of that situation that caused such strong emotions. She immediately said that it was not the content of the messages but their anonymity that felt most threatening. One of the respondents describes the situation of her friend who has made herself a photo while taking a bath. The film was stored in the phone. Then a girl was encouraged by some friends to lend a phone as they wanted to steal a video. Then the video was placed on the Net. Although it was published only in one portal and has been quickly withdrawn a girl was convinced that everybody has seen the video and she stopped attending to school for a longer period. Such mechanism is labeled by Danah Boyd (2007) as an invisible audience – a potentially huge, impossible to estimate number of people that may have seen a film clip.

The issue of intention is also ambiguous. As presented in the first part of the chapter some CMC features makes it difficult for the perpetrator to see the effects of his/her actions that makes it easier to engage in potentially harmful behavior even without negative intentions. Here in large number of cases we will have can have different opinions on intentionality depending on who will be asked: a victim or a perpetrator. For instance, a perpetrator, who once published a victim's photo on the Internet may resist the opinion that he did it in order to harm another person. However, for a victim who suffered a lot after such a publication has been viewed by a lot of people, the perpetrator's fault and negative intentions may be obvious.

Electronic Aggression Against the Vulnerable

Another electronic aggression type is targeted at "vulnerable" people, namely alcoholics, mentally disabled, mentally ill, etc. This type of aggression is often conducted by way of the unwanted filming of the victims and then the publication of the material on the Internet, mostly on the sites where the clips of the users may be uploaded. Victims are often depicted in a humiliating way and are probably unaware of the situation. The films seldom present physical aggression – they rather show actions like encouraging the victims to do "funny things" like undressing, showing physical imperfections, missing teeth, etc. Of course, such acts are harmful to the victims, even in cases when they are not aware that their human right are violated. On the other hand, a perpetrator of such acts becomes desensitized and probably more prone to repeat his/her behavior. It is worth underlying that this kind of aggression is an example of an overlapping of electronic aggression and traditional aggression. Perpetrators here conduct traditional aggression act, in the real world (e.g. insulting someone) and then record and publish the act. This resembles a phenomenon called *happy slapping*, which is conducted by young people; they approach another young person, conduct a simple physical aggression act, e.g. a kick, and then publish or disseminate electronically the recorded act (Saunders, 2005).

We should not forget that publicly presented aggressive media pictures showing hostile acts against the vulnerable persons may have a negative influence on all the viewers, particularly children and adolescents. It is worth underlying that in those cases, violence is usually presented in a way that is perceived as the most dangerous in terms of potential imitation. Those risks as meta-analysis show (Browne & Hamilton-Giachritsis, 2005) is higher when in a media picture there is no criticism or remorse to the violence, aggressive acts are often associated with humour and the violence is perceived as realistic. All those features are usuallly present in case of new media aggression against the vulnerable. Additionally, the aggressive acts are often conducted by the young people – that means those similar to the viewers what makes the risk of imitation even higher. Of course, those mechanism has been also present in some traditional media, e.g. reality news programs on TV (McCleneghan, 2002). The difference with the Internet is connected to scale – the number of users that can upload such kind of materials is really high – that means that the accessibility level is also very high – what is particularly dangerous in case of young audience. It is worth here to underline the active role of the viewers who by leaving they comments may reinforce the author of aggressive clips or on the contrary criticize his/her behavior. From this point of view, the activities of the audience can make the risk of negative media influence higher or lower; this active and influencing role of an audience seems to cause a substantial difference between traditional media (particularly television) and the new media. Moreover, this difference is also present in other electronic aggression types, not only that one against the vulnerable.

Aggression Against Random Victims (Random Electronic Aggression)

Cyber environment is a place that gives opportunity to contact a great variety of people, maintain existing relationships as well as start the new ones (Mishna at al. 2009). Young people meet the others while using synchronic and a-synchronic communication channels: chat rooms, discussion forums, instant messengers or e-mails. So it is very easy to start communication with totally unknown people, without the psychological restrictions that are normally present in the traditional offline communication. This may be perceived as an advantage for those from stigmatized groups (McKenna, 2008).

On the other hand, in the cyberspace it is convenient to attack verbally other people – particularly those that are not in a relationship with a perpetrator (neither online nor offline). Due to relative perceived anonymity is also much more safe to conduct such acts comparing to the similar acts in the offline environment. The potential consequences both legal as well as retaliation are more likely to occur in the real world comparing to the cyberspace.

Due to the fact that random aggression in the cyberspace takes place it is anyone present in the Internet (e.g. possessing a profile in social networking) may be victimized electronically.

Electronic Aggression Against Celebrities

Cyberspace, offers celebrities (e.g. actors, singers, etc.) and entertainment industry many channels to promote themselves. Of course this was also present in the traditional media e.g. newspapers. However, the Internet facilitates that kind of contact with general public, using also the advantages of interactivity that enables spectators to involve actively in communication acts. The same mechanisms may be also adopted to conduct harsh critics or electronic aggression against well-known people. In this case those attacked may be not treated by perpetrators as the actual individuals but more as a kind of a symbol. Electronic aggression against celebrities may be perceived as individual acts of the users as well as a part of "gossip industry" that makes profits from publishing "shocking" materials about the celebrities in a more organized way. Of course, both types of electronic aggression against celebrities are overlapping: gossip news portals are the preferred space for the users wanting to leave insulting or hostile comments.

Although in electronic aggression against celebrities victims are often perceived as "icons" or "symbols" sometimes the individual consequences of being a victim of electronic aggression are disastrous for them. The example of such situation is a suiceide of A Korean actress Jin-Sil who had been overwhelmed by the Internet users harm criticism (More on Jin-sil suicide, 2008).

Electronic Aggression Against Groups (Bias Cyberbullying)

Cyberspace is a place where the certain groups or individuals are represented through websites, forums, etc. Using those facilities, people representing certain ideas share their outlook with the general public or change information and ideas within their groups. This provides channels for potential aggressors to harass or insult groups. Indeed, vicious comments can be left in a portal's guest-book.

At first glance, this kind of electronic aggression may be perceived as less harmful since nobody is directly attacked. However, such aggression, as research shows, may be very frequent and the single act influences a great number of victims at the same time. The good example is a research by Tynes, et al. (2008) on behavior at chat rooms. Almost 60% of young people taking part in an unmoderated chat room face racist communications made by the other users. That means that everybody belonging to a racial minority may have been affected negatively by disparaging remarks. This kind of e-aggression is also difficult to target by way of legal intervention as there is no individual victim harassed by perpetrators.

Five electronic aggression types described above are not completely separated and may sometimes overlap. Despite this fact, it is clearly seen that there are substantial differences among them and that their potential consequences on the side of a victim vary substantially. Situation is completely different when a young person is victimized regularly by classmates (cyberbullying) comparing to a single attack by someone unknown in a chat room(random aggression).

Different Types of Electronic Aggression: A Quantitative Aspect

The involvement in different types of electronic aggression has been tested empirically within the study on the representative sample of 14-15 y.o. Polish adolescents (N=2143). The adolescents who revealed that in the last year they conducted at least one kind of specific electronic aggression acts (out of 20) were asked to indicate who their target was. The results reflecting their answers – the prevalence of different electronic aggression types are presented in the Table 2.

Analysis of the data in the table shows Polish adolescents were involved in the previous year in all types of electronic aggression as described above. The most prevalent was electronic aggression against the people a perpetrator knows only from cyberspace (42,5%). This is a qualitatively new situation as both: the relationship with a person and aggression are taking place in cyberspace. That means there are no "real" links between a perpetrator and a victim. Nevertheless, a perpetrator may have established a closer relationship with a victim online, e.g. they could have been friends from the same discussion forum. When the subsequent aggressive acts are regular, with accompanying intention to harm and abuse of power – then according to the definition this situation may be labeled as cyberbullying. (Wang, et al. 2009). Obviously, in many cases it will be hard to state that both sides of aggressive act belong to the same online group (comparable to the class or school in spatial world). Additionally, the hostile acts are not always intentional, regular and do not always overwhelm a victim. That means that electronic aggression acts against people known only from the Internet may take a specific form of cyberbullying only in some cases when the specific characteristics are present. Then we have peer aggression against young people known from "real" interactions – both close friends (26,8%) and young people who are known but not defined as close friends by perpetrators (39%). Similarly, as in case of aggression against people known only from the Internet, such acts may be in particular situations characterized by the distinctive cyberbullying features. Quite prevalent were also electronic aggression acts against totally random individuals – here neither online nor offline relationships between a perpetration and a victim were present. Aggression against a former boyfriend/all girlfriend (16,9%) may be interpreted as a specific kind of cyberbullying or a phenomenon referred in literature as cyberstalking (Spitzberg&Hoobler, 2002). The specifity of this electronic aggression type is connected to the formerly intimate relationship what can make the future attacks more damaging for a victim. Aggression against groups is also quite common and has been conducted by almost 16% of respondents in the previous year. One in nine adolescents revealed engagement in electronic aggression against celebrities or aggression against the vulnerable. About 9% of respondents attacked electronically teachers or other known adults – those acts may have been cyber bullying acts provided that all vital characteristics were present. In conclusion, it should be stated that although attacking people who are the same age (peer aggression) is the most prevalent, young people are involved in many other electronic aggression acts where the targets are the other people not those from a peer group.

SOLUTIONS AND RECOMMENDATIONS

The reflections and data presented above clearly show that electronic aggression is not a homogenous phenomenon. The characteristics that are referred as attributes of electronic aggression are present only in some electronic aggression acts making them sometimes more harmful for the victims. This should be reflected in the research tools in order to explore in depth the variety of electronic aggression acts in which young people may be involved as victims or perpetrators.

This is also true when we analyze different electronic aggression types depending on the relationship between a victim and perpetrator (e.g. is a victim a classmate or randomly chosen person on the Internet). If we do not take it into consideration asking our respondents whether they insulted someone on the Internet without precise distinction we will obtain only general statistics. They provide insufficient data on qualitatively different types of electronic aggression (e.g. aggression against the vulnerable, aggression against people known only from the Internet). This is an important issue also from the perspective of potential preventive measures. We have to analyze and check whether we have "one cure" for all the electronic aggression types or we have to plan and implement different measures against each type.

The involvement of young people in electronic aggression has been of growing concern in the recent years. A lot of action in this respect is asked from professional educators who should prevent electronic aggression as well as intervene when such aggression is conducted or experienced by their students.

Based on the empirical data and discussion presented in the chapter emerge a few practical issues that are important for professional prevention and intervention of electronic aggression among students.

The vital aspect is the knowledge of professionals concerning the "new" characteristics of electronic aggression (e.g. anonymity or unintentionality) and they awareness that those features are present only potentially in some electronic aggression acts. Educators who understand those features and accompanying psychological mechanisms may assess in a particular situation to what extent a particular act is distinctive and differs from typical traditional act. That should subsequently influence the type of intervention offered. For example, a student that unintentionally harmed someone else should be treated differently to someone who had hostile intentions. In the first case, awareness raising activities seem more reasonable than disciplinary consequences. Professionals ought to consider also new content of preventive programs where awareness concerning specific features of CMC will be raised. Knowledge on those features of CMC that may facilitate electronic aggression should act as a factor that discourage potential young perpetrators and helps the victims to cope with hostile acts.

On the other hand, it seems very rational to target all the electronic aggression types. That requires preparation of the specific programs targeting different electronic aggression types. For example, cyberbullying is usually connected to the offline relations of young people involved and all the actions may be conducted there, e.g. mediation sessions, etc. The situation seems complicated when the other types of electronic are analyzed. For instance, random aggression targeted against totally unknown victims met on the Internet, requires different measures, involving for example some online activities.

At this stage of research not enough is known to enable us to propose detailed actions tailored to the different types of electronic aggression. Still, the gathered data suggests that this "differential treatment" concept should be developed in the future.

FUTURE RESEARCH DIRECTIONS

Ellison&Akdeniz (1998) state that "the internet tends to produce extreme versions of problems, it rarely produces genuinely ones (1998, p. 29). This seems to have been proved by the data presented in the chapter. All the electronic aggression features may be also present in traditional aggression and all the electronic aggression types can have some parallel phenomena in the offline world. This gives a solid reason for research projects exploring both involvement in electronic aggression as well as traditional forms of this phenomenon. Some studies (e.g. Yan, 2005; Ybarra, et al., 2007) show that young people engaged in electronic aggression tend to exhibit other risk behaviors like substance use, sexual solicitation or school related problems. This involvement means both perpetration and victimization. It seems that exploring those online and offline risk behaviors together may be beneficial for future research projects. Here the relation between offline and online lives of young people should be the key problem needing exploration.

Serious consideration has to be given to research methodology on the variety of electronic aggression types. That is a big challenge when a construction of new tools is taken into account – they should reflect the richness of electronic aggression acts and at the same time be "plain" enough to be understood by young respondents.

Another issue is the developmental level of a young person. It influences both the involvement in electronic aggression as well as efficacy of any preventive activities is a developmental level of the child. Yan's research (2005) indicates that understanding of the complexity of the Internet is connected to the age of the child. An interesting finding here is that many of 11-12 years old adolescents understand the social aspects of the Internet only partially. This is absolutely important when we focus on risks and potential self -regulation of the internet behavior exhibited by a young user. The developmental aspect should be considered in terms of research (comparative aspects) as well as planning and implementing preventive strategies (e.g. educational programs). Thus, those aspects should be involved in the models so we can understand electronic aggression of young people from a broader perspective. Such approach takes into account the role of new media in young people' lives, development, and broadly understood health. Some researchers warn that the influence of the media has been neglected and often excluded from the significant big-scale projects relating to young people's health and wellbeing (Strasburger, 2009).

CONCLUSION

In the chapter, I have attempted to analyze two important issues concerning electronic aggression in young people – its distinction comparing to traditional aggression as well as diversity of electronic aggression acts.

The first issue has been analyzed with the help of ABACUS model that illustrates the potential character of features attributed to electronic aggression. Then the typology of electronic aggression types has been proposed using the identity of a target and his/her relationship with a perpetrator as a key criterion. All those reflections have been illustrated by research data, but we must remember that conceptual and theoretical clarity relating to electronic aggression has not been obtained, and still requires future research activities.

REFERENCES

Aricak, T., Siyahhan, S., Uzunhasanoglu, A., Saribeyoglu, S., Ciplak, S., Yilmaz, N., & Memmedov, C. (2008). Cyberbullying among Turkish adolescents. *Cyberpsychology & Behavior*, *3*, 253–261. doi:10.1089/cpb.2007.0016

Boyd, D. (2007). Why Youth (Heart) Social Network Sites: The Role of Networked Publics in Teenage Social Life. In Buckingham, D. (Ed.), *Mc Arthur Foundation on Digital Learning – Youth, Identity, and Digital Media Volume*. Cambridge, MA: MIT Press.

Browne, K., & Hamilton-Giachritsis, C. (2005). The influence of violent media on children and adolescents: a public-health approach. *Lancet*, *365*(9460), 702–710.

David-Ferdon, C., & Hertz, M. (2007). Electronic media, violence, and adolescents: an emerging public health problem. *The Journal of Adolescent Health*, *41*(6Suppl 1), S1–S5. doi:10.1016/j.jadohealth.2007.08.020

Dooley, J., Pyżalski, J., & Cross, D. (2009). (in press). Cyberbullying Versus Face-to-Face Bullying: A Theoretical and Conceptual Review. *Zeitschrift für Psychologie*. *The Journal of Psychology*, *217*(4), 182–188.

Due, P., Holstein, B., Lynch, J., Diderichsen, F., Gabhain, S., & Scheidt, P. (2005). Bullying and symptoms among school-aged children: international comparative cross sectional study in 28 countries. *European Journal of Public Health*, *15*(2), 128–132. doi:10.1093/eurpub/cki105

Ellison, L.,&Akdeniz, Y. (1998). Cyber-stalking: the regulation of harassment on the Internet. *Criminal Law review, December Special Edition: Crime, Criminal Justice and the Internet,* 29-48.

Griffin, R. S., & Gross, A. M. (2004). Childhood bullying:Current empirical findings and future directions for research. *Aggression and Violent Behavior*, *9*, 379–400. doi:10.1016/S1359-1789(03)00033-8

Hasset, J., & White, K. M. (1989). *Psychology in perspective*. New York: Harper&Row.

Hinduja, S., & Patchin, J. W. (2009). *Bullying Beyond the Schoolyard: Preventing and Responding to Cyberbullying*. Thousand Oaks, CA: Sage Publications (Corwin Press), Juvonen, J., & Gross, E. (2008). Extending the School Grounds?—Bullying Experiences in Cyberspace. *The Journal of School Health, 78*(9), 496–505.

Kowalski, R. M., & Limber, S. P. Agatson, P. W. (2008). *Cyber bullying. Bullying in the Digital Age.* New York: Blackwell Publishing Ltd.

Living and Learning with New Media (Report). (2008) McArthur Foundation. Retrieved November 25, 2009 from http://digitalyouth.ischool.berkeley.edu/files/report/digitalyouth-WhitePaper.pdf

McCleneghan, J. S. (2002). 'Reality violence' on TV NEWS: It began with Vietnam. *The Social Science Journal*, *39*, 593–598. doi:10.1016/S0362-3319(02)00232-X

McKenna, K., & Bargh, J. (2000). Plan 9 From Cyberspace: The Implications of the Internet for Personality and Social Psychology. *Personality and Social Psychology Review*, *4*(1), 57–75. doi:10.1207/S15327957PSPR0401_6

McKenna, K. Y. A. (2008). Influence on the nature and functioning of social groups. In Barak, A. (Ed.), *Psychological aspects of Cyberspace* (pp. 228–242). Cambridge, UK: Cambridge University Press.

Mishna, F., Saini, M., & Solomon, S. (2009). Ongoingand online: children's and youth perceptions of cyber bullying. *Children and Youth Services Review*, *31*, 1222–1228. doi:10.1016/j.childyouth.2009.05.004

Monks, C., Smith, P., Naylor, P., Barter, C., Ireland, J., & Coyne, I. (2009). Bullying in different contexts: Commonalities, differences and the role of theory. *Aggression and Violent Behavior*, *14*(2), 146–156. doi:10.1016/j.avb.2009.01.004

More on Choi Jin-shil's suicide (2008). Retrieved November 25, 2009 from http://www.dramabeans.com/2008/10/more-on-choi-jin-shils-suicide/.

Olweus, D. (1978). *Aggression in the schools. Bullies and whipping boys*. Washington, DC: Hemisphere.

Olweus, D. (1993). *Bullying at school. What we know and what we can do*. Oxford, UK: Blackwell.

Patchin, J. W., & Hinduja, S. (2006). Bullies move beyond the school yard: a preliminary look at cyberbullying. *Youth Violence and Juvenile Justice*, *4*, 48–69. doi:10.1177/1541204006286288

Pyżalski J. (2009) Agresja elektroniczna dzieci i młodzieży. Rożne wymiary zjawiska. *Dziecko krzywdzone. Teoria, badania, praktyka, 1* (26), 12-27.

Saunders R. (2005). Happy slapping: transatlantic contagion or home-grown, mass-mediated nihilism, *The London Consortium,* 1-11.

Silverblatt, A. (2004). Media as Social Institution. *The American Behavioral Scientist*, *48*(1), 35–41. doi:10.1177/0002764204267249

Spitzberg, B., & Hoobler, G. (2002). Cyberstalking and the technologies of interpersonal terrorism. *New Media & Society*, *4*(1), 71. doi:10.1177/14614440222226271

Stassen-Berger, K. (2007). Update on bullying at school: science forgotten? *Developmental Review*, *27*, 90–126. doi:10.1016/j.dr.2006.08.002

Strasburger, V. C. (2009). Why do adolescent health professionals ignore the impact of the media? *The Journal of Adolescent Health*, *44*, 203–205. doi:10.1016/j.jadohealth.2008.12.019

Subrahmanyam, K., Reich, M. S., Waechter, N., & Espinoza, G. (2008). Online and offline social networks: Use of social networking sites by emerging adults. *Journal of Applied Developmental Psychology*, *19*, 420–433. doi:10.1016/j.appdev.2008.07.003

Tynes, B. M., Giang, M. T., Williams, D. R., & Thopson, G. N. (2008). Online racial discrimination and psychological adjustment among adolescents. *The Journal of Adolescent Health*, *43*, 565–569. doi:10.1016/j.jadohealth.2008.08.021

Walrave, M., Heirman, W. (2009). Skutki cyberbullyingu – oskarżenie czy obrona technologii?, *Dziecko krzywdzone. Teoria, badania, praktyka, 1* (26), 78-89.

Wang, J., Iannotti, R., & Nansel, T. (2009). School bullying among adolescents in the United States: physical, verbal, relational, and cyber. *The Journal Of Adolescent Health: Official Publication Of The Society For Adolescent Medicine*, *45*(4), 368–375.

Williams, K. R., & Guerra, N. G. (2007). Prevalence and predictors of internet bullying. *The Journal of Adolescent Health*, *41*, S14–S21. doi:10.1016/j.jadohealth.2007.08.018

Wojtasik, Ł. (2009). Przemoc rówieśnicza a media elektroniczne. *Dziecko krzywdzone. Teoria, badania, praktyka, 1* (26), 12-27.

Wolak, J., Kimberly, J. M., & Finkelhor, D. (2007). Does online harassment Constitute bullying? An Exploration of Online Harassment by known peers and online-only contacts. *Journal of Adolescent Health, 41,* S51-S58. Vandebosch, H., & Van Cleemput, K. (2008). Defining cyberbullying: A qualitative research into the perceptions of youngsters. *Cyberpsychology & Behavior*, *11*, 499–503.

Yan, Z. (2005). Age differences in children's understanding the complexity of the Internet. *Applied Developmental Psychology*, *26*, 385–396. doi:10.1016/j.appdev.2005.04.001

Ybarra, L. M., Espelage, D. L., & Mitchell, K. J. (2007). The co-occurrence of Internet harassment and unwanted sexual solicitation victimization and perpetration. *The Journal of Adolescent Health*, *41*, S31–S41. doi:10.1016/j.jadohealth.2007.09.010

ENDNOTES

1 In the Project the data from two research projects was used:

1. Cyberbyllying as a new peer aggression type (Grant No. 106 067735) – a qualitative and quantitative research on a sample of 2143 Polish adolescents (14-15 y.o.). The grant is affiliated in Wyższa Szkoła Pedagogiczna w Łodzi. All the quotes of students' words in this chapter are taken from the data gathered during interviews in this study.
2. Electronic aggression as a new problem of teachers' occupational health (grant no. IMP 8.5). – a survey on 600 teachers. The grant is affiliated in Nofer Institute of Occupational Medicine. All the quotes of teachers' words In this chapter are taken from the data gathered during interviews in this study.

2 Those three characteristics are sometimes discussed in the literature and some authors raise doubts about them. However for the clarity of this chapter I adopt the most common definition (Monks, et al., 2009).

Chapter 17
Ways of ICT Usage among Mildly Intellectually Disabled Adolescents:
Potential Risks and Advantages

Piotr Plichta
The Pedagogy Academy in Lodz, Poland

ABSTRACT

The chapter explores patterns of ICT (Information and Communication Technologies) usage (particularly: cell phones and the Internet) among mildly intellectually disabled adolescents aged 13-17. Importance of leisure activity in life and rehabiliation of the disabled as well as a risk of digital exclusion are also underlined. The strong emphasis is put on an issue of victimization and perpetration of electronic aggression. Data were obtained from the unstructured interviews (qualitative approach). The research revealed several issues related to patterns of ICT usage. The data are presented as following categories: importance and declared range of ICT usage, ICT and its usage for communication purposes, awareness of the risks related to ICT usage, ICT usage and electronic aggression (divided in two categories: being a victim and/or being a perpetrator, ICT usage and issues related to family functioning). Such problem areas are not well recognized in either Polish or foreign literature.

INTRODUCTION: THE INTELLECTUALLY DISABLED AS USERS OF ICT

Undoubtedly, we are all living in the *Digital Age (Information Age or Computer Age)*. It is the effect of the considerable role of the New Media (Internet, cell phones) in education, work and leisure activities. The amount of time spend while using those media is still increasing. Therefore, new challenges for the education in terms of teaching and socialization appear. One of the most important tasks is to provide young people with support when concerning safe and responsible ICT usage. The cyberspace provides opportunities for acquiring positive experiences and contributes to personal development. On the other hand, it poses some risks like involvement in electronic aggression or other symptoms of *Problematic Internet Use* (Shapira et. al., 2000).

DOI: 10.4018/978-1-60960-209-3.ch017

Phenomena and problems related to presence of young people in VR (*Virtual Reality*) are common for all of those who have an access to modern technologies and possess even basic competencies. This is particularly true in case of the disabled, since their problems and limitations often result from the disabling environment not from their physical disability. Information and Communication Technologies such as the Internet and cell phones are favourable instruments to build and increase independence and self-empowerment of people with intellectual disabilities. Access to new technologies and relevant media education are the instruments of democratization and prevent disabled people from *digital exclusion as well as "media illiteracy"*.

According to an overview of the 2002 American Association on Mental Retardation's (AAMR's) *Definition, Classification, and Systems of Supports* an intellectual disability[1] is defined as a "disability characterized by significant limitations both in intellectual functioning and in adaptive behaviour as expressed in conceptual, social, and practical adaptive skills and that the basis for the disability has been present prior to age 18. (...) Five important assumptions are included as operational recommendations. These are: (1) limitations in present functioning must be considered within the context of community environments typical of the individual's age peers and culture; (2) valid assessment considers cultural and linguistic diversity as well as differences in communication, sensory, motor, and behavioural factors; (3) within an individual, limitations often coexist with strengths; (4) an important purpose of describing limitations is to develop a profile of needed supports; and (5) with appropriate personalized supports over a sustained period, life functioning of the person with ID generally will improve (Schalock & Luckasson, 2004, p. 4)".

Mildly intellectually disabled (IQ scores between 50 and 69) individuals comprise the largest part (approximately 85%) of the population of the intellectually disabled (the others levels, according to criteria of ICD-10, DSM IV-TR and categories of AAID are: moderate – IQ 35-49, severe – IQ 20-34 and profound - IQ below 20). "Mild Mental Retardation (MMR) was the official designation of a level of Mental Retardation (MR) that involved current intellectual functioning performance between 2 and 3 standard deviations below the population mean and significant limitations in some, but not all, facets of everyday adaptive functioning. (…) In contrast to the other levels, MMR typically is not diagnosed until the school age period, usually subsequent to a teacher referral due to severe and chronic school achievement problems. Persons with MMR do not exhibit the physical characteristics of many persons with MR at more severe levels, and they are not comprehensively impaired in the sense of requiring assistance with nearly all social roles and functions.(…) Persons with the least impairment, formerly called mild, are recognized now as needing intermittent levels of support as children and adults in coping with the everyday demands of living. Because there is no unequivocal sign or symptom of MMR, diagnosis is complex and sometimes controversial. Issues such as the relative importance of current intellectual functioning and adaptive behavior in diagnosis at different ages, appropriate cut scores or ranges to delineate MMR from borderline intellectual functioning, conceptions of adaptive behavior in terms of activities versus competencies, number and kind of normal, everyday functions that would rule out MMR are not resolved and are unlikely to be resolved unequivocally in the near future (Reschly, 2009, p. 125).

There is a great diversity in terms of questions which are taken up in professional literature on the topic of the intellectual disability. On the other hand, there are some knowledge gaps. One of them is a relatively small number of research projects exploring usage patterns of modern communication technologies[2], particularly with regards to potential risks such as involvement in cyber bullying or escape to "virtual world" and withdrawal from interpersonal contacts. Of course, there are also some big hopes related to using modern technologies in education and rehabilitation. According to Tamar Weiss, "Virtual environments provide patients with safe access to inter-active, true-to-life situations that would otherwise be inaccessible to them due to motor, cognitive and psychological limitations. The ability to change the virtual environment relatively easily, to grade task difficulty and to adapt it according to the patient's capabilities are important advantages of VR, since these features are essential to cognitive and motor remediation. (…). A review of the literature on leisure for people with disabilities highlights the need for increased exposure to augmented leisure opportunities and the positive correlation between life satisfaction, self-esteem, companionship, enjoyment and relaxation and leisure satisfaction. Greater involvement in leisure activities appears to improve the participant's coping skills, decreased stress, and adjustment to a life with disabilities (Weiss et.al., 2003, pp. 335-336). It seems that the intellectually disabled have the most difficult situation with the access to social life and possibility to make independent decisions.

Education and rehabilitation usage of ICT is beyond doubt (especially in terms of "*assistive technology*"). On the other hand, dysfunctional aspects of the Internet and ways of cell phones use are not well recognized within the field of special education. It is the case for Polish and foreign literature with the exception of Robert Didden's (2009) study on cyber bullying among students with intellectual and developmental disability in special education settings.

Unfortunately, sometimes in everyday life (e.g. in rehabilitation centers, at homes) television or computer become the „filling" of free time or even essence of the intellectually disabled life. This issue is underlined by Krause who describes a phenomenon observed in the disabled, namely addiction to TV - escape to the virtual world without efforts, rehabilitation etc. (Krause 2004, pp. 132-137). The great amount of free time that is not well organized, or without valuable activities, might be one of the risk factors in development of the intellectually disabled. Then, the following issue emerges - what could be the proper, valuable leisure activities' offer? It has been indicated by the conducted studies that: "about two-thirds of the activities chosen by the mentally retarded population were not directly related to cognitive ability, indicating that this ability does not specifically determine the extent in which one may participate or enjoy leisure activities (…). According to Falvey, 1989 and Schloss at al., 1986 the results of this investigations do indeed indicate that individuals with mental retardation have essentially the same interests and desires relative to recreation and leisure time (Whorton et.al., 1994, pp.177,180)".

Increasing presence of the intellectually disabled among the non-disabled entails mainly positive consequences but it is worth mentioning that some negative aspects and potential risks are possible as well. Both advantages and drawbacks should be considered by special educators and social scientists. According to Amadeusz Krause (2004, p. 176) "It is necessary to be aware of accumulating educational difficulties typical for the non-disabled (e.g. exposure to media manipulation, addiction and substance abuse) with problems characteristic for disability (low self-esteem, consequences of social isolation and exclusion, anxiety and uncertainty)".

METHODOLOGY

The presented data is a part of ongoing research (2-years internal grant of The Pedagogy Academy in Lodz; coordinator – Piotr Plichta, in cooperation with Jacek Pyżalski – projects FWN/8/2008-2010 and FWN 2/2009/2010). The main purpose of the research is to outline ICT (*Information and Communication Technology*) usage, especially: cell phones and the Internet, among mildly intellectually disabled adolescents aged 13-17 attending middle schools.

The strong emphasis was put on question of being a victim and/or a perpetrator of electronic aggression (*cyberbullying*). Data were obtained from the semi-structured interviews in the field of special education. The sample consists of 23 students (17 of them are boys) attending special schools (non-inclusive settings) for intellectually disabled students in Lodz. Gymnasium (pol. *gimnazjum*) in Poland is a compulsory middle school (comparable to English grammar schools or sixth form colleges and U.S. college preparatory high schools). It lasts three years (in non-inclusive schools for the disabled students it is possible to prolong this period with one additional year) and starts at the age of 13. Gymnasium is followed by six (or more in some cases) years of primary school. Further education is non-compulsory and for the mildly intellectually disabled students consists of 2 to 3 years attending vocational schools.

All the respondents have diagnosis of mental retardation in a mild degree (in Polish educational law, the term in force is "mental retardation" not "intellectual disability" but both of them are used interchangeably in common language) and kind of formal document: *special education needs statement*. Of course, there is possible for intellectually disabled students in Poland also attend to mainstream schools (inclusive settings) but the sample consists only of students from non-inclusive schools.

All the interviews were conducted by the members of research team (Piotr Plichta,Ph.D. and Jacek Pyzalski, Ph.D). The team obtained consent for the research proceedings from the headmasters and Parents' Boards of the schools. The parents of the potential respondents were informed (during the meetings with the school staff and through the announcement) about the purposes of the research. The parents as well as the students had the right to refusal but none of them exercised such right. The semi-structured interviews were conducted in school advisor room, temporarily made available for the research team. The interviews lasted between 20 and 80 minutes – mainly depended on the respondents level of willingness to participate and their experiences with ICT usage. The interviews were conducted in individual sessions. The answers were translated by the team members and corrected by certified translator. The responses were analyzed mainly with qualitative methodology.

The research was conducted between March and April 2009. The main research problems were forumlated into such questions like: What is declared importance and range of ICT usage among the respondents? What is the respondents point of view on ICT usage for communication purposes? What is respondents' awareness of the risks related to ICT usage? Are the respondents engaged in electronic aggression (*cyberbullying*), including being a victim and/or being a perpetrator? Is ICT usage related to family life of the respondents?

IMPORTANCE AND PURPOSES OF ICT USAGE

Do I have any other interests ? I don't. (...) I don't play football because I can't hit the ball properly. I always miss the ball and kick the legs of other players.

For the majority of the respondents usage of computers is a very important and enjoyable aspect of their life, especially in their leisure time. Five respondents out of the whole sample have not got computers at their homes and use it only occasionally at school or when visiting homes of their peers. Additionally, few students complain about not possessing their own computers ("*I have the computer but it is out of date*"), access to the Internet or having too slow Internet connection ("*Computer without access to the Internet connection is a piece of junk*").

The respondents were asked to evaluate the importance of using computers and the Internet in their leisure time. During the interviews, a picture of a sport podium was shown to the respondents. The interviewers asked the students to indicate which leisure time activity should be located on the relevant level of the podium. Which of them they would place on the highest step of podium, which on the step number two, etc. This method was used to estimate the role of the ICT usage in respondents' free time. Not surprisingly, the most frequent choice of the students was the highest platform of the podium. The students were also asked to imagine that computers and the Internet have disappeared and do not exist and they have to arrange their leisure activities without using ICT. One of the students stated that "*There is no life without Internet*". Another who was asked why computers are so important in his opinion, answered:

"Because it is possible to do anything you want to do".

I (Interviewer): *Do you think that computers are important in your life?*

R (Respondent): *It has to be important, because there is nothing to do without this technology. Instead of doing homework it is possible to download a movie or watching YouTube, football match of "Widzew" (one of the Polish football teams) or something.*

I: *So, computers are used...*

R: *To download movies, to read sport news, to know what's going on.*

I: *What do you download most often?*

R: *Songs, movies, skins for my cell phone – I must have it all.*

On the contrary, there are respondents who declare another point of view on importance of ICT: *"Computer for me? It is nothing important. I sit an hour or two or just a bit. School is the most important for me"* or next opinion: *"Computer is important but I prefer learning".*

The declared amount of time spent on using computers is approximately 3-4 hours a day. It seems that it consumes considerable part of the whole leisure time of the respondents. The amount of declared time is diversified and the exceptions to the rule are present. It occurs that some students do not possess their own computers, meanwhile another spend all the time sitting in front of the PC screens (*"I usually use the computer from 4 P.M. to midnight. I sleep off at school"* – informs one of the students). And continuation of this interview:

R (Respondent): *Days are good to sleep off.*

I (Interviewer): *You have said that you sit in front of computer till midnight... which means that you are sleepy.*

R: *Well, it happens.*

I: *But you probably have some breaks. Do you eat something?*

R: *I have supper while browsing websites.*

Another student describes wide range of his interests and possibilities in his leisure time: "*Computer? I just use it for an hour or hour and half and that's all. I can do many things*".

Astonishingly, the respondents use their PC's in a broad manner for several purposes. Views of one of the young people might be an illustration of this fact: "*It would be too long to explain what I can do with the computer*". In some cases students demonstrate decent knowledge and skills related to ICT, and according to one of respondents, including "*innards rummage*" (ability to replace and refit components and update computer sets).

Almost all of the participants possess their own profiles on the social network websites such as "*Nasza klasa*" or "*Fotka*". Simultaneously, most of them use one or more programs for instant messaging (mainly: "*Gadu Gadu*" and "*Skype*" - an instant messaging software) to get in touch with friends and even for starting new relationships. In two cases such "Internet acquaintance" is continued in "real life". According to the respondents, communication mediated via instant messaging has a lot of advantages. Due to one of the girls: "*It is better to communicate via the Internet than face to face, because more things happen (...) For example, my friend writes to me that her boyfriend has broke up with her and in the same time, that boy is writing to me too*"

In case of using computers for games (the most common - Counter Strike, Tibia, Grand Theft Auto etc.) some new issues emerge, e.g. striving for power "*I want to be a commander*". While describing significance of games, the respondents give some examples of thrill resulting from this activity: "*To understand it, one needs to be involved in game and emotions*" or "*I am the person who is chasing, shooting...*".

There are also a few (mostly positive) untypical forms of "web" activities. As the respondents revealed during the interviews: the more interests they had and the more they participated in varied forms of leisure activities, the less they seemed "immersed" in the virtual reality, especially in terms of negative or "purposeless" actions (e.g. "travelling" through the strangers' profiles on social network sites). Additionally, it is visible that in a few cases, ICT are utilized in connection with hobby or interests. The subsequent quotation illustrates that issue: One of the respondents – 17, girl, who likes to rewrite poems that are interesting to her:

R (Respondent): *I only visit the sites with poems*

I (Interviewer): *I don't get it. What poems?*

R: *Well, simply the poems. I love poems.*

I: *Is there any special portal with such poems?*

R: *Yes, one can type „poems" in Google and then look for love poems or friendship poems.*

I: *And do you only read them or do you write something as well and put on that website?*

R: *I read and then I sometimes rewrite them into my notebook.*

Another example of linking online and offline activities are: writing stories, creating music, downloading the music themes (backgrounds), searching for videos with dance moves:

I (Interviewer): *Do you search for something specific?*

R (Respondent): *Some hip-hop tricks.*

I: *Have you learned something from the Internet?*

R: *Thomas Flair.*

I: *What is this?*

R: *This is kind of dancing with hands.*

I: *And one day you were watching it and decided to do the same? Yes.*

The following respondent described the possibility to learn some elements of the so called "Parkour" in the Internet. *Le parkour* is the physical discipline of training to overcome any obstacle within one's path by adapting one's movements to the environment (http://en.wikipedia.org/wiki/Le_parkour). Two of the respondents have revealed watching *arranged 'fights'* between the football fans of mutually hostile football teams, as the examples of online activities. "*I want to watch how I performed in the fight*" – says one of the respondents while watching a video of the battle and his performance within. Given that fact, it seems obvious that such area of analysis is connected to problem of engagement in electronic aggression and cyber bullying. The photos of account owners (on social network sites) are a form of self-presentation, which is, of course, not only typical for respondents with intellectual disability. Nevertheless, it has special meaning among the intellectually disabled. As we can conclude from two of the opinions, there is a great involvement in searching for the most appropriate, attractive and private image of one's face on social websites. According to two respondents the reasons for deleting their accounts were "*ugly pictures of themselves*". The study provides evidence that the negative score resulting from the visitors evaluating respondents' profiles is something very important for the students.

Another interesting manifestation of online activities is "surfing" through the accounts on social networking sites and spending pretty much of their free time on this. One of the ways of interpreting such findings is the lack of knowledge on other people and the need to fill in the gap.

Despite the declaration of importance of ICT usage, some of the students have additional afterthoughts about it:

I (Interviewer): *How many hours do you spend in the Internet daily?*

R (Respondent): *Not much, I use Gadu-Gadu a little and then I log out.*

I: *You said that ICT are important, so you can use it more, maybe even the whole day...*

R: *Yes, once I sat the whole day and the following day I had a terrible headache ... In the beginning, I thought I would use and use computer and the Internet. Now I have an access to the Internet and I don't have to.*

I: *So have you become bored?*

R: *Yes.*

To summarize respondents statements, it is worth mentioning, that the biggest part of their online activities is devoted to communication purposes, which is in line with other findings of ICT usage among young people.

RESPONDENTS' NEEDS FOR ACQUIRING NEW ICT SKILLS

Almost half of the sample declares lack of need in the range of improving their skills. Such state may be interpreted in terms of using New Media in a limited way, as well as it may be a symptom of lacking the knowledge about the potential advantages of ICT utilization.

Particularly, the latter seems to be a starting point for education in the field of broadening the interests and building curiosity about unknown, so far, ways of using new technologies. The most often given answer to questions related to expanding knowledge and improving skills in using ICT was: "*no*" or "*nothing*". Another examples proving declared lack of needs in discussed area are: "*I can do everything I need*", "*It is enough for me*" or "*I don't need to learn more. I have what I want to have*".

One of the examined students stated that if he needs help with the computer he receives it from his sister, while three of the young people were not able to answer questions about their needs for improving their information skills. A few students revealed their ideas of what they could have learned but simultaneously some of them emphasized that it is nothing important. Such uncertainty seems to have a self-defending character and prevents them from suffering a possible defeat ("*I'd like to know how to create a web page but it is nothing important*"). Precisely, the skill of creating web pages is specified by three of the respondents as the most desirable, in their opinion. Additionally, in one case, there is a significantly stressed need to use a website as a tool for self-presentation ("*I'd create a website on myself*").

"*Hacking*" is perceived as a desirable skill to learn by one of the students. Other expressed needs are: "*I'd like to learn a lot*". For example: "*how to install, how to download safely from the Internet*", and "*I need to have new programs and desktop wallpapers for my cell phone" or "how to download music*". Another student, who was deeply engaged in online activities, expressed the need for improving his typing speed, because he noticed problems related to his limited efficiency while using the Internet.

ICT AND ITS USAGE FOR COMMUNICATION PURPOSES

Online communication is supposed to be easier than face-to-face one. This kind of contact may contribute to experiencing relief and decrease shyness (Skrzydlewski, 2005, p.32). Text-based communication also gives more control of passing on information (Krejtz & Zając, 2009, p.127) taking into account the fact of limited (comparing to non-disabled) abilities of the intellectually disabled to express themselves in writing. However, possibility of participation in activities typical for other young individuals like sending Short Text Messages or chatting may cause the feeling of belonging to the society.

Certainly, there are fairly many opinions on communication via ICT as only a substitute for "normal" conversation based on face-to-face contact. The Internet is a tool for equating the social status but also sometimes a substitute for relationships and illusion of the closeness (intimacy). The web activity is a chance for overcoming solitude and simultaneously risk and reason for social isolation and alienation (Cudak, 2007, p.43). Text based communication is a significant advantage especially for shy persons and for those who have speech impairments and language disorders.

I (Interviewer): *Why texting is better than talking on the phone?*

R (Respondent): *I know what to write but I do not know what to talk about.*

Sending an SMS is much cheaper for the participants but it is possible to regard them as a tool for augmentative communication. In the light of following statements, there are students who would not prefer phone calls even if they had been as cheap as SMSes: "*I prefer texting, because when I talk and hear people's voices, mhm... I am ashamed then, and by SMS it is possible to chat more comfortably, I don't need to hurry, or something*"; „*It is better to communicate through the Internet than face to face*" or: „*Talking is time consuming. I can send short text messages*".

The Internet has an impact on communication behaviour e.g. by greater openness and extremity in expressing the opinions (Krejtz & Zając, 2009, pp. 127-128). Moreover, there is significantly less pressure related to somebody's presence, which has its positive and negative impact, e.g. facilitates occurrence of disinhibition behaviors.

The respondents send approximately from several (10-20) to several dozens of short text messages each day. Of course, the deviations from the average score have been observed i.e. few students do not write SMS ("*I don't text because it is too tiresome for me*") or do it rarely. On the other hand, a few participants revealed that they even send several hundred messages a day. The latter means that such level of intensity must have significant influence on their daily activities and social life.

Remarkable engagement in communicating using an SMS is observed among students who do not have computers or the Internet access at home ("*The Internet is important but cell phone is more important for me*"). The following fragment of interview with "record holder" SMS sending (600 per day).

I (Interviewer): *How many SMS do you send every day?*

R (Respondent): *Oh my God! You'd better not ask me...*

I: *And why it is so important for you?*

R: *What? Are you asking me about 600 SMS per day? Simply speaking – at least I am not bored.*

I: *It would be dull without SMS?*

R: *Yes.*

As one can see, this is the kind of activity that is time-consuming and covers almost all of the free time.

One of the respondents (girl – 17 y.o.) is aware of her extra engagement in the cell phone usage and with a remarkable consciousness reported as follows: *"I can say, I am addicted to my cell phone (...). I cannot live without it (...) I have to watch it all the time, I have to keep it as close as possible (...) When I do not receive an SMS for more than a few minutes I feel dizzy* ". According to collected data, the cell phones are used to a higher degree by girls.

Ways of use of cell phones such as listening to music appeared only in few cases. Three of the students claim that they used their camera to take a picture of a classmate ("*because she is funny*"), a movie star, and the teacher who was not liked by the respondent. Coverage of the students' messages is related to such matters as school, relationships and others typical problems based on age of the students, which does not seem to be different compared to non-disabled peers.

I (Interviewer): *What are the matters discussed in SMS?*

R (Respondent): *e.g. "what happens", "shall we meet", "what classes are you attending now"?*

I: *Well, what else?*

R: *For example: that I was at school participating in mathematics classes and that the teacher is vicious.*

Subsequent pieces of interviews are displaying ways of use cell phones:

I (Interviewer): *What do you do during breaks at school?*

R (Respondent): *Most students keep and enjoy their cell phones. It is more common among girls (...) Boys prefer going to lavatory to smoke cigarettes. (...). With other girls we show each other messages from our boyfriends, favourite pictures and so on.*

According to Michel Walrave & Wannes Heirman (2009, p.32) "in many cases anonymous communication provides children and adolescents with some advantages. For instance, shy and withdrawn individuals seems to be more willing to leave their 'crusts', because anonymity increase their confidence".

Taking into account emotional problems accompanying intellectual disability, it is worth to consider the study of Shepherd & Edelman (2005) who proved relationship between Internet usage and social anxiety. According to them, it is easier for individuals afflicted by fear, it could be easier to establish and maintain online relationships than engage in face-to-face contacts. According to Campbell et al. (2006) intensive Internet usage may be good for some individuals for building self confidence and even play a therapeutic role.

AWARENESS OF THE RISKS RELATED TO ICT USAGE

The most common answers (seven respondents) to question whether somebody should pay attention when using computers and the Internet, are related do the "technical" side of the problem. The respondents almost unanimously indicate viruses as the most dangerous aspect of using New Media. One of the students, when asked to enumerate more risk factors multiplies technical dangers: "*What else? Trojans, electric shock and burning of the screen*".

Another respondents pointed that "*banners while downloading*" and "*hackers*" can be potentially dangerous. Two students do not know what are the dangers or risk factors related to using computers, especially with an access to the Internet. One of our participants mentioned: "*It has to be something dangerous but I do not remember what... The teacher told us something*". Next four of the respondents affirmed, that according to their experience, there is nothing dangerous related to online activities. The lack of conscience about potential dangers and having inappropriate knowledge in this area should be regarded as a threat for the respondents, both in individual and social dimension. The results show that 6 out of 23 students do not know or do not perceive any risks with online activities and 8 limit them only to the "technical", concrete dimension, therefore it should be a matter of concern for educators and researchers.

Three respondents revealed that while being online one should be concerned about revealing personal data. However, statements of the next four students exhibit the fact that intellectual disability does not have to be a definite limitation in understanding risk factors accompanying their online activity: "*There are people who lie about their age*"; "*Adults pretend to be children*"; "*One has to be aware when communicating with strangers*".

The following subgroup of statements refer video files as a reason of danger: "*I am not afraid but if I had invented the Internet I would have made more protectors to make it more safe for children*". When he was asked to elaborate on his statement he mentioned: "*sexual movies*". Another respondent says: "*It is advisable to watch out for awful movies in the Internet*" and gives an example: "*e.g when spiders emerge from the throat and so on*".

The most willingly and openly describing "*online life*" was respondent who revealed his knowledge about possibility of occurrence of destructive experiences on the Internet for example by putting and modifying content of the photo or video files.

***I (Interviewer):** What one should be aware of, when using the Internet in your opinion?*

***R (Respondent):** Personal photos, I think... For example, somebody can grab image of yourself and put... You know where?*

***I:** No I don't not know.*

***R:** On porn websites.*

***I:** Did it happen to somebody you know?*

***R:** Me?*

***I:** No, I mean anyone, whoever.*

***R:** No, It didn't happen to me, but it can happen to somebody else....*

ICT USAGE AND ELECTRONIC AGGRESSION[3]

Limited social skills and other features (e.g. lack of successes, strong need of acceptance, low self-esteem) accompanying intellectual disability, in connection with features of modern technologies, can cause serious consequences, both in individual and social sense. According to Samuel W. Flynt & Rhonda Collins Morton (2004) "children with mental retardation are prime candidates for being bullied. Additionally, such people with regard to probability of being potential victims are described as a "safe targets" (Williams, 2005). The examples of the researches on greater likelihood (for disabled people) to become a victim are also: Marini, Fairbairn & Zuber, 2001; Mishna, 2003; Whitney, Smith & Thompson, 1994, Little, 2004; Hershkowitz et al., 2007.

The intellectually disabled tend to have low self-esteem and to express "depending-on others" behaviour. They are often not aware of potentially dangerous situations. Children of moderate to low functioning intellectual disability are more likely to suffer from motor skill deficits, or physical and health impairments that make them easier marks for bullies seeking weaker victims" – also in "cyberspace" (Plichta, 2010). Subsequent problem, related to potential risk of being harmed, is a fact, that: "(…) disabled people have been regarded as unreliable witnesses (Gudjonsson, 2003; Hershkowitz et al., 2007) because of their poor memories, their susceptibility to suggestion, and their limited descriptive capacities (Perlman, Ericson, Esses, & Isaacs, 1994).

It occurs, that the intellectually disabled adolescents are also perpetrators of aggression. Beyond doubt, the mechanisms of aggression are common for people with or without intellectual disability. Even though, a few factors can be specific, to some degree, for intellectually disabled persons. For example, in agreement with Zdzisław Bartkowicz (1984) and Adam Mikrut (2000), aggression among intellectually disabled people is more related to the effect of instigation, peer pressure and willingness to attract somebody's attention. More frequently, than non-disabled individuals, they experience other people's aggression, therefore many of aggression acts are retaliation and may result from being attacked.

Using information and communication technologies by young people (such as The Internet and cell phones) as the means to conduct aggressive acts (e.g. cyber bullying) has received significant media attention driven by some recent cases resulting in criminal or civil lawsuits filed against the perpetrator as well as, in some incidences, the school (Dooley J., Pyżalski J. & Cross D., 2009). In this article, electronic aggression has a broader meaning and refers to all hostile acts of usage of modern technologies. Such meaning of electronic aggression "covers a wide range of behaviours that are qualitatively different with respect to a number of social aspects involved in particular situations. For example, the mechanisms of electronic aggression may differ when aimed at a victim known in the real world, at a person known only in cyberspace or a stranger (e.g. a celebrity person) or a group of people (e.g. racial/sexual minority). (…) Another important aspect is connected to the fact that some electronic aggression acts are targeted directly at a victim (e.g. someone sends vulgar e-mails, MNS messages) while the other harm a victim indirectly (e.g. someone posts unwanted visual material about a victim on the Internet or spreads rumours)" (Pyżalski, 2009, p.15).

BEING A VICTIM OF ELECTRONIC AGGRESSION

The respondents almost unanimously acknowledged, that usage of ICT gives them positive experiences. Nevertheless, it is worth mentioning, that almost half of the students admit they have unpleasant, negative experiences, even very serious ones, like receiving threats or sexual proposals ("*I was lured by a paedophile*"). A few respondents mentioned about their friends' nasty experiences, e.g. placing their personal photos on the Web without permission. Another of the students' friend received the following comment related (in perpetrators' point of view) to his appearance – "*you look gay*".

The examined students put strong emphasis on their "web image" – It is remarkably related to their personal photos on profiles available on social networking websites: "*I got one annoying comment about my appearance in the photo (...)*". How they are perceived, what people could think about them and how visitors evaluate their profiles - it is a matter of great importance, which is distinctively emphasised in respondents' opinions.

The following issue, derived from the interviews, is the jealousy as a main motive assigned by the victims (mainly girls) to the perpetrators. "*I got a nasty SMS from my best friend (...) She was jealous of my boyfriend. I wrote to her that R. is not her property, that she couldn't forbid me to meet him (...) and she stopped writing and offending me*". "*Via messenger I was insulted by my classmate (...). To pay him back, I did the same*". According to the respondent, she was attacked because the perpetrator was talked into doing it by jealous classmates. "*Nobody likes me in school. It might be, that they are jealous of my clothes. This jealousy is also of boys, that I have more colleagues from school, that I communicate via the Internet with them*". The respondent was also picked on by the cell the other student reported his way of responding to unpleasant situations on the Web: "*If slimy remark happens, then I remove it (...)*" and: "*I got such messages: for example: You are a f*****g Jew. When we catch you we beat you up. Remove your photos – you look like a muff. I don't care about their threats. We will see what happens if they find me*".

Several interviewees described examples of occurrences, which one of them interprets as "jokes" e.g. warnings or threats sent from unknown telephone number. "*Be ready. We'll be in a few minutes*" or "*I was insulted*". *Another example: "They called from the ex-directory phone number and they said, that they would break my bones and so on"*.

One of the respondents reported lack of fear despite the fact of being attacked via cell phone:

R (respondent): *I told him to shut up (...) If he phoned again, I would hung up.*

I (interviewer): *And? ...*

R: *...and he resigned and doesn't call me anymore.*

I: *Were you scared when he was threatening you?*

R: *Yes (irony)..., I can phone and scare somebody, too.*

Another example which proves receiving phone calls, including threats, is consecutive student's statement: "*I was intimidated two times*". The respondent was sure who was the perpetrator and she says "*It was my classmate. She did it because of boredom*". The next respondent describes what he had experienced: "*Somebody broke into my online game account*". Responding to interviewer question who could have done it, the respondent stated: "*It has to be an Englishman (...).People are angry with each other while they are playing net games*". Even though, some respondents have not experienced any difficult situation related to using computer or/and cell phone, they are often able to give the examples of unpleasant cases of their colleagues and friends.

To be a victim in the respondents statements, first of all, means receiving annoying messages, phone calls and bad remarks of their profiles on the social network websites. The respondents interpret this as it was done by chance, as a kind of jokes or youth boredom. The interviewed students generally ignore such events and do not respond. In two cases, reaction is an equivalent to the experienced offence. Regarding the girls, the motive of jealousy is very distinct.

BEING A PERPETRATOR OF ELECTRONIC AGGRESSION

Approximately, one third of the respondents (seven out of twenty three respondents) gives the examples of undertaking aggressive actions when using the ICT.

It is worth mentioning that such result is higher than in the research of typically developing (TD) adolescents (Pyzalski, 2009a), where 16 percent of respondents confess to be engaged in online aggression. Regarding the different way of obtaining the data and differences in the size of sample, the results of such comparison are limited and comparison has only "rough" character.

The cases are diversified in terms of the level of jeopardy. Some of them have relatively little annoying character, meanwhile some other should be interpreted as serious and potentially harmful to the victim. Also, some of the cases were continued in "real life". A few respondents should be regarded as both a victim and a perpetrator of online aggression. For example, one of the perpetrators describes: "*I was hooked in the Web*". According to him, it was caused by the photo of him in the t-shirt with logo of a football club. The respondent recognized who was the doer of such attack, found him and "*I bashed him in the face and he stopped bothering me*". As we can see, this is an example of "stretching" the online situation to the offline circumstances.

Another example of destructive use of the Internet is searching the Web to find profiles with the pictures of the males with long hair and making rude comments ("*Go to the hairdresser*") and other accusations (with offensive words) of being gay, related to the custom of having such hairstyle, in perpetrator's opinion. Some of the respondents indicate connection between "web aggression" and with everyday, "face to face", contacts. Sometimes, aggression activities have indirect character e.g. recording video materials with teachers or sending text messages among classmates with nasty comments about the teachers the perpetrators do not like.

In the subsequent dialogue with the interviewer, the respondent reveals attacking electronically somebody who had marked his profile with unsatisfactory rank:

I (Interviewer): *Have you ever written any hostile comments on somebody's profile, for example, on 'Fotka', that you didn't enjoy or done something similar?*

R (Respondent): *Yes, I posted a comment once. She gave me 'one'. She reduced my average score. So, I wrote to her: You are f*****g bitch. I had given you 10 points and you.... You know what I mean.*

Nevertheless, the respondent objected behaving in this manner more frequently. "*No, nowadays I don't write such things (...). It was my cousin who talked me into this*". One of the respondents, as he claimed, had received a message with sexual offer: "*Then I told him: F**k off. I am not gay. I don't like such people*".

The state of boredom as a reason for undertaking improper actions via the Internet or cell phones is revealed by one of the respondents. "*I was calling people I hadn't known (...) I simply made them up*", which means that sometimes the victim is chosen randomly.

Few respondents describe the situations which are not direct electronic aggression but have indirect character and are potentially harmful for the other people: "*I have taken a picture of my classmate – she is funny. I have it in my cell phone*". Another category revealed in interviews are "jokes". One of the respondents describes the following situation: "*We dialed our colleague's number and we said: You have ordered the pizza number seven. In a five minutes we will come*".

ICT USAGE AND FAMILY LIFE

In most cases, the participants do not report parents' attempts to limit the time they spend in the cyberspace.

I (Interviewer): *Do your parents have grudge against you for being online for so long?*

R (Respondent): *They sit all day in front of TV, so why couldn't I?*

In respondents' opinions their parents do not know what they do with computers.

I (Interviewer): *Do your parents know in details what you do on the Internet?*

R (Respondent): *No they don't.*

I: *Don't they talk to you about your activities on the Internet?*

R: *They ask what I do and I tell them:I'm playing or I am talking to a friend or I am listening to music...*

I: *You say and they believe...*

R: *Yes.*

One of respondents protects the PC from parents, not only due to the secrecy of correspondence, but because of fear that parents can break student's computer.

I (Interviewer): *Do you use the computer alone or with someone?*

R (Respondent): *I do it only myself. Parents haven't got the faintest idea what to do with a PC.*

I: *Do you help them with it?*

R: *Well, if they only touch computer, it will be destroyed.*

I: *Did it happen anytime?*

R: *Fortunately not, but they are not allowed even touch it?*

I: *Really?*

R: *I protected my PC with the password to prevent it from parents" control. They have not rights to my computer.*

The next two statements describe cases of parental supervising connected with spending leisure time on the Internet: "*My mother sometimes bawls: all the time only computer and nothing more! I don't accept it*" and another example:

I (Interviewer): *Do your parents mind that you spend four hours per day in front of PC? Don't they say: you are supposed to finish...*

R (Respondent): *Sometimes when it is getting late.*

I: *What do you mean saying "it is getting late"?*

R: *About 1 A.M.*

It would be naive to accuse the New Media of spoiling relations between parents and children. The way of usage and possible problems connected to this, are mainly the effects of previous contacts and the quality of bonds. The Internet is not "the fake world". There are the real people with their emotions, experiences which are transmitted to virtual environment (Szpunar, 2006, p. 166). In some cases, consent to unlimited ICT usage of young people is a choice of lesser evil for family members: "*Generally, grandfather enjoys the fact that I am at home*".

The piece of good news is the fact that in a few cases computer contributes to sharing interests understanding and is a common ground for adults and young people.

I (Interviewer): *Is your mother good at using a computer?*

R (Respondent): *She only enters "Fotka (social network site)" – nothing else (...) She sometimes asks me for help.*

As it was presented above, sometimes the respondents are more competent with ICT usage than their parents.

R (Respondent): *My mother rather doesn't have any skills to operate computers and browse the Internet.*

I (Interviewer): *Did she ask you for something to do for her on the Internet?*

R: *Well, e.g. to help her in searching for a job or to find the timetable of buses and trams.*

In one case online activity is a common mystery:

R (Respondent): *I write essays on the computer. I don' like writing on sheet of paper.*

I (Interviewer): *Do somebody know about such essays?*

R:*Nobody knows, except my mother.*

Having interests and important skills to organizing leisure activities might at least in part by explained by positive influence of parents:

I (Interviewer): *You have learned such things like putting together components to build a computer.*

R (Respondent):*It was always attractive to me when my father "fumbled" at electronic spare parts. When I was a kid I used to observe when he was mending, soldering etc., so it was something important to me.*

An illustration that suggests that some respondents have considerations of the world after and before computer appeared in their life, is the following statement:

I (Interviewer): *When I did not have a PC at home we (with family members) played cards, game boards.*

R (Respondent): *Mhm, and nowadays? No, Nowadays we do not do this (...) When I was a child I did it or I drew pictures.*

I: *You don't draw nowadays.*

R: *Now I have the computer. My sister brought it, needlessly".*

On the basis of the above facts it is possible to conclude that the ways of using ICT among interviewed students are often influenced by both cognitive impairments and in a larger scale by relationships with their families .

FUTURE RESEARCH DIRECTIONS

In further studies, the questions raised in the presented paper should be expanded with regards to the methodology, size of the sample, as well as in terms of range of undertaken problems. The next research should be based on both qualitative and quantitative methodology. Then the results of quantitative studies will contribute the data and in compliance with it, the possibility of making comparisons between disabled and non-disabled students (in terms of ICT usage with special regards to electronic aggression). Implementation of more variables (e.g. being a victim or/and perpetrator of "offline", traditional aggression, loneliness, social participation, educational achievements, leisure activities etc.) and examining relationships between them will bring answers in regards to such unrecognized issues as the intellectually disabled in the world of information and communication technology.

CONCLUSION

In regards to the chances and risks related to the participation in the ICT usage, as stated in the title, it is possible to derive following conclusions.

First of all, in some cases, one can observe the symptoms of *personal risk*, mainly related to possibility of becoming a victim of electronic aggression. Often this kind of danger is also connected with involvement in "traditional" peer aggression. The analysis of the gathered data, indicate that problematic Internet use among participants is mostly limited to their functioning as a member of peer groups. This means that on the one hand, the online aggression might be continuation of face-to-face contacts, on the other hand, the online aggression may by continued in the "real world". Another significant finding, distinctive for the respondents, shows that in a few most serious cases (in terms of engagement of destructive ICT use), such students are both victims and perpetrators of electronic aggression.

Secondly, it is possible to derive conclusion on occurrence of such dimensions as *risk of surrogate experiencing* versus *chance for success*. In some cases, it seems that the Internet as well as, the "old media" e.g. television, deliver substitute for important life experiences. In special education, the importance of experiencing and learning „life through life" is firmly underlined. On the contrary, one should not forget about the compensation and facilitating role of technology in helping intellectually disabled students to achieve in VR (virtual reality) what they cannot reach in the "*life offline*". It is worth mentioning that "(…) cognitive impairments of people with (…) cognitive disabilities impede the opportunity to influence their environment and to become aware of the results of their actions. Reduced participation in activities limits the opportunity to make choices, and even the awareness that one has the right and need to exert control over one's own lives. The provision of opportunities to exert control over one's environment and to influence and interact with other people is thought to help to reduce learned helplessness. The ability to make choices and to indicate preferences is a first step towards the achievement of a sense of independence and responsibility (Weiss, et.al., p. 339).

Thirdly, in case of people with emotional problems (e.g. unfulfilled need of power) it appears the *risk of emotional prosthesis* which manifests, for instance, by over-involvement in playing games with violent content. The evidence are statements of some respondents revealing that strong emotions, especially negative, emerge when using New Media. .

Fourthly, in case of people not having appropriate skills to shape their leisure activities in a positive way, there exists a serious *risk of immersion* – deep entrancing in a world of easier success resulting in withdrawal from more demanding *offline situations*. On the other hand, using ICT is also a hope for status equalization and a *chance for filling the void*, particularly with respect to broaden and sustain one's interpersonal relationships and finally enhancing the life chance for young people with intellectual disability.

I strongly believe that the modern information and communication technologies constitute a great hope for the people with disabilities in terms of greater social participation, but it seems unreasonable to think that only the technologies or disabled persons by themselves can change it. In this respect, the help has to be delivered and such help should be provided by emotionally significant people, present in the disabled people's life. As it was emerged from respondents' opinions, the patterns of ICT usage are inseparable from the everyday contacts with their families. Ways of using new technologies is often a consequence of weak bonds between family members, lack of common interests and typical ways of spending free time leisure activities.

The meaningful relationships with other people in the "real" world might be helpful, and also help building self-esteem. Another important factor which is worth consideration is the educators' presence in life of their pupils. Because of the increasing time devoted to New Media usage, the point is that the educators have to be present in the e-world, too. It would seem that the key point which might enable young people to safe, meaningful ICT usage is media education. Curriculum of such course should regard specificity of mediated communication - with its positive and negative consequences.

According to Wendy Seymour (2005, pp.195-196) use as well as under-use are lenses through which we can glimpse expressions of the self. Both the individuals strongly devoted to ICT usage and those who do not use it at all for several reasons (e.g. poverty, lack of self-confidence, negative parental attitudes towards media etc.) should be an object of educators' concern and interests. The former with regard to risk of over-involvement, the latter with regard to disappearing chances. However, "we should not question or accuse communication technologies, but the external reasons - namely the way the technologies are used by people. Such conclusion is supposed to be an inspiration in undertaking preventative actions which would underline the importance of media education" (Walrave & Heirman 2009, p.31).

The respondents participated in the interviews actively and with great involvement. Being an interviewee seemed very attractive to them – "my friend has told me that you wrote three pages after an interview with him" as one of respondents described. It means that one of the students noticed that the interviewer made three pages of notes. He evaluates this fact as a something to be proud of.

Accordingly to the gathered data, the young intellectually disabled students spend their free time with non-disabled students mainly. Unfortunately, in a few cases they participate in one of subculture groups - football fans and the way they meet their need for affiliation is supposed to have further consequences.

Quite broad knowledge and ICT skills and efficiency in dealing with everyday situations might be meaningful in considering the Paul Koegel & Robert Edgerton thesis that the intellectually disabled students are sometimes "*six-hour retarded children*" - only when they are at special schools. There are much more, than expected, similarities than exceptions with regards to using ICT between disabled and non-disabled students but it is only tentative finding that requires further exploration.

Of course, the presented data have only preliminary character (e.g. because of small sample and the fact that the research is still in progress) and therefore the range of conclusions is also limited. Nevertheless, the obtained data may be useful for both: practitioners (particularly, special educators) as well as researchers.

Educators can utilize presented findings e.g. for creating workshops aimed at preventing victimization among intellectually disabled students. While preparing such activities aimed at developing self-protective skills, it is worth to take the following features relevant to the intellectual disability into consideration:

1. Susceptibility to media manipulation (mainly addressed to consumption)
2. Acceptance and acquisition of negative and simplified patterns of behaviours
3. Inability to deep, critical analysis of the ambiguous, compound, provocative content
4. Difficulties in recognizing contextual, spatial and time relations
5. Limited ability to reflective and evaluative reception of selected aspects of New Media
6. Mechanism of seeking and modelling oneself behaviour upon persons without learning difficulties (Krause, 2004).

For researchers it is possible to make comparisons between the data obtained from special needs respondents and results of general population. Additional advantage of the research outcomes is its qualitative character with very interesting illustrative respondents' statements showing importance of New Media in respondents' life or describing advantages of using mediated communication.

One of the most important issues which arose from the research is to limit the risk of learning through substitute experiences. In process of education of the intellectually disabled it is crucial to exercise "life through life" which e.g. means experience of real significant contacts with other people. Sometimes New Media are a kind of escape from reality and are unbeneficial "filling" of their spare time. Certainly, ways of New Media usage should be included to "*absent plots*" in the field of special education. Especially, in the context of the choice-making, which has received relatively little attention in the education of the handicapped (Guess et al., 1998).

Obviously, modern technologies are civilization benefits and its usage always brings both advantages and risks. Then, let us examine the idea of "*razor's paradox*" described by Lech Witkowski (2007). Witkowski claims that within the education fields, the most evolutionary situations are these which have quite a lot dangers apart from profits.

REFERENCES

Bartkowicz, Z. (1984). *Nieletni z obniżoną sprawnością umysłową w zakładzie poprawczym*. Lublin, Poland: Wydawnictwo UMCS.

Campbell, A. J., Cumming, S. R., & Hughes, I. (2006). Internet use by the socially fearful: addiction or therapy? *Cyberpsychology & Behavior*, *9*, 69–81. doi:10.1089/cpb.2006.9.69

Cudak, E. (2007). Relacje wirtualne relacjami wirtualnymi? In Marek, Z., & Madej-Babula, M. (Eds.), *Bezradność wychowania* (pp. 37–48). Kraków, Poland: WAM.

Didden, R., Scholte, R. H. J., Korzilius, H., de Moor, J. M. H., Vermeulen, A., & O'Reilly, M. (2009). Cyberbullying among students with intellectual and developmental disability in special education settings. *Developmental Neurorehabilitation*, *12*(3), 146–151. doi:10.1080/17518420902971356

Dooley, J., Pyżalski, J., & Cross, D. (2009). Cyberbullying Versus Face-to-Face Bullying: A Theoretical and Conceptual Review. *Zeitschrift für Psychologie*. *The Journal of Psychology*, *217*(4), 182–188.

Falvey, M. (1986). *A Community-Based Curriculum: Instructional Strategies for Students with SevereHandicaps*. Baltimore, MD: Paul H. Brookes Publishing Company.

Flynt, S. W., & Morton, R. C. (2004). Bullying and Children with Disabilities. *Journal of Instructional Psychology*, *31*, 330–333.

Guess, D., Benson, H., & Siegel-Causey, E. (2008). Concepts and Issues Related to Choice Making and Autonomy Among Persons With Severe Disabilities. [Retrieved from Academic Search Complete database.]. *Research and Practice for Persons with Severe Disabilities*, *33*(1/2), 75–81.

Hershkowitz, I., Horovitz, D., & Lamb, M. (2007). Victimization of Children With Disabilities. *The American Journal of Orthopsychiatry*, *77*(4), 629–635. doi:10.1037/0002-9432.77.4.629

Krause, A. (2004). *Człowiek niepełnosprawny wobec przeobrażeń społecznych*. Kraków, Poland: Impuls.

Kreutz, K., & Zając, J. M. (2009). Psychologiczne aspekty wykorzystywania technologii internetowych w nauczaniu. In Nowak, A., Winkowska-Nowak, K., & Rycielska, L. (Eds.), *Szkoła w dobie Internetu* (pp. 121–138). Warszawa, Poland: Wydawnictwa Naukowe PWN.

Little, L. (2004). Victimization of children with disabilities. In Kendall-Tackett, K. A. (Ed.), *Health consequences of abuse in the family: A clinical guide for evidence-based practice* (pp. 95–108). Washington, DC: American Psychological Association. doi:10.1037/10674-006

Marini, Z. A., Fairbairn, L., & Zuber, R. (2001). Peer harassment in individuals with developmental disabilities: Towards the development of a multidimensional bullying identification model. *Developmental Disabilities Bulletin*, *29*, 170–195.

Mikrut, A. (2000). Próba wyjaśnienia związku między agresją i upośledzeniem umysłowym. In J. Pańczyk J. (Ed.), *Roczniki Pedagogiki Specjalnej*, 11, Warszawa, Poland: WSPS, 30-40.

Mishna, F. (2003). Learning disabilities and bullying: Double jeopardy. *Journal of Learning Disabilities*, *36*(4), 336–347. doi:10.1177/00222194030360040501

Perlman, N. B., Ericson, K. I., Esses, V. M., & Isaacs, B. J. (1994). The developmentally handicapped witness: Competency as a function of question format. *Law and Human Behavior*, *18*, 171–187. doi:10.1007/BF01499014

Plichta, P. (2010). Intellectualy disabled students – perpetrators and victims of electronic aggression. In Żółkowska, T., & Konopska, L. (Eds.), *Disability - the contextuality of its meaning* (p. 680). Szczecin, Poland: Wyd. Print Group.

Pyżalski, J. (2009). Agresja elektroniczna dzieci i młodzieży – różne wymiary zjawiska, *Dziecko krzywdzone. Teoria. Badania. Praktyka. „. Cyberprzemoc*, *1*(26), 12–26.

Pyżalski, J. (2009). Agresja elektroniczna wobec „pokrzywdzonych" – Inny jako ofiara agresji. In Chrzanowska, I., Jachimczak, B., & Podgórska-Jachnik, D. (Eds.), *Miejsce Innego we współczesnych naukach o wychowaniu. Trudy dorastania, trudy dorosłości* (pp. 54–59). Łódź: WSP/Edukacyjna Grupa Projektowa.

Reschly, D. (2009). Documenting the Developmental Origins of Mild Mental Retardation. *Applied Neuropsychology*, *16*(2), 124–134. .doi:10.1080/09084280902864469

Schalock, R., & Luckasson, R. (2004). American Association on Mental Retardation's Definition, Classification, and System of Supports and its relation to international trends and issues in the field of intellectual disabilities. *Journal of Policy and Practice in Intellectual Disabilities*, *1*(3-4), 136–146. .doi:10.1111/j.1741-1130.2004.04028.x

Schloss, P., Smith, M., & Kiehl, W. (1986). Rec Club: A community centered approach to recreational development for adults with mild to moderate retardation. [Retrieved from PsycINFO database.]. *Education and Training of the Mentally Retarded*, *21*(4), 282–288.

Seymour, W. (2005). ICTs and disability: Exploring the human dimensions of technological engagement. [Retrieved from Health Source: Nursing/Academic Edition database.]. *Technology and Disability*, *17*(4), 195–204.

Shapira, N. A., Goldsmith, T. D., & Keck, P. E. (2000). Psychiatric features of individuals with problematic internet use. *Journal of Affective Disorders*, *57*, 267–272. doi:10.1016/S0165-0327(99)00107-X

Shepherd, R., & Edelmann, R. (2005). Reasons for internet use and social anxiety. *Personality and Individual Differences*, *39*(5), 949–958. .doi:10.1016/j.paid.2005.04.001

Skrzydlewski, W. (2005). Kampanie społeczne w mediach-doświadczenia polskie. In Leppert, R., Melosik, Z., & Wojtasik, B. (Eds.), *Młodzież wobec niegościnnej przyszłości*. Wrocław, Poland: Wydawnictwo Naukowe DSW Edukacji TWP.

Szpunar, M. (2006). Społecznośći wirtualne – realne kontakty w wirtualnym świecie. In Haber, L., & Niezgoda, M. (Eds.), *Społeczeństwo informacyjne. Aspekty funkcjonalne i dysfunkcjonalne* (pp. 158–167). Kraków, Poland: Wydawnictwo UJ.

Walrave, M., & Wannes, H. (2009). Skutki cyberbullyingu – oskarżenie czy obrona technologii? *Dziecko krzywdzone. Teoria. Badania. Praktyka. „. Cyberprzemoc, 1*(26), 27–46.

Weiss, P., Bialik, P., & Kizony, R. (2003). Virtual reality provides leisure time opportunities for young adults with physical and intellectual disabilities. *Cyberpsychology & Behavior: The Impact Of The Internet, Multimedia And Virtual Reality On Behavior* [Retrieved from MEDLINE database.]. *Society, 6*(3), 335–342.

Whitney, I. (1994). Bullying and children with special needs. In Smith, P. K., & Sharp, S. (Eds.), *School bullying: Insights and perspectives* (pp. 213–240). London: Routledge.

Whorton, J., et al. (1994). A Comparison of Leisure and Recreational Activities for Adults with and without Mental Retardation. In D. Montgomery (Ed.), *Rural Partnerships: Working Together. Proceedings of the Annual National Conference of the American Council on Rural Special Education (ACRES)* (14th, Austin, Texas, March 23-26, 1994); see RC 019 557. Retrieved from ERIC database.

Williams, C. (1995). *Invisible victims: Crime and abuse against people with learning difficulties*. London: Jessica Kingsley.

Witkowski, L. (2007). *Edukacja i humanistyka. Nowe (Kon) teksty dla nowoczesnych nauczycieli*. Warszawa: IBE.

ENDNOTES

1. In older literature the term "mental retardation" was used to describe intellectually disabled people. Unfortunately, because of the editing limitations, it is impossible to provide here a broader discussion on definitional issues related to usage of the terms: "intellectual disability" and "mental retardation". In this chapter the terms are used interchangeably. Accordind to Reschly (2009, p. 125) one should note that "the name of the organization that deals primarily with mental retardation in the United States has changed over the last 50 years from the American Association on Mental Deficiency, to AAMR (American Association on Mental Retardation), and, since 2006, to the American Association on Intellectual and Developmental Disabilities".

2. The positive exceptions with respect to lack of literature of ICT usage among intellectually disabled are:
Löfgren-Mårtenson, L. (2008). Love in Cyberspace: Swedish Young People with Intellectual Disabilities and the Internet. *Scandinavian Journal of Disability Research, 10*(2), 125-138. doi:10.1080/15017410701758005. and: McClimens, A., & Gordon, F. (2009). People with intellectual disabilities as bloggers: what's social capital got to do with it anyway?. *Journal Of Intellectual Disabilities: JOID, 13*(1), 19-30. Retrieved from MEDLINE database.

3. This part of the paper is derived from the article: Plichta, P. (2010). Intellectualy disabled students – perpetrators and victims of electronic aggression. In T. Żółkowska & L. Konopska (Ed.), *Disability - the contextuality of its meaning* (pp. 680). Szczecin: Wyd. Print Group.

Chapter 18
Gaming and Aggression:
The Importance of Age-Appropriateness in Violent Video Games

Eva-Maria Schiller
University of Münster, Germany

Marie-Thérèse Schultes
University of Vienna, Austria

Dagmar Strohmeier
University of Vienna, Austria

Christiane Spiel
University of Vienna, Austria

ABSTRACT

Video games play an important role in the modern entertainment industry and determine the leisure time activities of many children and adolescents. A huge amount of video games are available, but many of them are not suitable for youth because of their violent content. Violent content in video games became an issue of public concern, not only in cases of extreme violent acts, such as school shootings (e.g. Littleton, Colorado, 1999; Kauhajoki, Finland, 2008; Winnenden, Germany, 2009) but also concerning the question of whether playing violent video games generally influences the development of aggressive behavior in children and adolescents. Considerable research showed that playing violent video games increases aggressive cognitions, and aggressive behavior (e.g., Anderson et al., 2010). A crucial issue in studies concerned with violent video games is the question of how to assess the presence of violent content in games. Most of the studies used expert ratings (e.g. Krahé & Möller, 2004), some studies asked study participants themselves (e.g., Gentile & Gentile, 2008; Wallenius, Punamäki, & Rimpelä, 2007), and only a few studies used categorizations directly displayed on games (e.g. Schiller, Strohmeier, & Spiel, 2009). In 2003, the Pan European Game Information (PEGI) system was established, aiming at the protection of children and adolescents from unsuitable video games. PEGI evaluates games according to five age categories (+3, +7, +12, +16, +18) and seven content descriptors (bad language, discrimination, drugs, fear, gambling, sexual content, and violence). These age categories and content descriptors are printed on games to inform customers about their appropriateness for children and adolescents.

DOI: 10.4018/978-1-60960-209-3.ch018

Although these descriptors are highly visible for parents and adults in 30 European countries, they are rarely used in research. The current chapter presents a study on pre-adolescents in which violent content of games was categorized based on PEGI descriptors. A distinction between playing age-appropriate violent video games and age-inappropriate violent games was made. The main goal of the study was to examine whether pre-adolescents who play non-violent or age-appropriate violent games systematically differ in aggression from youth who play age-inappropriate violent games. Gender differences were also examined. Conclusions for practical implications for adolescents and for parents are discussed.

It requires a tortuous logic to believe that children and adolescents are affected by what they observe in their living room, through the front window of their house, in their classroom, in their neighborhood, and among their peers but are not affected by what they observe in movies, on television, or in the video games they play. L. Rowell Huesmann, 2010, p. 179

VIDEO GAME USE IN CHILDREN AND ADOLESCENTS

The sale of video games is a booming industry. Video games are the most popular entertainment products available for adolescents, especially for boys (Buchmann & Funk, 1996; Klimmt, 2004). A study conducted by the Interactive Software Federation of Europe (ISFE) reported interactive software sale volumes of € 7.3 billions in nine European countries (United Kingdom, France, Germany, Spain, Italy, Netherlands, Switzerland, Sweden, Finland) in the year 2007 (Nielsen Games, 2008). In accordance to ISFE, it has been predicted that growth rates will further increase over the next years. The international *Health Behaviour in School-aged Children Study* (Currie et al., 2004) of the World Health Organization reported that, on average, 31% of eleven-year-olds, 35% of 13-year-olds and 31% of 15-year olds play video games for two or more hours on weekdays. Video game use varies among countries. For 11-year olds and 13-year-olds, the highest prevalence rates were reported in Israel and the lowest in Switzerland. For 15-year-olds, the highest prevalence rates were reported in Romania and the lowest in France.

Furthermore, consistent gender patterns were found in a considerable number of studies. Boys outperformed girls in frequency (Buchmann & Funk, 1996; Ferreira & Ribeio, 2001; Funk, Buchmann, Jenks, & Bechtoldt, 2003; Lucas & Sherry, 2004) and duration of video game use (Colwell & Payne, 2000; Colwell & Kato, 2003; Durkin & Barber, 2002; Gentile, Lynch, Linder, & Walsh, 2004; Möller & Krahé, 2009; Willoughby, 2008). Moreover, more boys than girls play video games with violent content (Dill & Dill, 1998; Polman, Orobio de Castro, & van Aken, 2008; Wallenius et al., 2007).

Violent content is an essential element of numerous video games. Thompson and Haninger (2001) revealed that 64% of available video games contained violent content. Smith, Lachlan, and Tamborini (2003) showed that 90% of games considered appropriate for teens and adults contained violent content. Games appropriate for children contained at least 57% violent content (Smith et al., 2003). Games that are appropriate for 17 year olds and above always contain violent content, according to Thompson, Tepichin, Haninger (2006). Furthermore, several violent video games contain violence in an extent that is not reasonable for children and adolescents. Therefore, the access to these games is controlled via age restrictions. Nevertheless, children and adolescents play games which are inappropriate for their age (Krahé &

Möller, 2004; Salisch et al., 2007). Studies indicated a peak in video game playing between 10 to 14 years (e.g., Gentile & Walsh, 2002) and this age group in particular plays a lot of games that are not suitable for them. After the age of 14 years, video game playing decreases again (Fromme, Meder, & Vollmer, 2000; Rideout, Roberts, & Foehr, 2005). The following two sections give an overview on terminologies and definitions of aggression and review relevant studies on playing violent games and aggression.

AGGRESSION: TERMINOLOGIES AND DEFINITIONS

Aggression is defined as 'any form of behavior directed toward the goal of harming or injuring another living being who is motivated to avoid such treatment' (Baron & Richardson, 1994, p. 7). In the current chapter, we only focus on weak aggression. Thus, we differentiated *violence* from aggression, what is described as 'physical aggression that is so severe that the target is likely to suffer serious physical injury' (Gentile, Saleem, & Anderson, 2007, p. 4).

In examining aggression it is important to differentiate *forms* of aggression (Little, Jones, Henrich, & Hawley, 2003). Forms of aggression refer to concrete aggressive behaviors. Forms of aggression are distinguished according overt and relational aggression (Card, Stucky, Sawalani, & Little, 2008). *Overt* aggression includes physical and verbal aggressive behavior that is aimed directly towards a person, like kicking, pushing, insulting, threatening or hitting. *Relational* aggression corresponds to social manipulation of peer relations in order to harm another person (Vitaro, Brendgen, & Barker, 2006). Relational aggression is aimed more indirectly towards a person, including destructing relationships or affecting acceptance in a group. For example, relational aggression includes behavior like spreading rumors or excluding others. A meta-analysis conducted by Card et al. (2008) found that boys show more overt aggression than girls, while gender differences in relational aggression were very small and trivial.

In examining aggression it is also important to differentiate *functions* of aggression (Little et al, 2003). Functions of aggression refer to underlying motives for aggressive behavior. In spite of contradicting positions (Bushman & Anderson, 2001), two concepts are differentiated as underlying motives for aggression for several decades: *reactive* and *instrumental* aggression (synonymous: proactive aggression; Card & Little, 2006; Dodge, 1991; Dodge & Coie, 1987; Vitaro & Brendgen, 2005; Vitaro, et al., 2006). The concept of reactive aggression has its roots in frustration-anger theory (e.g., Berkowitz, 1989, 1993; Dollard, Doob, Miller, Mowrer, & Sears, 1939). To harm somebody else occurs as a "reaction" to a (real or perceived) provocation, threat or frustration and is usually accompanied by strong feelings of anger. Reactive aggression is believed to develop from social experiences that are characterized by an unpredictable, rough and threatening environment or poor caring parenting (Vitaro et al., 2006). In contrast, the concept of instrumental aggression is based on social learning theory (Bandura, 1973, 1983). To harm somebody else is a premeditated, calculated behavior that is used as an "instrument" to reach particular goals. The dominant emotions involved in this type of aggression are pleasure and stimulation. Instrumental aggression is believed to develop from different social experiences than reactive aggression (Dodge, 1991). Instrumental aggression is believed to originate in an environment where aggression is a successful way to reach goals (Vitaro et al., 2006). Exposedness to role models who behave aggressively to reach goals, promotes the development of instrumental aggression, but not of reactive aggression (Dodge, Lochman, Harnish, Bates, & Pettit, 1997). Usually, boys score higher in both reactive and instrumental aggression than girls (e.g., Little, 2002).

Two important concepts of cognitions are related to aggression: hostile attributional style and normative beliefs about the acceptability of aggression.

The construct of hostile attributional style stands for the tendency to interpret ambiguous situations in a hostile way. If situations are interpreted as hostile, aggressive behavior is a more likely reaction. A variety of studies have indicated a relationship between hostile attributional style and physical aggression in children (e.g., Burks, Laird, Dodge, Pettit,& Bates, 1999; Dodge & Coie, 1987;) and in adolescents (VanOostrum & Horvath, 1997). Other studies reported that reactive aggression, not instrumental aggression, is related to hostile attributional style (e.g. Dodge & Coie, 1987; Dodge et al., 1997; Hubbard, Dodge, Cillessen, Coie, & Schwartz, 2001; Orobio de Castro, Veerman, Koops, Bosch, & Monshouver, 2002). Boys tend to interpret ambiguous situations more often in a hostile way than girls (Krahé & Möller, 2004) and men score higher in hostile attributional style than women (Anderson & Dill, 2000).

According to Huesmann and Guerra (1997, p. 409) normative beliefs are 'an individual's own cognition about the acceptability or unacceptability of a behavior'.

Normative beliefs about the acceptability of aggression are beliefs about the adequacy of behaving aggressively in specific situations. For example: if someone believes that it would be okay to kick or push others, he or she would be more likely to show aggressive behavior as a legitimate response. Huesmann and Guerra (1997) reported a positive relationship between normative beliefs and physical aggression. Also gender patterns were found. Boys reported higher acceptability of aggression than girls (Crick, Bigbee, & Howes, 1996; Krahé & Möller, 2004) and the correlations between normative beliefs and behaviors were higher for boys than for girls (Huesmann und Guerra, 1997).

However, aggression is a multidimensional construct which is predicted by various factors. One factor that could contribute to the development of aggression is *media violence*. Gentile et al. (2007, p.17), define media violence as 'depictions of aggressive and violent behavior directed at characters in the media story. Those characters can be human or nonhuman, cartoonish or visually realistic'. According to Barlett and Anderson (2009) media violence could be seen as a risk factor for the occurrence of aggressive behavior. Although media violence is neither the strongest, nor the most important risk factor for the development of aggressive behavior, it is one factor that can easily be controlled by parents and policy makers. In the last five decades,research on media violence centered mainly on violence in television (TV-violence). Substantial evidence supporting the effects of TV-violence exposure on aggressive behavior emerged from longitudinal studies conducted over several years (Eron, Huesmann, Lefkowitz, & Walder, 1972; Huesmann & Miller, 1994; Huesmann, Moise, & Podolski, 1997; Huesmann, Moise-Titus, Podolski, & Eron, 2003). Exposure to TV-violence during childhood increases later adult aggressive behavior, regardless of gender, socio-economic status or intellectual ability (Huesmann et al., 2003). Long-term effects of TV-violence on aggression are explained with the process of long-term observational learning of cognitions supporting aggression (social-cognitive observational learning theory, Berkowitz, 1993; Huesmann, 1988,1998). Cognitions supporting aggression are schemas about a hostile world, aggression-focused scripts for social problem solving, and normative beliefs about the acceptability of aggression (Bushman & Huesmann, 2001). According to the social-cognitive observational learning theory, long term effects of TV violence on aggression are caused by the development of aggressive problem-solving scripts, hostile attributional style, and normative beliefs endorsing aggression (Huesmann et al., 2003).

Violence in video games came into the focus of interest several years ago. The following chapter reviews empirical research on violent video game playing and aggression.

VIOLENT VIDEO GAME PLAYING AND AGGRESSION

Violent video game use is supposed to affect aggressive behavior more strongly than the reception of violence on television, because of three reasons: (1) violent scenes in violent games – especially scenes with physical aggression- are more frequent than in violent films, (2) dealing with characters in video games is much more interactive, and (3) aggressive behavior is often reinforced immediately in video games. Not surprisingly, there is ample evidence that exposure to violent video games increases aggressive cognitions and aggressive behavior (see meta-analysis by Anderson et al., 2010). Most studies found positive associations between playing violent games and aggression, irrespective of research design, e.g. in experimental studies (Anderson & Dill, 2000; Anderson, Gentile, & Buckley, 2007; Bartholow & Anderson, 2002; Eastin, 2006; Frindte & Obwexer, 2003; Polman et al., 2008), cross-sectional studies (Anderson et al., 2004; Bartholow, Sestir, & Davis, 2005; Gentile et al., 2004; Krahé & Möller, 2004) and also in longitudinal studies (Anderson et al., 2007; Möller & Krahé, 2009). Short-time effects of violent video game use are explained with priming mechanisms, increase of arousal, provocation of aggressive emotions, or imitation. Long-time effects of repeated violent video game use are explained by learning of new aggressive scripts and the development of aggression-endorsing beliefs or hostile schemas.

In the literature, contradictory positions have also been reported. Some studies indicated no association between violent video game use and aggressive behavior (e.g., Ferguson, 2007; Ferguson & Kilburn, 2009; Sherry, 2007). Furthermore, it is argued that effect sizes of violent video game playing on aggressive behavior in the most comprehensive meta-analysis conducted by Anderson et al. (2010) are very weak (Ferguson & Kilburn, 2010). Indeed, an overall effect of r =.24 for studies of high methodological quality was revealed (Anderson et al., 2010). In fact, this effect is classified as small, but the practical significance of this effect should not be neglected: When one takes into account how many youth are confronted with violent video games, even small effects could have considerable impact. Furthermore, effect sizes between r= .1 and r=.3 are most frequently reported in social psychology, due to the fact that behavior is a highly complex phenomenon and multiple causes for a certain behavior are possible (Bushman, Rothstein, & Anderson, 2010). In the following, exemplary studies concerning violent video game use, forms of aggressive behavior, functions of aggressive behavior and aggressive cognitions are outlined.

Aggressive Behavior: Overt and Relational Aggression

A longitudinal study on the relation between overt and relational aggression and violent video game playing was conducted by Anderson and colleagues (2007). The authors examined violent video game playing, overt (physical, verbal), and relational aggression in 3th-5th graders using students' peer nominations, self reports as well as teacher reports. Results showed that students who reported a higher extent in violent video game playing at the beginning of a school year scored higher in overt aggression and in both verbal aggression and physical aggression.

A correlational study with 200 male undergraduates by Bartholow et al. (2005) showed that violent video game playing is positively correlated with self reports of physical and verbal aggression. Also examining overt aggression, a correlational study conducted by Anderson et al. (2004) displayed a relationship between violent video game playing and verbal aggression (r =.20), weak physical aggression (r=.31) and severe physical aggression (r=.17). A meta-analysis by Anderson & Bushman (2001) revealed an effect size of r =.19 for the relationship between violent video game playing and aggressive behavior.

Anderson et al. (2010) meta-analyzed the relationship of aggressive behavior with violent video game playing. They reported an effect size of $r = .21$ in experimental studies, an effect size of $r = .17$ in cross-sectional studies, and a significant longitudinal effect $r = .075$, controlling for study quality and gender.*Functions of aggression: reactive and instrumental aggression*

In research on aggression, a distinction between reactive and instrumental aggression is often made. In research on violent video games, there is a lack of studies examining reactive and instrumental aggression. This might be due to the fact that the reactive versus instrumental aggression dichotomy was criticized (Anderson & Bushman, 2001). The authors supported their arguments with examples of severe violent acts. Anderson and Bushman argued that severe violent acts are often conducted because of a mixture of reactive and instrumental aggression. This approach might be accurate for severe forms of aggressive behavior, but not for mild forms of aggressive behavior, in which we are interested. In our opinion it is important to differentiate between reactive and instrumental aggression in video game research, mainly for two reasons. First, we assume different consequences of violent video game playing regarding reactive and instrumental aggression. Video game users are confronted with role models who use aggressive behaviors in order to reach goals. These role models use aggressive behavior as an instrument expecting positive rewarding outcomes for themselves. Second, it remains unclear until now whether violent video game playing increases instrumental aggression. Effects on reactive aggression are more often investigated. Anderson and Dill (2000) showed that violent video game playing in students increases reactive aggression. Male participants show more reactive aggression than females. Bartholow and Anderson (2002) found that violent video game playing only increases reactive aggression in men, not in women.

Aggressive Cognitions: Normative Beliefs and Hostile Attributional Style

There is consistent empirical support for an influence of violent video game playing on cognitions (e.g. Anderson, 2004; Möller & Krahé, 2009; Anderson et al., 2010). Two concepts of aggressive cognitions were considered in previous studies: Hostile attributional style and normative beliefs about the acceptability of physical aggression. Studies investigating the relationship between hostile attributional style and violent video game playing showed inconsistent results. Some surveys indicated a positive relationship between playing violent video games and hostile attributional style in children (Kirsh, 1998) and in college students (Anderson et al., 2007). No evidence for this relationship was found in Krahé & Möller (2004). Instead, violent video game playing was linked directly to normative beliefs about the acceptability of physical aggression, and only indirectly to hostile attributional style through normative beliefs. Krahé and Möller (2004) investigated normative beliefs in adolescents and found that the acceptability of physical aggression increased with increasing violent video game playing. According to Funk et al. (2004) only *violent* content in video games is linked with aggression supporting normative beliefs, not playing video games per se. No significant relationship was found when focusing on relational aggression - neither was violent video game use associated with normative beliefs concerning relational aggression nor was violent video game use associated with hostile attributional style in ambiguous relational situations (Krahé & Möller, 2004). Longitudinal studies on effects of violent video game playing on aggression are still rare. One longitudinal study that was conducted by Möller and Krahé revealed interesting findings. Adolescents' violent video game playing predicted physical aggression 30 months later via an increase of aggressive norms and hostile attribution bias. There was no relation between aggression at the beginning of the study and violent video game use 30 months later (Möller & Krahé, 2009).

CATEGORIZATION OF VIOLENT CONTENT IN GAMES

The categorization of violent content in video games is a considerable challenge. Previous studies used various categorization systems. Some used categorizations which were directly displayed on games, like age ratings or genres. However, these categorizations did not allow an exact division into violent and non violent games. Therefore, in some studies expert ratings were used (e.g. Krahé & Möller, 2004) or participants were asked to classify the games themselves (Gentile & Gentile, 2008; Wallenius et al., 2007). Expert ratings are considered to be highly reliable; but it is very time consuming to apply them. Participants' ratings of violence in video games are easy to get, but they are prone to biases and thus their reliability is hampered.

Since 2003, a video game rating system called Pan European Game Information (PEGI) has been established in Europe. PEGI combines age ratings with content descriptors including violent content as one of these descriptors. Thus, PEGI provides a clear classification of violent content in video games and offers a classification which is also easy to apply for research purposes. The PEGI system is presented in the following section.

THE PAN EUROPEAN GAME INFORMATION: A EUROPEAN VIDEO GAME RATING SYSTEM

Except for Germany[1], the *Pan European Game Information* system (PEGI) has been established in 30 European countries[2]. Comparable to PEGI, the *Entertainment Software Rating Board* (ESRB) evaluates games in North America. Both rating systems inform customers, especially parents, about age-appropriateness and potentially harmful content in video games. Pictorials about age ratings and content descriptors are placed on each game's cover. PEGI is developed by the *Interactive Software Federation of Europe* (ISFE). PEGI uses self-reports from video game publishers in order to evaluate age-appropriateness. Every game for 12, 16, and 18 year olds and above is evaluated and games for the age group of at least three to seven years are randomly checked. PEGI defines age categories and uses content descriptors to inform customers why a particular age-rating is given. PEGI provides five age categories that inform about the minimum age for which a game is considered suitable: 'suitable for at least three years', 'suitable for at least seven years', 'suitable for 12 years and above', 'suitable for 16 years and above' and 'suitable for 18 years and above' (PEGI, 2009). Content descriptors give information about potentially harmful content. Seven types of harmful content are distinguished: (1) frightening content, (2) occurrence of bad language, (3) encouragement to alcohol and drugs, (4) discrimination, (5) gambling, (6) sexual content or, (7) violent content.

In PEGI, violent video games for the age of 12 and above are defined as follows:

'Video games that show violence of a slightly more graphic nature towards fantasy character and/or non graphic violence towards human-looking characters or recognizable animals.' (PEGI, 2009)

Violence in games considered appropriate for 16 and 18 year olds and above is much more severe and realistic. Video games restricted for 16 years and above are described as follows:

'[...] depiction of violence reaches a stage that looks the same as would be expected in real life.' (PEGI, 2009)

Violent content in video games restricted for 18 years and above is defined as follows:

'[...] the level of violence reaches a stage where it becomes a depiction of gross violence and/or includes elements of specific types of violence. Gross violence is the most difficult to define since it can be very subjective in many cases, but in general terms it can be classed as the depictions of violence that would make the viewer feel a sense of revulsion.' (PEGI, 2009)

In existing studies, violent video games are hardly differentiated according to their age appropriateness *and* violent content. As described above, higher age ratings go along with increasingly realistic-looking violence. We consider this as a clear shortcoming in previous research. Thus we assume that it is very important to differentiate violent content in combination with age categories.

THE PRESENT STUDY

In the present study, we examined overt aggression (as a form of aggressive behavior), reactive and instrumental aggression (as functions of aggression), normative beliefs about the acceptability of physical aggression and hostile attributional style in three types of users: (1) in users who play video games that were rated as non-violent and appropriate for all age groups, (2) in users who play video games with violent content that were rated as suitable for up to 12 years, and (3) in users who play video games with violent content that were rated as suitable for 16 years and above. Like Möller and Krahé (2009), we focused on the age group of twelve-year olds, where the use of video games is at its peak.

To differentiate violent content in video games, a categorization based on PEGI content descriptors and age ratings was applied. Higher age ratings on games that contain violent content refer to a higher extent of violent content. The central hypothesis of the study was that users who are confronted with a higher extent of violent content in video games also score higher in aggression variables. Thus, the main goal of the study was to examine whether pre-adolescents who play non-violent or violent video games rated as suitable for up to 12 years systematically differ in aggression-related variables from youth who play violent video games that are suitable for 16 years and above. Gender differences were also examined.

Method

Sample and Procedure

The sample comprised 169 pre-adolescents (65 boys, 104 girls; 6th and 7th grade). The mean age was 12.24 years (*SD*= 0.73). The majority of students had German as mother tongue (75.7%). Participants took part in the study during regular school hours. After acceptance of the study by the local school council, parental consent was obtained. 16.15% of parents refused their children's participation in the study. 48.56% of non-participating pre-adolescents were boys.

Instruments

Pre-adolescents were administered a questionnaire including questions regarding (1) video game use, (2) self-ratings concerning violent content, (3) overt, reactive and instrumental aggression, (4) normative beliefs about aggression, and (5) attributional style.

(1) *Video game use* was measured by asking participants to indicate whether they play video games. Those who indicated video game playing were asked to write down the names of their five favorite video games. The same procedure to measure video game use was also applied in Anderson and Dill (2000) and Bartholow et al. (2005), for example.

(2) *Self-ratings concerning violent content* were measured by asking participants to rate their favorite video games for violent content. Participants could select between 'yes' (1) and 'no' (0) to indicate violent content.

(3) *Overt, reactive and instrumental aggression* were measured using Little's questionnaire about forms and functions of aggression (Little et al., 2003). Participants were administered three scales. First, overt aggression was measured with seven items about aggressive behavior (e.g. 'I'm the kind of person who threatens others', *a= .79)*. Second, reactive and instrumental aggression were measured with seven items respectively, containing questions regarding underlying motives for aggression (e.g., reactive aggression*:* 'When I'm threatened by someone, I often threaten back', *a= .83;* Instrumental aggression: 'I often threaten others to get what I want', *a= .86)*. These scales ranged from 'totally agree'(4) to 'totally disagree'(1).

(4) *Normative beliefs about aggression* were measured using a vignette (Möller & Krahé, 2009) and Möller's (2006) German adaptation of the Normative Beliefs About Aggression Scale (Huesmann & Guerra, 1997). The following vignette was presented:

Imagine you are extremely angry with a classmate because he/she treated you in a mean and unfair way in front of other classmates that morning. After school you meet him/her again and this time the two of you are alone. Immediately he/she starts quarrelling with you again, saying nasty things ...

Following the vignette, 12 items containing possible responses to provocation were presented: physically aggressive responses ('I think it would be okay to kick and push him/her'), verbally aggressive responses ('I think it would be okay to shout at him/her'), relationally aggressive responses ('I think it would be okay to spread rumors about him/her'). Participants assessed the acceptability of these responses on a five-point scale ranging from 'totally ok'(5) to 'not at all okay'(1).

(5) *Hostile attributional style* was assessed on the basis of reactions to a vignette (Möller, 2006) containing a hypothetical situation with ambiguous violent content in a situation that contained damage of property. Participants' hostile attributional style was measured by asking them if they perceived hostile intent ('Do you think the other person threw your things down on purpose*?'*). The following vignette was presented:

Imagine it is break time in school. You are sitting at your table in your classroom and you are chatting with your neighbor who is sitting next to the window, about the movie that both of you watched together yesterday. You have already placed your books and folders for the next lesson on the table. Suddenly, someone bumps into the table while passing so that the things fall off your table and lie scattered on the ground. When you bend down to pick them up, you notice that your new calculator is broken due to the fall.

Categorization of violent content

All games were categorized according to the norms of the *Pan European Game Information* (PEGI). In this study, video games were rated regarding (1) two age categories (from age 12 downwards '≤12', from age 16 onwards '≥16'), and (2) violent content indicated by PEGI. As pointed out above, violent content in games rated at ≥16 is much more severe and realistic than in ≤12 rated games. Based on the ratings of age levels in combination with violent content, three groups of game users were distinguished. Due to the fact, the average age of participants was 12.24 years, all games rated ≥16 were labeled as 'age-inappropriate'.

(1) Users of *non-violent* video games: These youth nominated games that were rated as non-violent and appropriate for all age groups according to PEGI. For example, users of non-violent video games indicated games like *Fifa, Gran Turismo* or *Singstar.*
(2) Users of *age-appropriate violent* video games: These youth nominated at least one game that was rated as suitable for up to 12 years and contained violent content according to PEGI. For example, users of age-appropriate violent video games named games like *The Sims, Age of Empires* or *James Bond.*
(3) Users of *age-inappropriate violent* video games: These youth nominated at least one game that was rated as suitable for 16 years and above and contained violent content according to PEGI. For example, users of age-inappropriate violent video games indicated games like *Grand Theft Auto, Counterstrike or Tomb Raider.*

Results

(1) Video game use

To obtain a picture of pre-adolescents' video game use, percentages of general video game use as well as percentages of users of non-violent, age-appropriate violent and age-inappropriate violent video games were computed.

Of all participants, 91% (154 participants) said that they play video games. Nine participants (5%) named games that could not be categorized according to PEGI questionnaires, so only 145 participants were used for further analyses.

42.6% of boys and 11.9% of girls played games that were categorized as violent and inappropriate for their age. Chi² -Tests revealed that the preference for age-inappropriate violent games was stronger among boys than girls (χ^2 (2) = 19.53, p<.001).

Age-appropriate violent video games were played by 34.4% of boys and 64.3% of girls. Almost 23% of boys and 23.8% of girls played non-violent games. No gender differences were found for users of age appropriate violent games and non-violent games.

Table 1 shows the number of nominations of the five most favorite games, their age appropriateness, their violent content and the corresponding user group according to PEGI. As shown in Table 1, the most frequently nominated games were *The SIMS, Singstar, Fifa, Need for Speed* and *Grand Theft Auto*.

(2) Self-ratings of violent content

Participants indicated a total of 282 different video games. Among all indicated games, violent content was rated by participants themselves as well as according to the PEGI content descriptor

Table 1. Most frequently indicated games

Game	Number of Nominations	Age Appropriateness	Violent Content	Categorization
The Sims	70	12+	Yes	Age-appropriate violent
Singstar	25	12+	No	Non-violent
Fifa	21	3+	No	Non-violent
Need for Speed	19	3+	No	Non-violent
Grand Theft Auto	11	18+	Yes	Age-inappropriate violent

for violent content. Overall, 376 ratings were given by participants. Participants indicated that 23.7% of video games they played contain violent content. According to PEGI, 50.3% of games contained violent content. The correlation between self-ratings and PEGI ratings was modest, ϕ = .37 (p <.001).

(3) Differences in aggression in three groups of game users

To find out whether non-violent or age-appropriate violent video game users differ regarding aggression related variables from age-inappropriate violent video game users, a 2x3 MANOVA with video game user group and gender as factors and aggression variables (overt, reactive, instrumental aggression, hostile attributional bias, normative beliefs about aggression) as dependent variables was conducted. We expected users of non-violent video games and age-appropriate violent video games to score lower in aggression related variables compared with age-inappropriate violent video game users. Thus, simple contrasts were computed.

Application of multivariate tests using Pillai's Criterion revealed a significant effect for video game user group, F (10, 272) = 2.42, p < .01, η^2 = .08 and for gender, F (5, 135) = 4.33, p < .001, η^2 = .14. There was no significant interaction between video game user group and gender, F (10, 272) = 1.11, p = n.s..

Video Game User Group

Univariate analyses showed an effect of video game use on overt aggression, F (2, 139) = 5.95, p < .01, η^2 = .08, reactive aggression, F (2, 139) = 3.42, p < .05, η^2 = .05, instrumental aggression F (2, 139) = 3.66, p < .05, η^2 = .05 and on normative beliefs about aggression F (2, 139) = 4.22, p < .05, η^2 = .06. There were no differences in attributional style F (2, 139) = 2.12, p = n.s., when comparing three groups of video game users.

As shown in Table 2, age-inappropriate violent video game users scored higher in reactive (p < .05) and overt aggression (p < .01) compared with non violent video game users. Users of age-inappropriate violent video games compared with users of age-appropriate violent video games scored higher in reactive (p < .05), overt (p < .01), and instrumental aggression (p < .01), respectively, as well as in normative beliefs about physical aggression (p < .01). Means and standard deviations are shown in Table 2.

Gender

Gender differences were found in reactive aggression, F (1, 139) = 6.20, p < .05, η^2 = .04, and in

Table 2. Means and standard deviations in aggression variables in three video game user groups

Video game user groups	Overt aggression	Reactive aggression	Instrumental aggression	Hostile attributional style	Normative beliefs
	M (*SD*)	*M* (*SD*)	*M* (*SD*)	*M* (*SD*)	*M* (*SD*)
Non-violent (*N*=34)	1.19 (.22)$_a$	1.53 (.54)$_a$	1.11 (.26)	2.17 (.83)	1.25 (.41)
Age-appropriate violent (*N*=75)	1.23 (.23)$_b$	1.51 (.45)$_b$	1.07 (.21)$_b$	2.65 (.96)	1.16 (.30)$_b$
Age-inappropriate violent(*N*=36)	1.46 (.57)$_c$	1.84 (.64)$_c$	1.20 (.39)$_c$	2.47 (.97)	1.51 (.73)$_c$

Note: Standard Deviations (*SD*) are reported in parenthesis, column means with different subscripts are significantly different at least at p < 0.05; Overt, reactive, instrumental aggression scales range from 1 to 4. Normative beliefs about aggression scale and hostile attributional style ranges from 1 to 5.

Table 3. Means and standard deviations in aggression variables in boys and girls

Video game user groups	Overt aggression	Reactive aggression	Instrumental aggression	Hostile attributional style	Normative beliefs
	M (*SD*)	*M* (*SD*)	*M* (*SD*)	*M* (*SD*)	*M* (*SD*)
Boys (*N*=34)	1.27 (.34)	1.78 (.47) $_a$	1.11 (.25)	2.61(.77)	1.45(.52) $_a$
Girls (*N*=75)	1.27 (.40)	1.46 (.65) $_b$	1.13 (.36)	2.41 (1.01)	1.1(.40) $_b$

Note: Standard Deviations (*SD*) are reported in parenthesis, column means with different subscripts are significantly different at least at $p < 0.05$; Overt, reactive, instrumental aggression scales range from 1 to 4. Normative beliefs about aggression scale and hostile attributional style ranges from 1 to 5.

normative beliefs about aggression $F(1, 139) = 6.67, p < 0.01, \eta^2 = .046$. No gender differences were found in the other variables. As shown in Table 3, boys scored higher in reactive aggression and in normative beliefs about aggression than girls.

DISCUSSION

In the present study, we examined overt, reactive and instrumental aggression, normative beliefs about the acceptability of physical aggression and hostile attributional style in three types of video game users: (1) users of *non violent video* games, (2) users of *age appropriate violent* video games and (3) users of *age inappropriate violent* video games.

To form these three groups, video games were categorized according to their violent content. To categorize violent content, a combination of violent content descriptors and age ratings given by the Pan European Games Information (PEGI) was used. This method was never used in research before and enables to distinguish between age-appropriate and age inappropriate violent video game players. This distinction is important, because users of age inappropriate violent video games are confronted with more violence in comparison to users of age appropriate and certainly of non-violent games. We expected non-violent video game users and age-appropriate violent video game users to score lower in aggression related variables compared to age-inappropriate violent video game users.

First, our study shows that video game playing is a widely spread leisure time activity in youth (91% of youth play video games), and *second* that a considerable number of youth play video games that are not appropriate for their age. Approximately 43% of boys and 12% of girls nominated favorite video games that were age-inappropriate and contained violent content. Moreover, the weak correlation between participant's self ratings of violent content and ratings given by PEGI showed that youth systematically underestimate the violent content of games. PEGI age ratings do not reflect pre-adolescents' estimations and vice versa. These results point out that violent content estimations depend on the rating method. Therefore, further research should contain a multi-informant rating system that combines different rating methods.

In line with previous studies, we found differences in violent video game users and non-violent video game users regarding aggression. In considering not only the violent content but also the age appropriateness, our study contributes to former findings. Previous studies reported higher overt aggression in users of violent video games (e.g. Anderson et al., 2007; Anderson et al., 2004; Bartholow et al., 2005).

The present study revealed that the age appropriateness of video games is an important variable to consider. We found higher levels of overt aggression only in users of age *in*appropriate violent video games and not in users of age appropriate violent games compared with players of non violent games.

Although prior research has consistently shown gender differences in overt aggression (see meta-analysis by Card et al., 2008), our findings suggest no gender differences in overt aggression between boys and girls. These findings however should be interpreted with caution. In our study, a considerable number of parents (16%) refused their children's participation. We suppose that parents' refusal of their children's participation could be driven by worries that their children's aggressive behavior becomes known. Thus it is possible that our study underestimates gender differences.

With respect to functions of aggression, users of age inappropriate games reported more reactive aggression compared to the users of the other two groups. This finding is of particular interest, because reactively aggressive adolescents often are confronted with severe consequences. Reactively aggressive adolescents suffer more often from rejection or victimization by their peers, and have fewer friends than non aggressive adolescents (Dodge et al., 1997; Price & Dodge, 1989; Prinstein & Cillessen, 2003). Thus every risk factor for reactive aggression should be minimized and this also applies to age inappropriate violent video games. Based on prior research, we expected gender differences in reactive aggression (e.g. Little et al., 2002) and our study yields consistent results. Boys scored higher in reactive aggression than girls.

The present study contributes to the existing literature in important ways by considering the relation between instrumental aggression and violent video game use. In comparison to users of age appropriate violent games, users of age *in*appropriate violent video games scored higher in instrumental aggression. We assumed that video game characters in violent games serve as role models and provide a learning environment for children and adolescents that promote aggressive behavior as an instrument to obtain goals. However, our results reveal that only users who are confronted with an age inappropriate extent of violence score higher in instrumental aggression. Because our study is cross sectional and data is based on self reports, we suggest addressing instrumental aggression in further studies. Inconsistent with existing literature (Little et al. 2002), we found no gender differences in instrumental aggression.

The results regarding hostile attributional style are in line with results obtained by Krahé and Möller (2004). In all three groups of video game users, no differences in interpreting ambiguous situations were found. Other studies indicated a positive relationship between violent video game playing and hostile attributional style (Anderson et al., 2007; Kirsh, 1998), however it should be noted that we measured hostile attributional style with only one vignette in the present study.

In line with previous studies (Funk et al., 2004; Krahé & Möller, 2004; Möller & Krahé, 2009) our results showed a positive relation between normative beliefs about the acceptability of physical aggression and violent video game playing. Users of age inappropriate violent video games reported a greater acceptance for physical aggression as an adequate response to provocations compared to users of age appropriate violent video games. This result points to the crucial role of cognitions in the development of aggression. Similar to previous studies, boys scored higher in normative beliefs about aggression than girls (e.g. Crick et al., 1996; Krahé & Möller, 2004).

In summary, results of the present study highlight the importance of age appropriateness in violent video game playing.

PRACTICAL IMPLICATIONS

First, *information about video game rating systems* should be given to youth.

Playing video games is most attractive for youth between the age of 10 and 14 years. Due to the fact that many video games also contain age-inappropriate violent content, this age group is often confronted with this kind of video games. Thus, it is important to educate adolescents how to deal with video games by drawing attention to the importance and meaning of the PEGI ratings printed on the games. Youth also should be informed about alternative video games that are non-violent and age-appropriate.

Second, *parental awareness and control* should be raised. Video game use is less supervised by parents during pre-adolescence. Due to the fact that pre-adolescents underestimated the violent content of video games in comparison with the categorizations provided by PEGI, our study points to the necessity of parental awareness and parental control of adolescents' video game use.

Parents as well as the broader public should be educated to consider the PEGI labels printed on the games before buying them for children and youth. We recommend PEGI as a rating system because it is characterized by high visibility in 30 European countries. Our recommendation is further supported by studies indicating that parents want to be informed about age ratings and potentially harmful content of a video game (Nikken, Jansz, & Schouwstra, 2007). However, 48% of parents with children between two and 17 years have never used a video game rating system before buying (Rideout et al., 2005). Approximately 90% of adolescents indicated that their parents never controlled age ratings before buying a video game (Rideout et al., 2005), and only 1% of parents didn't allow the purchase of a game because of age inappropriate ratings (Walsh, 2002).

FURTHER RESEARCH DIRECTIONS

Further research should address three issues. First, we plead for the *application of a PEGI based categorization method for violent content:* With the PEGI based categorization method; similar results compared to former studies were obtained in our study. We suggest using the PEGI based categorization as an alternative method to evaluate violent content in combination with self-ratings and expert ratings. A multi-rating approach allows for a systematical comparison between violent content ratings, helps to detect discrepancies in violent content estimations, and contributes to overcome estimation biases. Second, we suggest *conducting longitudinal studies on age-inappropriate video game use in adolescents.* Only longitudinal studies allow conclusions about causality and long-term effects of age inappropriate video game use on aggression. Third, we recommend research into *both reactive and instrumental aggression.* Although there are opposite opinions concerning the dichotomy of reactive and proactive aggression (e.g., Buchman & Anderson, 2001) we argue that a differentiation of functions of aggression is useful and should be maintained.

CONCLUSION

Most youth in the age between 10 and 14 years play video games on a regular basis. A considerable number of video games contain violent content. Violent content in video games has often been linked to aggression – in public debate as well as in research,Our study did not find that violent video game playing *per se* was associated with aggression. Our results show that *age-appropriateness* matters. Only playing *age-inappropriate* video games was linked with aggressive behavior and aggressive cognitions. Thus, we would like to draw the attention of parents and educators to one major point, namely to consider both violent content and age appropriateness of games when buying them for youth.

REFERENCES

Anderson, C. A. (2004). An update on the effects of playing violent video games. *Journal of Adolescence*, *27*, 113–122. doi:10.1016/j.adolescence.2003.10.009

Anderson, C. A., & Bushman, B. J. (2001). Effects of violent video games on aggressive behavior, aggressive cognition, aggressive affect, physiological arousal, and prosocial behavior: A meta-analytic review of the scientific literature. *Psychological Science*, *12*, 353–359. doi:10.1111/1467-9280.00366

Anderson, C. A., Carnagey, N. L., Flanagan, M., Benjamin, A. J., Eubanks, J., & Valentine, J. C. (2004). Violent video games: Specific effects of violent content on aggressive thoughts and behavior. *Advances in Experimental Social Psychology*, *36*, 199–249. doi:10.1016/S0065-2601(04)36004-1

Anderson, C. A., & Dill, K. E. (2000). Video games and aggressive thoughts, feelings, and behaviour in the laboratory and in real life. *Journal of Personality and Social Psychology*, *78*, 772–790. doi:10.1037/0022-3514.78.4.772

Anderson, C. A., Gentile, D. A., & Buckley, K. E. (2007). *Violent video game effects on children and adolescents. Theory, research, and public policy*. Oxford, UK: University Press. doi:10.1093/acprof:oso/9780195309836.001.0001

Anderson, C. A., Shibuya, A., Ihori, N., Swing, E. L., Bushman, B. J., & Sakamoto, A. (2010). Violent video game effects on aggression, empathy, and prosocial behavior in Eastern and Western countries. *Psychological Bulletin*, *136*(2), 151–173. doi:10.1037/a0018251

Bandura, A. (1973). *Aggression: a social learning analysis*. Englewood Cliffs, NJ: Prentice-Hall.

Bandura, A. (1983). Psychological mechanisms of aggression. In Green, R. G., & Donnerstein, E. I. (Eds.), *Aggression: Theoretical and empirical views* (*Vol. 1*, pp. 1–40). New York: Academic Press.

Barlett, C. P., & Anderson, C. A. (2009). Violent video games and public policy. In Bevc, T., & Zapf, H. (Eds.), *Wie wir spielen, was wir werden: Computerspiele in unserer Gesellschaft. Konstanz: UVK Verlagsgesellschaft*.

Baron, R. A., & Richardson, D. R. (1994). *Human aggression*. New York: Plenum Press.

Bartholow, B. D., & Anderson, C. A. (2002). Examining the effects of violent video games on aggressive behavior: Potential sex differences. *Journal of Experimental Social Psychology*, *38*, 283–290. doi:10.1006/jesp.2001.1502

Bartholow, B. D., Sestir, M. A., & Davis, E. B. (2005). Correlates and consequences of exposure to video game violence: Hostile personality, empathy, and aggressive behavior. *Personality and Social Psychology Bulletin*, *31*(11), 1573–1586. doi:10.1177/0146167205277205

Berkowitz, L. (1989). Frustration-aggression hypothesis: Examination and reformulation. *Psychological Bulletin*, *106*, 59–73. doi:10.1037/0033-2909.106.1.59

Berkowitz, L. (1993). *Aggression: Its causes, consequences, and control*. New York: Mc Graw-Hill.

Buchman, D. D., & Funk, J. B. (1996). Video and computer games in the '90s: Children's time commitment and game preference. *Children Today*, *24*, 12–16.

Burks, V. S., Laird, R. D., Dodge, K. A., Pettit, G. S., & Bates, J. E. (1999). Knowledge structures, social information processing, and children's aggressive behavior. *Social Development*, *8*, 220–236. doi:10.1111/1467-9507.00092

Bushman, B. J., & Anderson, C. A. (2001). Is it time to pull the plug on the hostile versus instrumental aggression dichotomy? *Psychological Review*, *108*, 273–279. doi:10.1037/0033-295X.108.1.273

Bushman, B. J., & Huesmann, L. R. (2001). Effects of televised violence on aggression. In Singer, D. G., & Singer, J. L. (Eds.), *Handbook of children and the media* (pp. 223–254). Thousand Oaks, CA: Sage.

Bushman, B. J., Rothstein, H. R., & Anderson, C. A. (2010). Much ado about something: Violent video game effects and a school of red herring: Reply to Ferguson und Kilburn (2010). *Psychological Bulletin*, *136*(2), 182–187. doi:10.1037/a0018718

Card, N. A., & Little, T. D. (2006). Proactive and reactive aggression in childhood and adolescence: A meta-analysis of differential relations with psychosocial adjustment. *International Journal of Behavioral Development*, *30*(5), 466–480. doi:10.1177/0165025406071904

Card, N. A., Stucky, B., Sawalani, G., & Little, T. D. (2008). Direct and indirect aggression during childhood and adolescence: A meta-analytic review of gender differences, intercorrelations, and relations to maladjustment. *Child Development*, *79*(5), 1185–1229. doi:10.1111/j.1467-8624.2008.01184.x

Colwell, J., & Kato, M. (2003). Investigation of the relationship between social isolation, selfesteem, aggression and computer game play in Japanese adolescents. *Asian Journal of Social Psychology*, *6*, 149–158. doi:10.1111/1467-839X.t01-1-00017

Colwell, J., & Payne, J. (2000). Negative correlates of computer game play in adolescents. *The British Journal of Psychology*, *91*, 295–310. doi:10.1348/000712600161844

Crick, N., Bigbee, M. A., & Howes, C. (1996). Gender differences in children's normative beliefs about aggression: How do I hurt thee? Let me count the ways. *Child Development*, *67*, 1003–1014. doi:10.2307/1131876

Currie, C., Roberts, C., Morgan, A., Smith, R., Settertobulte, W., & Samdal, O. (2004). *Young people's health in context. Health Behaviour in School-aged Children (HBSC) study: international report from the 2001/2002 survey*. Kopenhagen: World Health Organization Regional Office for Europe.

Dill, K. E., & Dill, J. C. (1998). Video game violence: A review of the empirical literature. *Aggression and Violent Behavior*, *3*, 407–428. doi:10.1016/S1359-1789(97)00001-3

Dodge, K. A. (1991). The structure and function of reactive and proactive aggression. In Pepler, D., & Rubin, K. (Eds.), *The development and treatment of childhood aggression* (pp. 201–218). Hillsdale, NJ: Erlbaum.

Dodge, K. A., & Coie, J. D. (1987). Social information processing factors in reactive and proactive aggression in children's playgroups. *Journal of Personality and Social Psychology*, *53*, 1146–1158. doi:10.1037/0022-3514.53.6.1146

Dodge, K. A., Lochman, J. E., Harnish, J. D., Bates, J. E., & Pettit, G. S. (1997). Reactive and proactive aggression in school children and psychiatrically impaired chronically assaultive youth. *Child Development*, *74*, 374–393. doi:10.1111/1467-8624.7402004

Dollard, D. J., Doob, L. W., Miller, N. E., Mowrer, O. H., & Sears, R. R. (1939). *Frustration and aggression*. New Haven, CT: Yale University Press. doi:10.1037/10022-000

Durkin, K., & Barber, B. (2002). Not so doomed: Computer game play and positive adolescent development. *Journal of Applied Developmental Psychology*, *23*(4), 373–392. doi:10.1016/S0193-3973(02)00124-7

Eastin, M. S. (2006). Video game violence and the female game player: Self- and opponent gender effects on presence and aggressive thoughts. *Human Communication Research*, *32*, 351–372. doi:10.1111/j.1468-2958.2006.00279.x

Eron, L. D., Huesmann, L. R., Lefkowitz, M. M., & Walder, L. O. (1972). Does television violence cause aggression? *The American Psychologist*, *27*, 253–263. doi:10.1037/h0033721

Ferguson, C. J. (2007). Evidence for publication bias in video game violence effects literature: A meta-analytic review. *Aggression and Violent Behavior*, *12*, 470–482. doi:10.1016/j.avb.2007.01.001

Ferguson, C. J., & Kilburn, J. (2009). The public health risk of media violence: A meta-analytic review. *The Journal of Pediatrics*, *154*, 759–763. doi:10.1016/j.jpeds.2008.11.033

Ferguson, C. J., & Kilburn, J. (2010). Much ado about nothing: The misestimation and overinterpretation of violent video game effects in Eastern and Western nations: Comment on Anderson et al.(2010). *Psychological Bulletin*, *136*(2), 174–178. doi:10.1037/a0018566

Ferreira, P. R., & Ribeiro, J. L. P. (2001). The relationship between playing violent electronic games and aggression in adolescents. In Martinez, M. (Ed.), *Prevention and control of aggression and the impact on its victims* (pp. 129–135). New York: Kluwer.

Frindte, W., & Obwexer, I. (2003). Ego-Shooter – Gewalthaltige Computerspiele und aggressive Neigungen. *Zeitschrift für Medienpsychologie*, *15*, 140–148. doi:10.1026//1617-6383.15.4.140

Fromme, J., Meder, N., & Vollmer, N. (2000). *Computerspiele in der Kinderkultur*. Opladen: Leske + Budrich.

Funk, J. B., Bechtoldt Baldacci, H., Pasold, T., & Baumgardner, J. (2004). Violence exposure in real-life, video games, television, movies, and the internet: Is there desensitization? *Journal of Adolescence*, *27*, 23–39. doi:10.1016/j.adolescence.2003.10.005

Funk, J. B., Buchman, D. D., Jenks, J., & Bechtoldt, H. (2003). Playing violent video games, desensitization, and moral evaluation in children. *Applied Developmental Psychology*, *24*, 413–436. doi:10.1016/S0193-3973(03)00073-X

Gentile, D. A., & Gentile, J. R. (2008). Violent video games as exemplary teachers: a conceptual analysis. *Journal of Adolescence*, *37*, 127–141.

Gentile, D. A., Lynch, P. J., Linder, J. R., & Walsh, D. A. (2004). The effects of violent game habits on adolescent hostility, aggressive behaviors, and school performance. *Journal of Adolescence*, *27*, 5–22. doi:10.1016/j.adolescence.2003.10.002

Gentile, D. A., Saleem, M., & Anderson, C. A. (2007). Public policy and the effects of media violence on children. *Social Issues and Policy Review*, *1*, 15–61. doi:10.1111/j.1751-2409.2007.00003.x

Gentile, D. A., & Walsh, D. A. (2002). A normative study of family media habitus. *Applied Developmental Psychology*, *23*, 157–178. doi:10.1016/S0193-3973(02)00102-8

Hubbard, J. A., Dodge, K. A., Cillessen, A. H., Coie, J. D., & Schwartz, D. (2001). The dyadic nature of social information processing in boys' reactive and proactive aggression. *Journal of Personality and Social Psychology*, *80*, 268–280. doi:10.1037/0022-3514.80.2.268

Huesmann, L. R. (1988). An information processing model for the development of aggression. *Aggressive Behavior*, *11*, 13–24. doi:10.1002/1098-2337(1988)14:1<13::AID-AB2480140104>3.0.CO;2-J

Huesmann, L. R. (1998). The role of social information processing and cognitive schema in the acquisition and maintenance of habitual aggressive behavior. In Geen, R. G., & Donnerstein, E. (Eds.), *Theories, research, and implications for social policy* (pp. 73–109). San Diego, CA: Academic Press.

Huesmann, L. R. (2010). Nailing the coffin shut on doubts that violent video games stimulate aggression: Comments on Anderson et al. (2010). *Psychological Bulletin, 136*(2), 179–181. doi:10.1037/a0018567

Huesmann, L. R., & Guerra, N. G. (1997). Children's normative beliefs about aggression and aggressive behavior. *Journal of Personality and Social Psychology, 72*, 408–419. doi:10.1037/0022-3514.72.2.408

Huesmann, L. R., & Miller, L. S. (1994). Long-term effects of repeated exposure to media violence in childhood. In Huesmann, L. R. (Ed.), *Aggressive behavior: Current perspectives* (pp. 153–186). New York: Plenum Press.

Huesmann, L. R., Moise, J. F., & Podolski, C.-L. (1997). The effects of media violence on the development of antisocial behavior. In Stoff, D. M., Breiling, J., & Maser, J. D. (Eds.), *Handbook of antisocial behavior* (pp. 181–193). New York: John Wiley.

Huesmann, L. R., Moise-Titus, J., Podolski, C.-L., & Eron, L. D. (2003). Longitudinal relations between childrens' exposure to TV violence and their aggressive and violent behavior in young adulthood: 1977-1992. *Developmental Psychology, 39*, 201–221. doi:10.1037/0012-1649.39.2.201

Kauhajoki/Finland. *Gunman sprayed bullets in classroom and corridor, and threw petrol bombs.* Retrieved 24.09.2009, from http://www.hs.fi/english/article/1135239693467

Kirsh, S. J. (1998). Seeing the world through mortal kombat-coloured glasses. Violent video games and the development of a short-term hostile attribution bias. *Childhood, 5*, 177–184. doi:10.1177/0907568298005002005

Klimmt, C. (2004). Computer- und Videospiele. In Mangold, R., Vorderer, P., & Bente, G. (Eds.), *Lehrbuch der Medienpsychologie* (pp. 696–716). Göttingen: Hogrefe.

Krahé, B., & Möller, I. (2004). Playing violent electronic games, hostile attributional style, and aggression-related norms in German adolescents. *Journal of Adolescence, 27*, 53–69. doi:10.1016/j.adolescence.2003.10.006

Little, T. D. (2002). *Pathways of the forms and functions of aggression during adolescence.* Paper presented at the biennial meeting of International Society for Research on Aggression, Montreal, QC.

Little, T. D., Jones, S. M., Henrich, C. C., & Hawley, P. H. (2003). Disentangling the "whys" from the "whats" of aggressive behavior. *International Journal of Behavioral Development, 27*, 122–133. doi:10.1080/01650250244000128

Littleton/Colorado. (n.d.). *Columbine killers planned to kill 500.* Retrieved 27.4.1999, from http://news.bbc.co.uk/2/hi/americas/329303.stm

Lucas, K., & Sherry, J. L. (2004). Sex differences in video game play: A communication-based explanation. *Communication Research, 31*(5), 499–523. doi:10.1177/0093650204267930

Möller, I. (2006). *Mediengewalt und Aggression. Eine längsschnittliche Betrachtung des Zusammenhangs am Beispiel des Nutzungs gewalthaltiger Bildschirmspiele.* Unpublished doctoral dissertation, University of Potsdam, Germany.

Möller, I., & Krahé, B. (2009). Exposure to violent video games and aggression in German adolescents: A longitudinal analysis. *Aggressive Behavior*, *35*, 75–89. doi:10.1002/ab.20290

Nielsen Games. (2008). *Video gamers in Europe -2008*. Brussles: Interactive Software Federation of Europe.

Nikken, P., Jansz, J., & Schouwstra, S. (2007). Parents' interest in videogame ratings and content descriptors in relation to game mediation. *European Journal of Communication*, *22*(3), 315–336. doi:10.1177/0267323107079684

Olweus, D. (1993). *Bullying at school: What we know and what we can do*. Oxford, UK: Blackwell.

Orobio de Castro, B., Veerman, J. W., Koops, W., Bosch, J. D., & Monshouwer, H. J. (2002). Hostile attribution of intent and aggressive behavior: A meta-analysis. *Child Development*, *73*, 916–934. doi:10.1111/1467-8624.00447

Pan European Game Information. (n.d.). Retrieved 15.12.2009, from http://www.pegi.info/de

Polman, H., Orobio de Castro, B., & van Aken, M. A. G. (2008). Experimental study of the differential effects of playing versus watching violent video games on children's aggressive behavior. *Aggressive Behavior*, *34*(3), 256–264. doi:10.1002/ab.20245

Price, J. M., & Dodge, K. A. (1989). Reactive and proactive aggression in childhood: Relations to peer status and social context dimensions. *Journal of Abnormal Child Psychology*, *17*, 455–471. doi:10.1007/BF00915038

Prinstein, M. J., & Cillessen, A. H. N. (2003). Forms and functions of adolescent peer aggression associated with high levels of peer status. *Merrill-Palmer Quarterly*, *49*, 310–342. doi:10.1353/mpq.2003.0015

Rideout, V., Roberts, D. F., & Foehr, U. G. (2005). *Generation M: Media in the lives of 8-18 year-olds*. Washington DC: The Henry J. Kaiser Family Foundation.

Schiller, E.-M., Strohmeier, D., & Spiel, C. (2009). Risiko Video- und Computerspiele? Eine Studie über Video- und Computerspielnutzung und Aggression bei 12- und 16-jährigen Jugendlichen. *Schweizerische Zeitschrift für Bildungswissenschaften*, *31*(1), 75–98.

Sherry, J. (2007). Violent video games and aggression: Why can't we find links? In Preiss, R., Gayle, B., Burrell, N., Allen, M., & Bryant, J. (Eds.), *Mass media effects research: Advances through meta-analysis*. Mahwah, NJ: Erlbaum.

Smith, S. L., Lachlan, K., & Tamborini, R. (2003). Popular video games: Quantifying the presentation of violence and its context. *Journal of Broadcasting & Electronic Media*, *47*, 58–76. doi:10.1207/s15506878jobem4701_4

Thompson, K. M., & Haninger, K. (2001). Violence in E-rated video games. *Journal of the American Medical Association*, *286*, 591–598. doi:10.1001/jama.286.5.591

Thompson, K. M., Tepichin, K., & Haninger, K. (2006). Content and ratings of mature-rated video games. *Archives of Pediatrics & Adolescent Medicine*, *160*, 402–410. doi:10.1001/archpedi.160.4.402

VanOostrum, N., & Horvath, P. (1997). The effects of hostile attribution on adolescents' aggressive responses to social situations. *Canadian Journal of School Psychology*, *13*, 48–59. doi:10.1177/082957359701300105

Vitaro, F., & Brendgen, M. (2005). Proactive and reactive aggression: A developmental perspective. In Tremblay, R. E., Hartup, W. W., & Archer, J. (Eds.), *Developmental origins of aggression*. New York: The Guilford Press.

Vitaro, F., Brendgen, M., & Barker, E. D. (2006). Subtypes of aggressive behaviors: A developmental perspective. *International Journal of Behavioral Development*, *30*(1), 12–19. doi:10.1177/0165025406059968

Von Salisch, M., Kristen, A., & Oppl, C. (2007). *Computerspiele mit und ohne Gewalt. Auswahl und Wirkung bei Kindern*. Stuttgart: Kohlhammer.

Wallenius, M., Punamäki, R.-L., & Rimpelä, A. (2007). Digital game playing and direct and indirect aggression in early adolescence: The roles of age, social intelligence, and parent-child communication. *Journal of Youth and Adolescence*, *36*, 325–336. doi:10.1007/s10964-006-9151-5

Winnenden/Germany. German school gun man 'kills 15'. Retrieved 11.3.2009, from http://news.bbc.co.uk/2/hi/europe/7936817.stm

ADDITIONAL READING

Anderson, C. A., & Bushman, B. J. (2002). Human aggression. *Annual Review of Psychology*, *53*, 27–51. doi:10.1146/annurev.psych.53.100901.135231

Björqvist, K., Lagerspetz, K. M., & Kaukiainen, A. (1992). Do girls manipulate and boys fight? Developmental trends in direct and indirect aggression. *Aggressive Behavior*, *18*, 117–127. doi:10.1002/1098-2337(1992)18:2<117::AID-AB2480180205>3.0.CO;2-3

Cantor, J., & Wilson, B. J. (2003). Media and violence: Intervention strategies for reducing aggression. *Media Psychology*, *5*, 363–403. doi:10.1207/S1532785XMEP0504_03

Crick, N. R. (1997). Engagement in gender normative versus nonnormative forms of aggression: Links to social-psychological adjustment. *Developmental Psychology*, *33*(4), 610–617. doi:10.1037/0012-1649.33.4.610

Crick, N. R., & Dodge, K. A. (1996). Social information-processing mechanisms in reactive and proactive aggression. *Child Development*, *67*, 993–1002. doi:10.2307/1131875

Erdley, C. A., & Asher, S. R. (1998). Linkages between children's beliefs about the legitimacy of aggression and their behavior. *Social Development*, *7*, 321–339. doi:10.1111/1467-9507.00070

Flanery, D. J., Vazsonyi, A. T., & Waldman, I. D. (2007). *The Cambridge Handbook of Violent Behavior and Aggression*. New York: Cambridge University Press.

Funk, J. B., Buchman, D. D., & Germann, J. N. (2000). Preference for violent electronic games, self-concept, and gender differences in young children. *The American Journal of Orthopsychiatry*, *70*(2), 233–241. doi:10.1037/h0087738

Gentile, D. A. (2003). *Media violence and children. A complete guide for parents and professionals*. Westport, CT: Praeger.

Guerra, N. G., Huesmann, L. R., & Hanish, L. (1995). The role of normative beliefs in children's social behavior. In Eisenberg, N. (Ed.), *Social development* (pp. 140–158). Thousand Oaks, CA: Sage.

Jansz, J. (2005). The emotional appeal of violent video games for adolescent males. *Communication Theory*, *15*(3), 219–241. doi:10.1111/j.1468-2885.2005.tb00334.x

Kirsh, S. J. (2003). The effects of violent video games on adolescents: The overlooked influence of development. *Aggression and Violent Behavior*, *8*, 377–389. doi:10.1016/S1359-1789(02)00056-3

Kirsh, S. J. (2006). *Children, adolescents, and media violence: A critical look at the research*. Thousand Oaks, CA: Sage Publications.

Krahé, B. (2001). *The social psychology of aggression*. Hove: Psychology Press.

Slater, M. D., Henry, K. L., Swaim, R. C., & Anderson, L. L. (2003). Violent media content and aggressiveness in adolescents: A downward spiral model. *Communication Research, 30*, 713–736. doi:10.1177/0093650203258281

Vorderer, P., & Bryant, J. (2006). *Playing Video Games: Motives, Responses, and Consequences*. Mahwah, NJ: Lawrence Erlbaum Associates.

KEY TERMS AND DEFINITIONS

Aggression: Aggression is defined as 'any form of behavior directed toward the goal of harming or injuring another living being who is motivated to avoid such treatment' (Baron & Richardson, 1994, p.7). The term 'aggression' describes a certain behavior and not emotions, motives or attitudes. Aggressive behavior is conducted intentionally and has the goal of damaging a victim. The damage can be either physical or psychological (Baron & Richardson, 1994).

Hostile Attributional Style: In ambiguous situations, people with a hostile attributional style tend to interpret other's intentions as hostile (Crick & Dodge, 1996). If a child is pushed by another child on the playground and it is not obvious if this was on purpose, a child with hostile attributional style will always impute hostile intentions to the 'perpetrator'.

Instrumental Aggression: The concept of instrumental aggression is based on social learning theory (Bandura, 1973, 1983). To harm somebody else is a premeditated, calculated behavior that is used as an 'instrument' to reach particular goals. The dominant emotions involved with this type of aggression are pleasure and stimulation.

Normative Beliefs: Normative Beliefs determine for the individual which forms of behavior are appropriate and which are not justifiable. Normative beliefs are learned by observation, experience, or direct instruction. They can be specific for certain situations, like 'It is OK to insult someone if they provoked you first' or they can have general validity for an individual, like 'It is always OK to insult someone' (Huesmann & Guerra, 1997).

Overt Aggression: Overt aggression includes direct aggressive behavior like hitting, pushing or insulting another person. Overt aggression can be divided into physical and verbal aggression. Physical aggressive behavior, like hitting or pushing, is conducted with the intention of harming another person physically. The goal of verbal aggressive behavior, like insulting someone, is to harm another person in a psychological way (Crick,1997).

Reactive Aggression: The concept of reactive aggression has its roots in frustration-anger theory (e.g., Berkowitz, 1989, 1993; Dollard et al., 1939). To harm somebody else occurs as a "reaction" to a (perceived) provocation, threat or frustration and is usually accompanied by strong feelings of anger.

PEGI: PEGI (Pan European Game Information) is an age rating system for electronic games which is implemented in almost all European countries. The PEGI-system rates electronic games according to their age appropriateness and provides information about critical content in games. PEGI's symbols can be found on the back side of electronic games and announce depictions of violence, sex, discrimination, drug use, bad language, or gambling.

Violence: Violence is described as 'physical aggression that is so severe that the target is likely to suffer serious physical injury' (Gentile, Saleem & Anderson, 2007, p. 4). Violent media contain depictions of individuals who intend to harm other people. These individuals can be real persons or cartoon characters (Anderson & Bushman, 2001).

ENDNOTES

1 In Germany the "Unterhaltungssoftware Selbstkontrolle" is established.

[2] PEGI is established in Austria Denmark, Hungary, Latvia, Norway, Slovenia, Belgium, Estonia, Iceland, Lithuania, Poland, Spain, Bulgaria, Finland, Ireland, Luxembourg, Portugal, Sweden, Cyprus, France, Israel, Malta, Romania, Switzerland, Czech Republic, Greece, Italy, the Netherlands, Slovak Republic and the United Kingdom.

[3] Bullying is defined as a subcategory of aggressive behavior, that is characterized by intentional, repetitive harm doing towards a person or a group. Bullying always includes imbalance of power between involved persons (e.g., Olweus, 1993).

Chapter 19
Young People, Sexual Content and Solicitation Online

Kareena McAloney
Queen's University Belfast, Northern Ireland

Joanne E. Wilson
Queen's University Belfast, Northern Ireland

ABSTRACT

Young people can potentially be exposed to sexual material from a variety of sources, both accidentally and purposefully. One such source, the internet, plays host to a vast array of information and imagery, among which sexually explicit material and pornography are in high concentration. Indeed within this virtual catalogue of material it is possible to find both adult and child pornography, particularly if one is aware of the correct methods of accessing such content. This chapter provides an in-depth overview of current knowledge regarding young people's exposure to and experiences of sexual material and sexual predators online, including those particular aspects of young people's online interactions that make them vulnerable to receive unwanted sexual material and solicitation. The authors then discuss the use of the Internet for the sexual exploitation of children and young people both in the nature of sexual material to which they are exposed to online including the transmission of images of child pornography and molestation, the processes by which young people access sexual material online, the solicitation of children by sexual predators in targeting young people and how young people in turn come to interact with sexual predators online. Finally they address current mechanisms designed to protect children and young people as they engage in online activities.

INTRODUCTION

The Internet can be viewed as the cornerstone of the modern world. When we talk about the "Internet" we are referring not only to the World Wide Web but also methods of electronic communication such as e-mails, chat-rooms and instant messaging. Undoubtedly the Internet is a valuable resource that puts people in connection with vast amounts of information (and in some cases, misinformation). The Internet is also a source of communication and support to many people and can facilitate

DOI: 10.4018/978-1-60960-209-3.ch019

cognitive, social and physical development (Guan & Subrahmanyam, 2009). Yet, it also opens the door to the transmission of harm and exploitation across communities and cultures, in no small part due to the lack of geographical restraints implicit in the World Wide Web. Despite the widespread acceptance and use of the internet within society, it is also shrouded in mystery and fear, particularly in relation to sex. This situation in not eased by a general lack of research and investigation in this area with much of the existing literature drawing from early research, which may be particularly problematic in an area with such a rapid developmental trajectory as that of the internet. Another particular difficulty which hampers the contemporary research and investigation of sex on the internet, and in particular on child pornography is the highly sensitive, and often illegal, nature of the information under study which restricts the ability of researchers to conduct a full investigation, and enforces a reliance on secondary sources of information.

Sex on the Internet

The growth of the Internet has allowed the rapid exchange of information over geographically diverse populations. The majority of children in the United Kingdom (UK) have access to the Internet either at home (75%) or at school (92%; Livingstone & Bober, 2005). Although most children spend less time on the Internet compared to watching television – less than one hour (Livingstone & Bober, 2005), the little time that they do spend *surfing* instills fear in many adults as there is the possibility that they may be exposed to unwanted sexually explicit material and solicitation from deviant individuals. There are several reasons as to why such concerns may be justified. First children are digital consumers – they have the skills and knowledge base to rapidly and extensively *surf the net* (Peter & Valkenburg, 2006). They are also capable and rather skilled at avoiding detection. For example, qualitative research by Cameron et al. (2005) found evidence that children can evade their parents' efforts at control: "…indirect monitoring (e.g. checking history files) was easily thwarted by adolescents" (p.537). While for many children their first exposure to adult sexually explicit material occurs during adolescence offline via television or magazines, even the lyrics in songs subconsciously expose children to sexual innuendos, the lack of control and responsibility bestowed on the Internet means that children are likely to encounter sexual material online irrespective of whether they intended to or not. Children are also at risk of exploitation through the transmission and viewing of child victimization in child pornography; and its use by predators to attract and engage children and young people as potential victims for both sexual pleasure and financial gain (Esposito, 1998). The Internet can be used to facilitate this abuse in a number of ways and four key areas have been identified by which child molesters/sexual predators use the Internet: to disseminate sexualized images of children; to establish online networks with individuals with similar interests in children; to engage with children in an inappropriate and sexual manner; and to locate potential child victims for their sexual overtures and attentions (Durkin, 1997).

Exposure to Adult Sexual Material Online

The Internet is a vast source of sexual material including sexually explicit pictures, sexually explicit movies/clips and erotic sites (Peter & Valkenburg. 2006). Given the widespread availability of and accessibility to sexually explicit material online, it is not surprising that the Internet has been described as a "sexual medium" (p.178) with mounting concerns regarding children's exposure to such material (Peter & Valkenburg, 2006). Such fears are not entirely unfounded. In a study of 2,880 children aged 10-16 years of age, Flander, Cosic & Profaca (2009) found that 27% had been exposed to human nudity and sexual activity online. Peter and Valkenberg (2006), in a survey of 745 Dutch adolescents aged between 13 to 18 years of age, also found that 71% of males and 40% of females were exposed to some form of sexually explicit material. One in five children in Canada report having found undesirable sexual material online (Media-Awareness, 2000), while one quarter of 7 – 16 year olds in the UK have been upset by online material (Wigley & Clark, 2000).

Children are often viewed as passive recipients to the information on the Internet; a view which fails to consider the active role children can play as digital consumers, actively seeking out such material online, whether to satisfy curiosity or a desire for sexual stimulation. Children are likely to be exposed to sexually explicit material through their own access, indeed, some of the sites most frequently visited by children and young people include "Yahoo.com, Google.com, Hotmail.com, eBay.com, ESPN.com, and MTV.com" all of which feature or facilitate the viewing of material which can be interpreted sexually (Escobar-Chaves et al., 2005, p.319). Ybarra and Mitchell (2005) reported that 68% of children from the Youth Internet Safety Survey intentionally sought to expose themselves to sexually explicit material both on- and off- line. Ybarra and Mitchell (2005) also reported evidence that those seeking out sexually explicit material were twice as likely to be older children (20% versus 8%; Ybarra & Mitchell, 2005). Where younger children did seek out sexually explicit material they were more likely to report traditional means of exposure lending support for their argument that those seeking out sexually explicit material "are simply age-appropriately curious about sex" (p.483). Part of the explanation is that there is a developmental curve such that the increase in age and biological changes that accompany puberty is matched with an increase in the young person's interest in sex and involvement in sexual activities. The teenage years represent an important and critical developmental period characterized by increased risk taking, experimentation and identify exploration (Erikson, 1968) particularly in relation to their sexuality and sexual identity (Arnett, 1995). The extent to which exposure to sexually explicit material online is detrimental to young people remains unclear. This may be due to varying definitions as to what constitutes sexually explicit material, lack of clarity as to what aspects of sexually explicit material are harmful, distressing and even illegal, and the lack of censorship regarding such material online (Livingstone, 2003). Indeed, while it is generally thought that exposure to sexually explicit material online during adolescence may have an adverse impact on a child's behaviour and sexual development, there is a lack of research accessing exposure to such material online with the focus very much on television exposure. Rideout (2001) suggests that any changes observed in older teenagers attitudes and beliefs towards sex are likely to be compounded by other media influences such as magazines and TV. Of this available evidence it would appear that adolescents' actual experience is positively related to their increased exposure particularly via the television. In a review of empirical research on the role of the entertainment media (i.e. TV, magazines, films, sitcoms, magazine advertisements, music videos) in the sexual socialisation of children in the USA, Ward (2003) found that exposure to media such as soaps and music videos was occasionally related to greater levels of sexual experience. Wider cultural shifts regarding roles and expectations of men and women are also likely to have an influence such as the shift towards open discussions about sex, engaging in promiscuous sexual acts, and acceptance of casual attitudes towards sex as the norm. However, evidence supporting this remains a *slow-burner* and much of the research evidence base linking exposure to sexually explicit material online to resulting behavioural change is anecdotal. However there are a few noteworthy exceptions (Ybarra & Mitchell, 2005; Peter & Valkenburg, 2006). For example, Braun-Courville and Rojas (2009) found that adolescents who were exposed to sexually explicit websites were more likely to have multiple lifetime sexual partners, and engage in substance use.

Exposure to Pornography Online

Pornography has existed for a very long time, reflecting human curiosity and interest in the human form, and in the display of sexuality. However, as Chatterjee (2000) points out the Internet has revolutionized the pornography industry by facilitating anonymity of sources and ease of access for purveyors. Jenkins (2001) tracks the evolution of pornography transmission from nudes in paintings, to photographic images, video footage, and the use of Bulletin Based Sites on modem connected computers in the late 80s, to the ever evolving use of the Internet. According to the 2006 Internet Filter Review (Ropelato, 2007), around 12% of websites contain pornography (approximately 4.2 million), and around 100,000 websites host child pornography. Yet, despite some notable exceptions, "Internet pornography, and especially child pornography is rarely the topic of academic discourse" (Adams, 2002, p.135). This is surprising given that one in three teens have viewed pornography online (Kaiser Foundation, 2000) and Wolak, Mitchell and Finkelhor (2007) report in their 2005 survey of 1500 American youth aged 10 - 17, 42% had been exposed to online pornography within the preceding twelve months; with the average age of first exposure is around 11 years of age (Ropelato, 2007). While Wolak et al. (2007) report that the majority (66%) of young people's exposure was unwanted, a sizeable minority (34%) actively sought out exposure to pornography online. Indeed, there is evidence that males are more likely to seek exposure to adult pornography (Ybarra & Mitchell, 2005; Peter & Valkenburg, 2006).

As noted by Carroll et al. (2008) Internet pornography has become an integral aspect of the lives of adolescents and emerging adults. However, children and young people may not always seek out sexually explicit material online and their exposure to such material instead may be unsolicited. In these instances the young people are exposed to unwanted sexual material. Mtichell, Finkelor and Wolak (2007a) examined trends in sexual solicitation, harassment and unwanted exposure to pornography online between 2000 and 2005. They found that the overall incidence has decreased but that an increase in unwanted exposure to pornography was apparent for 10-12 and 16-17 year olds, boys, whites and non-Hispanic youth. Unwanted exposure to adult pornography online occurs when pornography is distributed to the young person without them playing an active role in soliciting or receiving it. McAlinden (2006) notes that sex offenders often present potential child victims with pornographic material in a bid to sexualize their interactions. Similarly Langevin and Curnoe (2004) explored the use of pornography among over five hundred sex offenders and reported that 55% of offenders showed their victims pornographic material, and this was most likely among those with child victims rather than adult victims. Mitchell, Wolak and Finkelhor (2007) found a significant increase in the number of 10-12 year olds and 16-17 year olds reporting unwanted exposure to online pornography between 2000 and 2005. While many issues are raised regarding the viewing of adult material by youth, this is not the only type of sexual material available on the Internet. Child pornography represents a particularly disturbing niche within the pornography market, and one which is substantively distinct from adult pornography (Adams, 2002).

Child Pornography on the Internet

Edwards (2000) defines child pornography as a "record of the systematic rape, abuse and torture of children on film and photography and other electronic means" (p.1). Although the particular nature of the Internet makes it difficult to accurately assess the extent of child pornography on the Internet (Taylor & Quayle, 2003), the existing evidence reveals a substantial volume of child focused pornographic imagery and content online. The extent of child pornography online rose drastically during the 1990s (Taylor, et al, 2001). Yet between 2000 and 2005, the overall incidence of sexual solicitation, harassment and unwanted exposure to pornography online has decreased (Wolak, et al, 2007). Undoubtedly, the early rise resulted in public angst and concern with implications for legislation, government enforcement and policy to manage and control the growing trade of online child pornography and despite the lack of empirical evidence, policy regarding child pornography online continues to emerge and the issue continues to remain top of the government agenda. For example, over the last two decades there has been a dramatic surge in (1) the number of laws and (2) changes to legislation to address child abuse across countries including the development of: the Audiovisual Media Services Directive (adopted in 2007), the Council of Europe Convention on the protection of children against sexual exploitation and sexual abuse (2007), the European Union 'Guidelines for the Promotion and Protection of the Rights of the Child (2007), the Revised Council Framework decision on combating the sexual abuse, sexual exploitation of children and child pornography (2009) and the European Convention on Human Rights (ECHR).

Generally speaking online child pornography exists in three forms – 'barely legal' pornography which features adults portrayed as teenagers; 'falsified' child pornography, which consists of sexualized images of children which are either computer-generated or those which are created from the imposing of a child's image onto a second image; and authentic child pornography which features children and underage teenagers in sexualized contexts or situations. All three types of child pornography raise significant concerns regarding their dissemination within the public sphere, and regarding the well-being of children. While 'barely legal' pornography is often considered legal, given that the subject can be verified as an adult, 'falsified', and authentic child pornography, are illegal in the majority of Western societies due to the representation of victimization of children.

Child pornography is not restricted to one particular age group of children, images of infants, toddlers, school-age children and teens can all be found in abundance on the Internet, the content varying from everyday images of bathing to images of children being sexually assaulted by adults and involved in bestiality (Beech, et al, 2008). Mitchell, Wolak and Finkelhor's (2005) study of online sex offenders found that the majority of child pornographic images collected by these individuals concerned children aged between 6 and 12 years (82%), although images of younger children were also present. Images tended to show explicit images of genitalia or specific acts (92%), penetration of the child (80%), and other sexual contact with adults (71%), in a significant minority of cases (21%) images of bondage, rape, torture or other violence were recorded. Frei, Erenay, Dittmann and Graf's (2005) study of child pornography involving 33 men convicted of child pornography offences revealed that 45% of images documented a serious assault and 27% of images depicted sadism or bestiality.

Kierkgaard (2008) identifies three main methods which facilitate the distribution of sexualized child content online - commercial websites, personal websites, and peer-to-peer file sharing. Jenkins (2001) counters claims by law enforcement and government agencies that paedophiles trade images in chat rooms and by email stating that such mechanisms are counterproductive to the nature of child pornography networks, but instead identifies four main mechanisms - "newsgroups (Usenet); corporate-linked "communities"; web-based bulletin boards; and closed groups"(p.53). Newsgroups are essentially commercial websites which host large volumes of material along central themes. Jenkins (2001) identified two main types of content on newsgroups - images and story boards hosting written descriptions of sexual fantasies and actual actions on children. Newsgroups are by nature difficult to access, not normally identified in a standard search on a search engine, and in most cases require some form of payment or subscription in order to access the content. Given the paper trail left by a credit card payment to such an organization it is difficult to imagine how law enforcement agencies are not better able to access complete lists of paying paedophiles, however as Jenkins (2001) describes, people with an interest in this material become very skilled at side-stepping processes which lead to their identification, with the creation of false identities and email addresses, and the development of hacking skills to allow them to access material for free.

Bulletin boards appear to have been central to the organization of child pornography. These boards do not provide pornographic images, but instead provide Uniform Resource Locators (URLs) with descriptions of hosted content, followed by passwords, so that other members of the board can access material on an external site. As Jenkins (2001) describes them "Though the boards forbid the posting of visual materials, they nonetheless act as guideposts to actual images, operating on a global scale and freely crossing international boundaries and jurisdictions" (p.67). In addition these boards provide an open forum for those with an interest in child pornography to discuss their particular sexual interests, to network with likeminded individuals, to access support for sexual interests in children, and to request links which display images which are of specific interest to them.

Closed groups or *paedo* rings are described by Jenkins (2001) as the networks of the elite, and among these circles peer-to-peer file sharing is most likely. While normally such activity is shunned by child pornographers, the members of closed groups tend to be experienced members of bulletin boards, with established histories of participating and posting material, thus making them less likely to be masquerading law enforcement personnel, and also tend to have extreme *hard core* pornographic tastes, often involved directly in the molestation, abuse and rape of children and subsequent production of child pornography. Several examples of these paedo rings have been identified in recent years, among the most infamous being the Wonderland Club, an international ring spanning thirteen countries. The Wonderland Club was the target of an international policing operation which uncovered 200 individuals trading in child pornography, with 750,000 images and 1,800 video-taped scenes of no less that 1,200 children ranging in age from infancy to teens. The Wonderland Club was tightly administered and guarded by its members, with potential new members strictly vetted, and asked to provide references and a personal, available catalogue of at least 10,000 images of interest (Panorama, 1998 - 2001).

In contrast to child pornography distribution which is facilitated in covert networks, solicitation and predation requires direct access to children, occurring in chat rooms and social network sites, as well as instant or private messaging and emails. However child pornography can be employed to seduce or blackmail a child in order to make contact with them (Frei et al., 2005). Mitchell, Wolak and Finkelhor (2005) report in their study of online sex offenders that 52% of offenders who solicited a child and 41% of offenders who solicited an undercover police or law enforcement agent possessed child pornography.

Solicitation and Perpetration of Sex Offences Online

Much of the growing concern around child and teenage Internet use is the potential for exploitation either through unwanted or inappropriate sexual contact made by predatory adults. In theory the internet provides access to an infinitely large number of potential victims that can be *groomed* for sexual contact offline (See Buschman &

Bogaerts, 2009; Wolak et al., 2004; Wolak et al., 2003). Mechanisms such as chat rooms offer a way to meet potential victims. The *assurance* of anonymity allows such individuals to pose as children and communicate with their victims in turn establishing trust (Quayle & Taylor, 2001).

According to the research literature adolescents are more likely than younger children to be solicited online (Finkelhor, Mitchell & Wolak, 2002), particularly as they are more likely to instigate or participate in an online discussion threads about sex or relationships (Magid, 1998). Ybarra and Mitchell (2008) report that 15% of 10 - 15 year olds surveyed had received an unwanted sexual solicitation whilst online. Mitchell, Finkelhor and Wolak (2007) report that in a national sample of American youth 1 in 25 had received an online request to provide a sexual picture of themselves in the year prior to the survey, and such requests were more likely to be made during online communications with adults. Between 1st July 2000 and 30th June 2001 the US law enforcement agencies reported almost 1,000 arrests for Internet sex offences against minors (Wolak, Mitchell & Finkelhor, 2003). It is worth noting that between 2000 and 2005 in the US there was an overall decrease in youth solicitation (Mitchell, Wolak and Finkelhor, 2007b), and more recent findings from The Growing Up with Media survey revealed that most adolescents, contrary to popular belief, are not involved in harassment or unwanted sexual solicitation as either victim or perpetrator (Ybarra, Espelage & Mitchell, 2007). Yet there is evidence from both the UK and the US that crimes against children are increasing, and that adult sex offenders are using the Internet as a viable means to gain access to their potential victims. Ybarra and Mitchell (2008) found that among 10 to 15 year old American youths who had received an online solicitation, instant messaging was the most common method of solicitation (43%) followed by chat rooms (32%), social networking sites (27%) and emails (22%), only a small minority of youth reported being the target of a sexual solicitation in a blog (6%). There is evidence that adolescent *bloggers* are at increased risk of harassment online regardless as to whether or not they interacted with others online which suggests that simply being online and blogging renders individuals susceptible to online harassment (e.g. Mitchell, Wolak & Finkelhor, 2008). However there are some studies which indicate that social networking sites are not associated with unwanted sexual solicitation or harassment, suggesting that interventions should focus on the youth as opposed to specific internet sites such as Facebook (Ybarra & Mitchell, 2008). For example, in a sample of 1,588 youth, Ybarra and Mitchell (2008) found that 15% reported unwanted sexual solicitation and 33% reported online harassment but only 4% of unwanted sexual solicitation and 9% of online harassment reported an incident on a specific social networking site.

Perhaps more worryingly however, are the intentions of young people when they go online and how certain actions make them more vulnerable to solicitation by potential predators. Subrahmanyam et al. (2000) report that adolescents often visit chat rooms with the intention of engaging with strangers. This is particularly concerning when one considers that chat room users are four times more likely than non-users to be exposed to an unwanted online sexual advance (Mitchell, Finkelhor & Wolak, 2001). Furthermore, online interactions are particularly problematic as they generally involve revelations of highly personal information to unknown individuals. Particularly striking is the apparent naivety of young children in disclosing information regarding their personal identity. Livingstone (2003) reported evidence from the NOP Kids.net survey which found that 20% of children would give out their home address and 14% would give out their email address. Wolak, Mitchell and Finkelhor (2003b) report 14% of 10 to 17 year old Americans surveyed had formed a close romantic relationship with someone they met online. While Internet relationships may be beneficial to teens in allowing them to explore their sexuality and emotional commitment while retaining their anonymity and detachment (Clark, 1998) they can present an opportunity for potential predators to engage with young people and forge relationships with them.

Mitchell, Finkelhor and Wolak (2007b) suggest that adolescents were 1.7 times more likely to report more aggressive solicitations particularly if they were female, talking to people they met online, including talking about sex, using chat rooms to name a few. Unwanted sexual solicitation has been linked with both depression and offline victimisation and higher rates of substance use (Mitchell, Finkelhor, and Wolak, 2007; Mitchell, Finkelhor & Wolak, 2001; Mitchell, Ybarra & Finkelhor, 2007). For those who are molested as a result of their solicitation the consequences again are profound for their well-being, with higher rates of anxiety, depression, suicide ideation and substance use often reported (Dombrowski, Ahia & McQuiillan, 2003; Oddone, Genuis & Violato, 2001). Victims of online solicitation are also more than two times likely to report depression and high levels of substance use (Mitchell, Ybarra & Finkelhor, 2007). Inevitably all youth are susceptible to solicitation online however more research is needed to understand what factors place individuals at greater risk and to help identify effective interventions (Guan & Subrahmanyam, 2009).

The Perpetrators

"Sexual predators are a heterogeneous group, and as a result it is difficult to define a typology of the sexual predator" (Dombrowski, LeMasney, Ahia & Dickson, 2004, p. 66). This becomes somewhat more difficult when one considers the different categories of perpetrators, and the different victims they target. We have concentrated on two main types – the child pornographer, and the online predator who solicits young people. Although it is worth bearing in mind that these categories are by no means mutually exclusive, on a superficial level there are some distinguishing characteristics in terms of their characteristics, their victims, and the mechanisms by which they use the internet to victimise children.

Describing a 'typical' child pornographer is not an easy task, in part because the existing research base examining child sex offenders does not tend to distinguish between pornographers and molesters and also because as Jenkins (2001) points out accessible information "tells us about those inept and seemingly atypical offenders who fail to take the obvious precautions and who get caught" (p.13). According to some researchers the profile of online predators and their victims is inaccurate and internet crimes involving children and young people are more consistent with a model of rape than child molestation (Wolak, Finkelhor, Mitchell & Ybarra, 2008). Media reports often depict apprehended sex offenders, both pornographers and molesters, as 'loners', 'misfits' and intellectually slow (Panorama, 1998 - 2001). Indeed much of the existing research literature on sex offenders (pornographers and molesters) does suggest that in addition to being predominantly male, sex offenders tend to have a lower than average IQ (Kalichman, 1991), be in lower socio-economic status groups (Abel et al., 2001), and be victims themselves. However, online child pornographers may not be well represented by such a typology. Jenkins (2001) asserts that the world of online child pornography is one of substantial technological sophistication of which the purveyors of images and network members must be cognizant in order to survive and evade detection. Frei et al.'s (2005) study of 33 individuals convicted of online child pornography offences revealed a substantially different typology among this group. While all were adult males, ranging between 25 and 69 years of age, almost all were employed, a third of whom were in respectable, professional occupations. Forty percent had children themselves and almost three quarters (70%) had no previous offending histories. Furthermore while one third of these men had never been in an intimate romantic relationship in their lifetime two thirds did report having been in an intimate relationships including marriage.

Evidence from qualitative work suggests that the standards of the pedophile culture justify their behaviour and attitudes to sexual relationships with children and young people both off- and online (Holt, Blevins & Burkert, 2010). One potential source of relationship for the child pornographer is other pornographers, or the 'loli/boy lover' community. Loli-lover is a particular term of reference to those with an interest in young girls, while 'boy-lover' describes those with an interest in young boys. Many have identified a tendency among child molesters and child sex-offenders to develop relationships with similarly inclined individuals who can support their beliefs and actions (Durkin & Bryant, 1999; Ward & Hudson, 2000). Malesky and Ennis (2004) covertly studied interactions between members of a 'boy love' network over a seven day period and reported that over one fifth of all comments contained justification for their attitudes, beliefs and actions towards children. In discussing the group trading in the Wonderland Club Adams describes

the problem was not just the trading of images, but also the way that paedophiles had an easy way to contact each other and to reinforce their beliefs that sex with children was not wrong, to promote the ghastly idea that somehow these children were 'in relationships' with adults. (Adams, 2002, p.140).

Of particular note is the 'collector' trait of many paedophilies, with images much sought after in the loli-lover and boy-lover communities and members striving to complete particular collections, bargaining and exchanging with fellow network members to complete the series. Taylor, Holland and Quayle (2001) suggest that collections can be characterized in one of two ways - a thematic series of images depicting a particular theme, or act, and a narrative series dedicated to a particular child or group of children. Indeed some of the most infamous collections have been narrative in nature such as the KG and KX collections - KG depicts sexualized nude images of toddlers at a kindergarten/pre-school, and KX includes more explicit images of these toddlers being sexually abused by adults (Taylor, Holland and Quayle, 2001). Similarly, the infamous 'He-lo' (Helena -lolita) series portrays images of seven year old Helena, nude and engaged in sexual intercourse with a young boy of similar age. As well as images of sex involving both children and an adult man. The He-lo series is unfortunately thought to be only a small section of a larger collection which documents Helena's abuse from a toddler to around the age of 12 (Jenkins, 2001).

Elliott, Browne and Kilcoyne (1995) report that sex offenders range in age from the young adult to the elderly, but with a particularly high proportion in the thirty and early forty age bands. Traditionally the majority of sex offences against children are perpetrated by someone known to them or their family (Fieldman & Crespi, 2002), with comparatively fewer acts by strangers (Snyder, 2000). However the global connectivity of the Internet appears to facilitate greater 'stranger danger' with 97% of youth in Mitchell, Finkelhor and Wolak's (2001) study reporting solicitation by a stranger. While most sexual solicitations and offences against children are perpetrated by adults, it is also important to recognize the potential of children to sexually abuse their peers, and younger children. Estimates in the United States suggest that as many as one third of child molestations are perpetrated by children (Davis & Leitenberg, 1987; Fieldman & Crespi, 2002).

The Child Victims

Notwithstanding the implications of the availability of child pornography for viewing by other children and adolescents in their online interactions, it is important to frame child pornography on the Internet, not just as a problem for those using the website, but also (excepting those cases where the images are computer generated) as a visual record of an act or multiple acts of abuse against a child or children. While many purveyors of child pornography may defend their actions in an 'at least I'm not touching' mantra, it is inescapable that for such material to be available a child must be harmed and abused. A report by the IWF (2009) stated that 69% of those being exploited were between the ages of 0 and 10 years. Furthermore, for 58% the severity of the abuse displayed reached level 4 (penetrative sexual activity involving a child/children or both children and adults) or 5 (sadism or penetration of or by an animal) according to the Sentencing Guidelines Council (IWF, 2009).

Sexual abuse has considerable ramifications for the health and well-being of victims, both in the short- and long- term (Dombrowski, Ahia & McQuiillan, 2003; Oddone, Genuis & Violato, 2001). As Cicetti and Toth (1995) have pointed out sexual victimization of a child can result in significant developmental problems which can have lasting implications. Victims of sexual abuse are also more likely to report mental health disorders and substance use disorders (Dombrowski, Ahia & McQuiillan, 2003; Oddone, Genuis & Violato, 2001). However, for those who have their abuse documented and made public to others there may be even further consequences, Lanning (1984) reports that the permanency of pornographic images may hold considerable repercussions for the child victim. Adams (2002) cites a television interview between a reporter and police personnel following the discovery and arrest of paedophiles involved in the 'Wonderland' series in which the effect of this permanence is discussed, the reporter states "The policemen who patrol the Internet still see the faces of hundreds of Wonderland children. They are out there forever" and police representatives comment "their abuse is going to continue for the rest of their life. That documentation of their abuse is going to be part of their life forever" (Panorama, Transcript of 'The Wonderland Club' p. 12, cited by Adams, 2002, p. 46). The literature consistently reports that victimized youth are significantly more likely to experience further victimization in their lives, and this is also true in online interactions. Among online solicited youth there is a higher prevalence of prior offline sexual and physical abuse (Mitchell, Finkelhor and Wolak, 2007b). As is so often the case those children most at risk of such victimization are less likely to have mechanisms within their lives which can identify and address the victimization (Glasser, et al, 2001).

As noted by Wolak, Finkelhor, Mitchell and Ybarra (2008) Internet initiated sex crimes against children generally concerns a more restricted age range of children than in offline child molestation. Victims of online solicitation are generally in or approaching their teens (Wolak, Finkelhor & Mitchell, 2004). Aside from the nature of adolescent activities online which may act to facilitate contact with potentially abusive adults there are a number of characteristics which appear to increase a particular teenager's vulnerability to sexual solicitation. Wolak et al. (2008) have identified females, particularly those who are sexually active early; homosexual or sexually unsure males; and victimized young people. Female youth tend to receive online sexual solicitations more than males (Mitchell, Wolak & Finkelhor, 2007). Wolak, Mitchell and Finkelhor (2004) reported that 75% of victims in their study of 129 Internet initiated sex crimes were female teens between 13 and 15 years of age. Contrastingly boys appear to be less likely to be sought out for sexual solicitation 25% of victims in the Wolak, Mitchell and Finkelhor (2004) study and Wolak et al. (2008) suggest that the activity of these young boys online, suggests that they were either homosexual or sexually confused.

Younger children are currently much less likely to be using Internet technology, and subsequently at less at risk for online solicitation, but rather face risk of molestation from offline sources. The fact that most recipients of online sexual solicitations are adolescent has some important implications for our understanding of the processes of online sexual solicitation and 'relationships' with adults. It is very important to appreciate that young people are not necessarily passive recipients of sexual overtures online, but rather much of their online behaviour can facilitate contact with potential sex offenders. According to work by Subrahmanyam and Greenfield (2008) the potential for contact with strangers is highly dependent on the type of online communication and the context of use, with chat room interactions, and social networking sites most likely to involve strangers while instant messaging more commonly used to connect with known offline contacts. However it is important to bear in mind that children's technological skills are becoming more advanced at earlier ages, and that this may substantially impact on the profile of young victims of solicitation. It remains to be seen if younger children's growing presence on and familiarity with the internet facilitates greater solicitation of this age group by predators.

Protecting Children Online

Sexually abusive and exploitative practices towards children (e.g. child prostitution, child sex tourism and child pornography), are not new and remain prevalent on a worldwide scale. For example, approximately one million children, of whom the majority is girls, are victims of sex trafficking every year (UNICEF, 2008). Protecting children as they engage online is an important part of the United Nations Convention on the Rights of the Child (UNCRC) which requires governments to safeguard children from all forms of sexual abuse and exploitation including child pornography and trafficking (UNICEF, 2008). Yet, the criminalization of such material and the prosecution of those involved are problematic and difficult to execute. The main difficulty is that "the production of child abuse material may take place in one country, the distribution may be hosted in a second and the material may be downloaded from all other the world" (European Commission, 2008, p.4). Furthermore, the surge in legislation has done little to facilitate apprehension and prosecution of criminals. For example, from all 187 countries of Interpol, 93 countries have no legislation targeting child pornography whilst a further 36 do not criminalise possession irrespective of their intention to distribute such material (ICMEC, 2008). Furthermore many legal issues centre on possession, distribution and production of child pornography (Taylor et al, 2001). Other difficulties hampering the criminalization of child pornography and sexual solicitation of children on an international scale include the lack of a global consensus on when childhood ends and adulthood begins. The result is that there exist different ages of consent, and different age thresholds for the prosecution of child pornography and solicitation offences both within and between countries. Another concern relates to the way in which documents operationalise sexual abuse and exploitation. Usually, such documents state that maltreatment involves (1) engagement of the child in real or simulated sexual activities, (2) representation of sexual parts of the child for sexual purposes (e.g. UN optional protocol and sentencing guidelines). This definition suggests that such material is harmful, criminal and deserving of prosecution. However, it ignores large amounts of other potentially harmful material to which children may be exposed to on the Internet and other audio/visual mediums which should also be criminalized (Quayle, Loof & Palmer, 2008).

In their desire to unite countries in combating the problem, the UN has organized a number of world Conferences against the sexual exploitation of children and adolescents, to mobilize countries into taking action to safeguarding the rights and needs of the child against such acts of abuse. A number of events have been organized worldwide to target child abuse and exploitation. For example, in June 2009, the 'Nobody's Children Foundation' initiated and coordinate a joint venture between ECPAT (End Child Prostitution, Child Pornography and the Trafficking of Children for

Sexual Purposes) and a number of organizations responsible for assisting abused children and children at risk in Poland (CANEE, 2009a). In Poland a number of organizations have joined forces to establish a Polish network against the commercial sexual exploitation of children (CANEE, 2009b). Finally the Safer Internet Centers in Poland and Germany are worked together to organize the 3rd International Conference entitled 'Keeping Children and Young People Safe Online' in Warsaw in September 2008 (CANEE, 2008a).

Work from the second World Congress on sexual exploitation of children recommended that more hotlines are established to curb potential harm and victimization online. Since then the International Association of Internet Hotlines (INHOPE) has grown from 15 members in 2001 across 12 countries to 28 by 2008 (Quayle et al., 2008). Furthermore, INHOPE appears to be making progress on tackling the issue of the sexual exploitation of children having made 6,400 reports to the police in the October to December of 2006, but reported receiving little feedback from the law enforcement agencies and some stated that reports were not forwarded as they lay outside their control (Quayle et al., 2008). There have also been changes in practice and in the UK, the Internet Watch Foundation (IWF), a founding member of INHOPE, has been established to enable the public and IT professionals to report potentially criminal and reprehensible material encountered on the Internet. To date, IWF have successfully reduced images of child sexual abuse (of what is known) that were hosted in the UK from 18% in 1997 to less than 1% since 2003 (IWF, 2009). However, there is still much work to be done. For instance, 81% of domains depicting images of child sex abuse are live for less than 100 days. Frequent hopping across servers and countries to avoid detection and potential prosecution for those involved (much like a game of cat and mouse) continues to pose great challenges in managing the sexual abuse and exploitation of children and young people online. Furthermore, this process is further hampered by different jurisdictions and laws and thus requires international cooperation if it is to be a success since the IWF have no remit in countries outside of the UK and thus can only share its relevant intelligence with the hotline in the hosting country (IWF, 2009).

Fuelled in part by a moral obligation and in part public panic, organizations have started to produce material to educate young children about how to be safe online. In 1999 the European Commission developed the 'Safer Internet Programme' which is designed to protect children online by raising awareness through various initiatives and combating the display of illegal material. The UK has endorsed the EU Safer Internet Programme and on February 10th 2009, the Nobody's Children Foundation and NASK organized the Safer Internet Day in Poland (i.e. SaferInternet.pl) after implementing the EU's project (CANEE, 2009c). The Council of Europe has also developed an online game to teach children (7-10 years of age) in a fun and friendly way how to be safe online. Entitled 'Wild Web Woods' it is currently available in 20 languages such as Russian, Hungarian and Lithuanian (CANEE, 2008b). Education packages have been developed for different age groups for use in schools across Europe, USA, Australia and New Zealand. However, while much work and effort has gone into promoting safety online, many of these tools have yet to be evaluated (Quayle et al., 2008) and yet policy should be based on evidence. Furthermore, the extent to which they actually impact on and influence behavior, remains to be seen as they may just impact on attitudes and knowledge (Quayle et al., 2008). There is a need for future research to produce reliable and systematic evaluations of educational and awareness programmes which chart their effectiveness at behavior change in the youth (Quayle et al., 2003).

On a practical level Dombrowski et al. (2004) suggest a number of techniques which can be employed to assist in securing children's safety online. First, firewall and anti-virus software should be installed to prevent infiltration by potential predators seeking information which they can exploit in their interactions with children. There is evidence that preventive software was associated with significantly reduced risk of unwanted exposure for 10-12 year olds and 13-15 year olds, although not for 16-17 year olds (Ybarra, Finkelhor, Mtichell & Wolak, 2009) when stratified by age. Caregivers of younger children who want to reduce the incidence of unwanted exposure should consider preventative software. Second, they suggest a need for adults to better understand children's screen names, and the messages that these may send to adults with a sexual interest in children, a responsibility which should be carried by all in roles of responsibility towards children. They also emphasize the need for understanding between the child and care-giver as to acceptable online behaviour, and the creation of *contracts* which stipulate the responsibility of each in relation to online behaviour and relationships. The placing of the computer in a public place can be particularly important in assisting parents in monitoring children's online risk exposure, although as previously mentioned, children appear to be adept at evading such forms of monitoring and regulation. Finally Dombrowski et al. (2004) stress the need for action on discovering inappropriate actions by an adult towards a child, and advise for parents, care-givers and teachers to contact available hotlines to alert agencies to child pornography and episodes of solicitation, and to contact the Internet server who have a responsibility to protect their child users. To facilitate the reporting of abuse, the CEOP introduced an 'abuse button' in the UK and the Norwegian police have implemented a similar devise. Microsoft carries the button on Windows Live Messenger which allows individuals to report any suspicious and malicious behavior encountered when online (Quayle et al., 2008). Belguim and the Netherlands have both established websites on which concerned individuals can lodge concerns or complaints about child pornography. Networking sites are also becoming more responsible using moderators in chat rooms to block improper conduct (Quayle et al., 2008). There is evidence that monitored chats environments contain less explicit sexuality and fewer obscenities (Subrahmanyam, Smahel & Greenfield, 2006). Perhaps most importantly however, is that age and developmentally appropriate strategies should be developed which target children and young people directly and particularly those in high risk groups (Wolak et al., 2008).

CONCLUSION

This chapter sought to address several questions regarding young people's exposure to and experiences of sexual content and solicitation online including the nature of sexual material to which they are exposed to online; the processes by which young people access sexual material online; the nature by which potential predators may target young people and how young people in turn come to interact with sexual predators online. There is much debate as to the nature and extent of regulation (e.g. government, commercial, private) needed to protect children from exposure to sexually explicit material and from potential exploitation by deviant individuals. While the need to protect children from harm yet simultaneously allow their freedom of expression is juggled, there remains nonetheless a moral obligation that children are protected from predatory individuals and consequently legislation has arisen to manage this. However, to fully uphold children's rights and ensure the continued safety of children and young people as they participate online will require the commitment of everyone from the family to schools and other institutions. The simple reason being that for "as long as there is a sexual interest in a behavior or type of person, that will lead to commodification and commercialization, in the form of prostitution and pornography" Jenkins (2001, p.30).

As already noted despite the widespread acceptance and use of the internet within society, it is also shrouded in mystery and fear, particularly in relation to sex. Undoubtedly, the area is hampered by outdated, unsystematic research that relies largely on small samples and cross-sectional design (Escobar-Chaves et al., 2005). Use of robust measures and longitudinal studies are warranted if understanding as to the effectiveness of interventions and if the findings from studies are to be used as evidence for policy (Escobar-Chaves et al., 2005)

REFERENCES

Abel, G. A., Jordan, A., Hand, C. G., Holland, C. A., & Phipps, A. (2001). Classification models of child molesters utilizing the Abel Assessment for Sexual Interest. *Child Abuse & Neglect*, *25*, 703–718. doi:10.1016/S0145-2134(01)00227-7

Adams, A. (2002). Cyberstalking and Internet pornography: gender and the gaze. *Ethics and Information Technology*, *4*(2), 133–143. doi:10.1023/A:1019967504762

Arnett, J. J. (1995). Adolescents' use of the media for self-socialization. *Journal of Youth and Adolescence*, *24*, 519–533. doi:10.1007/BF01537054

Beech, A. R., Elliott, I. A., Birgden, A., & Findlater, D. (2008). The Internet and child sexual offending: A criminological review. *Aggression and Violent Behavior*, *13*, 216–218. doi:10.1016/j.avb.2008.03.007

Braun-Courville, D. K., & Rojas, M. (2009). Exposure to sexually explicit web sites and adolescent sexual attitudes and behaviours. *The Journal of Adolescent Health*, *45*(2), 156–162. doi:10.1016/j.jadohealth.2008.12.004

Buschman, J., & Bogaerts, S. (2009). Polygraph testing internet offenders. In D.T. Wilcox (Ed.). *The use of the polygraphy in assessing, treating and supervising sex offenders: A practitioner's guide* (pp.113-128). West Sussex, UK: John Wiley & Sons LTd.

Cameron, K. A., Salazar, L. F., Bernhardt, J. M., Burgess-Whitman, N., Wingwood, G. M., & DiClemente, R. J. (2005). Adolescents' experience with sex on the web: Results from online focus groups. *Journal of Adolescence*, *28*, 535–540. doi:10.1016/j.adolescence.2004.10.006

CANEE. (2008a). *2nd International Conference "Keeping Children and Young People Safe Online", 18th -19th September 2008, Warsaw.* Available at: http://www.canee.net/poland/2nd_international_conference_keeping_children_and_young_people_safe_online_18th_19th_september_2008_warsaw

CANEE. *(2008b)*. Internet safety through Council of Europe online game. *Available at:*http://www.canee.net/child_and_internet/internet_safety_through_council_of_europe_online_game

CANEE. (2009a). *Keeping children and young people safe online.* Available at: http://www.canee.net/poland/keeping_children_and_young_people_safe_online

CANEE. (2009b). *Polish network against commercial sexual exploitation of children*. Available at: http://www.canee.net/poland/polish_network_against_commercial_sexual_exploitation_of_children

CANEE. (2009c). *Safer internet day 2009 celebrated in Poland.* Available at: http://www.canee.net/poland/safer_internet_day_2009_celebrated_in_poland

Carroll, J. S., Padilla-Walker, L. M., Nelson, L. J., Olson, C. D., Barry, C. M., & Madsen, S. D. (2008). Generation XXX: Pornography acceptance and use among emerging adults. *Journal of Adolescent Research*, *23*, 6–30. doi:10.1177/0743558407306348

Chatterjee, B. (2000). Cyberpornography, cyberidentities, and Law. *International Review of Law Computers & Technology*, *14*(1), 89–93. doi:10.1080/13600860054926

Ciccetti, D., & Toth, S. L. (1995). A developmental psychopathology perspective on child abuse and neglect. *Journal of the American Academy of Child and Adolescent Psychiatry*, *34*, 541–565. doi:10.1097/00004583-199505000-00008

Clark, L. (1998). Dating on the Net: Teens and the rise of 'pure' relationships. In Jones, S. (Ed.), *Cybersociety: Revisiting computer-mediated communication and community* (pp. 159–183). Thousand Oaks, CA: Sage.

Davis, G. E., & Leitenberg, H. (1987). Adolescent sexual offenders. *Psychological Bulletin*, *101*, 417–427. doi:10.1037/0033-2909.101.3.417

Dombrowski, S. C., Ahia, C. E., & McQuillan, K. (2003). Protecting children through mandated child abuse reporting. *The Educational Forum*, *67*(2), 76–85. doi:10.1080/00131720308984549

Dombrowski, S. C., LeMasney, J. W., Ahia, C. E., & Dickson, S. A. (2004). Protecting children from online sexual predators: technological, psychoeducational, and legal considerations. *Professional Psychology, Research and Practice*, *35*(1), 65–73. doi:10.1037/0735-7028.35.1.65

Durkin, K. F. (1997). Misuse of the Internet by pedophiles: implications for law enforcement and probation practice. *Federal Probation*, *61*, 14–18.

Durkin, K.F. & Bryant, C.D. (1999). Propagandizing pederasty: A thematic analysis of the online exculpatory accounts of unrepentant pedophiles. *Deviant Behavior: An interdisciplinary journal, 20,* 103 - 207.

Edwards, S. S. M. (2000). Prosecuting 'child pornography': Possession and taking of indecent photos of children. *Journal of Social Welfare and Family Law*, *22*, 1–21. doi:10.1080/014180300362732

Elliott, M., Browne, K., & Kilcoyne, J. (1995). Child sexual abuse prevention: what offenders tell us. *Child Abuse & Neglect*, *19*, 579–594. doi:10.1016/0145-2134(95)00017-3

Erikson, E. (1968). *Identity: Youth and crisis*. New York: Norton.

Escobar-Chaves, S. L., Tortolero, S. R., Markham, C. M., Low, B. J., Eitel, P., & Thickstun, P. (2005). Impact of the Media on Adolescent Sexual Attitudes and Behaviors. *Pediatricsm*, *116*(1), 303–326.

Esposito, L. C. (1998). Regulating the Internet: The new battle against child pornography. *Case Western Reserve Journal of International Law*, *3*(213), 541–567.

Fieldman, J. P., & Crespi, T. D. (2002). Child sexual abuse: offenders, disclosure, and school-based initiatives. *Adolescence*, *37*, 151–161.

Finkelhor, D., Mitchell, K.J., & Wolak, (2002). Online victimization: A report on the nation's youth. University of New Hampshire, Crimes Against Children research Center. Available at http://www.missingkids.com (accessed 18 July 2008).

Flander, G. B., Cosic, I., & Profaca, B. (2009). Exposure of children to sexual content on the Internet in Croatia. *Child Abuse & Neglect*, *33*(12), 849–856. doi:10.1016/j.chiabu.2009.06.002

Frei, A., Erenay, N., Dittmann, V., & Graf, M. (2005). Paedophilia on the Internet: A study of 33 convicted offenders in the Canton of Lucerne. *Swiss Medical Weekly*, *133*, 488–494.

Glasser, M., Kolvin, I., Campbell, D., Glasser, A., Leitch, I., & Farrell, S. (2001). Cycle of child sexual abuse: links between being a victim and becoming a perpetrator. *The British Journal of Psychiatry*, *179*, 482–494. doi:10.1192/bjp.179.6.482

Guan, S. S., & Subrahmanyam, K. (2009). Youth Internet use: risks and opportunities. *Current Opinion in Psychiatry*, *22*(4), 351–356. doi:10.1097/YCO.0b013e32832bd7e0

Holt, T. J., Blevins, K. R., & Burkert, N. (2010). Considering the pedophile subculture online. *Sexual Abuse*, *22*(1), 3–24. doi:10.1177/1079063209344979

IWF. (2009). *2008 Annual and charity report*. Cambridge, UK: Internet Watch Foundation.

IWF. (2009). *2008 Annual and charity report*. Cambridge, UK: Internet Watch Foundation.

Jenkins, P. (2001). *Beyond Tolerance: Child Pornography on the Internet*. New York: New York University Press.

Kaiser Foundation. (2000) US Adults and Kids on New Media Technology.InC. Von Feilltzen &U Carlsson (eds). *Children in the New Media Landscape* (pp 349 – 350). Goteborg: UNESCO/ Nodricom.

Kalichman, S. C. (1991). Psychopathology and personality characteristics of criminal sex offenders as a function of victim age. *Archives of Sexual Behavior*, *20*(2), 187–197. doi:10.1007/ BF01541943

Kierkegaard, S. (2008). Cybering, online grooming and ageplay. *Computer Law & Security Report*, *24*, 41–55. doi:10.1016/j.clsr.2007.11.004

Lanning, K. V. (1984). Collectors. In A.W. Burgess &M.L. Clark (Eds). *Child Pornography and Sex Rings* (pp. 83 - 92). Toronto: Lexington.

Livingstone, S. (2003). Children's use of the Internet: Reflections on the emerging research agenda. *New Media & Society*, *5*, 147–166. doi:10.1177/1461444803005002001

Livingstone, S., & Bober, M. (2005). UK children go online: listening to young people's experiences. London: London School of Economics. Retrieved from http://news.bbc.co.uk/1/shared/ bsp/hi/pdfs/28_04_05_childrenonline.pdf on 26th February 2009.

Magid, L. (1998). *Teen safety on the information highway*. National Centre for Missing and Exploited Children. Available at http://www.missingkids.com (accessed 18 July 2008).

Malesky, L. A. Jnr, & Ennis, L. (2004). Supportive Distributions: An analysis of posts on a pedophile Internet message boards. *Journal of Addictions & Offender Counseling*, *24*(2), 92–101.

Media-Awareness. (2000). Canada's Children in a Wired World: The parents' view. A survey of Internet use in Canadian families. *Media Awareness*, *20*(2), 17–18.

Mitchell, K., Finkelhor, D., & Wolak, J. (2001). Risk factors for and impact of online sexual solicitation of youth. *Journal of the American Medical Association*, *285*(23), 3011–3014. doi:10.1001/ jama.285.23.3011

Mitchell, K., Finkelhor, D., & Wolak, J. (2007a). Online requests for sexual pictures from youth: risk factors and incident characteristics. *The Journal of Adolescent Health*, *41*(2), 196–203. doi:10.1016/j.jadohealth.2007.03.013

Mitchell, K., Finkelhor, D., & Wolak, J. (2007b). Youth Internet users at risk for the most serious online solicitations. *American Journal of Preventive Medicine*, *32*(6), 532–537. doi:10.1016/j. amepre.2007.02.001

Mitchell, K. J., Wolak, J., & Finkelhor, D. (2005). Internet sex crimes against minors. In Kendall-Tackett, K., & Giacomoni, S. (Eds.), *Child victimization* (pp. 2.1–2.17). Kingston, NJ: Civic Research Institute.

Mitchell, K. J., Wolak, J., & Finkelhor, D. (2007). Trends in youth reports of sexual solicitations, harassment and unwanted exposure to pornography on the Internet. *The Journal of Adolescent Health*, *40*, 116–126. doi:10.1016/j. jadohealth.2006.05.021

Mitchell, K. J., Wolak, J., & Finkelhor, D. (2008). Are blogs putting youth at risk for online sexual solicitation or harassment? *Child Abuse & Neglect*, *32*(2), 277–294. doi:10.1016/j.chiabu.2007.04.015

Mitchell, K. J., Ybarra, M., & Finkelhor, D. (2007). The relative importance of online victimization in understanding depression, delinquency, and substance use. *Child Maltreatment*, *12*(4), 314–324. doi:10.1177/1077559507305996

Oddone, E., Genuis, M. L., & Violato, C. (2001). A meta-analysis of the published research on the effects of child sexual abuse. *The Journal of Psychology*, *135*, 17–36. doi:10.1080/00223980109603677

Panorama, (1998 - 2001). *The Wonderland Club.* Available at http://news.bbc.co.uk/1/hi/programmes/panoram/archive/1166945.stm

Peter, J., & Valkenburg, P. M. (2006). Adolescents exposure to sexually explicit material on the Internet. *Communication Research*, *33*, 178–204. doi:10.1177/0093650205285369

Quayle, E., Loof, L., & Palmer, T. (2008). *Child Pornography and Sexual Exploitation of Children Online: A Contribution of ECPAT International to the World Congress III against Sexual Exploitation of Children and Adolescents.* Bangkok: ECPAT International.

Quayle, E., & Taylor, M. (2001). Child Seduction and Self-representation on the internet. *Cyberpsychology & Behavior*, *4*(5), 597–610. doi:10.1089/109493101753235197

Rideout, V. (2001). *Generation Rx.com: How young people use the Internet for health information.* Available at http://www.kff.org/entmedia/upload/Toplines.pdf Accessed on 30th December 2009.

Ropelato, J. (2007). *Internet pornography statistics.* Available from http://Internet-filter-review.toptenreviews.com/Internet-pornography-statistics.html Accessed 23/12/09.

Sabina, C., Wolak, J., & Finkelhor, D. (2008). The nature and dynamics of Internet pornography exposure for youth. *Cyberpsychology & Behavior*, *11*(6), 691–693. doi:10.1089/cpb.2007.0179

Snyder, H. N. (2000). *Sexual assault of young children as reported to law enforcement: victims, incident, and offender characteristics. Report NCJ 182990.* Washington, DC: US Department of Justice, National Center for Juvenile Justice, Bureau of Statistics.

Subrahmanyam, K., & Greenfield, P. (2008). Online communication and adolescent relationships. *The Future of Children*, *18*(1), 119–146. doi:10.1353/foc.0.0006

Subrahmanyam, K., Greenfield, P. M., & Tynes, B. (2004). Constructing sexuality and identity in an online teen chat room. *Applied Developmental Psychology*, *25*, 651–666. doi:10.1016/j.appdev.2004.09.007

Subrahmanyam, K., Kraut, R. E., Greenfield, P. M., & Gross, E. F. (2000). The impact of home computer use on children's activities and development. *The Future of Children*, *10*, 123–144. doi:10.2307/1602692

Subrahmanyam, K., Smahel, D., & Greenfield, P. M. (2006). Connecting developmental processes to the Internet: Identity presentation and sexual exploration in online teen chatrooms. *Developmental Psychology*, *42*, 1–12. doi:10.1037/0012-1649.42.3.395

Taylor, M., Holland, G., & Quayle, E. (2001). Typology of paedophile picture collections. *The Police Journal*, *74*(2), 97–107.

Taylor, M., & Quayle, E. (2003). *Child Pornography: An Internet Crime*. Hove, UK: Brunner-Routledge.

UNICEF. (2008). *Convention on the Rights of the Child: An introduction.* Available at http://www.unicef.org/crc/index_30160 accessed on 5th February 2009.

Ward, L. M. (1995). Talking about sex: common themes about sexuality in the prime-time television programs children and adolescents view most. *Journal of Youth and Adolescence*, *24*, 595–615. doi:10.1007/BF01537058

Ward, L. M. (2003). Understanding the role of entertainment media in the sexual socialization of American youth: A review of empirical research. *Developmental Review*, *23*, 347–388. doi:10.1016/S0273-2297(03)00013-3

Ward, T., & Hudson, S. M. (2000). Sexual offenders' implicit planning: a conceptual model. *Sexual Abuse*, *12*, 189–202. doi:10.1177/107906320001200303

Wigley, K., & Clark, B. (2000) *Kids.net London National Poll.* http://www.nop.co.uk

Wolak, J., Finkelhor, D., & Mitchell, K.J. (2004). Internet-initiated sex crimes against minors: implications for prevention based on findings from a national study. *Journal of Adolescent Health, 35(5),* 424.e11 - 424.e20.

Wolak, J., Finkelhor, D., Mitchell, K. J., & Ybarra, M. L. (2008). Online "Predators" and their victims: myths, realities and implications for prevention and treatment. *The American Psychologist*, *63*, 111–128. doi:10.1037/0003-066X.63.2.111

Wolak, J., Mitchell, K., & Finkelhor, D. (2003). *Internet sex crimes against minors: The response of law enforcement.* Alexandria, VA: National Centre for Missing and Exploited Children, Publication No. 10-03-022.

Wolak, J., Mitchell, K., & Finkelhor, D. (2003b). Escaping or connecting? Characteristics of youth who form close online relationships. *Journal of Adolescence*, *26*(1), 105–119. doi:10.1016/S0140-1971(02)00114-8

Wolak, J., Mitchell, K., & Finkelhor, D. (2004). Internet-initiated sex crimes against minors: Implications for prevention based on findings from a national study. *The Journal of Adolescent Health*, *35*, 424–433. doi:10.1016/j.jadohealth.2004.05.006

Wolak, J., Mitchell, K., & Finkelhor, D. (2007). Unwanted and wanted exposure to online pornography in a national sample of youth Internet users. *Pediatrics*, *119*(2), 247–257. doi:10.1542/peds.2006-1891

Ybarra, M. L., Espelage, D. L., & Mitchell, K. J. (2007). The co-occurrence of Internet harassment and unwanted sexual solicitation victimization and perpetration: associations with psychosocial indicators. *The Journal of Adolescent Health*, *41*(6), S31–S41. doi:10.1016/j.jadohealth.2007.09.010

Ybarra, M. L., Finkelhor, D., Mitchell, K. J., & Wolak, J. (2009). Associations between blocking, monitoring and filtering software on the home computer and youth-reported unwanted exposure to sexual material online. *Child Abuse & Neglect*, *33*(12), 857–869. doi:10.1016/j.chiabu.2008.09.015

Ybarra, M. L., & Mitchell, K. J. (2005). Exposure to Internet pornography among children and adolescents; a national survey. *Cyberpsychology & Behavior*, *8*(5), 473–486. doi:10.1089/cpb.2005.8.473

Ybarra, M. L., & Mitchell, K. J. (2008). How risky are social networking sites? A comparison of places online where youth sexual solicitation and harassment occurs. *Pediatrics*, *121*(2), e350–e357. doi:10.1542/peds.2007-0693

Ybarra, M. L., & Mitchell, K. J. (2008). How risky are social networking sites? A comparison of places Online where youth sexual solicitation and harassment occurs. *Pediatrics*, *121*(2), E350–E357. doi:10.1542/peds.2007-0693

Chapter 20
Spirituality in Cybercrime (Yahoo Yahoo) Activities among Youths in South West Nigeria

Agunbiade Ojo Melvin
Obafemi Awolowo University, Nigeria

Titilayo Ayotunde
Obafemi Awolowo University, Nigeria

ABSTRACT

This chapter explores the relevance and adoption of spirituality in cybercrime; the roles of spiritualists; experiences of self-confessed youths that are involved in 'yahoo yahoo' activities and the future intentions of youths to engage in cybercrime. This was with a view to providing a socio-cultural analysis of the influence of spirituality in cybercrime ('yahoo yahoo') activities among Nigerian youths. Vignette based focus group discussions were held with male and female youths (18-35 years), in-depth interviews with 'yahoo yahoo' youths and some spiritualists. Findings showed that spirituality attracts high cultural relevance in life achievements and the conduct of cybercrimes. Perceptions on youths' involvement in cybercrime activities attracted mixed reactions. To the 'Yahoo yahoo' youths, they are playing a game, to other participants; 'yahoo yahoo' was a criminal act. Cybercrime among the youths have received the support of some spiritualists within a political economy that creates an enabling environment for cybercrimes and related activities. A few participants indicated future interests in cybercrime if their economic conditions remain unchanged or worsen. In conclusion, we argued that a holistic approach grounded in the cultural system would be more effective in re-orientating and empowering the youths to positively utilizes their internet skills. Thus, curbing cybercrimes would require a process that would not rely exclusively on legal and policing frameworks.

DOI: 10.4018/978-1-60960-209-3.ch020

INTRODUCTION

The web of social interactions has widened with the emergence of the Internet (Castells, 2001). Historical origins of the Internet and related technologies are properly documented (Norman, 2005). The relevance of the internet in the social arrangements of global events has become more evident but with inherent challenges. With the increase in social interactions through the cyber-space (Hughes, 1995), both the latent and manifest functions of the internet have evolved and may increase in dimension. A re-occurring latent function associated with social interactions through the cyber space is the phenomenon of cybercrime. Attempts by Criminologists and scholars from other fields at providing theoretical explanations for the nature and the complexity surrounding cybercrime have yielded contentious positions (Jaishankar, 2008; Mcquade, 2009). There are those who believed in the potentiality of traditional criminological theories in exploring the reality of cybercrime; while other scholars support new theoretical positions (Jaishankar, 2008; Mcquade, 2009). Among the latter are the Integrated theory; the theory of Technology-Enabled Crime, Policing, and Security (Mcquade, 2009); and the Space Transition Theory of Cybercrimes (Jaishankar, 2008). Despite the Inherent differences in these theoretical positions, a common stance is that all forms of crimes are explainable.

Cybercrimes are driven by intentions or motivations formed within space and time. Irrespective of the nature of the intentions, each type of cybercrime requires a particular set of skills, knowledge, resources, and access to particular data or information systems (Moore, 2009: 173). With increase in internet access and the likely occurrence of cybercrimes, there is need to understand the patterns and practices common with the various types of cybercrime especially in those targeted against persons. Various skills, knowledge, resources, and the type of data or information systems that facilitate particular cybercrimes are context driven within space and time. Our theoretical stance is influenced by Berger and Luckmann's(1966) position that knowledge production occurs within context. Thus, cultural beliefs system would affect both knowledge production and obtainable dominant patterns of cybercrime within time and space. This chapter situates Cybercrime targeted against persons within a cultural context by exploring the philosophical and psychosocial basis influencing the usefulness of spirituality- a culturally and psychologically rooted social phenomenon in the conduct of cybercrime activities like email frauds, fake identities, online dating, and online social networking among youths in south western Nigeria. In pursuing these objectives, this chapter did not examine the various processes involved in carrying out such activities. Rather, it focuses on the cultural beliefs and the knowledge systems that support the achievement of desired results in cybercrimes against persons. There is a void in the literature on the relevance of spirituality in the social construction of success by cybercrime perpetrators and the future intentions of other youths in relation to cybercrime in Nigeria.

The chapter starts with an overview of online interactions and the phenomenon of cybercrime; the universality of cybercrime; psychosocial cause-effects of cybercrime; cybercrime activities among Nigerian and Ghanaian youths; and the social nature of spirituality among the Yoruba people. The Yoruba cultural framework was adopted mainly for analytical purposes.

While the chapter's general objective is to explore the social perception of spirituality as a potent ideology and framework for negotiating successful interactions in '*yahoo yahoo*'(a social tag for cybercrime) activities among Nigerian Youths. The specific objectives include: exploring the perspectives of youths (18-35 years) and spiritualists on the relevance of spirituality in achieving desired expectations in life and in cybercrimes; investigate the future intentions of youths to engage in cybercrime activities. Examine the various social interactions and actors involved in the use of spirituality especially the role and position of spiritualists. And document the experiences of youths that are involved in '*yahoo yahoo*' activities in relation to spirituality.

ONLINE INTERACTIONS AND THE PHENOMENON OF CYBERCRIME

The advent of the internet and its wide spread has created new opportunities in almost every field of life with overt and covert revolutions on business, politics, education, security and other areas of human endeavour (Wellenius & Townsend, 2005). With the internet, the social order that bind time and space have become disembodied and distanced (Bottoms & Wiles, 1996). Through the creation of borderless social interactions, the internet has become a veritable tool for knowledge production, distribution, acquisition and utilization. The Internet as a social phenomenon has both manifest and latent functions on individuals, groups, societies and the global world. A re-occurring latent function of the internet is Cybercrime- a global phenomenon associated with the widened web of social relations in the virtual space. All forms of crime including Cybercrime are social in orientation and practice (Ballantine & Roberts, 2008). However, the multicultural nature of human brings diversity and uniqueness in several dimensions, including crime related activities. This diversity also extends into what constitute Cybercrime and the predisposing factors.

From the position of the Council of Europe's Cybercrime Treaty, cybercrimes include criminal activities against data to content and copyright infringement (Krone, 2005). This definition in Zeviar-Geese opinion is broader as it includes activities such as fraud, unauthorized access, child pornography, and cyber stalking (Zeviar-Geese, as cited in Gordon & Ford, 2006). In a restricted sense, Gordon and Ford (2006) describe cybercrime as any crime facilitated or committed using a computer, network, or hardware device. In this wise, the computer acts as the agent of the crime, facilitator, or taker of the crime in virtual or other non-virtual locations (Gordon & Ford, 2006), and the target could be individuals using the computer or the computer itself. Cybercrime is global but with peculiarities in practice, orientation, and patterns. Like the traditional crime, cybercrime is multidimensional as it occurs in different social context (Gordon & Ford, 2006). The transnational nature of cybercrime is supported by the growing relevance of computer knowledge, increasing global access to the internet and the endorsement of the virtual space as an online platform for social interactions (Burden, Palmer, Lyde & Gilbert, 2003; Embar-Seddon, 2002; Gies, 2008; Gordon & Ford, 2006).

In terms of functions, Cyberspace or virtual space perform similar functions to other physical space like bus stop, campus or social clubs (Grabosky, 2001). Also, it possesses equal potentials for both beneficial and unintended consequences including criminal activities. Durkheim opined that crime is a normal phenomenon in any society or social context, but with variations in typologies (Simpson, 1963). There are variations in Cybercrime. Gordon & Ford (2006) identify two types of cybercrime, type I and Type II. The type I is characterized by the following:

1. When a victim of an online interaction defines a singular or discrete event as fraudulent (*a single or discrete event could occur when the user goes online to perform a task and takes action which allows the criminal access to information that can be used in defrauding the user; when the information is used by the attacker; and the user or victim becomes conscious of the crime and the crime is investigated and resolved*).
2. When crime ware programmes such as keystroke loggers, viruses, root kits or

Trojan horses are introduced into the user's computer system.

3. The introductions can or may not be facilitated by vulnerabilities (Gordon & Ford, 2006).

"The type II Cybercrime include but not limited to activities such as cyber stalking and harassment, child protection, extortion, blackmail, stock market manipulation, complex corporate espionage, and planning or carrying out terrorist activities online"(Gordon & Ford, 2006). In Gordon and Ford's (2006) opinion, these typologies are relevant in clarifying the confusion on the real extent of Cybercrime. However, our focus in this chapter is on Cybercrimes targeted against humans, a characteristic of the type I described by Gordon and Ford (2006). Common examples of Cybercrimes targeted at persons include crimes like transmission of child-pornography, harassment of any one with the use of a computer such as e-mail, the trafficking, distribution, posting, and dissemination of obscene material including pornography and indecent exposure. This is not an exhaustive list.

Cybercrimes targeted against humans like other forms of criminal acts vary within time and space. A likely rationale behind the variations may be found in the cultural beliefs, prevailing forms of knowledge, existing social structures, and the obtainable patterns of social relations. This form of Cybercrime involves a level of social interaction between two or more social actors through the virtual space or non-virtual space. Like other forms of social relations, virtual interactions are contracted on trust (Lahno, 2004; McGeer, 2004). Trust, a subjective reality entails an expectation that the virtual interaction with others will be mutually exclusive. Such expectations are subject to the influence of intentions that may be overt to other actors that are involved. Intentionality is derived from the Latin word *intentio*, which in turn derives from the verb *intendere.*

Woolfolk (1990) describes intention as 'those mental states that relate human subjectivity to those corporal events and entities outside the mind'. In Watson's (1999: 119) opinion, Intentionality is "the projection of awareness, with purpose and efficacy, toward some object or outcome. Intentionality represents both philosophical and methodological stance. As a Philosophically stance, intentionality consists of consciousness about something or some content of consciousness, such as belief, volition, expectation, attention, action, and even the unconscious" (Dennett, 1987; Searle, 1983). In contemporary usage, intentionality encompasses issues relating to the nature of the mind-body interactions and how this relationship influences ontological and metaphysical questions that addresses the fundamental nature of mental states: such as perceiving, remembering, believing, desiring, hoping, knowing, intending, feeling, experiencing, and so on (Stanford Encyclopaedia, 2003).

Often social actors' intentions are concealed in thoughts, but may be observed in overt behaviours. Normatively, all social interactions are conducted in trust and good faith. Pragmatically, trusts is compromised at times as social actors' intentions are hidden expect with careful observations and with the outcomes, which can be judged as fraudulent or otherwise by those involved (Bierhoff & Vornefeld, 2004; Lahno, 2004; McGeer, 2004). Social actors using the Cyberspace as a platform for online social interactions have experienced both trust and distrust from other actors at various especially with the increasing rate of Cybercrimes.

The Universality of Cybercrime

The emergence of the internet has increased the web of possible social interactions through the cyberspace. The Cyberspace and its borderless nature of interactions have its challenges among which include Cybercrimes. The universality of Cybercrime is been strengthening by the increasing adoption and access to the internet. Benedikt

(1991) describes the cyberspace as a "new universe, a parallel universe created and sustained by the world's computers and communication lines". In a slightly different position, Hughes (1995) conceives cyberspace as "an interconnected, computer- mediated environment in which all prior media are represented." The Internet provides an environment with psycho-social and socio-political implications on virtual or non-virtual social interactions (Castells 2001; Jordan 1999).

Theoretical positions on the nature and complexity of Cybercrime have remained diverse (Jaishankar, 2008; Mcquade, 2009). There are scholars who believed in the potentiality of traditional criminological theories in explaining and addressing cybercrimes and those with contrary positions (Jaishankar, 2008; Mcquade, 2009). In both groups, there are inherent diversity methodologies, questions being addressed and the expected outcomes. While taking positions may amount to producing partial knowledge on the reality, appreciable contributions have emerged from this debate. Researchers now have the opportunity to explore the reality of Cybercrime from diverse perspectives. While our interest in this chapter is not to join issues on this theoretical debate, but we intend to explore the usefulness of conceptualizing cybercrimes and the various typologies as identified in the literature within a cultural context. Thus, situating the reality of cybercrime targeted humans within a cultural framework could be a fruitful approach towards building relevant cultural theories for cybercrime typologies. Hence, we explore the psychosocial variables and the philosophical positions that may be supporting the obtainable practices and patterns of Cybercrime targeted against persons within the framework of Yoruba Cultural system.

Psychosocial Cause-Effects of Cybercrime

Social actors' expectations and engagement in virtual and physical interactions as well as the means found useful cannot be divorced from the actors' social cultural environment. An insight into social actors' intentions governing social interactions in the cyber space or the physical space could provide an opportunity for understanding and formulating measures that may be useful in preventing or reducing cybercrimes targeted against other social actors in the virtual space or otherwise. We are in consonance with Mead's(Kinloch,1977: 146) position that intentions are formed and modulated within social cultural processes and that social actors through the socialization process imbibe one form of cultural belief or the other. Such beliefs are overt or covert in relationships as well as measures that are adopted in resolving challenges, attaining pleasures, satisfactions or relieving painful experiences. It is not our position that crime oriented intentions is solely facilitated by cultural beliefs. There are other contributing factors like the political economy, structural constraints and opportunities, personality and psychological effects of previous success of others in related activities.

Social actors are largely motivated by the level of rewards received from engaging in an activity or previous success of others and the likely success of those currently involved in related activities (Makarenko, 2004). Classical criminologists conceive punishment as potential deterrent to crime prevention (Stevens & Payne, 1999). While this position has been contested, the need for social control measures and process that could discourage crime including cybercrimes cannot be over emphasized.

From the structuralism perspective, social structures perform greater roles than personal characteristics of individuals when addressing issues relating to trend and depth of crime related activities (Sampson & Wilson, 2005). Similarly, influence of social structures on crime occurrence can also be observed in the light of opportunities and supporting environment that emphasises 'quick money' than reward through hard work. The increasing materialistic culture in many parts of the global and the shrinking opportunities to increase wealth is another reality. Even with the global recession, Harwood (2009), in a recent interview with David Dewalt the CEO and president of McAfee, argued that:

"Cyber criminals are exploiting the global recession by luring in susceptible victims through the promise of easy money. While governments and law enforcement have their attention diverted by the economy, the door is left open for cyber criminals to continue targeting bank balances and potentially damage consumer confidence, which is essential to the economic recovery." (http://www.security-management.com/news/cybercrime-trends-will-worsen-2009-according-forecasts-004969).

Structural influence on cybercrime is interconnected with psychological variables. So also is the psychosocial cause-effect relationship of cybercrime, which consists of macro and micro dimensions. At the macro level, interactions among the various social structures through the approved social agencies have effects on the lives of the social actors within that environment. When the activities of these agencies are corrupt, the effect becomes enormous, as other social actors may tend to repeat similar actions at various levels. The same holds at the micro level where social actors daily interactions are meaning laden. Youths that are involved in criminal activities with relative level of success and without any threat may serve as a source of inspiration to other youths. It is easy for an individual or group to rationalize their involvement in criminal acts like cybercrime after repeated failures or little success in other socially approved activities. Furthermore, crime may also become rational when there are other actors that engage in criminal activities with little societal disapproval or approval.

Thus, irrespective of the space and time, social actors' context has influence on available knowledge, reasoning, and intentions while interacting with members of their group or outsiders. More so, knowledge is context relevant, culturally constructed, and shared (Banks, 1993; Berger & Luckmann, 1966; Meyer & Jepperson, 2000).

With this presumption, it will be out of place to expect fundamental variations in the choice of suitable means considered effective by actors involved in cybercrime targeted against humans in protecting their interest, identity and achieving their desired goals in cyberspace interactions.

Cybercrime Activities among Nigerian and Ghanaian Youths

Beyond access and knowledge of the internet, effective monitoring of internet users activities and patterns in an environment where there are limited employment opportunities is another issue in cybercrime patterns and practices. Cybercrime has drastic implications on the perpetrators, victims and the society at large (Aghatise, 2006; Longe, Ngwa, Wada, & Mbarika, 2009; Saban, McGivern & Saykiewicz, 2002; Philippsohn, 2001; Smith, 2004). With these implications in mind, stakeholders across the globe have embarked on series of campaigns in dissuading, preventing, and punishing cybercrime perpetrators (Ribadu, 2004). Similarly, the increasing monitoring of cybercrime activities by some governmental agencies in Nigeria such as the Economic and Financial Crimes commission, the Nigerian Police, are good developments (Ribadu, 2004). However, despite the measures being put in place, some youths have persisted in cybercrime and related activities by employing unconventional mean to achieve their goals. Such means are unconventional to the cyber space but are part of the social realities of these youths.

Robert Merton, a renowned Sociologist in his Anomie theory (1968) explained the influence of social structures on social actors' capability in accepting, creating, or substituting old measures with modified or new ones that could facilitate the actualization of socially approved goals within a cultural system. Merton (1968) argued that Anomie would emerge due to disjunction between

Table 1.

Mode of adaptation	*Relationship to norms*
I. Conformity	+ +
II. Innovation	+ -
III. Ritualism	- +
IV. Retreatism	- -
V. Rebellion	Xx

goals and means available to social actors within a cultural system. Merton developed a typology of adaptation modes in relation to norms as depicted in Table 1.

The symbol (+) designates people's relationship to norms about goals; the second symbol (-) designates their relationship to norms about the means of achieving those goals. In this diagram, a "+" means acceptance, a "-" signifies rejection, and an "x" means rejection of prevailing values and substitution of new ones. Many African youths including their Nigerian counterparts have embraced the Internet as a useful platform for social interactions and other profitable activities (Oyelaran-Oyeyinka & Adeya, 2004). Some other youths have adopted the internet for both fraudulent and profitable activities or vice versa like their counterparts in other African societies (Ayoku, 2005; Lu, Jen, Chang & Chou, 2006; Makarenko, 2004).

In Nigeria and Ghana, Internet services at inception were largely provided by cybercafés located in commercially viable cities. Even with the emergence of the Global System for Mobile Communication, internet service provision and consumption in both countries is largely an urban phenomenon. The cost of Internet access in Africa and Nigeria is still high (Oyelaran-Oyeyinka & Adeya, 2004). This may be related to the fact that internet service provision is still largely controlled by 'state-owned monopolies- or their privatised successors - which still enjoy near monopolies' (The Association for Progressive Communications, 2006). While the internet has become a global platform for profitable and fraudulent practices; Merton's anomie theory recognises the peculiarity of crime patterns and practices found within context. For instance, in the literature, the type II cybercrime characteristics described by Gordon and Ford (2006) are more predominant in Western Cultures compare with emerging evidence from some African countries. The Internet and cell phones have been tools used for identity theft, e-mail fraud, trafficking, sexual exploitation, and prostitution. However, E-mail fraud and phishing have remained predominant among Nigerians and in other African countries (Longe, Ngwa, Wada, & Mbarika, 2009). For culturally explainable reasons, common patterns and practices of internet fraud in Nigeria and Ghana have taken a new dimension with the introduction of spirituality and occultism into the execution of cybercrime targeted against persons. In both countries, some youths involved in such cybercrimes have fortified themselves by employing diabolic means in the conduct of their activities. Spirituality as a social phenomenon is common in organizing and executing traditional crimes in Nigeria and some other African cultures.

The subculture of cybercrime that has emerged among some Nigerian and Ghanaian youths may have been fueled by the relative absence of effective monitoring of cybercrime and the ease at which corrupt practices have thrived over the years in both countries. 'Yahoo yahoo' as a cybercrime practice is common among Nigerian youths (Adeniran, 2008), while the Ghanaian version is 'Sakawa' (The Ghanaian Journal, 2009). The term 'yahoo yahoo' connotes a social label used in describing youths that searches the internet for individuals or network of relations that could be deceptively manipulated for personal or group gains with negative consequences on other participating members. Majority of the Cybercrimes against persons are targeted at individual's emotions and psychological makeup that requires less technical expertise on the part of these criminals. Generally, human weaknesses such as greed and gullibility are exploited with psychological and financial implications on the victims of such acts. These crimes are just like theft and other means of swindling victims that has existed offline before the advent of the internet (Ayoku, 2005; Longe & Chiemeke, 2007; Longe, Ngwa, Wada, & Mbarika, 2009; Longe, Mbarika, Kourouma, et al, 2009).

Sakawa is a name given to cybercrimes among Ghanaians. It is a way of defrauding many unsuspecting individuals on the internet. A common pattern is for a Ghanaian woman to pose nude in search of husbands and other forms of relationship on the internet. The unsuspecting victims would do anything, including the doling out of gifts and large sums of money to win the internet suitors. Over time, the reality became manifest on these Foreigners that many of the attractive nude women were either fronting for their husbands, boyfriends or other male friends. Major benefits from the elicit business therefore goes to the males with the women only acting as pawns in the larger business plan (The Ghanaian Journal, 2009).

With increased awareness and campaign against this type of crime, many previous victims became aware and reluctant in succumbing to their tricks. However, successes in previous activities and the psychological effects of 'easy money' without much resistance have encouraged many of these youths to devise other means of achieving their desires by relying on occultisms and spiritualism to maintain the ostentatious lifestyles they have acquired earlier from the internet business. With their beliefs in such practices, 'Sakawa' has produced many instant youth billionaires through cybercrimes targeted at humans. On several occasions, victims of cybercrimes that had previously refused suspicious proposals end up yielding unconsciously to 'Sakawa'. To maintain their level of results, Sakawa youths are ordered by their spiritualists to sleep in coffins and with corpses; walk bare-footed, eat from refuse dumps and any other weird prescriptions of their spiritualists or occult masters (The Ghanaian Journal, 2009).

Spirituality is socially perceived as potent measure criminals can adopt in reducing their vulnerability to law enforcement agencies and increasing their chances in achieving their desired results. Social commentators, researchers, social, and religious groups have used the term spirituality in several ways. The concept of spirituality goes beyond religious and cultural boundaries (Delgado, 2005). Historically, spirituality has been used by the twelfth century Christians who associated the term with the 'subjective life of faith'. It is different from religion, defined as the outward and objectified elements of a tradition (Roof, 2003 as cited in Olupona, 2008). The use of the term has changed in recent times. Roof (2003) argued that the term has transcends its traditional use to 'refer to the presence of the human spirit or soul and the human quest for meaning and experiential wholeness. From the humanistic psychology position, it is a search on the part of an individual for reaching some regimen of self-transformation and one's greatest potential'. In contemporary times, spirituality encompasses the inner life that is embedded within religious forms. Or much more loosely in keeping with the way the term spirituality is currently been used, spirituality indicates some qualities which makes it possible for individuals and groups to participate in multiple religious traditions without having to claim to –or be claimed by any of them (Olupona, 2008). Spirituality is common with many cultures. Delgado (2005) asserts that 'spirituality is an inherent human quality, and may manifest in various degrees influenced in part by the social and cultural environment. Spirituality for many involves faith or the willingness to believe, a search for meaning and purpose in life, a sense of connection with others, and a transcendence of the self, resulting in a sense of inner peace and well-being'.

In Nigeria, spirituality has been a critical factor in the activities of criminals involved in both organized and non-organized crimes. For instance, in the Anini Saga (a late-armed robber in Nigeria),

Marening (1987) portrayed how Anini and his gang explored spirituality in terrorizing the Nigerian state and the citizens until the gang eventually met their waterloo. Again, the values placed on spirituality are situated in cultural symbols with explicit meanings. These explicit meanings are not static but varied according to the status and knowledge of the member of the group concerned (Firth, 1973: 82). Social analysts have also toed this line of reasoning by arguing for the adoption of spirituality in solving the contemporary socio-economic and political problems in Nigeria. In the area of security problems, Fourchard, (2008) and Pratten (2008) noted that vigilante groups rely largely on traditional methods of crime control including spirituality in fighting crime and improving their social recognition.

SPIRITUALITY IN YORUBA CULTURAL BELIEFS

The Yoruba people share the philosophical position that binary relationships exist between the spiritual and the physical world. God is seen as the source and ultimate controller of the vital forces regulating the universe while the deities are the intermediaries between man and God. The invisible and the visible world are two dimensions of the same universe. Generally, it may be safe to say that Yoruba people believe in God; divinities; spirits; ancestors; mysterious powers, destiny (*ayanmo*), and inner head (*Ori inu*). In most social interpretations, the duality between the physical and the spiritual is often shared. For instance, there is the perception that the harmony between the physical and the spiritual can only be sustained with *Iwapele* on the part of the social actor in relation to their ancestors and the living. *Iwapele* as espoused by Abimbola (1975) deals with good character. In Ifa literary corpus, the concept of *Iwapele* (good character) is a normative and religious term that emphasis building and maintaining good social relations as requisite to attaining success in life and enhance better relations with others.

Also, the Yoruba people share the worldview that whatever happens to an individual in life is a function of one's ayanmo, ori and *iwapele*. However, this will not come on its own volition; the individual must be well fortified after arriving into this world of evil and good. Such fortifications could be in form of inquiring from those who possess the right knowledge such as the Babalawo (Diviner) and take precautions by performing the necessary *etutu* (sacrifice) or prayers. Conscious efforts in form of prayers, rituals, and sacrifices should be taken on a regular basis as they are part of the cultural ways of protecting one's *ayanmo* and *ori* from evil machinations (see: Abimbola 1967; Jegede, 2002; Karade, 1994; Morakinyo 1983 for details).

The underlying assumption for the continuous relevance of fortification throughout the life course is the philosophy of '*ogun laiye*' (that the world is full of battles and hurdles) and all social beings must be spiritually prepared to reduce or avoid mishaps in life. Even with the embracement of Christianity and Islam among the Yoruba cultural group, Jegede (2002) asserts that 'the thoughts of the people about life and their attitudes to it are still shaped by the old worldview. This, however, they exhibit in their day-to-day interpersonal interactions, within and outside the churches and mosques'. Hence, it will be safe to infer that spirituality occupies a germane position in the worldview of the Yoruba people especially in achieving personal or group goals.

Methodology

Study Design

A mixed qualitative research design consisting of Vignette based Focus Group Discussion and in-depth interview was adopted in generating relevant data. Mixed methods refer to the use of diverse procedures that combine and synthesize methods (Creswell, 2003; Creswell, Plano Clark, Gutmann, & Hanson, 2003; Tashakkori & Teddlie, 2003), or triangulate methods. Triangulation is anchor on the logic that no single method ever adequately presents a total picture of a reality (Patton, 1999; Schulze, 2003), and that the weakness in each single method will be complimented by the counter-balancing strengths of another (Morse & Chung, 2003; Hugentobler, Israel, & Schurman, 1992). In this study, the first level of data collection was the vignette based focus group discussions with male and female participants. This was followed by in-depth interviews.

The Study Setting

The study was conducted in Bariga and Ojo communities in Lagos State Nigeria. Bariga play host to the University of Lagos, while the Lagos State University is situated in Ojo. The latter operates a multi-campus system with four fully owned campuses. Lagos State University's main campus is at Ojo (along the Badagry Expressway), with other campuses at Epe (where the Engineering Faculty is located), Ikeja and Surulere as well as six external/affiliated campuses (Lagos State University: http://www.lasunigeria.org/).

The study areas are similar in many respects. Both Communities are suburb in orientation with a high proportion of University students as residents. There are functional Cyber cafes, commercial banks and other commercial organisations in both communities. As suburb areas, majority of the residents may be classified as low and middle income earners. Largely, Lagos State is one of the most commercially viable states in Nigeria. The presence of the universities in both communities has somewhat contributed to the influx of thousands of people from different ethnic inclinations to the areas. The Gown and Town interactions have increased the population of students' residents in both communities. Although, the University of Lagos operates functional on-campus accommodation, but the available students' hostels are inadequate for the student population. A high proportion of the students reside off- campus. On the other hand, Lagos State University does not provide on-campus accommodation for the students. Apartments for rent are available to willing students within Ojo and other locations in Lagos state.

Population and Sample

The study population consisted of three categories of participants. The first category consists of youths (18-35years) that are residents in *Bariga* and *Iyana-Iba Ojo* communities (suburb areas), the second category, 'yahoo yahoo' youths and the third, spiritualists. The study areas were purposively selected based on their functions as university communities, the presence of a number of cybercafés; high number of students as residents; and previous reported cases of cybercrimes among university students in Nigeria. Although this does not imply that such activities are limited to students of both universities and Lagos State as a geographical location. Similar cybercrimes have been reported else where among Nigerian youths (Longe, & Chiemeke, 2008).

Pre-field visits were made to the study locations prior to the actual data collection. Members of the two communities affirmed their awareness of the involvement of some youths in '*yahoo yahoo'* activities. Thus, based on our familiarity with some of the community's members and their knowledge of our identities it was easier to gain entry and their support.

All the three categories of participants that featured in this study were recruited using purposive sampling techniques. The first category of participants (Male and female youths aged 18-35 years) were recruited within the communities for the FGD. The inclusion criteria for the FGD include religious affiliation, age (18-35 years) and ethic background (Yoruba). The next two categories of informants were youths that are involved in '*yahoo yahoo'* activities while the third category was spiritualists. These last two groups were recruited through snowballing technique. A total of 10 self confessed and known youths involved in 'y*ahoo yahoo'* volunteered to share their experiences. The recruitment of the 'Yahoo yahoo' using snowballing technique was made possible with the support of four key informants. These informants are friends to some of the self confessed and known *'yahoo yahoo'* youths in the study locations.

Ten male and female spiritualists were also recruited based on the information provided by the 'yahoo yahoo' youths that were interviewed. Recruitment of the spiritualists was facilitated by two male adult informants who had close links with traditional medical practitioners within and outside Lagos State. In selecting the spiritualists, conscious efforts were made to ensure the recruitment of spiritualists with background in any of the three common religions (Christianity, Islam and Traditional) obtainable among the Yoruba people in Southwest Nigeria. Based on the diversity in doctrines and practices of Christianity in Nigeria, spiritualist from the Celestial Church and the Cherubim and Seraphim faith were recruited for the Christian category, *Alfa* (Islamic Cleric) for the Islamic faith and Babalawo (Diviners) with Onisegun (Herbalist) were recruited for the traditional Yoruba religion.

Data Collection Methods

The data collection lasted between August and September, 2009. Vignette based focus group discussion and in-depth interviews were the techniques employed in collecting the data. Vignettes are short descriptions of persons or social situations (Torres, 2009). Vignettes can be used in exposing personal matters and experiences indirectly through text, images, songs, or other forms of information framed within a scenario or story about a hypothetical situation (Hughes & Huby, 2002; Spalding & Phillips, 2007). Primarily, this method reckons with social definition of reality and promotes context-based understanding of lived experiences (Finch, 1987 as cited in Torres, 2009). In investigating sensitive issues, vignettes allow participants to express their ideas at any time without feeling personally exposed to the interviewer or other participants (Huges, 1998; Barter & Renold, 1999; Schoenberg & Ravdal, 2000).

For this study, we used the vignette method as the lead data collection technique. The vignettes employed in this study were developed by the authors alongside with two psychologists and an anthropologist whose research interests were in youth and internet behaviours. The vignettes stories were in two parts (a) the story of *'yahoo yahoo'* youth that ended on a "happy note" and (b) the story of a *'yahoo yahoo'* youth that ended on a "tragic note." The vignettes were used in the Focus Group Discussion. The inclusion criteria for the FGD include religious affiliation, age (18-35 years) and ethic background (Yoruba). For analytical purposes, the FGD participants were recruited on the basis of the dominant religions (Christianity and Islam). The FGDs were held with male and female youths (18-35years) as defined in this study. In each of the focus group discussion session, all the participants were exposed to both the "happy" and the "tragic" vignettes.

Focus Group Discussion (FGD) as a method has the capacity to unravel critical constructs that may be lost with individually generated data. In addition, it is reactionary to cultural values and norms (Kingry, Tiedje, & Friedman, 1990; Kitzinger, 1995). As a method, it does not only facilitate the elicitation of information from participants, it also encourages group interaction among participants (Morgan, 1988). In all the focus group discussions, the content of the vignettes were used to prompt participants to discuss and consider factors responsible for the 'successes' in *'yahoo yahoo'* (CyberCrime) and perceived influence of spirituality in particular. Before the vignettes were used, a pre-test of it was done among selected participants who met the study inclusion criteria. For the pre-test, two focus group discussion sessions were held with males and females not included in this study. After the pre-test, we restructured some aspects of the vignettes for clarity. In each FGD session a minimum of eight to ten discussants as recommended by Casey and Kreuger (2000), and O'Donnell, Lutfey, Marceau, & Mckinlay (2007), participated. In general a total of four FGD sessions were held with a total number of 37 male and female discussants. Each FGD session lasted for an average of one and half hours.

Two categories of informants were recruited for the in-depth interviews. The first category consisted of youths involved in *'yahoo yahoo'* activities while the second category of informants was spiritualists. In-depth Interview (IDI) allows for the collection of more meanings that informants give to events and the complexities of their attitudes, behaviour and experiences (Gubrium & Holstein, 2003). The In-depth interviews in this study were held with 10 self confessed and known youths involved in *'yahoo yahoo'* who volunteered to share their experiences. After the interviews held the *'yahoo yahoo'* youths, ten in-depth interviews were subsequently held with purposively selected spiritualists (males and females). The information provided by the 'yahoo yahoo' youths were acted on in the recruitment of the spiritualists. The interviews were conducted in either English or Yoruba language as preferred by the informants. Each interview session lasted between 45minutes to one and half hours. The first author facilitated the FGDs and the in-depth interviews, while the second author and two experienced social researchers acted as note takers and taped recorded all the discussions.

Data Analysis

A qualitative content analysis which entailed coding responses, creating categories, and identifying recurring themes (Hsieh & Shannon, 2005; Graneheim & Lundman, 2004), was used in analysing the data generated from the recorded Vignettes based FGDs. All interviews conducted in Yoruba language were translated, transcribed verbatim and checked for accuracy. However, the initial transcription of all the interviews held in Yoruba was done in Yoruba language and were later translated to English language. Thereafter, the transcriptions and translations were given to an expert in Yoruba and English Language to ensure proper and accurate representation of participants' views. On the overall, all data collected from all sources were transcribed before being coded using Nvivo8 qualitative software. A triangulation approach was adopted in the analysis and presenting of the salient themes that emerged from the data and the literature. Salient themes that emerged from the issues raised and discussed by the participants were noted and developed into themes and analysed.

Ethical Considerations

The audiotape recordings were done with the consent of all the participants. Before each discussion, written informed consent was obtained where possible, while uneducated participants were asked to give verbal consent. All the discussions took place at preferred suitable locations suggested by the Participants. Through out the interaction periods, participants were informed of their rights to decline participation at any point in the discussions.

Findings

Findings from the study revealed five interrelated themes under the following: Spirituality and life achievements; 'yahoo yahoo' is a game; spirituality in 'yahoo yahoo' activities; deceptions and identity falsification; gender and future intentions on 'yahoo yahoo'

Profiles of Participants

There were 18 males and 19 females among the focus group participants. Among the males, 10 of them were Christians and 8 were Muslims. Among the females, 9 were Christians while 10 were Muslims. By age category, 11 of the males in the FGDs were within ages 18-26 while 7 were the age bracket of 27-35 years. A high proportion (19) of the FGD participants (males and females) was students. Similarly, 14 of them were unemployed, only 4 of them were employed. Educational status of the FGD participants also showed that 28 of them had post-secondary qualifications while 9 of them had only secondary school education.

All the 'yahoo yahoo' youths that were interviewed were university students and all of them were below age 30. Among these youths, 7 of them claimed they were Christians while 3 of them claimed they were Muslims. All the 'yahoo yahoo' youths had one time or the other visited a spiritualist within the last month before the interview. Three among the 'yahoo yahoo' youths at the point of interview claimed they had travelled outside the country courtesy of their 'maga' or 'maye'(both terms refer to '*yahoo yahoo'* victim). Information on their socioeconomic background showed that 6 of them were from low-socioeconomic background while the remaining 4 had parents that were above average in socioeconomic status. Among the spiritualists interviewed, 3 were from the Celestial Church of Christ (CCC) and the Cherubim and Seraphim(C&S), 2 were *Alfas* (Islamic Cleric) 3 were Babalawos (Diviners) and 2 were Oniseguns (Herbalists). Only two of the spiritualists were females and the remaining were males. Educational background of the informants showed that all of them had a minimum of secondary education and all had practiced one occupation before their present occupation as spiritualist. None among the informants had spent less than 5 years in their present occupation. Three among the informants had spent more than 25 years in the occupation.

Spirituality and Life Achievements

The participants accorded a high relevance to spirituality. In the focus group discussions, both male and female discussants considered spirituality as inherent part of life that is inseparable from meaningful achievements in life. Both the spiritualists and the 'yahoo yahoo' youths expressed a similar same view. In addition, the participants accorded high importance to hard work as a requisite to success; nevertheless, they maintained the position that there must be harmony between activities in the physical and the spiritual before meaningful achievements could occur. In the position of another spiritualist (Babalawo) interviewed, *ori eni ni gbe ire ko nio* (it is one inner head that brings good) it is not hard work but good inner head. It was also interesting to observe that religious variations were not significant in participants' worldview and their interpretations of life events. The extracts below reveal some of their views:

There are certain events in life that hard work or physical strength cannot achieve except one understands and possess some spiritual powers (Babalawo aged 72)

We have witnessed situations in this country that only prayers, sacrifices, and rituals saved us (FGD with males)

Without spirituality, your level of success in 'yahoo yahoo' will remain very low. Many 'maga' are now recalcitrant but spirituality will give you an easy breakthrough. 'This moves you to the next level in the game' (male 'yahoo yahoo' youth aged 25)

In my secondary school days, my academic performance was too low but after my mother took me to a spiritualist in Ilorin, I can tell you today that I am among the best student in my department ('Yahoo yahoo' youth aged 23).

However, some of the male and female discussants in the focus group discussions cautioned against the use of spiritual powers in attaining success in some activities that are dubious in orientation. Armed robbery, money rituals, stealing and other related criminal activities were the most cited. In the same vein, three spiritualists (Prophets, *Alfa, and Babalawo*) emphasized the risks involved in employing spiritual powers for criminal activities as against God's will and commandments. On the over all, the participants maintained that spiritual powers are available for any interested person to explore, but only a few possess the knowledge. Spirituality cuts across all religions of the *'yahoo yahoo'* informants. For instance, among them, those who claimed Christianity also confessed seeking spiritual supports from Babalawos and Alfas. To them, no form of contradictions exists between their primary faith and the religion of their spiritual supporters. What remains most important is the faith in the powers gained from these spiritualists:

I am a Christian, but I have an Alfa and a Babalawo that I consult regularly. Initially when I started this 'yahoo yahoo' of a thing, things were very difficult and little results were achieved. The moment I started using one black soap and the power given to me by my spiritualist, things changed for better. Now, with that power whatever I ask my 'maga' it must be granted ('yahoo yahoo' youth aged 27)

Four male and three female Participants in the FGDs also expressed a similar position that individuals are free in life to search for solutions to life challenges anywhere even though there may be religious divergences. However, the belief that irrespective of religious divergences people should be free to explore different avenues to their problems was boldly contested by other participants in the FGDs arguing that such approach are synonymous to 'spiritual adultery'.

'Yahoo Yahoo' is a Game

The FGD discussants and the 'yahoo yahoo' informants expressed contradictory views on whether *'yahoo yahoo'* is a crime or a game, while six spiritualists described 'yahoo yahoo' as similar to robbery. The position of the *'yahoo yahoo'* youths is that they are playing game and not stealing. For a player to succeed in the game certain skills were considered necessary. Seven out of the Ten 'yahoo yahoo' youths expressed the view that Knowledge of the internet and possession of adequate skills about the computer, ability to communicate in a foreign language (English language in particular being the Lingua Franca in Nigeria), and proper induction into 'Yahoo yahoo' activities were essential skills a new entrant in *'yahoo yahoo'* activities must possess. These factors are important, as the new entrant has to undergo on-the-job training usually under the tutelage of a 'master':

No one can jump into 'yahoo yahoo' without learning from a master. They provide you with the links; teach you the language and show you how it works. From there, a newcomer can then use his initiative, and move forward ('Yahoo yahoo' youth aged 24)

Another informant argued that:

Appropriate induction into the 'game' cannot be sacrificed. To make it in this game one must follow a master and strive to outshine the master ('Yahoo yahoo' youth aged 22)

'Yahoo yahoo' is a game of intellectuals. You have to be sound and bold ('Yahoo yahoo' youth aged 29)

Spirituality in 'Yahoo Yahoo' Activities

Some of the male and female discussants in the FGD further extend the position that spirituality has promoted the level of success 'yahoo yahoo' youths have enjoyed in their activities. These participants argued that total reliance on personal skills and networks alone could not guarantee the level of success 'yahoo yahoo' youths are recording now. Two female discussants in the FGD argued intently that:

Yahoo yahoo youths are using spiritual powers it can't just be their brainwork and networks alone. There are occasions some of these guys have been caught with some diabolic materials in their rooms and some of them have even confessed to using spiritual powers in conducting their cybercrimes (FGD with Female youths).

Within the community, vigilante groups have arrested three self-confessed "yahoo yahoo' youths coming out of the Cemetery after performing a ritual. Some of the materials they used for the rituals were still with them at the point of their arrest (FGD with Male youths).

I am aware of a spiritualist who specializes in providing spiritual supports for 'yahoo yahoo' youths and those youths have made it in terms of material wealth (A male spiritualist)

On the part of the 'yahoo yahoo' youths, a consensus that emerged was the position that to move to the next phase in the game, vigilance, and pro-activeness in the sense of employing spirituality in various forms are essential. Fasting and praying, sacrificing to gods, confession of sins, incantations, and rituals were mentioned as common practices. Eight among the informants argued that 'yahoo yahoo' youths that adopt spiritual measures would have an edge over those relying on physical skills and connections with their 'magas'. They further argued that the swift competition among 'yahoo yahoo' youths and the monitoring of cybercrime activities at the local and international levels requires spiritual powers as against the use of brain and networking alone.

Using their personal experiences, the informants ('yahoo yahoo' youths) argued that the level at which 'good money' could be made in *'yahoo yahoo'* game and without being caught in the process requires using spiritual powers. Such powers will serve as a means of fortification and guaranteeing success in the activity. Below are extracts from two of the informants(Yahoo yahoo) youths on how the use of spiritual powers have worked for them so far in attaining success in life and 'yahoo yahoo' activities in particular:

I started with a master and learnt so much from him. Now I am the master, I explored spiritual powers and now things have really changed for me. 'Moti lo owo' (I am rich now ('Yahoo yahoo' youth aged 23)

Before I speak or mail my maga (yahoo yahoo victim) there are certain rituals I must perform to ensure success. Most maga's are now recalcitrant so I have the ones for speech (afose, mayehun) and others that I use. Believe me the maga has no option than to give me whatever I ask ('Yahoo yahoo' youth aged 26)

Many of the male and female discussants in the Focus group discussions also supported the view that spirituality is essential in achieving remarkable feet in the face of stiff competition in life. However, three among the spiritualist cautioned on the dangers associated with the use of spiritual powers in attaining success by indulging in dubious activities. One among the three spiritualist also narrated an experience a late client of his went through before eventually given up the ghost. The spiritualist narrated how the former client obtained a spiritual power from him under false pretence but promised to use the power as instructed. He later defaulted by using the powers to rob and kill without knowing that the consequences would be a miserable death within seven days after his actions.

Deceptions and Identity Falsification

All the informants ('yahoo yahoo' youths) narrated their experiences and their activities as a game of deceit and greed. One of the 'yahoo yahoo' youths argued emphatically that 'Nothing in this game is real'. Even in dating relations, the youths disclosed that extortion of money and using the networks of their maga as the major objective:

I have a 'maga' that I worked on for three months before she eventually came to meet me. We spent two weeks together and I pretended to be in love just to get her money ('Yahoo yahoo' youth aged 26)

Another informant shared how he used deceit in getting some powers with the support of his unaware mother from a spiritualist in Ogbomosho (a town in Osun State, Southwest Nigeria):

Initially, when I started yahoo game things were not glowing for me. The little money I made then I reinvested it into getting more spiritual powers and little additional skills. I specialized in dating and issuing fake cheques. All I need to do is to gain the maga's ('yahoo yahoo' victim) emotions. Once that is achieved, it becomes easier for spiritual powers to work. Therefore, I lied to the 'Baba' (spiritualist) that some people owe me some money but have refused to pay me. The spiritualist performed some rituals and gave me a black smelling soup and another object that I cannot describe. Immediately I started using them, things turned around for me. That was how I got some powers ('Yahoo yahoo' youth aged 28).

In the FGD sessions, many of the male and female discussants confirmed the use of deceits by both the *'yahoo yahoo'* youths and some spiritualists. To the discussants, there are some spiritualists that are aware of the activities of the *'yahoo yahoo'* youths and other criminals but who still go ahead and give them the needed spiritual powers due to their financial abilities to pay for such services. They would even tell them in details the taboos attached to the powers and the renewal processes involved. In addition, Moneymaking rituals cases were cited as good examples of how the activities of some spiritualists have supported criminal activities in some Nigerian communities. A spiritualist also corroborated the above position:

You see, Ifa stands for truth and hard work. However, some spiritualists have defiled themselves by supporting evil activities such as money rituals and providing supports for thieves in general. They provide charms and conduct rituals to aide criminal activities and other related activities (Male Herbalist aged 67)

The political economy of the country and corrupt practices of some Nigerian police officers were also implicated in the event of things. Some of the FGD discussants (males and females) cited the political economy as frustrating and largely corrupt. This in their view has also influenced the disposition of some spiritualists and the youths in particular into criminal activities. Two of the spiritualists interviewed also supported this position:

Alfas are not supposed to support people with Tira (charms) and other spiritual powers to dupe people. I am aware of some alfas, especially the young ones who are desperate at making quick money but they have forgotten that there are consequences for participating in such acts and those using Tiras in unguarded manners (Male Alfa aged 70).

Many people are out there due to economic and financial needs. Some spiritualists are still sincere with their practice and are godly. Nevertheless, a high proportion out there has deviated because of money. Such are the wolves among us (Prophetess aged 58)

The high level of unemployment in the land, the high activity and awareness rates of 'yahoo yahoo' have influenced the involvement of some young people into these scam activities which had paid off handsomely for a good number of them(FGD with male youths and females)

Some police officers have also supported the activities of these 'yahoo yahoo' youths. Five among the 'yahoo yahoo' youths narrated how some police officers who were aware of their activities had used that as an avenue to extort money from them on several occasions. In some instances, these youths have also narrated how some banks officials have assisted on some occasions in transacting some illegal business.

A spiritualist (Prophet) argued that not all of the youths that are involved in 'yahoo yahoo' have eventually made it. However, some youths have been influenced by the success some 'yahoo yahoo' youths have recorded. The spiritualist decried the situation as destructive for youths that are involved in the crime as well as those with future intentions. Although the 'yahoo yahoo' youths preferred the tag game to stealing in describing their activities, but six out of these youths also expressed concerns over their ways of life. They argued passively that other activities that can make their lives better but such opportunities are restricted to youths that have powerful and influential parents. Similar fears were also expressed in their use of spiritual powers in getting desired results from their 'maga'. An informant narrated how a spiritualist invoked the spirit of a *maga* and through this medium he (the 'yahoo yahoo' youths) requested for a car and $ 10,000 US Dollars from his *maga.* Prior to this point, the 'maga' has refused to yield for over a year and within three days the money the youth requested sent was sent and two weeks after, the car was also sent by the *'maga'*('Yahoo yahoo' victims). While 'yahoo yahoo' youths have achieved a level of material success through this means, there are consequences attached. Three among the 'yahoo yahoo' youths also narrated how they stopped using spiritual powers after a while on their '*magas*'. An explanation for this was that their spiritualists have warned them on the repercussions and they do not want that. One of them lamented that business have changed since he stopped using such powers:

Since I stopped using my magical ring and black powder, it has been very difficult. I am only living on my previous proceeds and now concentrating on low returns activities just to survive ('Yahoo yahoo' youth aged 27).

Gender and Future Intentions on 'Yahoo Yahoo'

It was not a surprise that some youths that claimed non-involvement in 'yahoo yahoo' activities nurse future intentions about it. Four males indicated their willingness to participant in 'yahoo yahoo' if they cannot secure any meaningful employment and there is opportunity for 'yahoo yahoo'. They tied their willingness to indulge in the activity to the easy at which 'yahoo yahoo' youths are making it and the poor situation of the Nigerian economy. Among the male discussants, five of them owned up to be friends of some 'yahoo yahoo' youths but and that their friend's involvement in the activity has not affect their friendship. While this was not a general census, a common position was that the society has encouraged 'yahoo yahoo' through corrupt practices. Similarly, the participants expressed dismay in the prevailing political logjam in Nigeria and youth unemployment in the absence of social security programme for youths.

For the female discussants, all of them expressed pessimistic positions on their likely involvement in 'yahoo yahoo' activity in the nearest future. However, a slight shift from this position was observed when three female discussants in the FGD admitted their willingness to date a 'yahoo yahoo' male. They explained that 'yahoo yahoo' youths spend extravagantly on girls and they would prefer dating *'guys'* that are not stingy and asocial:

Yahoo yahoo boys? Why not, I will definitely date him o o o!!! They are very outgoing and can spend their last penny on a girl. Why wouldn't I (FGD with female Christian)

DISCUSSION AND CONCLUSION

Previous findings reveal a high magnitude of youths involvement in cybercrime activities in other parts of the world and Nigeria as a nation (Adeniran, 2008; Aghatise, 2006; Ayoku, 2005; Longe, Ngwa, Wada, & Mbarika, 2009; Longe, & Chiemeke, 2008). Similarly, such studies have also shown the level of efforts governments and international organizations are employing in reducing cybercrimes among Nigerian youths but without looking at the efforts and initiatives by 'yahoo yahoo' youths in safe guarding their identities, protecting themselves and improving their earnings. Our findings suggest that ''yahoo yahoo' youths, as social actors are proactive in utilizing cultural symbols and values in forming potential strategies that could protect their interests and attract other interested entrants into their activities. Our findings in this direction suggest that spirituality accords high relevance among the study participants. To the participants' spirituality is an inherent part of life inseparable from meaningful achievements in life. This is line with findings from the literature on the nature of spirituality and its relevance (Delgado, 2005; Winterowd, Harrist, Thomason, Worth, & Carlozzi, 2005). The adoption of shortcuts and quick rich measures as shown in some of the 'yahoo yahoo' youths calls for concerns. This fear has been raised by other researchers as detrimental not only to the image of the Nigerian nation but the future of these youths (Aghatise, 2006; Longe, & Chiemeke, 2008; Ribadu, 2004). They have already tasted the good side of cybercrime and hard work is no longer cherished among this category of youths. Some of the youths that are involved in this type of cybercrime might have suffered negatively, but may enjoy recounting positive experiences than the negative ones.

Despite the cultural beliefs that adopting spirituality in criminal activities has its consequences a high proportion of the 'yahoo yahoo', youths in this study for personal reasons have continuously employed spirituality in their activities. These youths are not the only ones involved, there are other social actors providing the needed support for these youths, and the spiritualists are in the forefront. This contradicts the manifest roles of spiritualists in the society. Social structures and their agents possess the potentials to be functional or dysfunctional. There are possibilities that agents of religious institutions could at times act against the interest of the social system. In addition, the internet as a medium of communication and interactions could function as a medium of crimes (Rustad, 2001).

Our findings also underscore divergent positions in the label attributed to 'yahoo yahoo' activity. To these youths, they are not thieves but game players and only greed is what is making their victims to fall. For every entrant and the old members, the game must be played according to the rules. This game also required certain skills that may be learned without being a member. Knowledge of the internet and possession of adequate skills about the computer, ability to communicate in a foreign language (English language in particular being the Lingua Franca in Nigeria), and proper induction into Yahoo yahoo activities were essential for a new entrant in *'yahoo yahoo'* activities. Nevertheless, the language or slangs of the activity must be learned through tutelage under a master. The 'yahoo yahoo' culture and induction process may be likening to some findings on gay culture (Fraley, Mona, & Theodore, 2007).

Beyond the primary skills and slangs needed to operate, spirituality was widely considered paramount in achieving success in all endeavours and 'yahoo yahoo' in particular. Such attribution may be traceable to the cultural beliefs of the people. From the experiences of the '*yahoo yahoo*' youths, the views of the spiritualists and the FGD discussants, the 'yahoo yahoo' youths involved in cybercrime have greatly explored this avenue. The consensus that emerged was that the level of success the youths involved in cybercrime are recording might be attributed to the influence of spirituality. While the participants' foresaw dangers in the use of spirituality in criminal activities, it is obvious that quite a number of the '*yahoo yahoo*' youths still believed that they would quickly get their share of foreign currency before nemesis catch up with them. Such beliefs are traceable to cultural values and belief systems. It also confirms a Yoruba proverb which says: *Abẹ́rẹ́ á lọ kí ọ̀nà okùn tó dí* (The needle will pass before the way of the thread is blocked). While some of these youths are taught the virtues of hard work, there are indications that some of them have refused to conform to such teachings. Another reason may be that some parents and guardians are themselves indulging in occultism, corrupt practices, or too busy to attend to the moral upbringing of their children.

Our findings also support other scholars' arguments that cybercrime activities targeted against humans are governed by greed, deceptions, and falsehood (Burden, Palmer, Lyde, & Gilbert, 2003; Embar-Seddon, 2002; Gordon & Ford, 2006). All the participants in this study affirmed the position especially the youths currently involved in cybercrime. Similar deceptions have been extended in the search for spiritual powers by the 'yahoo yahoo' youths. A high proportion of the participants regretted this position and some spiritualists as there are consequences attached to such measures.

Similar to findings in the literature, the political economy situation in Nigeria was also implicated as a contributor to cybercrime. These factors have been associated as contributors to cybercrime especially in an economy where there are large social restrictions on peoples' ability to meet socially approved goals (Adeniran, 2008; Aghatise, 2006; Longe & Chiemeke, 2008; Teh, 2009).

The willingness of some male participants to engage in cybercrime activities in the nearest future and their female counterparts' intention to date a 'yahoo yahoo' male indicates a form of acceptance of 'yahoo yahoo' culture and as a means of survival among some of the youths. This is also in line with some studies that investigated trend and use of internet among youths in Nigeria (Adeniran, 2008; Longe, Ngwa, & Wada, et al, 2009). The future intentions to engage in cybercrime activities among the youths indicate that more sensitization and empowerments of youths in particular should be continuous and relevant to the youths than putting control measures to arrest youths that are presently involved in cybercrime activities.

Notwithstanding the strengths of our study, our findings are limited in a few respects. To start with, the use of spirituality in cybercrime is restricted in many respects. Cybercrimes targeted against computers would require high technical skills than those targeted against individuals. The study also focused on cybercrimes targeted against humans and not the computer. Certain cybercrimes may be common among Nigerian youths (Adeniran, 2008; Aghatise, 2006; Longe, Mbarika, Kourouma, et al, 2009); it does not eliminate the plausibility of other forms of cybercrimes among Nigerian youths. Furthermore, the sample size of this study being a qualitative study is also limited. Despite the depth of understanding achieved through the findings, generalizing the findings to include all patterns and practices of cybercrime among Nigerian youths is limited. Hence, it will be relevant if other studies could investigate the cultural differences in patterns and practices of cybercrimes among youths in Nigeria. This is essential as cultural and ethnic variations in work attitudes have been confirmed among Nigerian workers (Aluko, 2003).

In conclusion, cybercrime is a torn in the flesh of Nigerians and the rest of the world. To the youths involved in this crime, 'yahoo yahoo' has been functional in providing for their present needs but are afraid of their future as youths. This present predicament must not be allowed to remain as many youths may become frustrated if the political and economic situation of Nigeria does not improve. The corrupt practices of law enforcement agencies, spiritualists and the moral decadence in the society are contributing towards the trend of cybercrime. Similarly, the use of spirituality in some of the cybercrime activities calls for a more holistic approach in the search for solutions. All stakeholders must be involved in the quest for answers to the cybercrimes. In the final analysis, in the literature, there are evidences that the best police in the world cannot effectively fight crime in a culture that is materialistic and where the social arrangements and structural factors are constraining social actors from attaining socially approved goals (Teh, 2009). Thus, addressing cybercrimes would require more than legal framework and effective policing. These frameworks are relevant, but the social conditions constraining people from achieving socially approved goals must be addressed in a holistic manner backed with high political will. With these measures in place, culturally grounded strategies would be available in re-orientating the majority of youths connected with cybercrimes and other criminal activities in the Nigerian.

REFERENCES

Abimbola, W. (1967). *Ifa: an exposition of the Ifa literary corpus*. Ibadan, Nigeria: Oxford University Press.

Abimbola, W. (1975): Iwapele: The concept of good character in Ifa literary corpus. In: Abimbola, W. (ed.), *Yoruba oral tradition.* (Ile-Ife: Department of African Languages and Literature University of Ife), 389-420.

Adeniran, A. I. (2008). The Internet and Emergence of Yahooboys sub-Culture in Nigeria. *International Journal of Cyber Criminology*, *2*(2), 368–381.

Aghatise, E. J. (2006). Cybercrime definition. *Computer Crime Research Center*. Retrieved from http://www.crime-research.org/articles/joseph06/2

Aluko, M. A. O. (2003). The impact of culture on organizational performance in selected textile firms in Nigeria. *Nordic Journal of African Studies*, *12*(2), 164–179.

Ayoku, A. O. (2005). The evolving sophistication of Internet abuses in Africa. *The International Information & Library Review*, *37*(1), 11–17. doi:10.1016/j.iilr.2005.01.002

Ballantine, J. H., & Roberts, K. A. (2008). *Our Social World* (2nd ed.). London: Sage Publications.

Banks, J. A. (1993). The Canon Debate, Knowledge Construction, and Multicultural Education. *Educational Researcher*, *22*(5), 4–14.

Barter, C., & Renold, E. (1999). The Use of Vignettes in Qualitative Research. [University of Surrey]. *Social Research Update*, *25*, 1–6.

Benedikt, M. (Ed.). (1991). *Cyberspace First Steps*. Cambridge, MA: MIT Press.

Berger, P. L., & Luckmann, T. (1966). *The Social Construction of Reality: A Treatise its the Sociology of Knowledge*. New York: Anchor Books.

Bierhoff, H., & Vornefeld, B. (2004). The Social Psychology of Trust with Applications in the Internet. *Analyse & Kritik*, *26*(1), 48–62.

Bottoms, A., & Wiles, P. (1996). Understanding crime prevention in late modern societies. In Bennett, T. (Ed.), *Preventing crime and disorder: Targeting strategies and responsibilities*. Cambridge, UK: University of Cambridge.

Burden, K., Palmer, C., & Lyde, B., & Gilbert. (2003). Internet crime: Cyber Crime — A new breed of criminal? *Computer Law & Security Report, 19*(3), 222–227. doi:10.1016/S0267-3649(03)00306-6

Casey, M. A., & Kreuger, R. A. (2000). *Focus groups: A practical guide for applied research* (3rd ed.). Thousand Oaks, CA: Sage.

Castells, M. (2001). Lessons from the History of Internet. In *The Internet Galaxy* (pp. 9–35). New York: Oxford University Press.

Creswell, J. W. (2003). *Research design: qualitative, quantitative, and mixed methods approaches* (2nd ed.). Thousand Oaks, CA: Sage Publications.

Creswell, J. W., Plano Clark, V. L., Gutmann, M. L., & Hanson, W. E. (2003). Advanced mixed methods research designs. In Tashakkori, A., & Teddlie, C. (Eds.), *Handbook of mixed methods in the social and behavioral research. (pp. 209e240)*. Thousand Oaks, CA: Sage Publications.

Delgado, C. (2005). A Discussion of the Concept of Spirituality. *Nursing Science Quarterly, 18*, 157–162. doi:10.1177/0894318405274828

Dennett, D. C. (1987). *The intentional stance. Cambridge, MA*. Bradford: MIT Press.

Dennett, D. C. (2003). *Freedom Evolves*. New York: Viking Press.

Embar-Seddon, A. (2002). Cyberterrorism: Are We Under Siege? *The American Behavioral Scientist, 45*(6), 1033–1043. doi:10.1177/0002764202045006007

Firth, R. (1973). *82). Symbols: Public and Private*. Ithaca, NY: Cornell Univ. Press.

Fourchard, L. (2008). A New Name for an Old Practice: Vigilantes in South-Western Nigeria. *Africa, 78*(1), 16–40. doi:10.3366/E000197200800003X

Fraley, S. S., Mona, L. R., & Theodore, P. S. (2007). The Sexual Lives of Lesbian, Gay, and Bisexual People With Disabilities: Psychological Perspectives. *Sexuality Research & Social Policy, 4*(1), 15–26. doi:10.1525/srsp.2007.4.1.15

Gies, L. (2008). How material are cyberbodies? Broadband Internet and embodied subjectivity *Crime Media Culture,* 4(3), 311-330. (References and further reading may be available for this article. To view references and further reading you must purchase this article).

Gordon, S., & Ford, R. (2006). On the definition and classification of cybercrime. *Journal in Computer Virology, 2*(1), 13–20. doi:10.1007/s11416-006-0015-z

Grabosky, P. N. (2001). Virtual Criminality: Old Wine in new bottles? *Social & Legal Studies, 10*(2), 243–249.

Graneheim, U. H., & Lundman, B. (2004). Qualitative content analysis in nursing research: Concepts, procedures and measures to achieve trustworthiness. *Nurse Education Today, 24*, 105–112. doi:10.1016/j.nedt.2003.10.001

Gurbrium, J. F., & Holstein, J. A. (2003). *Postmodern interviewing*. Thousand Oaks, CA: Sage Publications.

Harwood, M. (2009). *Cybercrime Trends Will Worsen in 2009*, According to Forecasts (2009). Retrieved May 15, 2010 from http://www.securitymanagement.com/news/cybercrime-trends-will-worsen-2009-according-forecasts-004969.

Hsieh, H., & Shannon, S. E. (2005). Three Approaches to Qualitative Content Analysis. *Qualitative Health Research, 15*(9), 1277–1288. doi:10.1177/1049732305276687

Hugentobler, M. K., Israel, B. A., & Schurman, S. J. (1992). An action research approach to workplace health: integrating methods. *Health Education Quarterly*, 19(1), 55e76.

Hughes, K. (1995). *From Webspace to Cyberspace*. Retrieved May 20 2010, from http://tecfa.unige.ch/guides/vrml/hughes/cspace.1.1.pdf.

Hughes, R. (1998). Considering the Vignette Technique and its Application to a Study of Drug Injecting and HIV Risk and Safer Behaviour. *Sociology of Health & Illness*, *20*, 381–400. doi:10.1111/1467-9566.00107

Hughes, R., & Huby, M. (2002). The Application of Vignettes in Social and Nursing Research. *Journal of Advanced Nursing*, *37*(4), 382–386. doi:10.1046/j.1365-2648.2002.02100.x

Jaishankar, K. (2008). Space Transition Theory of Cyber Crimes. In Schmallager, F., & Pittaro, M. (Eds.), *Crimes of the Internet* (pp. 283–301). Upper Saddle River, NJ: Prentice Hall.

Jegede, A. S. (2002). The Yoruba cultural construction of health and illness. *Nordic Journal of African Studies*, *11*(3), 322–335.

Jordan, T. (1999). *Cyberpower: The culture and politics of cyberspace and the Internet*. London: Routledge. doi:10.4324/9780203448632

Karade, I. (1994). *The handbook of Yoruba religious concepts*. York Beach: Weiser Books.

Kingry, M. J., Tiedje, L. B., & Friedman, L. L. (1990). Focus groups: a research technique for nursing. *Nursing Research*, *39*(2), 124–125.

Kinloch, G. C. (1977). *Sociological Theory its Development and Major Paradigms*. New York: McGraw- Hill, Inc.

Kitzinger, J. (1995). Qualitative research: introducing focus groups. *British Medical Journal*, *311*, 299–302.

Krone, T. (2005). *High tech crime brief.* Canberra, Australia:Australian Institute of Criminology, Lagos State University.(n.d.). Retrieved July, 2009 http://www.lasunigeria.org/

Lahno, B. (2004). Three Aspects of Interpersonal Trust. *Analyse & Kritik*, *26*(1), 30–47.

Langhout, R. D. (2003). Reconceptualizing quantitative and qualitative methods: a case study dealing with place as an exemplar. *American Journal of Community Psychology*, *32*(3/4), 229e244.

Leininger, M. (1970). *Nursing and Anthropology: Two Worlds to Blend*. New York: John Wiley & Sons.

Longe, O., Mbarika, V., Kourouma, M., Wada,F. & Isabalija, R. (2009). Seeing beyond the surface: Understanding and Tracking Fraudulent Cyber Activities. *International Journal of Computer Science and Information Security*, 6(3,124-135.

Longe, O., Ngwa, O., Wada, F., & Mbarika, V. (2009). Criminal Uses of Information and Communication Technologies in Sub-Saharan Africa: Trends, Concerns and Perspectives. *Journal of Information Technology Impact*, *9*(3), 155–172.

Longe, O. B., & Chiemeke, S. C. (2007). Information and communication technology penetration in Nigeria: Prospects, challenges and metrics. *Asian Journal of Information Technology*, *6*(3), 280–287. Retrieved November 7, 2009 from http://www.medwellonline.net/fulltext/ajit/2007/280-287.pdf/

Longe, O. B., & Chiemeke, S. C. (2008). Cyber Crime and Criminality in Nigeria – What Roles are Internet Access Points in Playing? *European Journal of Soil Science*, *6*(4), 132–139.

Lu, C., Jen, W., Chang, W., & Chou, S. (2006). Cybercrime & Cybercriminals: An Overview of the Taiwan Experience. *Journal of Computers*, *6*(1), 11–18.

Makarenko, T. (2004). The Crime–Terror Continuum: Tracing the Interplay between Transnational Organised Crime and Terrorism. *Global Crime*, *6*(1), 129–145. doi:10.1080/1744057042000297025

Marening, O. (1987). The Anini Saga: Armed Robbery and the Reproduction of Ideology in Nigeria. *The Journal of Modern African Studies*, *25*(2), 259–281. doi:10.1017/S0022278X00000380

McGeer, V. (2004). Developing Trust on the Internet. *Analyse & Kritik*, *26*(1), 91–107.

Mcquade, S. C. (2009). Theories of Cybercrime. In Mcquade, S. C. (Ed.), *Encyclopedia of Cybercrime* (pp. 179–181). Westport, CT: Greenwood Press.

Merton, R. K. (1968). *Anomie theory in Social theory and Social structure*. New York: free press.

Meyer, J. W., & Jepperson, R. L. (2000). The "Actors" of Modern Society: The Cultural Construction of Social Agency. *Sociological Theory*, *18*(1), 100–120. doi:10.1111/0735-2751.00090

Moore, J. (2009). Technologies commonly used for Cybercrime. In Mcquade, S. C. (Ed.), *Encyclopedia of Cybercrime* (pp. 173–179). Westport, CT: Greenwood Press.

Morakinyo, O. (1983). The Yoruba ayanmo myth and mental health care. *West Africa Journal of Cultures and Ideas*, *1*(1), 61–92.

Morgan, D. L. (1988). *Focus Group as Qualitative Research*. Newbury Park, CA: Sage.

Morse, J. M., & Chung, S. E. (2003). Toward holism: the significance of methodological pluralism. [Article 2]. *International Journal of Qualitative Methods*, *2*(3). Available from http://www.ualberta. ca/w;iiqm/backissues/ 2_3final/html/morsechung.html.

Norman, J. M. (2005). *From Gutenberg to the Internet: A source Book on the History of Information Technology.* Novato, CA: Historyofscience.com.

O'Donnell, A. B., Lutfey, K. E., Marceau, L. D., & McKinlay, J. B. (2007). Using focus groups to improve the validity of cross-national survey research: A study of physician decision making. *Qualitative Health Research*, *7*, 971–981. doi:10.1177/1049732307305257

Olupona, J. K. (2008). Sacred Ambiguity: Global African Spirituality, Religious Tradition, Social Capital and Self-Reliance. In Babawale, T., & Alao, A. (Eds.), *Global African Spirituality Social Capital and Self-Reliance in Africa* (pp. xvii–xxxii). Lagos, Nigeria: Malthouse Press Limited.

Oyelaran-Oyeyinka, B., & Adeya, C. N. (2004). Internet Access in Africa: Empirical Evidence from Kenya and Nigeria. *Telematics and Informatics*, *21*(1), 67–81. doi:10.1016/S0736-5853(03)00023-6

Patton, M. Q. (1999). Enhancing the quality and credibility of qualitative analysis. (Part II). *Health Services Research, 34*(5), 1189e1208.

Philippsohn, S. (2001). Trends in Cybercrime - an overview of current financial crimes on the Internet. *Computers & Security*, *20*(1), 53–69. doi:10.1016/S0167-4048(01)01021-5

Plog, F. (1980). *Anthropology: Decisions, Adaptation, and Evolution*. New York: Alfred A. Knopf.

Pratten, D. (2008). 'The Thief Eats His Shame': Practice and Power in Nigerian Vigilantism. *Africa*, *78*(1), 64–83. doi:10.3366/E0001972008000053

Ribadu, N. (2004). *Obstacles to the effective prosecution of corrupt practices and financial crime cases in Nigeria.* Paper Presented at the Summit on Corrupt Practices and Financial Crimes in Nigeria, Kaduna, Nigeria.

Rustad, M. (2001). Private Enforcement of Cybercrime on the Electronic Frontier. *Southern California Interdisciplinary Law Journal*, *11*(1), 63–116.

Saban, K. A., McGivern, E., & Saykiewicz, J. N. (2002). A critical look at the impact of cyber crime on consumer internet behavior. *Journal of Marketing Theory and Practice*, *10*(2), 29–37.

Sampson, R. J., & Wilson, W. J. (2005). Toward a theory of race, crime and urban inequality. In Gabbidon, L. S., & Greene, H. T. (Eds.), *Race, Crime and Justice: A Reader* (pp. 37–54). New York: Routledge.

Schoenberg, N. E., & Ravdal, H. (2000). Using vignettes in awareness and attitudinal research. *International Journal of Social Research Methodology, 3*(1), 63–74. doi:10.1080/136455700294932

Schulze, S. (2003). Views on the combination of quantitative and qualitative research approaches. *Progressio, 25(*2), 8e20.

Searle, J. (1983). *Intentionality: An essay in the philosophy of mind*. Cambridge, UK: Cambridge University Press.

Simpson, G. (1963). *Emile Durkheim: Selections from His Work*. New York: Thomas Y. Crowell Co.

Smith, A. D. (2004). Cyberciminal impacts on online business and consumer confidence. *Online Information Review, 28*(3), 224–234. doi:10.1108/14684520410543670

Spalding, N. J., & Phillips, T. (2007). Exploring the Use of Vignettes: From Validity to Trustworthiness. *Qualitative Health Research, 17*(7), 954–962. doi:10.1177/1049732307306187

Stanford Encyclopaedia. *Intentionality* (2003). Retrieved January 12, 2010 from http://plato.stanford.edu/entries/intentionality/.

Stevens, E. D., & Payne, B. K. (1999). Applying deterrence theory in the context of corporate wrongdoing: Limitations on punitive damages. *Journal of Criminal Justice, 27*(3), 195–207. doi:10.1016/S0047-2352(98)00060-9

Tashakkori, A., & Teddlie, C. (2003). *Handbook of mixed methods in the social and behavioral research*. Thousand Oaks, CA: Sage Publications.

Teh, Y. K. (2009). The best police force in the world will not bring down a high crime rate in a materialistic society. *International Journal of Police Science and Management, 11*(1), 1–7. doi:10.1350/ijps.2009.11.1.104

The Association for Progressive Communications (APC). (2006). *Warning over African internet cable*. Retrieved from http://news.bbc.co.uk/2/hi/africa/4787422.stm.

The Ghanaian Journal. (2009). *Top Pastors in Ghana Patronise Sakawa Boys*. Retrieved May 29, 2010 from http://www.theghanaianjournal.com/2009/05/04/top-pastors-in-ghana-patronize-sakawa-boys/.

Torres, S. (2009). Vignette methodology and culture-relevance: lessons learned from a study on Iranian's understandings of successful aging. *Journal of Cross-Cultural Gerontology, 24*(1), 93–114. doi:10.1007/s10823-009-9095-9

Watson, J. (1999). *Postmodern nursing and beyond*. Edinburgh, UK: Churchill Livingstone.

Wellenius, B., & Townsend, D. (2005). Telecommunications and Economic Development. In Majumdar, S. K., Vogelsang, I., & Cave, M. E. (Eds.), *Handbook of Telecommunications Economics Technology Evolution and the Internet* (*Vol. 2*, pp. 557–621). London: Emerald Group Publishing.

Winterowd, C., Harrist, S., Thomason, N., Worth, S., & Carlozzi, B. (2005). The Relationship of Spiritual Beliefs and Involvement with the Experience of Anger and Stress in College Students. *Journal of College Student Development, 46*(5), 515–529. doi:10.1353/csd.2005.0057

Woolfolk, R. L. (1990). Intentional Explanation and Its Limits. *Psychological Inquiry, 1*(3), 273–274. doi:10.1207/s15327965pli0103_24

Compilation of References

Aapola, S., Gonick, M., & Harris, A. (2005). *Young Feminity. Girlhood, Power and Social Change*. New York: Palgrave Macmillan.

Abanihe – isiugo, U.C. (2002). *Currents and Perspectives in Sociology*.

Abel, G. A., Jordan, A., Hand, C. G., Holland, C. A., & Phipps, A. (2001). Classification models of child molesters utilizing the Abel Assessment for Sexual Interest. *Child Abuse & Neglect*, *25*, 703–718. doi:10.1016/S0145-2134(01)00227-7

Abimbola, W. (1967). *Ifa: an exposition of the Ifa literary corpus*. Ibadan, Nigeria: Oxford University Press.

Abimbola, W. (1975): Iwapele: The concept of good character in Ifa literary corpus. In: Abimbola, W. (ed.), *Yoruba oral tradition.* (Ile-Ife: Department of African Languages and Literature University of Ife), 389-420.

Adams, G. R., & Montemayor, R. (1983). Identify Formation During Early Adolescence. *The Journal of Early Adolescence*, *3*(3), 193–202. doi:10.1177/0272431683033002

Adams, A. (2002). Cyberstalking and Internet pornography: gender and the gaze. *Ethics and Information Technology*, *4*(2), 133–143. doi:10.1023/A:1019967504762

Adeniran, A. I. (2008). The Internet and Emergence of Yahooboys sub-Culture in Nigeria. *International Journal of Cyber Criminology*, *2*(2), 368–381.

AfDB/OECD (African Development Bank/ Organisation for Economic Co-operation and Development). (2008) '*Mauritius*'. Retrieved December 10 2008 from ww.oecd.org/dataoecd/13/7/40578285.pdf

Aghatise, E. J. (2006). Cybercrime definition. *Computer Crime Research Center*. Retrieved from http://www.crime-research.org/articles/joseph06/2

Ajaheb-Jahangeer, S., & Jahangeer, A. C. (2004). School Culture in a Private Secondary Institution in Mauritius. *International Education Journal*, *5*(2), 247–254.

Alanen, L. (1992). *Modern Childhood? Exploring the 'Child Question' in sociology*. Jyväskylä: University of Jyväskylä.

Albright, J. M., & Conran, T. (2003). Desire, Love and Betrayal: Constructing and Deconstructing Intimacy Online. *Journal of Systemic Therapies*, *22*(3), 42–53. doi:10.1521/jsyt.22.3.42.23352

Alderson, P. (2005). Generation Inequalities. *UK Health Watch*, *2005*, 47–52.

Allison, B. N., & Schultz, J. B. (2001). Interpersonal Identity Formation During Early Adolescence. *Adolescence*, *36*(143), 509–523.

Altheide, D., & Michalowski, S. (1999). Fear in the News: A Discourse of Control. *The Sociological Quarterly*, *40*(3), 475–503. doi:10.1111/j.1533-8525.1999.tb01730.x

Altheide, D. (1987). Ethnographic Content Analysis. *Qualitative Sociology*, *10*(1), 65–77. doi:10.1007/BF00988269

Althusser, L. (1965). *For Marx*. London: Verso.

Aluko, M. A. O. (2003). The impact of culture on organizational performance in selected textile firms in Nigeria. *Nordic Journal of African Studies*, *12*(2), 164–179.

Alwood, E. (1996). *Straight news: Gays, lesbians and the news media*. New York: Columbia University Press.

Alzae, H. (1978). Sexual Behaviour of Columbia Female University Students. *Archives of Sexual Behavior*, *7*(1), 43–53. doi:10.1007/BF01541897

Ambjörnsson, F. (2004). *I en klass för sig. Genus, klass och sexualitet bland gymnasietjejer*. Stockholm, Sweden: Ordfront.

American Psychological Association. (2002). *Developing Adolescents: A Reference for Professionals*. Washington, DC: American Psychological Association.

Anderson, B., & Tracey, K. (2001). Digital Living: The Impact (or Otherwise) of the Internet on Everyday Life. *The American Behavioral Scientist*, *45*(3), 456–475. doi:10.1177/00027640121957295

Anderson, B. (1983). *Imagined communities: Reflections on the origin and spread of nationalism*. London: Verso.

Anderson, C. A. (2004). An update on the effects of playing violent video games. *Journal of Adolescence*, *27*, 113–122. doi:10.1016/j.adolescence.2003.10.009

Anderson, C. A., & Bushman, B. J. (2001). Effects of violent video games on aggressive behavior, aggressive cognition, aggressive affect, physiological arousal, and prosocial behavior: A meta-analytic review of the scientific literature. *Psychological Science*, *12*, 353–359. doi:10.1111/1467-9280.00366

Anderson, C. A., Carnagey, N. L., Flanagan, M., Benjamin, A. J., Eubanks, J., & Valentine, J. C. (2004). Violent video games: Specific effects of violent content on aggressive thoughts and behavior. *Advances in Experimental Social Psychology*, *36*, 199–249. doi:10.1016/S0065-2601(04)36004-1

Anderson, C. A., & Dill, K. E. (2000). Video games and aggressive thoughts, feelings, and behaviour in the laboratory and in real life. *Journal of Personality and Social Psychology*, *78*, 772–790. doi:10.1037/0022-3514.78.4.772

Anderson, C. A., Gentile, D. A., & Buckley, K. E. (2007). *Violent video game effects on children and adolescents. Theory, research, and public policy*. Oxford, UK: University Press. doi:10.1093/acprof:oso/9780195309836.001.0001

Anderson, C. A., Shibuya, A., Ihori, N., Swing, E. L., Bushman, B. J., & Sakamoto, A. (2010). Violent video game effects on aggression, empathy, and prosocial behavior in Eastern and Western countries. *Psychological Bulletin*, *136*(2), 151–173. doi:10.1037/a0018251

Ang, I. (1985). *Watching Dallas: soap opera and the melodramatic imagination* (Couling, D., Trans.). New York: Routledge.

Apell, S. (2008). YouTube - Le Parkour Sweden Simon Apell interview. Retrieved August 26, 2009 from http://www.youtube.com/watch?v=K_C4oCDAXsg

Aricak, T., Siyahhan, S., Uzunhasanoglu, A., Saribeyoglu, S., Ciplak, S., Yilmaz, N., & Memmedov, C. (2008). Cyberbullying among Turkish adolescents. *Cyberpsychology & Behavior*, *3*, 253–261. doi:10.1089/cpb.2007.0016

Arkhem, H. (2005) *Jätten i spegeln*. Richters förlag.

Arnett, J. J. (2006). *Emerging Adulthood The Winding Road from the Late Teens Through the Twenties*. New York: Oxford University Press.

Arnett, J. J. (1995). Adolescents' use of the media for self-socialization. *Journal of Youth and Adolescence*, *24*, 519–533. doi:10.1007/BF01537054

Ashcroft, B., Griffiths, G., & Tiffin, H. (2007). *Post-Colonial Studies: The Key Concepts* (2 ed.). Oxon, UK: Routledge.

ATA News (2009). *RCMP and CTF join forces to fight cyberbullying, 43*(10), 6.

Attewell, P., Suazo-Garcia, B., & Battle, J. (2003). Computers and Young Children: Social Benefit or Social Problem? *Social Forces*, *82*(1), 277–296. doi:10.1353/sof.2003.0075

Attwood, F. (2009). Deepthroatfucker' & `Discerning Adonis': Men & cybersex. *International Journal of Cultural Studies, 12*(3), 279–294. doi:10.1177/1367877908101573

Aumeerally, N. L. (2007). The Ambivalence of Postcolonial Mauritius. *International Journal of Cultural Policy, 11*(3), 307–323. doi:10.1080/102866304200031254 3

Avisen.dk. (2007, January 15). Unge piger smider tøjet. *Nyhedsavisen*. Retrieved from http:/www.avisen.dk

Ayoku, A. O. (2005). The evolving sophistication of Internet abuses in Africa. *The International Information & Library Review, 37*(1), 11–17. doi:10.1016/j.iilr.2005.01.002

Bäckman, M. (2003). *Kön och känsla: samlevnadsundervisning och ungdomars tankar om sexualitet* [Sex and emotion: Education in sexual matters and young peoples thoughts about sexuality]. Stockholm, Sweden: Makadam.

boyd, d. (2008). Why youth love social network sites: The role of networked publics in teenage social life. In Buckingham, D. (Ed.) *Youth, identity, and digital media*, pp. 119-142. The John D. and Catherine T. MacArthur Foundation Series on Digital Media and Learning. Cambridge, MA: The MIT Press.

Baker, M., & Brooks, K. (1998). *Knowing Audiences: Judge Dredd, its friends, fans and foes*. Luton, UK: University of Luton Press.

Bakhtin, M. M. (1968). *Rabelais and his world*. Cambridge, MA: M.I.T. Press.

Ballantine, J. H., & Roberts, K. A. (2008). *Our Social World* (2nd ed.). London: Sage Publications.

Bandura, A. (1973). *Aggression: a social learning analysis*. Englewood Cliffs, NJ: Prentice-Hall.

Bandura, A. (1983). Psychological mechanisms of aggression. In Green, R. G., & Donnerstein, E. I. (Eds.), *Aggression: Theoretical and empirical views* (*Vol. 1*, pp. 1–40). New York: Academic Press.

Banks, J. A. (1993). The Canon Debate, Knowledge Construction, and Multicultural Education. *Educational Researcher, 22*(5), 4–14.

Barack, L. (2007). Schools Mute Cellphones: Teachers nationwide hear one ringy-dingy too many. *School Library Journal, 10*(1), 1–2.

Barak, A., & Sadovsky, Y. (2008). Internet use and personal empowerment of hearing impaired adolescents. *Computers in Human Behavior, 24*(5), 1802–1815. doi:10.1016/j.chb.2008.02.007

Barbara, S. M., Daniel, B., Wesley, H. C., & Binka, P. (1999). The Changing Nature of Adolescence in the Kassena-Nankana District of Northern Ghana. *Studies in Family Planning, 30*(2), 95–111. doi:10.1111/j.1728-4465.1999.00095.x

Barlett, C. P., & Anderson, C. A. (2009). Violent video games and public policy. In Bevc, T., & Zapf, H. (Eds.), *Wie wir spielen, was wir werden: Computerspiele in unserer Gesellschaft. Konstanz: UVK Verlagsgesellschaft.*

Baron, R. A., & Richardson, D. R. (1994). *Human aggression*. New York: Plenum Press.

Baron, N. (n.d.). *The Myth of Impoverished Signal: Dispelling the Spoken-Language Fallacy for Emoticons in Online Communication*. Retrieved December 5, 2009, from: http://www1.american.edu/tesol/Scholarly%20Documents,%20Articles,%20Research/Baron_Emoticons%5B1%5D.pdf.

Barron, C. (2000). *Giving Youth a Voice: A Basis for Rethinking Adolescent Violence*. Halifax, Nova Scotia: Fernwood Publishing.

Barter, C., & Renold, E. (1999). The Use of Vignettes in Qualitative Research. [University of Surrey]. *Social Research Update, 25*, 1–6.

Bartholow, B. D., & Anderson, C. A. (2002). Examining the effects of violent video games on aggressive behavior: Potential sex differences. *Journal of Experimental Social Psychology, 38*, 283–290. doi:10.1006/jesp.2001.1502

Bartholow, B. D., Sestir, M. A., & Davis, E. B. (2005). Correlates and consequences of exposure to video game violence: Hostile personality, empathy, and aggressive behavior. *Personality and Social Psychology Bulletin, 31*(11), 1573–1586. doi:10.1177/0146167205277205

Bartkowicz, Z. (1984). *Nieletni z obniżoną sprawnością umysłową w zakładzie poprawczym*. Lublin, Poland: Wydawnictwo UMCS.

Baucom, H. D., Epstein, N., Rankin, L. A., & Burnett, C. K. (1996). Assessing Relationship Standards: The Inventory of Specific Relationship Standards. *Journal of Family Psychology*, *10*(1), 72–88. doi:10.1037/0893-3200.10.1.72

Baum, R. (2003). *Female absence: women, theatre and other metaphors*. New York: P.I.E.-Peter Lang.

Baumeister, R. F., & Tice, D. M. (2001). *The Social Dimension of Sex*. Boston: Allyn & Bacon.

Baumeister, R. F., & Leary, M. R. (1995). The Need to Belong: Desire for interpersonal Attachment as a Fundamental Human Motivation. *Psychological Bulletin*, *117*, 497–529. doi:10.1037/0033-2909.117.3.497

Bay-Cheng, L. Y. (2001). SexEd.com: Values and norms in Web-based sexuality education. *Journal of Sex Research*, *38*(3), 241–251. doi:10.1080/00224490109552093

Baym, N. K. (1998). The emergence of on-line community. In Jones, S. G. (Ed.), *Cybersociety 2.0: Revisiting computer-mediated communication* (pp. 35–68). Thousand Oaks, CA: Sage.

Baym, N. K. (2007). The new shape of online community: The example of Swedish independent music fandom. *First Monday, 12*(8). Retrieved from URL: http://firstmonday.org/issues/issue12_8/baym/index.html

Bayne, S., & Ross, J. (2007). *The 'digital native' and 'digital immigrant': a dangerous opposition*. Paper presented at the Annual Conference of the Society for Research into Higher Education (SRHE) December 2007.

Beale, A., & Hall, K. (2007). Cyberbullying: What school administrators (and parents) can do. *Clearing House (Menasha, Wis.)*, *81*(1), 8–12. doi:10.3200/TCHS.81.1.8-12

Beaty, L. A. (1999). Identity development of homosexual youth and parental and familial influences of the coming out process. *Adolescence*, *34*(135), 597–601.

Beech, A. R., Elliott, I. A., Birgden, A., & Findlater, D. (2008). The Internet and child sexual offending: A criminological review. *Aggression and Violent Behavior*, *13*, 216–218. doi:10.1016/j.avb.2008.03.007

Beentjes, J. W. J., Koolstra, C. M., Marseille, N., & van der Voort, T. H. A. (2001). Children's use of different media: For how long and why? In Livingstone, S., & Bovill, M. (Eds.), *Children and their changing media environment: A European comparative study* (pp. 85–112). Mahwah, NJ: Lawrence Erlbaum.

Beggan, J. K., & Allison, S. T. (2003). Reflexivity in the pornographic films of Candida Royalle. *Sexualities*, *6*(3-4), 301–324. doi:10.1177/136346070363003

Bell, R. (1985). *Holy Anorexia*. Chicago: University of Chicago Press.

Benedikt, M. (Ed.). (1991). *Cyberspace First Steps*. Cambridge, MA: MIT Press.

Benkler, Y. (2007). *The Wealth of Networks: How Social Production Transforms Markets and Freedom*. New Haven, CT: Yale University Press.

Bennett, S., Maton, K., & Kervin, L. (2008). The 'digital natives' debate: A critical review of the evidence. *British Journal of Educational Technology*, *39*(5), 775–786. doi:10.1111/j.1467-8535.2007.00793.x

Ben-Ze'ev, A. (2004). *Love Online*. Cambridge, UK: Cambridge University Press. doi:10.1017/CBO9780511489785

Berger, P., & Luckmann, T. (1966). *The Social Construction of Reality. A Treatise in the Sociology of Knowledge*. New York: Anchor books.

Bergström, A. (2001). Internet - från revolution till vardagsanvändning. In S. Holmberg & L. Weibull (Eds.). *Land, Du Välsignade?: SOM-Undersökningen 2000* (pp. 283-296). Göteborg: SOM-institutet, Univ.

Berkowitz, L. (1989). Frustration-aggression hypothesis: Examination and reformulation. *Psychological Bulletin*, *106*, 59–73. doi:10.1037/0033-2909.106.1.59

Berkowitz, L. (1993). *Aggression: Its causes, consequences, and control*. New York: Mc Graw-Hill.

Betterton, R. (Ed.). (1987). *Looking on: images of femininity in the visual arts and media*. New York: Pandora Press.

Bhat, C. (2008). Cyber Bullying: Overview and Strategies for School Counsellors Guidance Officers and All School Personnel. *Australian Journal of Guidance & Counselling*, *18*(1), 53–66. doi:10.1375/ajgc.18.1.53

Bierhoff, H., & Vornefeld, B. (2004). The Social Psychology of Trust with Applications in the Internet. *Analyse & Kritik*, *26*(1), 48–62.

Bjørnstad and Ellingsen. (2004). *Onliners: A report about youth and the Internet*. Norwegian Board of Film Classification.

Blau, P. M. (1977). A Macrosociological Theory of Social Structure. *American Journal of Sociology*, *83*(1), 26–54. doi:10.1086/226505

Bledsoe, C. H., & Cohen, B. (1993). *Social Dynamics of Adolescents Fertility in Sub-Sahara Africa*. Washington, D.C: National Academy Press.

Blood, R. (2000). *Weblogs: a History and Perspective*. Retrieved 13 September 2007, found at http://www.rebeccablood.net/essays/weblog_history.html.

Bloustien, G. (2003). *Girl making: a cross-cultural ethnography on the processes of growing up female*. New York: Berghahn Books.

Blumer, H. (1969). *Symbolic Interactionism. Perspective and Method*. Berkley, CA: University Press.

Boase, J., Horrigan, J. B., Wellman, B., & Rainie, L. (2006). *The strength of internet ties*. Pew Internet and American Life Project. Retrieved March 10, 2010 from http://www.pewinternet.org/~/media//Files/Reports/2006/PIP_Internet_ties.pdf.pdf

Boies, S. C., Knudson, G., & Young, J. (2004). The Internet, Sex, and Youths: Implications for Sexual Development. *Sexual Addiction & Compulsivity*, *11*, 343–363. doi:10.1080/10720160490902630

Bordo, S. (1993). *Unbearable Weight. Feminism, Western Culture and the Body*. Berkeley, CA: California University Press.

Bortree, D. S. (2005). Presentation of self on the Web: an ethnographic study of teenage girls' weblogs. *Education Communication and Information*, *5*(1), 25–39.

Bottoms, A., & Wiles, P. (1996). Understanding crime prevention in late modern societies. In Bennett, T. (Ed.), *Preventing crime and disorder: Targeting strategies and responsibilities*. Cambridge, UK: University of Cambridge.

Bourdieu, P. (1977). *Outline of a Theory of Practice*. Cambridge, UK: Cambridge University Press.

Bourdieu, P. (1998). *Pracrical reason: On the theory of action*. Stanford, CA: Stanford University Press.

Bowker, N. I., & Tuffin, K. (2007). Understanding positive subjectivities made possible online for disabled people. *New Zealand Journal of Psychology*, *36*(2), 63–71.

Boyd, D. (2007). Why Youth (Heart) Social Network Sites: The Role of Networked Publics in Teenage Social Life. In Buckingham, D. (Ed.), *Mc Arthur Foundation on Digital Learning – Youth, Identity, and Digital Media Volume*. Cambridge, MA: MIT Press.

boyd, d. (2008). *Taken Out of Context American Teen Sociality in Networked Publics*. Unpublished doctoral dissertation, University of California, Berkley.

boyd, D., & Ellison, N. B. (2007). Social Network Sites: Definition, History, and Scholarship. *Journal of Computer-Mediated Communication, 13*(1), 210-230.

Brannon, R. (1976). The male sex role: Our culture's blueprint of manhood, and what it's done for us lately. In Brannon, R., & David, D. (Eds.), *The forty-nine percent majority: The male sex role. Reading, MA.: Addison-Wesley. Buckingham, D. (2009). Video Cultures in Technology and Everyday Creativity*. New York: Palgrave Macmillan.

Bratton, J. (2007). *Work and Organizational Behaviour*. New York: Palgrave Macmillan.

Braun-Courville, D. K., & Rojas, M. (2009). Exposure to sexually explicit web sites and adolescent sexual attritudes and behaviours. *The Journal of Adolescent Health, 45*(2), 156–162. doi:10.1016/j.jadohealth.2008.12.004

Brembeck, H., Johansson, B., & Kampman, J. (2004). *Beyond the competent child: Exploring contemporary childhoods in the Nordic welfare societies*. Fredriksberg, Denmark: Roskilde University Press.

Bridgewater, D. (1997). Effective coming out: Self-disclosure strategies to reduce sexual identity bias. In Sears, J. T., & Williams, W. L. (Eds.), *Overcoming heterosexism and homophobia: Strategies that work* (pp. 65–75). New York: Columbia University Press.

Bristow, J. (1997). *Sexuality*. London: Routledge.

Brown, K., Jackson, M., & Cassidy, W. (2006). Cyber-bullying: Developing policy to direct responses that are equitable and effective in addressing this special form of bullying. [), HTTP://UMANITOBA.CA/PUBLICATIONS/CJEAP/]. *Canadian Journal of Educational Administration and Policy, 57*,

Brown, J. S., & Duguid, P. (2002). *The Social Life of information*. Boston: Harvard Business School publishing Corporation

Brown, L. (2007). Cellphone policy needed, teachers says. *The Star, July 14*, 1-3

Browne, K., & Hamilton-Giachritsis, C. (2005). The influence of violent media on children and adolescents: a public-health approach. *Lancet, 365*(9460), 702–710.

Brumberg, J. (1988). *Fasting Girls. The Emergence of Anorexia Nervosa as a Modern Disease*. Cambridge, MA: Harvard University Press.

Buchman, D. D., & Funk, J. B. (1996). Video and computer games in the '90s: Children's time commitment and game preference. *Children Today, 24*, 12–16.

Buchmann, M. (2004) Sociology of Youth Culture. *International Encyclopedia of the Social & Behavioral Sciences*, 1: *16660-16664*

Buckingham, D. (1999). Children, Young People and Digital Technology. *In Convergence*, *5*(4).

Buckingham, D. (2008). *Youth, Identity, and Digital Media*. Cambridge, MA: Massachusetts Institute of Technology.

Buckingham, D. (2000). *After the death of childhood. Growing up in the age of electronic media*. Cambridge, UK: Polity Press.

Buckingham, D. (2002). The Electronic Generation? Children and New Media. In Lievrouw, L., & Livingstone, S. (Eds.), *The Handbook of New Media* (pp. 77–89). London: Sage.

Bunwaree, S. (1999). Gender inequality: the Mauritian experience. In Heward, C., & Bunwaree, S. (Eds.), *Gender, Education and Development: Beyond Access to Empowerment*. London: Zed Books.

Burden, K., Palmer, C., & Lyde, B., & Gilbert. (2003). Internet crime: Cyber Crime — A new breed of criminal? *Computer Law & Security Report, 19*(3), 222–227. doi:10.1016/S0267-3649(03)00306-6

Burgess, J. E., & Green, J. B. (2009). *YouTube: Online Video and Participatory Culture*. Cambridge, UK: Polity Press.

Burks, V. S., Laird, R. D., Dodge, K. A., Pettit, G. S., & Bates, J. E. (1999). Knowledge structures, social information processing, and children's aggressive behavior. *Social Development, 8*, 220–236. doi:10.1111/1467-9507.00092

Burr, V. (2003). *Social Constructionism*. London: Routledge.

Burton, G. (2004). *Media and Society: Critical Perspectives*. Berkshire/GBR: McGraw-Hill Education.

Busby, L. (1975). Sex-role research on the mass media. *The Journal of Communication, 25*(4), 107–131. doi:10.1111/j.1460-2466.1975.tb00646.x

Buschman, J., & Bogaerts, S. (2009). Polygraph testing internet offenders. In D.T. Wilcox (Ed.). *The use of the polygraphy in assessing, treating and supervising sex offenders: A practitioner's guide* (pp.113-128). West Sussex, UK: John Wiley & Sons LTd.

Bushman, B. J., & Anderson, C. A. (2001). Is it time to pull the plug on the hostile versus instrumental aggression dichotomy? *Psychological Review*, *108*, 273–279. doi:10.1037/0033-295X.108.1.273

Bushman, B. J., Rothstein, H. R., & Anderson, C. A. (2010). Much ado about something: Violent video game effects and a school of red herring: Reply to Ferguson und Kilburn (2010). *Psychological Bulletin*, *136*(2), 182–187. doi:10.1037/a0018718

Bushman, B. J., & Huesmann, L. R. (2001). Effects of televised violence on aggression. In Singer, D. G., & Singer, J. L. (Eds.), *Handbook of children and the media* (pp. 223–254). Thousand Oaks, CA: Sage.

Butcher, H., Coward, R., Evaristi, M., Garber, J., Harrison, R., & Winship, J. (1974). *Images of women in the media*. Birmingham, West Midlands: Centre for Contemporary Cultural Studies, University of Birmingham.

Butler, J. (1990). *Gender trouble. Feminism and the subversion of identity*. New York: Routledge.

Butler, J. (1993). Critically queer. *GLQ: A Journal of Lesbian and Gay Studies*, *1*(1), 17-32.

Callaghan, D. (1989). *Woman and gender in Renaissance tragedy: a study of King Lear, Othello, The Duchess of Malfi and The White Devil*. New York: Harvester Wheatsheaf.

Cameron, K. A., Salazar, L. F., Bernhardt, J. M., Burgess-Whitman, N., Wingood, G. M., & DiClemente, R. J. (2005). Adolescents' experience with sex on the web: results from online focus groups. *Journal of Adolescence*, *28*(4), 535–540. doi:10.1016/j.adolescence.2004.10.006

Campbell, M. (2005). Cyberbullying: An older problem in a new guise? *Australian Journal of Guidance & Counselling*, *15*(1), 68–76. doi:10.1375/ajgc.15.1.68

Campbell, A. J., Cumming, S. R., & Hughes, I. (2006). Internet use by the socially fearful: addiction or therapy? *Cyberpsychology & Behavior*, *9*, 69–81. doi:10.1089/cpb.2006.9.69

CANEE. (2008a). *2nd International Conference "Keeping Children and Young People Safe Online", 18th -19th September 2008, Warsaw*. Available at: http://www.canee.net/poland/2nd_international_conference_keeping_children_and_young_people_safe_online_18th_19th_september_2008_warsaw

CANEE. *(2008b)*. Internet safety through Council of Europe online game. *Available at:*http://www.canee.net/child_and_internet/internet_safety_through_council_of_europe_online_game

CANEE. (2009a). *Keeping children and young people safe online*. Available at: http://www.canee.net/poland/keeping_children_and_young_people_safe_online

CANEE. (2009b). *Polish network against commercial sexual exploitation of children*. Available at: http://www.canee.net/poland/polish_network_against_commercial_sexual_exploitation_of_children

CANEE. (2009c). *Safer internet day 2009 celebrated in Poland*. Available at: http://www.canee.net/poland/safer_internet_day_2009_celebrated_in_poland

Card, N. A., & Little, T. D. (2006). Proactive and reactive aggression in childhood and adolescence: A meta-analysis of differential relations with psychosocial adjustment. *International Journal of Behavioral Development*, *30*(5), 466–480. doi:10.1177/0165025406071904

Card, N. A., Stucky, B., Sawalani, G., & Little, T. D. (2008). Direct and indirect aggression during childhood and adolescence: A meta-analytic review of gender differences, intercorrelations, and relations to maladjustment. *Child Development*, *79*(5), 1185–1229. doi:10.1111/j.1467-8624.2008.01184.x

Carlsson, U. (2010). *Mediesverige 2010*. Göteborg, UK: Nordicom.

Carroll, J. S., Padilla-Walker, L. M., Nelson, L. J., Olson, C. D., Barry, C. M., & Madsen, S. D. (2008). Generation XXX: Pornography acceptance and use among emerging adults. *Journal of Adolescent Research*, *23*, 6–30. doi:10.1177/0743558407306348

Carvalheira, A., & Gomes, F. A. (2003). Cybersex in portuguese chatrooms: A study of sexual behaviors related to online sex. *Journal of Sex & Marital Therapy*, *29*(5), 345–360. doi:10.1080/00926230390224729

Case, S.-E. (1988). *Feminism and theatre*. New York: Routledge.

Casey, M. A., & Kreuger, R. A. (2000). *Focus groups: A practical guide for applied research* (3rd ed.). Thousand Oaks, CA: Sage.

Cass, V. C. (1979). Homosexual identity formation: A theoretical model. *Journal of Homosexuality*, *4*(3), 219–235. doi:10.1300/J082v04n03_01

Cassell, J., & Cramer, M. (2008). High Tech or High Risk: Moral Panics about Girls Online. In McPherson, T. (Ed.), *Digital Young, Innovation, and the Unexpected. The John D. and Catherine T. MacArthur Foundation Series on Digital Media and Learning*. Cambridge, MA: MIT Press.

Cassidy, W., & Bates, A. (2005). "Drop-outs" and "push-outs": finding hope at a school that actualizes the ethic of care. *American Journal of Education*, *112*(1), 66–102. doi:10.1086/444524

Cassidy, W., Jackson, M., & Brown, K. (2009). Sticks and stones can break my bones, but how can pixels hurt me? Students' experiences with cyber-bullying. *School Psychology International*, *30*(4), 383–402. doi:10.1177/0143034309106948

Cassidy, W., Brown, K., & Jackson, M. (2010). *Redirecting students from cyber-bullying to cyber-kindness.* Paper presented at the 2010 Hawaii International Conference on Education, Honolulu, Hawaii, January 7 to 10, 2010.

Castells, M. (1996). The Information Age. Economy, Society and Culture.: *Vol. 1. The Rise of the Network Society*. Oxford, UK: Blackwell.

Castells, M. (1997). *The power of identity. Volum 2, information age*. Oxford, UK: Blackwell.

Castells, M. (2001). Lessons from the History of Internet. In *The Internet Galaxy* (pp. 9–35). New York: Oxford University Press.

Cerpas, N. (2002). Variation in the Display and Experience of Love between College Latino and Non-Latino Heterosexuals Romantic Couples. *The Berkley McNair Research Journal*, *10*, 173–189.

Chaney, D. (2002). *Cultural Change and Everyday Life*. Basingstoke, UK: Palgrave.

Charon, J. M. (2007). *Symbolic Interactionism. An Introduction, an Interpretation, an Integration*. London: Pearson Prentice Hall.

Chatterjee, B. (2000). Cyberpornography, cyberidentities, and Law. *International Review of Law Computers & Technology*, *14*(1), 89–93. doi:10.1080/13600860054926

Cheng, X., Dale, C., & Liu, J. (2008). Statistics and social network of YouTube videos. In Proceedings of the 16th International Workshop on Quality of Service (pp. 229–238). Enschede: IWQoS.

Chou, C., Yu, S., Chen, C., & Wu, H. (2009). Tool, Toy, Telephone, Territory, or Treasure of Information: Elementary school students' attitudes toward the Internet. *Computers & Education*, *53*(2), 308–316. doi:10.1016/j.compedu.2009.02.003

Ciccetti, D., & Toth, S. L. (1995). A developmental psychopathology perspective on child abuse and neglect. *Journal of the American Academy of Child and Adolescent Psychiatry*, *34*, 541–565. doi:10.1097/00004583-199505000-00008

Clark, L. (1998). Dating on the Net: Teens and the rise of 'pure' relationships. In Jones, S. (Ed.), *Cybersociety: Revisiting computer-mediated communication and community* (pp. 159–183). Thousand Oaks, CA: Sage.

Cline-Cole, R., & Powell, M. (2004). ICTs, 'Virtual Colonisation' and Political Economy. *Review of African Political Economy*, *31*(99), 5–9. doi:10.1080/0305624042000258388

Cohen, S. (1972). *Folk Devils and Moral Panics: the Creation of the Mods and Rockers*. London: MacGibbon and Kee.

Coleman, E. (1982). Developmental stages of the coming-out process. *Journal of Homosexuality*, *7*(2/3), 31–43. doi:10.1300/J082v07n02_06

Coleman, J. C., & Hendry, L. B. (1999) *The Nature of Adolescence, (3rd Ed).* London: Routledge.

Colnerud, G. (1997). I de MÖRKASTE vrårna av skolans värld. *Pedagogiska Magasinet*, *1*(4), 61–65.

Colnerud, G. (2002). Den kollegiala paradoxen. *Pedagogiska Magasinet*, *4*(4), 24–30.

Colwell, J., & Kato, M. (2003). Investigation of the relationship between social isolation, selfesteem, aggression and computer game play in Japanese adolescents. *Asian Journal of Social Psychology*, *6*, 149–158. doi:10.1111/1467-839X.t01-1-00017

Colwell, J., & Payne, J. (2000). Negative correlates of computer game play in adolescents. *The British Journal of Psychology*, *91*, 295–310. doi:10.1348/000712600161844

Connell, R. W. (1995). *Masculinities*. Cambridge, UK: Polity Press.

Connolly, R. (2001). The rise and persistence of the technological community ideal. In Werry, C., & Mowbrey, M. (Eds.), *Online community: Commerce, community action and the virtual university* (pp. 317–364). Upper Saddle River, NJ: Prentice Hall.

Cook-Sather, A. (2002). Authorizing students' perspectives: Toward trust, dialogue, change in Education. *Educational Researcher*, *24*(June-July), 12–17.

Cooley, C. H. (1909). *Social Organization*. New York: Charles Scribner's Sons.

Cooley, C. H. (1922). *Human Nature and the Social Order* (Revised Edition). New York: Charles Scribner's Sons.

Coombe, R. (1998). *The cultural life of intellectual properties – Authorship, Appropriation, and the Law*. Durham, NC: Duke University Press.

Cooper, A., Mcloughlin, I. P., & Campbell, K. M. (2000). 'Sexuality in Cyberspace: Update for the 21st Century'. *Cyberpsychology & Behavior*, *3*(4), 521–536. doi:10.1089/109493100420142

Cooper, A., Scherer, C. R., Boies, S. C., & Gordon, B. L. (1999). Sexuality on the internet: rom sexual exploration to pathological expression. *Professional Psychology, Research and Practice*, *30*(2), 154–164. doi:10.1037/0735-7028.30.2.154

Cooper, A., & Griffin-Shelley, E. (2002). The internet: The next sexual revolution. In Cooper, A. (Ed.), *Sex & the internet: A guidebook for clinicians*. New York: Brunner-Routledge.

Creswell, J. W. (2003). *Research design: qualitative, quantitative, and mixed methods approaches* (2nd ed.). Thousand Oaks, CA: Sage Publications.

Creswell, J. W., Plano Clark, V. L., Gutmann, M. L., & Hanson, W. E. (2003). Advanced mixed methods research designs. In Tashakkori, A., & Teddlie, C. (Eds.), *Handbook of mixed methods in the social and behavioral research. (pp. 209e240)*. Thousand Oaks, CA: Sage Publications.

Crick, N., & Grotpeter, J. (1995). Relational aggression, gender, and social-psychological adjustment. *Child Development*, *66*, 710–722. doi:10.2307/1131945

Crick, N., Bigbee, M. A., & Howes, C. (1996). Gender differences in children's normative beliefs about aggression: How do I hurt thee? Let me count the ways. *Child Development*, *67*, 1003–1014. doi:10.2307/1131876

Crick, N., Werner, N., Casas, J., O'Brien, K., Nelson, D., Grotpeter, J., & Markon, K. (1999). Childhood aggression and gender: A new look at an old problem. In Bernstein, D. (Ed.), *Nebraska symposium on motivation* (pp. 75–141). Lincoln, NE: University of Nebraska Press.

Critcher, C. (2003). *Moral panics and the media. Issues in cultural and media studies*. Buckingham, UK: Open University Press.

Critcher, C. (2008). Making Waves: Historical Aspects of Public Debates about Children and Mass Media. In *The International Handbook of Children, Media and Culture* (pp. 91-104). Los Angeles, CA: SAGE.

Croteau, D., & Hoynes, W. (2000). *Media society: Industries, images, audiences*. Thousand Oaks, CA: Pine Forge.

CSO (Central Statistics Office). (2000). *Population Census 2000*. Retrieved December 23, 2009, from www.gov.mu/portal/sites/ncb/cso/report/.../census5/index.htm

Cudak, E. (2007). Relacje wirtualne relacjami wirtualnymi? In Marek, Z., & Madej-Babula, M. (Eds.), *Bezradność wychowania* (pp. 37–48). Kraków, Poland: WAM.

Currie, C., Roberts, C., Morgan, A., Smith, R., Settertobulte, W., & Samdal, O. (2004). *Young people's health in context. Health Behaviour in School-aged Children (HBSC) study: international report from the 2001/2002 survey*. Kopenhagen: World Health Organization Regional Office for Europe.

D'Augelli, A. R., Pilkington, N. W., & Hershberger, S. L. (2002). Incidence and mental health impact of sexual orientation victimization of lesbian, gay, and bisexual youths in high school. *School Psychology Quarterly*, *17*(2), 148–167. doi:10.1521/scpq.17.2.148.20854

Daher-Larsson, A. (1990). *Ormen och jag. Om en kamp mot anorexia*. Södra Sandby: Atlantis.

Dahlberg, C. (2000). *I himlen får jag äta*. St. Petersburg: Ord & Visor förlag.

Daneback, K., Cooper, A., & Månsson, S.-A. (2005). An internet study of cybersex participants. *Archives of Sexual Behavior*, *34*(3), 321–328. doi:10.1007/s10508-005-3120-z

Daneback, K., Ross, M. W., & Månsson, S.-A. (2006). Characteristics and behaviors of sexual compulsives who use the internet for sexual purposes. *Sexual Addiction & Compulsivity*, *13*(1), 53–67. doi:10.1080/10720160500529276

Daneback, K. (2006). *Love and sexuality on the internet* [Doctoral dissertation]. Göteborg, Sweden: Göteborg University.

Daneback, K., & Månsson, S-A. (2009). Kärlek och sexualitet på internet år 2009

Danet, B. (2001). *Cyberpl@y: Communicating Online*. Oxford, UK: Berg.

Danmarks Statistik. (2009). *Befolkningens brug af internet 2009*. Danmarks Statistik. Retrieved from http://www.dst.dk/publikation.aspx?cid=14039

Darren Star Productions, D. Star, M. P. King & S. J. Parker (Producer). (1998-2004) *Sex and the City*.New York: Home Box Office.

David-Ferdon, C., & Hertz, M. (2007). Electronic media, violence, and adolescents: an emerging public health problem. *The Journal of Adolescent Health*, *41*(6Suppl 1), S1–S5. doi:10.1016/j.jadohealth.2007.08.020

Davies, J. (2004). Negotiating femininities online. *Gender and Education*, *16*(1), 35–49. doi:10.1080/0954025032000170327

Davis, G. E., & Leitenberg, H. (1987). Adolescent sexual offenders. *Psychological Bulletin*, *101*, 417–427. doi:10.1037/0033-2909.101.3.417

de Beauvoir, S. (1949/1972). *The second sex* (Parshley, H. M., Trans.). Harmondsworth, UK: Penguin.

Delgado, C. (2005). A Discussion of the Concept of Spirituality. *Nursing Science Quarterly*, *18*, 157–162. doi:10.1177/0894318405274828

Denks, D., Bos, A. & von Grumbkow (2008). Emoticons and Online Message Interpretation.

Dennett, D. C. (1987). *The intentional stance. Cambridge, MA*. Bradford: MIT Press.

Dennett, D. C. (2003). *Freedom Evolves*. New York: Viking Press.

Deschamps, Y. C., & Devos, T. (1998). Regarding the Relationship between social identity and Personal Identity. In Worchel (Ed.), *Social Identity: International Perspectives* (pp1- 12). London: Sage Publications

Dias, K. (2003). The Ana Sanctuary: Women's Pro-Anorexia Narratives in Cyberspace. *Journal of International Women's Studies*, *4*(2).

Didden, R., Scholte, R. H. J., Korzilius, H., de Moor, J. M. H., Vermeulen, A., & O'Reilly, M. (2009). Cyberbullying among students with intellectual and developmental disability in special education settings. *Developmental Neurorehabilitation*, *12*(3), 146–151. doi:10.1080/17518420902971356

Diepold, J., & Young, R. D. (1979). Empirical Studies of Adolescents Sexual Behaviour: A Critical Review. *Adolescence*, *16*(53), 45–64.

Digitala infödingar (n.d.). Google search for "Digitala infödingar", March 3, 2010.

Dijck van, J. (2005). Composing the Self: Of Diaries and Lifelogs. *Fibreculture*, (3).

Dill, K. E., & Dill, J. C. (1998). Video game violence: A review of the empirical literature. *Aggression and Violent Behavior*, *3*, 407–428. doi:10.1016/S1359-1789(97)00001-3

Directive 2006/24/EC. Directive 2006/24/EC of the European Parliament and of the Council on the retention of data generated or processed in connection with the provision of publicly available electronic communications services or of public communications networks and amending Directive 2002/58/EC.

Doane, M. A. (1987). *The desire to desire: the woman's film of the 1940s*. Bloomington, IN: Indiana University Press.

Dobson, A. S. (2008). Femininities as commodities: cam girl culture. In Harris, A. (Ed.), *Next wave cultures: feminism, subcultures, activism* (pp. 123–148). New York: Routledge.

Dobson, A. S. (2010). *Bitches, Bunnies and BFF's: a feminist analysis of young women's performance of contemporary popular femininities on MySpace*. Unpublished doctoral dissertation. Monash University, Melbourne.

Dodge, K. A., & Coie, J. D. (1987). Social information processing factors in reactive and proactive aggression in children's playgroups. *Journal of Personality and Social Psychology*, *53*, 1146–1158. doi:10.1037/0022-3514.53.6.1146

Dodge, K. A., Lochman, J. E., Harnish, J. D., Bates, J. E., & Pettit, G. S. (1997). Reactive and proactive aggression in school children and psychiatrically impaired chronically assaultive youth. *Child Development*, *74*, 374–393. doi:10.1111/1467-8624.7402004

Dodge, K. A. (1991). The structure and function of reactive and proactive aggression. In Pepler, D., & Rubin, K. (Eds.), *The development and treatment of childhood aggression* (pp. 201–218). Hillsdale, NJ: Erlbaum.

Doise, W. (1998). Social Representations in Personal Identity. In Worchel, Y. (Ed.), *Social Identity: International Perspectives* (pp. 13–23). London: Sage Publications.

Dollard, D. J., Doob, L. W., Miller, N. E., Mowrer, O. H., & Sears, R. R. (1939). *Frustration and aggression*. New Haven, CT: Yale University Press. doi:10.1037/10022-000

Dombrowski, S. C., Ahia, C. E., & McQuillan, K. (2003). Protecting children through mandated child abuse reporting. *The Educational Forum*, *67*(2), 76–85. doi:10.1080/00131720308984549

Dombrowski, S. C., LeMasney, J. W., Ahia, C. E., & Dickson, S. A. (2004). Protecting children from online sexual predators: technological, psychoeducational, and legal considerations. *Professional Psychology, Research and Practice*, *35*(1), 65–73. doi:10.1037/0735-7028.35.1.65

Donath, J., Karahalios, K., & Viegas, F. (1999). Visualizing conversation. *Journal of Computer-Mediated Communication, 4*(4). Retrieved March 10, 2010 from http://jcmc.indiana.edu/vol4/issue4/donath.html

Dooley, J., Pyżalski, J., & Cross, D. (2009). (in press). Cyberbullying Versus Face-to-Face Bullying: A Theoretical and Conceptual Review. *Zeitschrift für Psychologie. The Journal of Psychology*, *217*(4), 182–188.

Döring, M. D. (2009). The Internet's impact on sexuality: A critical review of 15 years of research. *Computers in Human Behavior*, *25*, 1089–1101. doi:10.1016/j.chb.2009.04.003

Döring, N. (2000). Feminist Views of Cybersex: Victimisation, Liberation, & Empowerment. *CyberPsychology & Behaviour*, *3*(5), 863–884. doi:10.1089/10949310050191845

Dowell, E., Burgess, A., & Cavanaugh, D. (2009). Clustering of Internet Risk Behaviors in a Middle School Student Population. *The Journal of School Health, 79*(11), 547–553. doi:10.1111/j.1746-1561.2009.00447.x

Dresang, E. T. (1999). More research needed: Informal information-seeking behavior of youth on the internet. *Journal of the American Society for Information Science American Society for Information Science, 50*(12), 1123–1124. doi:10.1002/(SICI)1097-4571(1999)50:12<1123::AID-ASI14>3.0.CO;2-F

Drotner, K. (1999). Dangerous Media? Panic Discourses and Dilemmas of Modernity. *Paedagogica Historica, 35*(3), 593–619. doi:10.1080/0030923990350303

Drotner, K., & Rudberg, M. (Eds.). (1993). *Dobbeltblikk på det moderne: Unge kvinners hverdagsliv og kultur i Norden.* Oslo, Sweden: Universitetforlaget.

Drotner, K. (1992). Modernity and Media Pancis. In Skovmand, M., & Schrøder, K. C. (Eds.), *Media Cultures. Reappraising Transnational Media* (pp. 42–62). London: Routledge.

Drotner, K. (1991). *At skabe sig - selv: ungdom, æstetik, pædagogik* (1 ed.). Copenhagen, Denmark: Gyldendal.

Due, P., Holstein, B., Lynch, J., Diderichsen, F., Gabhain, S., & Scheidt, P. (2005). Bullying and symptoms among school-aged children: international comparative cross sectional study in 28 countries. *European Journal of Public Health, 15*(2), 128–132. doi:10.1093/eurpub/cki105

Duffy, M. E. (2003). Web of Hate: A Fantasy Theme Analysis of the Rhetorical Vision of Hate Groups Online. *The Journal of Communication Inquiry, 27*, 291–312.

Dunkels, E. (2008). Children's Strategies on the Internet. *Critical Studies in Education, 49*(2). doi:10.1080/17508480802123914

Dunkels, E. (2009). När kan vi tala om nätmobbning? [When can we speak of cyber bullying?]. *Locus (Denton, Tex.), 2009*(2).

Dunkels, E. (2007). *Bridging the Distance.* Umeå, Sweden: Umeå University.

Dunkels, E., & Enochsson, A. (2008). Interview with young people using online chat. In Quigley, M. (Ed.), *Encyclopaedia of Information Ethics and Security.* Hershey: Idea Media Group.

Dunkels, E. (2005). Nätkulturer - Vad gör barn och unga på Internet? (Net cultures - What do Children and Young People do on the Internet?). *Tidskrif för lärarutbildning och forskning, 1-2*, 41-49.

Dunkels, E. (2008). Children's Strategies on the Internet. In *Critical Studies in Education*, vol. 49 no. 2 Sep. 2008.

Durkin, K., & Barber, B. (2002). Not so doomed: Computer game play and positive adolescent development. *Journal of Applied Developmental Psychology, 23*(4), 373–392. doi:10.1016/S0193-3973(02)00124-7

Durkin, K. F. (1997). Misuse of the Internet by pedophiles: implications for law enforcement and probation practice. *Federal Probation, 61*, 14–18.

Durkin, K.F. & Bryant, C.D. (1999). Propagandizing pederasty: A thematic analysis of the online exculpatory accounts of unrepentant pedophiles. *Deviant Behavior: An interdisciplinary journal, 20,* 103 - 207.

Dyer, R. (2002). *The matter of images: Essays on representation.* London: Routledge.

Eastin, M. S. (2006). Video game violence and the female game player: Self- and opponent gender effects on presence and aggressive thoughts. *Human Communication Research, 32*, 351–372. doi:10.1111/j.1468-2958.2006.00279.x

Edgar, P., & McPhee, H. (1974). *Media she.* Melbourne, Victoria: Heinemann.

Edwards, S. S. M. (2000). Prosecuting 'child pornography': Possession and taking of indecent photos of children. *Journal of Social Welfare and Family Law, 22*, 1–21. doi:10.1080/014180300362732

Ehn, A. (1995). *Vårfrost.* Stockholm, Sweden: Norstedt.

Elliott, M., Browne, K., & Kilcoyne, J. (1995). Child sexual abuse prevention: what offenders tell us. *Child Abuse & Neglect*, *19*, 579–594. doi:10.1016/0145-2134(95)00017-3

Ellison, L.,&Akdeniz, Y. (1998). Cyber-stalking: the regulation of harassment on the Internet.*Criminal Law review, December Special Edition: Crime, Criminal Justice and the Internet,* 29-48.

Ellison, N. B. (2008). Introduction: Reshaping Campus Communication and Community through Social Network Sites. In *The ECAR Study of Undergraduate Students and Information Technology, 2008*, ECAR Research Study Vol. 8, pp. 19-32. EDUCAUSE Center for Applied Research. Retrieved from http://www.educause.edu/ecar.

Ellison, N. B., Steinfield, C., & Lampe, C. (2007). The benefits of Facebook "friends:" Social capital and college students' use of online social network sites. *Journal of Computer-Mediated Communication, 12*(4), article 1.

Elvstrand, H. (2009). *Delaktighet i skolans vardagsarbete.* Linköping, Sweden: Linköping universitet.

Embar-Seddon, A. (2002). Cyberterrorism: Are We Under Siege? *The American Behavioral Scientist*, *45*(6), 1033–1043. doi:10.1177/0002764202045006007

En enkätundersökning bland män och kvinnor 18-24 år [Love and sexuality on the internet 2009: A survey among men and women 18-24]. I Ungdomsstyrelsen, *Se mig. Unga om sex och internet [See me. Young people on sex and the internet]*, pp 182-237. Ungdomsstyrelsen 2009:9.

Erikson, E. H. (1968). *Identity, youth, and crisis*. New York: W. W. Norton.

Eriksson, M. (2008). *Barns röster om våld: att tolka och förstå [[Children's vioces about violence: to interpret and understand*]]. Malmö, Sweden: Gleerups.

Eron, L. D., Huesmann, L. R., Lefkowitz, M. M., & Walder, L. O. (1972). Does television violence cause aggression? *The American Psychologist*, *27*, 253–263. doi:10.1037/h0033721

Escobar-Chaves, S. L., Tortolero, S. R., Markham, C. M., Low, B. J., Eitel, P., & Thickstun, P. (2005). Impact of the Media on Adolescent Sexual Attitudes and Behaviors. *Pediatricsm*, *116*(1), 303–326.

Esposito, L. C. (1998). Regulating the Internet: The new battle against child pornography. *Case Western Reserve Journal of International Law*, *3*(213), 541–567.

Etienne, P. (2007). *'Sex Education is Becoming Urgent for Self-Protection'*. Retrieved May 05, 2007 from www.lexpress.mu/display_article_sup.php?news

Eurobarometer. (2007). *Safer Internet for Children, Qualitative study in 29 European countries, National Analysis: Denmark*. Hellerup: European Commission, Directorate-General Information Society and Media. Retrieved from http://andk.medieraadet.dk/upload/denmark_report.pdf

Fafunwa, A. B. (1974). *History of Education in Nigeria.* London: George Allen and union Ltd.

Faludi, S. (1992). *Backlash: the undeclared war against women*. London: Chatto & Windus.

Falvey, M. (1986). *A Community-Based Curriculum: Instructional Strategies for Students with Severe Handicaps*. Baltimore, MD: Paul H. Brookes Publishing Company.

FDIM. (2009). Charmærket. *Foreningen af Danske Interaktive Medier*. Retrieved January 15, 2010, from http://www.fdim.dk/?pageid=48

Featherstone, M. (1982). The Body in Consumer Culture'. *Theory, Culture & Society*, *1*(2), 18–33. doi:10.1177/026327648200100203

Feingold, A. (1992). Good-looking people are not what we think. *Psychological Bulletin*, *111*, 304–341. doi:10.1037/0033-2909.111.2.304

Ferguson, C. J. (2007). Evidence for publication bias in video game violence effects literature: A meta-analytic review. *Aggression and Violent Behavior*, *12*, 470–482. doi:10.1016/j.avb.2007.01.001

Ferguson, C. J., & Kilburn, J. (2009). The public health risk of media violence: A meta-analytic review. *The Journal of Pediatrics*, *154*, 759–763. doi:10.1016/j.jpeds.2008.11.033

Ferguson, C. J., & Kilburn, J. (2010). Much ado about nothing: The misestimation and overinterpretation of violent video game effects in Eastern and Western nations: Comment on Anderson et al.(2010). *Psychological Bulletin*, *136*(2), 174–178. doi:10.1037/a0018566

Ferreira, P. R., & Ribeiro, J. L. P. (2001). The relationship between playing violent electronic games and aggression in adolescents. In Martinez, M. (Ed.), *Prevention and control of aggression and the impact on its victims* (pp. 129–135). New York: Kluwer.

Ferris, L. (1990). *Acting women: images of women in theatre*. Basingstoke, UK: Macmillan.

Fieldman, J. P., & Crespi, T. D. (2002). Child sexual abuse: offenders, disclosure, and school-based initiatives. *Adolescence*, *37*, 151–161.

Figueroa Sarriera, H. (2006). Connecting the Selves Computer-Mediated Identification Processes. In Silver, D. (Ed.), *Critical Cyberculture Studies*. New York: NYU Press.

Fikar, C. R., & Keith, L. (2004). Information needs of gay, lesbian, bisexual, and transgender health care professionals: Results of an internet survey. *Journal of the Medical Library Association*, *92*(1), 56–65.

Findahl, O. (2001). *Svenskarna och Internet 2000*. World Internet Institute. Retrieved from http://www.worldinternetinstitute.org.

Findahl, O. (2008). *Svenskarna och Internet 2008*. World Internet Institute. Retrieved from http://www.worldinternetinstitute.org.

Findahl, O. (2009*). Internet 15 år [Internet 15 years]*. Stockholm, Sweden: Stiftelsen för internetinfrastruktur.

Findahl, O. (2009). *Svenskarna och Internet 2009*. Stockholm.

Finkelhor, D., Mitchell, K.J., & Wolak, (2002). Online victimization: A report on the nation's youth. University of New Hampshire, Crimes Against Children research Center. Available at http://www.missingkids.com (accessed 18 July 2008).

Firat, A. F. (1987). Towards a deeper understanding of consumption experiences: The underlying dimensions. *Advances in Consumer Research. Association for Consumer Research (U. S.)*, *14*(1), 342–346.

Firth, R. (1973). *82). Symbols: Public and Private*. Ithaca, NY: Cornell Univ. Press.

Fish, S. M. (1976). Interpreting the "Variorum.". *Critical Inquiry*, *2*(3), 465–485. Retrieved from http://www.jstor.org/stable/1342862. doi:10.1086/447852

Fisher, W.A. (2001). *'Internet Pornography: A Social Psychological Perspective on Internet Sexuality'*. Retrieved December 12 2009 from http://www.findarticles.com

Flander, G. B., Cosic, I., & Profaca, B. (2009). Exposure of children to sexual content on the Internet in Croatia. *Child Abuse & Neglect*, *33*(12), 849–856. doi:10.1016/j.chiabu.2009.06.002

Flum, H. (1994). Styles of Identity Formation in Early and Middle Adolescence. *Genetic, Social, and General Psychology Monographs*, *120*(4), 435–468.

Flynt, S. W., & Morton, R. C. (2004). Bullying and Children with Disabilities. *Journal of Instructional Psychology*, *31*, 330–333.

Foucault, M., & Gordon, C. (Eds.). (1980). *Power/Knowledge. Selected Interviews and other Writings 1972-1977*. Brighton, UK: Harvester Press.

Foucault, M. (1976). The history of sexuality: *Vol. I. The will to know*. London: Penguin.

Foucault, M. (1972). *The archeology of knowledge and the discourse on language*. New York: Pantheon. (Original work published 1969)

Foucault, M. (1978). *The history of sexuality 1: An introduction*. New York: Random House. (Original work published 1976)

Fourchard, L. (2008). A New Name for an Old Practice: Vigilantes in South-Western Nigeria. *Africa, 78*(1), 16–40. doi:10.3366/E000197200800003X

Fraley, S. S., Mona, L. R., & Theodore, P. S. (2007). The Sexual Lives of Lesbian, Gay, and Bisexual People With Disabilities: Psychological Perspectives. *Sexuality Research & Social Policy, 4*(1), 15–26. doi:10.1525/srsp.2007.4.1.15

Frånberg, G., & Gill, P. (2009). Vad är mobbning? [What is bullying?] In *På tal om mobbning – och det som görs*. Stockholm, Sweden: Skolverket.

Francis, DA. (in press). Sexuality Education in South Africa: Three Essential Questions. *International Journal of Educational Development*, in Press, Corrected Proof, Available online 31 December 2009

Francis, J. (producer) (1998-present) *Girls Gone Wild.* Los Angeles, CA: Mantra Entertainment.

Franck, M. (2009). *Frigjord oskuld. Heterosexuellt mognadsimperativ i svensk ungdomsroman.* Åbo, Finland: Åbo Akademi University Press.

Frei, A., Erenay, N., Dittmann, V., & Graf, M. (2005). Paedophilia on the Internet: A study of 33 convicted offenders in the Canton of Lucerne. *Swiss Medical Weekly, 133*, 488–494.

Friedan, B. (1963). *The feminine mystique*. London: Gollancz.

Friends. (2009). *Koll på nätet: En bok om att skapa samtal mellan barn, unga och vuxna om nätet*. Stockholm, Sweden: Friends.

Frietas, D., & Buckenmeyer, J. (2009). Cell Phones in American High Schools: 21st Century Connections. *FSC News, January 19,* 1-3.

Frindte, W., & Obwexer, I. (2003). Ego-Shooter – Gewalthaltige Computerspiele und aggressive Neigungen. *Zeitschrift für Medienpsychologie, 15*, 140–148. doi:10.1026//1617-6383.15.4.140

Froese-Germain, B. (2008). Bullying in the Digital Age: Using Technology to Harass Students and Teachers. *Our Schools. Our Selves, 17*(4), 45.

Fromme, J., Meder, N., & Vollmer, N. (2000). *Computerspiele in der Kinderkultur*. Opladen: Leske + Budrich.

Funk, J. B., Bechtoldt Baldacci, H., Pasold, T., & Baumgardner, J. (2004). Violence exposure in real-life, video games, television, movies, and the internet: Is there desensitization? *Journal of Adolescence, 27*, 23–39. doi:10.1016/j.adolescence.2003.10.005

Funk, J. B., Buchman, D. D., Jenks, J., & Bechtoldt, H. (2003). Playing violent video games, desensitization, and moral evaluation in children. *Applied Developmental Psychology, 24*, 413–436. doi:10.1016/S0193-3973(03)00073-X

Furman, W., & Wehner, E. A. (1993). Romantic Views: Toward a Theory of Adolescent Romantic Relationships. In Montemayor, R., Adams, G. R., & Gullota, G. P. (Eds.), *Advances in Adolescent Development* (pp. 168–195). Thousand Oaks, CA: Sage Publications.

Gamble, T. K., & Gambel, M. (2002). *Communication works* (7th ed.). Boston: McGraw Hill.

Ganetz, H., & Lövgren, K. (Eds.). (1991). *Om unga kvinnor*. Lund, Sweden: Studentlitteratur.

Garnets, L. D., & Kimmel, D. C. (2003). *Psychological perspectives on lesbian, gay and bisexual experiences.* New York: Columbia University Press.

Garnets, L. D., Herek, G. M., & Levy, B. (2003). Violence and victimization of lesbians and gay men: Mental health consequences. In Garnets, L. D., & Kimmel, D. C. (Eds.), *Psychological perspectives on lesbian, gay and bisexual experiences* (pp. 188–206). New York: Columbia University Press.

Gavin, J., Rodham, K., & Poyer, H. (2008, March). The Presentation of "Pro-Anorexia" in Online Group Interactions. *Qualitative Health Research, 18*(3), 325–333. doi:10.1177/1049732307311640

Gentile, D. A., & Gentile, J. R. (2008). Violent video games as exemplary teachers: a conceptual analysis. *Journal of Adolescence*, *37*, 127–141.

Gentile, D. A., Lynch, P. J., Linder, J. R., & Walsh, D. A. (2004). The effects of violent game habits on adolescent hostility, aggressive behaviors, and school performance. *Journal of Adolescence*, *27*, 5–22. doi:10.1016/j.adolescence.2003.10.002

Gentile, D. A., Saleem, M., & Anderson, C. A. (2007). Public policy and the effects of media violence on children. *Social Issues and Policy Review*, *1*, 15–61. doi:10.1111/j.1751-2409.2007.00003.x

Gentile, D. A., & Walsh, D. A. (2002). A normative study of family media habitus. *Applied Developmental Psychology*, *23*, 157–178. doi:10.1016/S0193-3973(02)00102-8

Genuis, S., & Genuis, S. (2005). Implications for Cyberspace Communication: A Role for Physicians. *Southern Medical Journal*, *98*(4), 451–455. doi:10.1097/01.SMJ.0000152885.90154.89

Gibbs, G. R. (2002). *Qualitative data analysis: explorations with NVivo. Understanding social research*. Buckingham, UK: Open University Press.

Gibson, J. J. (1979). *The Ecological Approach to Visual Perceptions*. Boston, MA: Houghton Mifflin.

Gibson, P. (1994). Gay male and lesbian youth suicide. In Remafedi, G. (Ed.), *Death by denial* (pp. 15–64). London: Alyson Publication.

Giddens, A. (1992). *The transformation of intimacy: sexuality, love and eroticism in modern societies*. Stanford, CA: University.

Gies, L. (2008). How material are cyberbodies? Broadband Internet and embodied subjectivity *Crime Media Culture*, 4(3), 311-330. (References and further reading may be available for this article. To view references and further reading you must purchase this article).

Giles, D. (2006). Constructing identities in cyberspace: The case of eating disorders. *The British Journal of Social Psychology*, *45*, 463–477. doi:10.1348/014466605X53596

Gill, R. (2003). From sexual objectification to sexual subjectification: the resexualisation of women's bodies in the media. *Feminist Media Studies*, *3*(1), 100–106.

Gill, R., & Arthurs, J. (2006). Editors' Introduction – New Femininities? *Feminist Media Studies*, *6*(4), 443–451.

Glaser, B. G. (1965). The constant comparative method of qualitative analysis. *Social Problems*, , 436–445. doi:10.1525/sp.1965.12.4.03a00070

Glasser, M., Kolvin, I., Campbell, D., Glasser, A., Leitch, I., & Farrell, S. (2001). Cycle of child sexual abuse: links between being a victim and becoming a perpetrator. *The British Journal of Psychiatry*, *179*, 482–494. doi:10.1192/bjp.179.6.482

Globe & Mail. (2007-04-07). *'Star Wars Kid' cuts a deal with his tormentors*. www.theglobeandmail.com/servlet/story/RTGAM.20060407.wxstarwars07/BNStory/National/home

Goffman, E. (1969). *The presentation of self in everyday life*. London: Allen Lane.

Goffman, E. (1963). *Stigma: Notes of the management of spoiled identity*. Upper Saddle River, NJ: Prentice-Hall.

Goffman, E. (1983). The interaction ritual. *Americal Sociological Review*, (48), 1-19.

Golding, P., & Middleton, S. (1979). Making claims: news media and the welfare state. *Media Culture & Society*, *1*(1), 5–21. doi:10.1177/016344377900100102

Gordon, S., & Ford, R. (2006). On the definition and classification of cybercrime. *Journal in Computer Virology*, *2*(1), 13–20. doi:10.1007/s11416-006-0015-z

Göthlund, A. (1997). *Bilder av tonårsflickor. Om estetik och identitetsarbete*. Tema Kommunikation: Linköpings universitet.

Gould, M. S., Munfakh, J. L. H., Lubell, K., Kleinman, M., & Parker, S. (2002). Seeking help from the internet during adolescence. *Journal of the American Academy of Child and Adolescent Psychiatry*, *41*(10), 1182–1189. doi:10.1097/00004583-200210000-00007

Gourevitch, P., & Morris, E. (2008). *STANDRAD OPERATING PROCEDURE: A WAR STORY*. London: Pan Macmillian.

Grabosky, P. N. (2001). Virtual Criminality: Old Wine in new bottles? *Social & Legal Studies*, *10*(2), 243–249.

Gradinger, P., Strohmeier, D., & Spiel, C. (2009). Traditional Bullying and Cyberbullying: Identification of Risk Groups for Adjustment Problems. *The Journal of Psychology*, *217*(4), 205–213.

Graneheim, U. H., & Lundman, B. (2004). Qualitative content analysis in nursing research: Concepts, procedures and measures to achieve trustworthiness. *Nurse Education Today*, *24*, 105–112. doi:10.1016/j.nedt.2003.10.001

Granit, E., & Nathan, L. (2000). Virtual communities: A new social structure? [Hebrew]. *Megamot*, *40*(2), 298–315.

Granovetter, M. S. (1973). The Strength of Weak Ties. *American Journal of Sociology*, *78*(6), 1360–1380. doi:10.1086/225469

Gray, L. (2008). A Fangirl's Crush. *Houston Chronicle*. Retrieved from http://www.chron.com/disp/story.mpl/moms/5611220.html.

Green, J., & Jenkins, H. (2009). The Moral Economy of Web 2.0: Audience Research and Convergence Culture. In Holt, J., & Perren, A. (Eds.), *Media Industries: History, Theory, and Method*. Boston: Blackwell.

Green, H., & Hannon, C. (2007). *TheirSpace: Education for a digital generation*. Demos, London. Retrieved July 6, 2009 from: http://www.demos.co.uk/publications/theirspace

Greer, G. (1970). *The female eunuch*. London: MacGibbon & Kee.

Gregory, M. (2000). Care as a goal of democratic education. *Journal of Moral Education*, *29*(4), 445–461. doi:10.1080/713679392

Gregson, K. (2005). What if the Lead Character Looks Like Me? Girl Fans of Shoujo Anime and Their Web Sites. In S. Mazzarella, *Girl Wide Web. Girls, the Internet, Negotioation of Identity*. New York: Peter Lang.

Griffin, R. S., & Gross, A. M. (2004). Childhood bullying:Current empirical findings and future directions for research. *Aggression and Violent Behavior*, *9*, 379–400. doi:10.1016/S1359-1789(03)00033-8

Grindal, B. T. (1982). *Growing Up in Two Worlds: Education and Transition Among the Sisala of Northern Nigeria*. New York: Irvington Publisher.

Grisso, A. D., & Weiss, D. (2005). What are gURLS talking about? Adolescent girls' construction of sexual identity on gURL.com. In Mazzarella, S. R. (Ed.), *Girl wide web: girls, the Internet, and the negotiation of identity* (pp. 31–49). New York: Peter Lang.

Gross, E. F., Juvonen, J., & Gable, S. L. (2002). Internet use and well-being in adolescence'. *The Journal of Social Issues*, *58*(11), 75–90. doi:10.1111/1540-4560.00249

Grotevant, H. D., Thorbecke, W., & Meyer, M. L. (1982). An extension of Marcia's Identity Status Interview into the interpersonal domain. *Journal of Youth and Adolescence*, *11*(1), 33–47. .doi:10.1007/BF01537815

Grov, C., Bimbi, D. S., Nanin, J. E., & Parsons, J. T. (2006). Race, ethnicity, gender and generational factors associated with the coming-out process among gay, lesbian, and bisexual individuals. *Journal of Sex Research*, *43*(2), 115–121. doi:10.1080/00224490609552306

Grov, C., Debusk, J. A., Bimbi, D. S., Golub, S. A., Nanin, J. E., & Parsons, J. T. (2007). Barebacking, the internet, and harm reduction: An intercept survey with gay and bisexual men in Los Angeles and New York City. *AIDS and Behavior*, *11*(4), 527–536. doi:10.1007/s10461-007-9234-7

Grusin, R. (2009). In Snickars, P., & Vonderau, P. (Eds.), *YouTube at the End of New Media* (pp. 60–67). Stockholm, Sweden: National Library of Sweden.

Guan, S. S., & Subrahmanyam, K. (2009). Youth Internet use: risks and opportunities. *Current Opinion in Psychiatry*, *22*(4), 351–356. doi:10.1097/YCO.0b013e32832bd7e0

Guess, D., Benson, H., & Siegel-Causey, E. (2008). Concepts and Issues Related to Choice Making and Autonomy Among Persons With Severe Disabilities. [Retrieved from Academic Search Complete database.]. *Research and Practice for Persons with Severe Disabilities*, *33*(1/2), 75–81.

Gupta, G. R., & Weiss, E. (2005). Women's lives and sex: Implications for AIDS prevention. *Culture, Medicine and Psychiatry*, *17*(4), 399–412. doi:10.1007/BF01379307

Gurbrium, J. F., & Holstein, J. A. (2003). *Postmodern interviewing*. Thousand Oaks, CA: Sage Publications.

Haddon, L., & Stald, G. (2009). *A cross-national European analysis of press coverage of children and the internet. LSE*. London: EU Kids Online.

Hake, K. (1999). Barneperspektivet - en forskningsstrategi. In *Børn, unge og medier. Nordiske forskningsperspektiver* (pp. 193–208). Göteborg, Sweden: Nordicom.

Hall, M. (1999). Virtual Colonization. *Journal of Material Culture*, *4*(1), 39–55. doi:10.1177/135918359900400103

Hall, G. S. (1904). *Adolescence*. New York: Appleton.

Hällgren, C. (2006). *Researching and developing Swedkid: a Swedish case study at the intersection of the web, racism and education /Diss./*. Fakulteten för lärarutbildning vid Umeå universitet.

Hardey, M. (2004). Mediated Relationships: Authenticity and the Possibility of Romance. *Information Communication and Society*, *7*(2), 207–222. doi:10.1080/1369118042000232657

Harel, A. (2000). The rise and fall of the Israeli gay legal revolution. *Columbia Human Rights Law Review*, *31*(2), 443–471.

Harjunen, H. (2009). *Women and Fat. Approaches to the Social Study of Fatness*. Jyväskylä, Finland: Jyväskylä University Press.

Harris, A. (2004). *Future Girl*. New York: Routledge.

Harter, S. (1990). Self and identity development. In S. S. Feldman & G. R. Elliott, *At the threshold: The developing adolescent* (1 ed., pp. 352-387). Cambridge, MA: Harvard University Press.

Hartmann, M. (2005). The Discourse of the Perfect Future - Young People and New Technologies. In R. Silverstone (Ed.), *Media, Technology, and Everyday Life in Europe: From Information to Communication* (pp. 141-158). Aldershot, Hants, England: Ashgate.

Harwood, M. (2009).*Cybercrime Trends Will Worsen in 2009*, According to Forecasts (2009). Retrieved May 15, 2010 from http://www.securitymanagement.com/news/cybercrime-trends-will-worsen-2009-according-forecasts-004969.

Hasset, J., & White, K. M. (1989). *Psychology in perspective*. New York: Harper&Row.

Haythornthwaite, C. (2001). Introduction: The Internet in Everyday Life. *The American Behavioral Scientist*, *45*(3), 363–382. doi:10.1177/00027640121957240

Hedin, U.-C. Månson. S-A., & Tikkanen, R. (2008). *När man måste säga ifrån: Om kritik och whistleblowing i offentliga organisationer.* Stockholm, Sweden: Natur och Kultur.

Hellekson, K., & Busse, K. (2006). *Fan fiction and fan communities in the age of the Internet: new essays*. Jefferson, NC: McFarland & Co.

Henderson, L. (2008). Slow love. *Communication Review*, *11*(3), 219–224. doi:10.1080/10714420802306650

Hendriks, A.The right to health promotion and protection of women's rights to sexual & reproductive health under international law. The economic covenant and the women's convention. *The American University Law Review*, *4*, 1123–1134.

Herdt, G. (1999). Clinical ethnography and sexual study. *Annual Review of Sex Research*, *10*, 100–119.

Herdt, G., & Boxer, A. M. (1993). *Children of horizons: How gay and lesbian teens are leading a new way out of the closet*. Boston: Beacon.

Herring, S. C. (2004). Slouching toward the ordinary: Current trends in computer-mediated communication. *New Media & Society*, *6*(1). doi:10.1177/1461444804039906

Herring, D. (2008). Questioning the Generational Divide: Technological Exoticism and Adult Constructions of Online Youth Identity. In Buckingham, D. (Ed.), *Youth, Identity, and Digital Media, The John D. and Catherine T. MacArthur Foundation Series on Digital Media and Learning* (pp. 71–92). Cambridge, MA: The MIT Press.

Hershkowitz, I., Horovitz, D., & Lamb, M. (2007). Victimization of Children With Disabilities. *The American Journal of Orthopsychiatry*, *77*(4), 629–635. doi:10.1037/0002-9432.77.4.629

Hesse-Biber, S. (1996). *Am I Thin Enough Yet? The Cult of Thinness and the Commercialization of Identity*. New York: Oxford University Press.

Hey, V. (1997). *The company she keeps: an ethnography of girls' friendships*. Buckingham, Milton Keynes, UK: Open University Press.

Hilbert, P. (2007). *'Academic Success with Special Attention to Moral Values'*, Retrieved May 05, 2007 from www.lexpress.mu/display_article_sup.php?news

Hillcoat-Nalletamby, S., & Ragobur, S. (2005). 'The Need for Information on Family Planning Among Young, Unmarried Women in Mauritius'. *Journal of Social Development in Africa*, *20*(2), 39–63.

Hillcoat-Nallétamby, S., & Dharmalingam, A. (2005). *The influence of historical cultural identity in shaping contemporary reproductive behaviour in Mauritius*. Paper presented at the International Union for the Scientific Study of Population, XXV International Population Conference, Tours, France, July 18-23, 2005.

Hills, M. (2002). *Fan Cultures*. London: Routledge.

Hinduja, S., & Patchin, J. (2008). Personal information of adolescents on the Internet: A quantitative content analysis of MySpace. *Journal of Adolescence*, *31*(1), 125–146. doi:10.1016/j.adolescence.2007.05.004

Hinduja, S., & Patchin, J. W. (2009). *Bullying Beyond the Schoolyard: Preventing and Responding to Cyberbullying*. Thousand Oaks, CA: Sage Publications (Corwin Press),

Juvonen, J., & Gross, E. (2008). Extending the School Grounds?—Bullying Experiences in Cyberspace. *The Journal of School Health*, *78*(9), 496–505.

Hine, C. (2000). *Virtual Ethnography*. London: Sage Publications.

Hine, C. (1998). Virtual Ethnography. In *IRISS '98: Conference Papers*. Retrieved from http://www.intute.ac.uk/socialsciences/archive/iriss/papers/paper16.htm

Holloway, S. L., & Valentine, G. (2003). *Cyberkids, children in the information age*. London: Routledge.

Holt, T. J., Blevins, K. R., & Burkert, N. (2010). Considering the pedophile subculture online. *Sexual Abuse*, *22*(1), 3–24. doi:10.1177/1079063209344979

Honowar, V. (2007). Cellphones in Classrooms Land Teachers on Online Video Sites. *Education Week*, *27*(11), 1–12.

Hopkins, S. (2002). *Girl heroes: the new force in popular culture*. Annandale, Australia: Pluto Press.

Horowitz, B. (2006). Creators, Synthezisers and Consumers. *Elatable*. Retrieved from http://blog.elatable.com/2006/02/creators-synthesizers-and-consumers.html.

Hsieh, H., & Shannon, S. E. (2005). Three Approaches to Qualitative Content Analysis. *Qualitative Health Research*, *15*(9), 1277–1288. doi:10.1177/1049732305276687

Hubbard, J. A., Dodge, K. A., Cillessen, A. H., Coie, J. D., & Schwartz, D. (2001). The dyadic nature of social information processing in boys' reactive and proactive aggression. *Journal of Personality and Social Psychology*, *80*, 268–280. doi:10.1037/0022-3514.80.2.268

Huesmann, L. R. (1988). An information processing model for the development of aggression. *Aggressive Behavior*, *11*, 13–24. doi:10.1002/1098-2337(1988)14:1<13::AID-AB2480140104>3.0.CO;2-J

Huesmann, L. R. (2010). Nailing the coffin shut on doubts that violent video games stimulate aggression: Comments on Anderson et al. (2010). *Psychological Bulletin*, *136*(2), 179–181. doi:10.1037/a0018567

Huesmann, L. R., & Guerra, N. G. (1997). Children's normative beliefs about aggression and aggressive behavior. *Journal of Personality and Social Psychology*, *72*, 408–419. doi:10.1037/0022-3514.72.2.408

Huesmann, L. R., Moise-Titus, J., Podolski, C.-L., & Eron, L. D. (2003). Longitudinal relations between childrens' exposure to TV violence and their aggressive and violent behavior in young adulthood: 1977-1992. *Developmental Psychology*, *39*, 201–221. doi:10.1037/0012-1649.39.2.201

Huesmann, L. R. (1998). The role of social information processing and cognitive schema in the acquisition and maintenance of habitual aggressive behavior. In Geen, R. G., & Donnerstein, E. (Eds.), *Theories, research, and implications for social policy* (pp. 73–109). San Diego, CA: Academic Press.

Huesmann, L. R., & Miller, L. S. (1994). Long-term effects of repeated exposure to media violence in childhood. In Huesmann, L. R. (Ed.), *Aggressive behavior: Current perspectives* (pp. 153–186). New York: Plenum Press.

Huesmann, L. R., Moise, J. F., & Podolski, C.-L. (1997). The effects of media violence on the development of antisocial behavior. In Stoff, D. M., Breiling, J., & Maser, J. D. (Eds.), *Handbook of antisocial behavior* (pp. 181–193). New York: John Wiley.

Hugentobler, M. K., Israel, B. A., & Schurman, S. J. (1992). An action research approach to workplace health: integrating methods. *Health Education Quarterly*, 19(1), 55e76.

Hughes, R. (1998). Considering the Vignette Technique and its Application to a Study of Drug Injecting and HIV Risk and Safer Behaviour. *Sociology of Health & Illness*, *20*, 381–400. doi:10.1111/1467-9566.00107

Hughes, R., & Huby, M. (2002). The Application of Vignettes in Social and Nursing Research. *Journal of Advanced Nursing*, *37*(4), 382–386. doi:10.1046/j.1365-2648.2002.02100.x

Hughes, B. (2009). What is Web 2.0, and what's not: A Roadmap for Research Relevance. Presented at *2009 EURAM (European Academy of Management) Conference on Renaissance and Renewal in Management Studies*, Liverpool.

Hughes, K. (1995). *From Webspace to Cyberspace*. Retrieved May 20 2010, from http://tecfa.unige.ch/guides/vrml/hughes/cspace.1.1.pdf.

Hunt, A. (1999). *Governing Morals: A Social History of Moral Regulation. Cambridge studies in law and society*. Cambridge, UK: Cambridge University Press.

Hunter, K., Hari, S., Egbu, C., & Kelly, J. (2005). 'Grounded Theory: Its Diversification & Application Through two Examples from Research Studies on Knowledge & Value Management'. *The Electronic Journal of Business Research Methodology*, *3*(1), 57–68.

Hybels, S., & Weaver, R. L. (2001). *Communicating Effectively* (6th ed.). Boston: McGraw Hill.

Infomedia. (2010). *Infomedia*. Retrieved January 15, 2010, from http://www.infomedia.dk

Ins@fe. (2007). Safer Internet Day 2007. *Safer Internet Day 2007*. Retrieved January 15, 2010, from http://old.saferinternet.org/ww/en/pub/insafe/mediaroom/sid2007.htm

Internet World Stats. (2009). *List of countries classified by internet penetration rates*. Retrieved March 10, 2010 from http://www.internetworldstats.com/list4.htm

Ito, M. (2008). Mobilizing the Imagination in Everyday Play: The Case of Japanese Media Mixes. In *The International Handbook of Children, Media and Culture* (pp. 397-412). Los Angeles, CA: SAGE.

Ito, M., Baumer, S., & Bittanti, M. boyd, D., Cody, R., Herr, B., Horst, H. A., et al. (2010). *Hanging Out, Messing Around, Geeking Out: Living and Learning with New Media*. The John D. and Catherine T. MacArthur Foundation Series on Digital Media and Learning. Cambridge, MA: MIT Press.

IWF. (2009). *2008 Annual and charity report*. Cambridge, UK: Internet Watch Foundation.

Jackson, C. (2006). Wild girls? An exploration of ladette cultures in secondary schools. *Gender and Education*, *18*(4), 339–360. doi:10.1080/09540250600804966

Jackson, C., & Tinkler, P. (2007). 'Ladettes' and 'Modern Girls': 'troublesome' young femininities. *The Sociological Review*, *55*(2), 251–272. doi:10.1111/j.1467-954X.2007.00704.x

Jackson, M., Cassidy, W., & Brown, K. (2009a). Out of the mouths of babes: Students "voice" their opinions on cyber-bullying. *Long Island Education Review*, *8*(2), 24–30.

Jackson, P. W., Hansen, D., & Boomstrom, R. (1993). *The moral life of schools*. San Francisco: Jossey-Bass Publishers.

Jackson, M., Cassidy, W., & Brown, K. (2009b). *"You were born ugly and youl die ugly too": Cyber-bullying as relational aggression. In Education (Special Issue on Technology and Social Media (Part 1)*, 15(2), December 2009.http://www.ineducation.ca/article/you-were-born-ugly-and-youl-die-ugly-too-cyber bullying-relational-aggression

Jacquemot, N. (2002). *Inkognito. Kärlek, relationer & möten på Internet* [Incognito. Love, relations & encounters on the internet]. Stockholm, Sweden: Bokförlaget DN.

Jaishankar, K. (2008). Space Transition Theory of Cyber Crimes. In Schmallager, F., & Pittaro, M. (Eds.), *Crimes of the Internet* (pp. 283–301). Upper Saddle River, NJ: Prentice Hall.

James, W. (2003). *Media Communication: An Introduction to Theory and process* (2nd ed.). New York: Palgrave Macmillan.

James, S.L., Ode, I., & O., Soola (1990). *Introduction to Communication: For Business and Organisations*. Ibadan: Spectrum Books Ltd.

Jegede, A. S. (2002). The Yoruba cultural construction of health and illness. *Nordic Journal of African Studies*, *11*(3), 322–335.

Jenkins, H. (1992). *Textual poachers: television fans & participatory culture*. New York: Routledge.

Jenkins, H. (2006). *Convergence Culture: Where old and new media collide*. New York: New York University Press.

Jenkins, H. (2006). *Fans, Bloggers, and Gamers: Exploring Participatory Culture*. New York: New York University Press.

Jenkins, H. (2006). *Convergence culture: Where old and new media collide*. New York: New York University Press.

Jenkins, P. (2001). *Beyond Tolerance: Child Pornography on the Internet*. New York: New York University Press.

Jenkins, H. (2007): Nine Propositions Towards a Cultural Theory of YouTube. http://henryjenkins.org. Blog post, May 28.

Jimenez, R. T. (2000). Literacy and the Identity Development of Latina/o Students. *American Educational Research Journal*, *37*(4), 971–1000. doi:. doi:10.3102/00028312037004971

Jones, R. H., & Norris, S. (2005). *Discourse in Action: Introducing Mediated Discourse Analysis* (1st ed.). New York: Routledge.

Jones, R. H., & Norris, S. (2005). Discourse as action/discourse in action. In Norris, S., & Jones, R. H. (Eds.), *Discourse in Action. Introducing mediated discourse analysis* (pp. 3–14). London: Routledge.

Jones, S., & Madden, M. (2002). The Internet Goes to College: how students are living in the future with today's technology\.*Pew Internet and American Life Project*. Retrieved December 31, 2009, from http://www.pewinternet.org/~/media//Files/Reports/2002/PIP_College_Report.pdf.pdf.

Jonsson, L., Warfvinge, C., & Banck, L. (2009). *Children and Sexual Abuse via IT*. Linköping, Sweden: BUP-Elefanten.

Jordan, T. (1999). *Cyberpower: The culture and politics of cyberspace and the Internet*. London: Routledge. doi:10.4324/9780203448632

Josselson, R., Greenberger, E., & McConochie, D. (1977). Phenomenological aspects of psychosocial maturity in adolescence. Part I. Boys. *Journal of Youth and Adolescence*, *6*(1), 25–55. doi:10.1007/BF02138922

Josselson, R., Greenberger, E., & McConochie, D. (1977). Phenomenological aspects of psychosocial maturity in adolescence. Part II. Girls. *Journal of Youth and Adolescence*, *6*(2), 145–167. doi:10.1007/BF02139081

Joyce, E., & Kraut, R. E. (2006). Predicting continued participation in newsgroups. *Journal of Computer-Mediated Communication*, *11*(3), 723–747. doi:10.1111/j.1083-6101.2006.00033.x

Juvonen, J., & Gross, E. F. (2008). Extending the school grounds? Bullying experiences in cyberspace. *The Journal of School Health*, *78*(9), 496–505. doi:10.1111/j.1746-1561.2008.00335.x

K.L. Bishop, A., Chang, R. & Churchward A. (2007). The net generation are not big users of Web 2.0 technologies: preliminary findings. In *ICT: Providing choices for learners and learning. Proceedings ascilite Singapore 2007*. Available from http://www.ascilite.org.au/conferences/singapore07/procs/kennedy.pdf

Kaiser Foundation. (2000) US Adults and Kids on New Media Technology.InC. Von Feilltzen &U Carlsson (eds). *Children in the New Media Landscape* (pp 349 – 350). Goteborg: UNESCO/Nodricom.

Kalichman, S. C. (1991). Psychopathology and personality characteristics of criminal sex offenders as a function of victim age. *Archives of Sexual Behavior*, *20*(2), 187–197. doi:10.1007/BF01541943

Kama, A. (2002). The quest for inclusion: Jewish-Israeli gay men's perceptions of gays in the media. *Feminist Media Studies*, *2*(2), 195–212. doi:10.1080/14680770220150863

Kama, A. (2007). Israeli gay men's consumption of lesbigay media. In Barnhurst, K. G. (Ed.), *Media Q, Media\ Queered: Visibility and its discontents* (pp. 125–142). New York: Peter Lang.

Kama, A. (2005). GLBT issues in Israel. In Sears, J. T. (Ed.), *Youth, education, and sexualities: An international encyclopedia* (pp. 448–453). Westport, CT: Greenwood.

Kampmann, J. (1998). *Børneperspektiv og børn som informanter*. København, Denmark: Børnerådet.

Karade, I. (1994). *The handbook of Yoruba religious concepts*. York Beach: Weiser Books.

Kauhajoki/Finland. *Gunman sprayed bullets in classroom and corridor, and threw petrol bombs*. Retrieved 24.09.2009, from http://www.hs.fi/english/article/1135239693467

Kawaura, Kawakami & Yamashita. (1998). Keeping A Diary in Cyberspace. *The Japanese Psychological Research*, *40*(4), 234–245. doi:10.1111/1468-5884.00097

Keane, H. (2002). *What's Wrong with Addiction?* Victoria, Australia: Melbourne University Press.

Kearney, M. (2009). Coalescing: The Development of Girls' Studies. *NWSA Journal*, *21*(1), 1–28.

Kearney, M. C. (2006). *Girls make media*. New York: Routledge.

Keesing, R. M & Keesing, F.M. (1971). *New Perspectives in Cultural anthropology*. New York: Holt, Rinehart and Winston.

Kelly, D. M., Pomerantz, S., & Currie, D. H. (2006). "No boundaries"? Girls' interactive, online learning about femininities. *Youth & Society*, *38*(1), 3–28. doi:10.1177/0044118X05283482

Kennedy, G., Dalgarno, B., Gray, K., Judd, T., Waycott, J., Bennett, S., Maton, K., Krause,

Kenway, J., & Willis, S. (1990). *Hearts and minds: self-esteem and the schooling of girls*. New York: Falmer Press.

Kierkegaard, S. (2008). Cybering, online grooming and ageplay. *Computer Law & Security Report, 24*, 41–55. doi:10.1016/j.clsr.2007.11.004

Kiesler, S., Siegel, J., & McGuire, T. W. (1984). Social psychological aspects of computermediated communication. *The American Psychologist, 39*, 1123–1134. doi:10.1037/0003-066X.39.10.1123

King, S. A. (1996). Researching Internet communities: Proposed ethical guidelines for the reporting of results. *The Information Society, 12*(2), 119–128. doi:10.1080/713856145

Kingry, M. J., Tiedje, L. B., & Friedman, L. L. (1990). Focus groups: a research technique for nursing. *Nursing Research, 39*(2), 124–125.

Kinloch, G. C. (1977). *Sociological Theory its Development and Major Paradigms*. New York: McGraw-Hill, Inc.

Kinsey, A. C. (1998). *Sexual behavior in the human male*. Indiana: Indiana University Press. (Original work published 1948)

Kiriakidis, S. P., & Kavoura, A. (2010). Cyberbullying: a review of the literature on harassment through the internet and other electronic means. *Family & Community Health, 33*(2), 82–93.

Kirsh, S. J. (1998). Seeing the world through mortal kombat-coloured glasses. Violent video games and the development of a short-term hostile attribution bias. *Childhood, 5*, 177–184. doi:10.1177/0907568298005002005

Kitzinger, J. (1995). Qualitative research: introducing focus groups. *British Medical Journal, 311*, 299–302.

Klein, F. (1993). *The bisexual option*. New York: Harrington Park.

Kleven, K. (1992). *Jentekultur som kyskhetsbelte. Om kuturelle, samfunnsmässige og psykologiske endringer i unge jenters verden*. Oslo, Sweden: Universitetsforlaget.

Klimmt, C. (2004). Computer- und Videospiele. In Mangold, R., Vorderer, P., & Bente, G. (Eds.), *Lehrbuch der Medienpsychologie* (pp. 696–716). Göttingen: Hogrefe.

Kowalski, R., & Limber, S. (2007). Electronic Bullying Among Middle School Students. *The Journal of Adolescent Health, 41*(6), S22–S30. doi:10.1016/j.jadohealth.2007.08.017

Kowalski, R. M., Limber, S. E., & Agatston, P. W. (2008). *Cyber bullying: Bullying in the digital age*. Malden, MA: Blackwell Publishers. doi:10.1002/9780470694176

Kozinets, R. (1999). E-Tribalized Marketing?: The Strategic Implications of Virtual Communities of Consumption. *European Management Journal, 17*(3), 252–264. doi:10.1016/S0263-2373(99)00004-3

Krahé, B., & Möller, I. (2004). Playing violent electronic games, hostile attributional style, and aggression-related norms in German adolescents. *Journal of Adolescence, 27*, 53–69. doi:10.1016/j.adolescence.2003.10.006

Krause, A. (2004). *Człowiek niepełnosprawny wobec przeobrażeń społecznych*. Kraków, Poland: Impuls.

Kreutz, K., & Zając, J. M. (2009). Psychologiczne aspekty wykorzystywania technologii internetowych w nauczaniu. In Nowak, A., Winkowska-Nowak, K., & Rycielska, L. (Eds.), *Szkoła w dobie Internetu* (pp. 121–138). Warszawa, Poland: Wydawnictwa Naukowe PWN.

Krippendorff, K. (1980). *Content analysis: An introduction to its methodology*. London: Sage.

Krone, T. (2005). *High tech crime brief.* Canberra, Australia:Australian Institute of Criminology, Lagos State University.(n.d.). Retrieved July, 2009 http://www.lasunigeria.org/

Krueger, R. A., & Casey, M. A. (2009). *Focus Groups* (4th ed.). Thousand Oaks, CA: SAGE Inc.

Kryger, N. (2004). Childhood and "New Learning" in a Nordic Context. In H. Bremback, B. Johansson, & J. Kampman (Eds.), *Beyond the competent child: Exploring contemporary childhoods in the Nordic welfare societies.* (pp. 153-176). Fredriksberg, Denmark: Roskilde University Press.

Kumble, R. (Director)(2002). *The Sweetest Thing.*. New York: Sony Pictures Entertainment.

Küng, L. (2008). *Strategic Management in the Media: Theory to Practice*. London: SAGE Publications.

Kuttler, A. F., & La Greca, A. M. (2004). 'Linkages among adolescent girls' romantic relationships, best friendships, and peer networks'. *Journal of Adolescence*, *27*, 395–414. doi:10.1016/j.adolescence.2004.05.002

Kvale, S. (1997). *Den kvalitativa forskningsintervjun*. Lund, Sweden: Studentlitteratur.

L'Express. (2006a). *'More schoolgirl porn "shows" under scrutiny.'* Retrieved March 03 2006 from http://www.lexpress.mu

L'Express. (2006b). *'Opinions De Jeunes: Réaction au "Show": "Nous ne sommes pas des enfants modele'*. Retrieved March 03 2006 from http://www.lexpress.mu

LaBelle, B. (2006). *Background noise: perspectives on sound art*. New York: Continuum International.

Lagos, N. Malthouse Press Limited Adedayo, O. (2009). *The internet as a medium of communication: A uses and Gratificationtheory amongst* Bowen University Students. Unpublished Undergraduate Long essay, Bowen University, Iwo Nigeria.

Lahno, B. (2004). Three Aspects of Interpersonal Trust. *Analyse & Kritik*, *26*(1), 30–47.

Lampe, C., Ellison, N., & Steinfield, C. (2006). A face(book) in the crowd: social Searching vs. social browsing. In *Proceedings of the 2006 20th Anniversary Conference on Computer Supported Cooperative Work*. Association for Computing Machinery.

Lamy, M. N. Hampel. R. (2007). *Online communication in language learning & teaching*. London: Palgrave Macmillan.

Lange, P. G. (2008). Publicly private and privately public: Social networking on YouTube. *Journal of Computer-Mediated Communication*, *13*(1), 361–380. doi:10.1111/j.1083-6101.2007.00400.x

Langellier, K., & Peterson, E. (2004). *Storytelling in Daily Life: Performing Narrative*. Philadelphia, PA: Temple University Press.

Langellier, K. (1998). Voiceless Bodies, Bodiless Voices: The Future of Personal Narrative Performance. In Dailey, S. J. (Ed.), *The Future of Performance Studies: Visions and Revisions* (pp. 207–213). Annandale, VA: National Communication Association.

Langhout, R. D. (2003). Reconceptualizing quantitative and qualitative methods: a case study dealing with place as an exemplar. *American Journal of Community Psychology, 32*(3/4), 229e244.

Lanning, K. V. (1984). Collectors. In A. W. Burgess &M. L. Clark (Eds). *Child Pornography and Sex Rings* (pp. 83 - 92). Toronto: Lexington.

Lärarförbundet., & Lärarnas Riksförbund. (2002). *Lärares Yrkesetik*. Stockholm, Sweden: Lärarförbundet, Lärarnas Riksförbund.

Larsen, M. (2009). Sociale netværkssider og digital ungdomskultur: Når unge praktiserer venskab på nettet [Social networking sites and digital youth culture: When young people practice friendship online]. *MedieKultur*, *47*, 45–65.

Larsen, M. (2008). *Online Social Networking: From Local Experiences to Global Discourses*. Paper presented at Internet Research 9.0: Rethinking Community, Rethinking Place, The IT University, Copenhagen, Denmark 20081015–18.

Larsen, M. C. (2005). *Ungdom, venskab og identitet - en etnografisk undersøgelse af unges brug af hjemmesiden Arto* (Upubliceret specialeafhandling). Institut for Kommunikation, Aalborg Universitet. Retrieved from http://www.ell.aau.dk/fileadmin/user_upload/documents/staff/Malene_Larsen_-_Documents/Ungdom__venskab_og_identitet_Malene_Charlotte_Larsen.pdf

Larsen, M. C. (2007a). Kærlighed og venskab på Arto.dk (Love and friendship on Arto.dk). *ungdomsforskning*, *6*(1), 11.

Larsen, M. C. (2007b). Understanding Social Networking: On Young People's Construction and Co-construction of Identity Online. In *Proceedings from the conference Internet Research 8.0: Let's Play, Association of Internet Researchers, Vancouver*. Association of Internet Researchers.

Larson, D. G., & Chastain, R. L. (1990). Self-concealment: Conceptualization, measurement, and health implications. *Journal of Social and Clinical Psychology*, *9*(4), 439–455.

Lasala, M. C. (2000). Lesbians, gay men and their parents: Family therapy for the coming-out crisis. *Family Process*, *39*(1), 67–81. doi:10.1111/j.1545-5300.2000.39108.x

Laurendeau, J. (2003). Gender and the culture of skydiving: Misogyny, trivialization, and sexualization in a "gender-neutral" sport. Paper presented at the 2003 Annual Meeting of the American Sociological Association, Atlanta, GA.

Lave, J., & Wenger, E. (1991). *Situated Learning: Legitimate Peripheral Participation*. Cambridge, UK: Cambridge University Press.

Lee, S. (2009). Online Communication and Adolescent Social Ties: Who benefits more from Internet use? *Journal of Computer-Mediated Communication*, *14*(3), 509–531. doi:10.1111/j.1083-6101.2009.01451.x

Lefebvre, H. (1974). *The production of space*. Oxford, UK: Basil Blackwell.

Leininger, M. (1970). *Nursing and Anthropology: Two Worlds to Blend*. New York: John Wiley & Sons.

Lemke, J. L. (2005). Multimedia genres and traversals. *Folia Linguistica*, *39*(1-2), 45–56. doi:10.1515/flin.2005.39.1-2.45

Lenhart, A. (2009). *Adults and social network websites*: Pew Internet. Available from http://www.pewinternet.org/Reports/2009/Adults-and-Social-Network-Websites.aspx

Lenhart, A., & Madden, M. (2007). *Social networking websites and teens*: Pew Internet. Available fromhttp://www.pewinternet.org/Reports/2007/Social-Networking-Websites-and-Teens.aspx

Lenhart, A., & Madden, M. (2007). Teens, Privacy & Online Social Networks: How teens manage their online identities and personal information in the age of MySpace. *Pew Internet & American Life Project*. Retrieved from: http://www.pewinternet.org

Lenhart, A., Madden, M., & Macgill, A. (2007). Teens and Social Media: The use of social media gains a greater foothold in teen life as they embrace the conversational nature of interactive online media. *Pew Internet & American Life Project*. Retrieved from: http://www.pewinternet.org

Lenhart, A., Purcell, K., Smith, A., & Zickuhr, K. (2010). *Social media and mobile internet use among teen and young adults*. Retrieved March 10, 2010 from http://pewinternet.org/~/media//Files/Reports/2010/PIP_Social_Media_and_Young_Adults_Report.pdf

LeParkour Sweden. (n.d.). Retrieved from http://www.le-parkour.se/

Lessig, L. (2008). *Remix: Making Art and Commerce Thrive in the Hybrid Economy*. New York: Penguin Press.

Levine, D. (2000). Virtual Attraction: What Rocks Your Boat. *CyberPsychology & Behaviour*, *3*(4), 565–573. doi:10.1089/109493100420179

Levy, A. (2005). *Female Chauvinist Pigs*. Melbourne: Schwartz Publishing.

Lewis, L. A. (1992). *The Adoring Audience: Fan Culture and Popular Media*. London: Routledge. doi:10.4324/9780203181539

Li, Q. (2006). Cyberbullying in Schools: A research of Gender Differences. *School Psychology International*, *27*(2), 157–170. doi:10.1177/0143034306064547

Li, Q. (2007). New bottle but old wine: A research of cyberbullying in schools. *Computers in Human Behavior*, *23*(4), 1777–1791. doi:10.1016/j.chb.2005.10.005

Liau, A., Khoo, A., & Ang, P. (2008). Parental awareness and monitoring of adolescent Internet use. *Current Psychology (New Brunswick, N.J.)*, *27*(4), 217–233. doi:10.1007/s12144-008-9038-6

Lincoln, Y. S., & Guba, E. G. (1985). *Naturalistic inquiry*. London: Sage.

Lindgren, S., & Lelievre, M. (2009). In the Laboratory of Masculinity: Renegotiating Gender Subjectivities in MTV's Jackass. *Critical Studies in Media Communication*, *26*(5), 393–410. doi:10.1080/15295030903325313

Lindgren, S. (2009). YouTube Gunmen? Mapping participatory media discourse on school shooting videos. Paper presented at Violence and Network Society: School Shootings and Social Violence in Contemporary Public Life, Helsinki, Finland, November 6-7.

Lindlof, T. R., & Shatzer, M. (1998). Media ethnography in virtual space: Strategies, limits, and possibilities. *Journal of Broadcasting & Electronic Media*.

Lister, M., Dovey, J., Giddings, S., Grant, I., & Kelly, K. (2009). *New Media: A critical Introduction* (2nd ed.). London: Routledge.

Little, T. D., Jones, S. M., Henrich, C. C., & Hawley, P. H. (2003). Disentangling the "whys" from the "whats" of aggressive behavior. *International Journal of Behavioral Development*, *27*, 122–133. doi:10.1080/01650250244000128

Little, L. (2004). Victimization of children with disabilities. In Kendall-Tackett, K. A. (Ed.), *Health consequences of abuse in the family: A clinical guide for evidence-based practice* (pp. 95–108). Washington, DC: American Psychological Association. doi:10.1037/10674-006

Little, T. D. (2002). *Pathways of the forms and functions of aggression during adolescence.* Paper presented at the biennial meeting of International Society for Research on Aggression, Montreal, QC.

Littleton/Colorado. (n.d.). *Columbine killers planned to kill 500.* Retrieved 27.4.1999, from http://news.bbc.co.uk/2/hi/americas/329303.stm

Liu, H. (2007). Social network profiles as 'taste performances'. *Journal of Computer-Mediated Communication*, *13*(1), 252–275. doi:10.1111/j.1083-6101.2007.00395.x

Living and Learning with New Media (Report). (2008) McArthur Foundation. Retrieved November 25, 2009 from http://digitalyouth.ischool.berkeley.edu/files/report/digitalyouth-WhitePaper.pdf

Livingstone, S. (2002). *Young People and New Media.* London: SAGE Publications.

Livingstone, S. (2008). Taking risky opportunities in youthful content creation: teenagers' use of social networking sites for intimacy, privacy and self-expression. *New Media & Society*, *10*(3), 393–411. doi:10.1177/1461444808089415

Livingstone, S., & Helsper, E. J. (2007). Taking risks when communicating on the Internet: the role of offline social-psychological factors in young people's vulnerability to online risks. *Information Communication and Society*, *10*(5), 619. doi:10.1080/13691180701657998

Livingstone, S. (2003). Children's use of the Internet: Reflections on the emerging research agenda. *New Media & Society*, *5*, 147–166. doi:10.1177/1461444803005002001

Livingstone, S., & Bober, M. (2005). *UK Children Go Online. Final report of key project findings*. Department of Media and Communications, The London School of Economics and Political Science. Retrieved from www.children-go-online.net

Livingstone, S., & Bober, M. (2005). UK children go online: listening to young people's experiences. London: London School of Economics. Retrieved from http://news.bbc.co.uk/1/shared/bsp/hi/pdfs/28_04_05_childrenonline.pdf on 26th February 2009.

Löfberg, C. (2008). *Möjligheternas arena? Barns och ungas samtal om tjejer, killar och sexualitet på en virtuell arena.* [Arena of Possibilities? Children's and Young People's conversations about Girls, Boys, Emotions and Sexuality on a Virtual arena] [Doctoral dissertation] Stockholm, Sweden: Pedagogiska institutionen, Stockholms universitet.

Long, B., & Baecker, R. (1997). A taxonomy of internet communications tools. Retrieved April 1, 2008 from http://www.dgp.toronto.edu/people/byron/webnet/Taxonomy.html

Longe, O., Ngwa, O., Wada, F., & Mbarika, V. (2009). Criminal Uses of Information and Communication Technologies in Sub-Saharan Africa: Trends, Concerns and Perspectives. *Journal of Information Technology Impact*, *9*(3), 155–172.

Longe, O. B., & Chiemeke, S. C. (2008). Cyber Crime and Criminality in Nigeria – What Roles are Internet Access Points in Playing? *European Journal of Soil Science*, *6*(4), 132–139.

Longe, O. B., & Chiemeke, S. C. (2007). Information and communication technology penetration in Nigeria: Prospects, challenges and metrics. *Asian Journal of Information Technology, 6*(3), 280–287. Retrieved November 7, 2009 from http://www.medwellonline.net/fulltext/ajit/2007/280-287.pdf/

Longe, O., Mbarika, V., Kourouma, M., Wada, F. & Isabalija, R. (2009). Seeing beyond the surface: Understanding and Tracking Fraudulent Cyber Activities. *International Journal of Computer Science and Information Security*, 6(3,124-135.

Longmore, M. A. (1998). Symbolic Interactionism and the Study of Sexuality - The Use of Theory in Research and Scholarship on Sexuality. *Journal of Sex Research*, *35*(1), 44–58. doi:10.1080/00224499809551916

Loret, A. (1995). *Génération glisse. Dans l'eau, l'air, la neige... la revolution des sports dans les années fun.* Paris: Autrement.

Lu, C., Jen, W., Chang, W., & Chou, S. (2006). Cybercrime & Cybercriminals: An Overview of the Taiwan Experience. *Journal of Computers*, *6*(1), 11–18.

Lucas, K., & Sherry, J. L. (2004). Sex differences in video game play: A communication-based explanation. *Communication Research*, *31*(5), 499–523. doi:10.1177/0093650204267930

Lüders, M., Bae Brandtzæg, P., & Dunkels, E. (2009). Risky Contacts. In Livingstone, S., & Haddon, L. (Eds.), *Kids Online*. London: Policy Press.

Lull, J., & Hinerman, S. (1997). The Search for the Scandal. In Lull, J., & Hinerman, S. (Eds.), *Media Scandals: Morality and Desire in the Popular Culture Marketplace*. Cambridge, UK: Polity Press.

Lund, J. (2009). *Digital view: Life on the Danish Internet, August 17-23 2009 – arto.com vs facebook*. København, Denmark: Jon Lund. Retrieved from http://jon-lund.com/main/digital-view-life-on-the-danish-internet-august-17-23-2009-artocom-vs-facebook/

Lybarra, M., Leaf, P. J., & Diener-West, M. (2004). Sex Differences in Youth-Reported Depressive Symptomatology & Unwanted Internet Sexual Solicitation. *Journal of Medical Internet Research*, *6*(1), 10–22.

Madden, M., Fox, S., Smith, A., & Vitak, J. (2007). *Digital Footprints – Online identity management and search in the age of transparency*. Washington: PEW/Internet.

Magid, L. (1998). *Teen safety on the information highway*. National Centre for Missing and Exploited Children. Available at http://www.missingkids.com (accessed 18 July 2008).

Makarenko, T. (2004). The Crime–Terror Continuum: Tracing the Interplay between Transnational Organised Crime and Terrorism. *Global Crime*, *6*(1), 129–145. doi:10.1080/1744057042000297025

Malesky, L. A. Jnr, & Ennis, L. (2004). Supportive Distributions: An analysis of posts on a pedophile Internet message boards. *Journal of Addictions & Offender Counseling*, *24*(2), 92–101.

Mallon, G. P. (1998). Knowledge for practice with gay and lesbian persons. In Mallon, G. P. (Ed.), *Foundations of social work practice with lesbian and gay persons* (pp. 1–30). New York: Haworth Press.

Månsson, S.-A., Daneback, K., Tikkanen, R., & Löfgren-Mårtenson, L. (2003). *Kärlek och sex på internet* [Love and sex on the internet]. Göteborg University and Malmö University.

Månsson, S.-A., & Löfgren Mårtenson, L. (2007). Let's talk about porn: On youth, gender and pornography in Sweden. In Knudsen, S. V., Löfgren-Mårtenson, L., & Månsson, S.-A. (Eds.), *Generation P? Youth, gender and pornography* (pp. 241–258). Copenhagen: Danish School of Education Press.

Marcoccia, M. (2004). On-line polylogues: Conversation structure and participation framework in internet newsgroups. *Journal of Pragmatics*, *36*(1), 115–145. doi:10.1016/S0378-2166(03)00038-9

Marening, O. (1987). The Anini Saga: Armed Robbery and the Reproduction of Ideology in Nigeria. *The Journal of Modern African Studies*, *25*(2), 259–281. doi:10.1017/S0022278X00000380

Marini, Z. A., Fairbairn, L., & Zuber, R. (2001). Peer harassment in individuals with developmental disabilities: Towards the development of a multidimensional bullying identification model. *Developmental Disabilities Bulletin*, *29*, 170–195.

Martin, A. D., & Hetrick, E. S. (1988). The stigmatization of the gay and lesbian adolescent. In Ross, M. W. (Ed.), *Psychopathology & psychotherapy in homosexuality* (pp. 163–184). New York: Haworth Press.

Marwick, A. (2008). To catch a predator? *The MySpace moral panic*. *First Monday*, *13*(6).

Masuda, Y. (1983). *The information Society as Post-Industrial Society.* Bethesda,MD: World future society.

Maticka-Tyndale, E., & Smylie, L. (2008). Sexual Rights: Striking a Balance. *International Journal of Sexual Health*, *20*(1&2), 7–24. doi:10.1080/19317610802156996

McCabe, J. (2009). Resisting Alienation: The Social Construction of Internet Communities Supporting Eating Disorders. *Communication Studies*, *60*(1), 1–16. doi:10.1080/10510970802623542

McCleneghan, J. S. (2002). 'Reality violence' on TV NEWS: It began with Vietnam. *The Social Science Journal*, *39*, 593–598. doi:10.1016/S0362-3319(02)00232-X

McDonald, K. (2004). At the UN: Mainstreaming Sexual Rights. *Sexual Reproductive Health & Rights*, *1*(1), 1–4.

McDonald, J. G. (1982). Individual differences in the coming-out process for gay men: Implications for theoretical models. *Journal of Homosexuality*, *8*(1), 47–60. doi:10.1300/J082v08n01_05

McGeer, V. (2004). Developing Trust on the Internet. *Analyse & Kritik*, *26*(1), 91–107.

McKenna, K., & Bargh, J. (2000). Plan 9 From Cyberspace: The Implications of the Internet for Personality and Social Psychology. *Personality and Social Psychology Review*, *4*(1), 57–75. doi:10.1207/S15327957PSPR0401_6

McKenna, K. Y. A. (2008). Influence on the nature and functioning of social groups. In Barak, A. (Ed.), *Psychological aspects of Cyberspace* (pp. 228–242). Cambridge, UK: Cambridge University Press.

McMillan, S. J., & Morrison, M. (2006). Coming of age with the internet: A qualitative exploration of how the internet has become an integral part of young people's lives. *New Media & Society*, *8*(1), 73–95. doi:10.1177/1461444806059871

McNeill, L. (2003). Teaching an Old Genre New Tricks: The Diary on the Internet. *Biography*, *26*(1). doi:10.1353/bio.2003.0028

Mcquade, S. C. (2009). Theories of Cybercrime. In Mcquade, S. C. (Ed.), *Encyclopedia of Cybercrime* (pp. 179–181). Westport, CT: Greenwood Press.

McRobbie, A., & Thornton, S. L. (1995). Rethinking 'Moral Panic' for Multi-Mediated Social Worlds. *The British Journal of Sociology*, *46*(4), 559–574. doi:10.2307/591571

McRobbie, A. (1991). *Feminism and Youth Culture: From "Jackie" to "Just Seventeen."*. Boston: Unwin Hyman.

McRobbie, A. (2007). Top Girls? Young women and the post-feminist sexual contract. *Cultural Studies*, *21*(4-5), 718–737. doi:10.1080/09502380701279044

McRobbie, A. (2008). Pornographic Permutations. *Communication Review*, *11*(3), 225–236. doi:10.1080/10714420802306676

McRobbie, A., & Garber, J. (1976). Girls and subcultures: an exploration. In Hall, S., & Jefferson, T. (Eds.), *Resistance through rituals: youth subcultures in post-war Britain* (pp. 209–222). London: Hutchinson.

McRobbie, A. (2004). Notes on postfeminism and popular culture: Bridget Jones and the new gender regime. In Harris, A. (Ed.), *All about the girl: power, culture, and identity* (pp. 3–14). New York: Routledge.

McRobbie, A. (1999). *More*! New sexualities in girls and women's magazines. In McRobbie, A. (Ed.), *In the culture society: art, fashion, and popular music* (pp. 46–61). New York: Routledge.

McRobbie, A. (1978). 'Working Class Girls and the Culture of Femininity'. In Women'sStudies Group, Centre for Contemporary Cultural Studies (eds). *Women Take Issue: Aspects of Women's Subordination*, pp. 96–108. London: Hutchinson.

Mead, G. H. (1967). *Mind, Self and Society*. Chicago: The University of Chicago Press.

Media, R. D. F. (Producer). (2005-present) *Ladette to Lady*. UK: ITV (UK) and Channel 9 (Australia).

Media-Awareness. (2000). Canada's Children in a Wired World: The parents' view. A survey of Internet use in Canadian families. *Media Awareness, 20*(2), 17–18.

Mediarådet (The Swedish Media Council). (2008). *Ungar och medier 2008*. Elanders.

Medierådet. (2007). *Tips till vuxna om barn/unga och mobiltelefoner*. Stockholm, Sweden: Regeringskansliet.

Mehra, B., Merkel, C., & Bishop, A. P. (2004). The internet for empowerment of minority and marginalized users. *New Media & Society, 6*(6), 781–802. doi:10.1177/146144804047513

Meilman, P. W. (1979). Cross-sectional age changes in ego identity status during adolescence. *Developmental Psychology, 15*(2), 230–231. doi:10.1037/0012-1649.15.2.230

Meixner, E. (2006). Teacher agency and access to LGBTQ young adult literature. *Radical Teacher, 76*, 13–19.

Merton, R. K., Fiske, M., & Kendall, P. L. (1990). *The Focused Interview: A Manual of Problems and Procedures* (2nd ed.). New York: The Free Press.

Merton, R. K. (1968). *Anomie theory in Social theory and Social structure*. New York: free press.

Meyer, J. W., & Jepperson, R. L. (2000). The "Actors" of Modern Society: The Cultural Construction of Social Agency. *Sociological Theory, 18*(1), 100–120. doi:10.1111/0735-2751.00090

Meyer, I. H. (2003). Minority stress and mental health in gay men. In Garnets, L. D., & Kimmel, D. C. (Eds.), *Psychological perspectives on lesbian, gay and bisexual experiences* (pp. 699–731). New York: Columbia University Press.

MFPA (Mauritius Family Planning Association). (1993). *Research Report on Young Women, Work & AIDS-related Risk Behaviour in Mauritius*. Port Louis: Mauritius Family Planning Association.

MFPA (Mauritius Family Planning Association). (2005) *'Mauritius Reproductive health Education for Women Workers'*. Retrieved December 21, 2009 from http://www.icomp.org.my/South-South/S-S-Catalogue.htm

MIH (Mauritius Institute of Health). (1996). *Research Report on National Survey on Youth Profile*. Pamplemousses, Mauritis: MIH

Mikrut, A. (2000). Próba wyjaśnienia związku między agresją i upośledzeniem umysłowym. In J. Pańczyk J. (Ed.), *Roczniki Pedagogiki Specjalnej*, 11, Warszawa, Poland: WSPS, 30-40.

Milne, J. (2008). What have we got to be scared of? *Times Educational Supplement, 25*(January), 1–5.

Mishna, F., Saini, M., & Solomon, S. (2009). Ongoingand online: children's and youth perceptions of cyber bullying. *Children and Youth Services Review, 31*, 1222–1228. doi:10.1016/j.childyouth.2009.05.004

Mishna, F. (2003). Learning disabilities and bullying: Double jeopardy. *Journal of Learning Disabilities, 36*(4), 336–347. doi:10.1177/00222194030360040501

Mitchell, K. J., Finkelhor, D., & Wolak, J. (2001). Risk Factors for & Impact of Online Sexual Solicitation of Youth. *Journal of the American Medical Association*, *285*, 3011–3014. doi:10.1001/jama.285.23.3011

Mitchell, K., Finkelhor, D., & Wolak, J. (2001). Risk factors for and impact of online sexual solicitation of youth. *Journal of the American Medical Association*, *285*(23), 3011–3014. doi:10.1001/jama.285.23.3011

Mitchell, K., Finkelhor, D., & Wolak, J. (2007a). Online requests for sexual pictures from youth: risk factors and incident characteristics. *The Journal of Adolescent Health*, *41*(2), 196–203. doi:10.1016/j.jadohealth.2007.03.013

Mitchell, K., Finkelhor, D., & Wolak, J. (2007b). Youth Internet users at risk for the most serious online solicitations. *American Journal of Preventive Medicine*, *32*(6), 532–537. doi:10.1016/j.amepre.2007.02.001

Mitchell, K. J., Wolak, J., & Finkelhor, D. (2007). Trends in youth reports of sexual solicitations, harassment and unwanted exposure to pornography on the Internet. *The Journal of Adolescent Health*, *40*, 116–126. doi:10.1016/j.jadohealth.2006.05.021

Mitchell, K. J., Wolak, J., & Finkelhor, D. (2008). Are blogs putting youth at risk for online sexual solicitation or harassment? *Child Abuse & Neglect*, *32*(2), 277–294. doi:10.1016/j.chiabu.2007.04.015

Mitchell, K. J., Ybarra, M., & Finkelhor, D. (2007). The relative importance of online victimization in understanding depression, delinquency, and substance use. *Child Maltreatment*, *12*(4), 314–324. doi:10.1177/1077559507305996

Mitchell, K. J., Wolak, J., & Finkelhor, D. (2005). Internet sex crimes against minors. In Kendall-Tackett, K., & Giacomoni, S. (Eds.), *Child victimization* (pp. 2.1–2.17). Kingston, NJ: Civic Research Institute.

Moinian, F. (2007). *Negotiating Identities*. Stockholm, Sweden: Stockholm Institute of Education Press.

Möller, I., & Krahé, B. (2009). Exposure to violent video games and aggression in German adolescents: A longitudinal analysis. *Aggressive Behavior*, *35*, 75–89. doi:10.1002/ab.20290

Möller, I. (2006). *Mediengewalt und Aggression. Eine längsschnittliche Betrachtung des Zusammenhangs am Beispiel des Nutzungs gewalthaltiger Bildschirmspiele*. Unpublished doctoral dissertation, University of Potsdam, Germany.

Monks, C., Smith, P., Naylor, P., Barter, C., Ireland, J., & Coyne, I. (2009). Bullying in different contexts: Commonalities, differences and the role of theory. *Aggression and Violent Behavior*, *14*(2), 146–156. doi:10.1016/j.avb.2009.01.004

Montgomery, M. J., & Sorell, G. T. (1998). Love and Dating Experience in Early and Middle Adolescence: Grade and Gender Comparisons. *Journal of Adolescence*, *21*, 677–689. doi:10.1006/jado.1998.0188

Moore, J. (2009). Technologies commonly used for Cybercrime. In Mcquade, S. C. (Ed.), *Encyclopedia of Cybercrime* (pp. 173–179). Westport, CT: Greenwood Press.

Morakinyo, O. (1983). The Yoruba ayanmo myth and mental health care. *West Africa Journal of Cultures and Ideas*, *1*(1), 61–92.

More on Choi Jin-shil's suicide (2008). Retrieved November 25, 2009 from http://www.dramabeans.com/2008/10/more-on-choi-jin-shils-suicide/.

Morgan, D. L. (1988). *Focus Group as Qualitative Research*. Newbury Park, CA: Sage.

Morgan, R. (1980). Theory and Practice: Pornography and Rape. In Lederer, L. (Ed.), *Take Back the Night: Women on Pornography* (pp. 134–140). New York: William Morrow.

Moriel, L. (2000). Israel and Palestine. In Haggerty, G. E. (Ed.), *Gay histories and cultures: An encyclopedia* (pp. 481–484). New York: Garland Publishing.

Morse, J. M., & Chung, S. E. (2003). Toward holism: the significance of methodological pluralism. [Article 2]. *International Journal of Qualitative Methods*, *2*(3). Available from http://www.ualberta. ca/w;iiqm/backissues/ 2_3final/ html/morsechung.html.

Mudrey, R., & Medina-Adams, A. (2006). Attitudes, perceptions, and knowledge of pre service teachers regarding the educational isolation of sexual minority youth. *Journal of Homosexuality*, *51*(4), 63–90. doi:10.1300/J082v51n04_04

Mulvey, L. (1975/1989). Visual Pleasure and Narrative Cinema. In Mulvey, L. (Ed.), *Visual and other pleasures* (pp. 14–26). Houndmills, UK: Palgrave.

Murthy, D. (2008). Digital ethnography: an examination of the use of new technologies for social research. *Sociology*, *42*(5), 837–855. doi:10.1177/0038038508094565

Mustanski, B. S. (2001). Getting wired: Exploiting the Internet for the collection of valid sexuality data. *Journal of Sex Research*, *38*, 292–302. doi:10.1080/00224490109552100

MYS (Ministry of Youth & Sports). (2009). National [Port Louis: Government of Mauritius]. *Youth & Policy*, 2010–2014.

Näre, S., & Lähteenmaa, J. (1992). *Letit liehumaan! Tyttökulttuuri*. Helsinki: Suomalaisen Kirjallisuuden Seura.

Nielsen Games. (2008). *Video gamers in Europe -2008*. Brussles: Interactive Software Federation of Europe.

Nikken, P., Jansz, J., & Schouwstra, S. (2007). Parents' interest in videogame ratings and content descriptors in relation to game mediation. *European Journal of Communication*, *22*(3), 315–336. doi:10.1177/0267323107079684

Nir, L. (1998, July). *A site of their own: Gay teenagers' involvement patterns in IRC and newsgroups*. Paper presented at the international communication conference, Jerusalem, Israel.

Nishida, K. (1958). *Intelligibility and the Philosophy of Nothingness*. Tokyo: Maruzen Co. Ltd.

Nishimura, Y. H., Ono-Kihara, M., Mohith, J. C., Ngmansun, R., Homma, T., & Diclemente, R. J. (2007). Sexual behaviors and their correlates among young people in Mauritius: a cross-sectional study. *BMC International Health and Human Rights*. Published online 2007 October 5. .doi:10.1186/1472-698X-7-8

Noddings, N. (2002). *Educating moral people: A caring alternative to character education*. New York: Teachers College Press.

Noddings, N. (2005). *The challenge to care in schools: An alternative approach to education*. New York: Teachers College Press.

Norman, J. M. (2005). *From Gutenberg to the Internet: A source Book on the History of Information Technology*. Novato, CA: Historyofscience.com.

Norring, D., & Engström, I. (2002). Ätstörningarnas förekomst. In Clinton, C. (Ed.) *Ätstörningar. Bakgrund och aktuella behandlingsmetoder*. Stockholm: Natur och Kultur.

Norris, S., & Jones, R. H. (Eds.). (2005). *Discourse in Action: Introducing Mediated Discourse Analysis*. London: Routledge.

NOVA. (2006). *Ungdom som selger eller bytter sex– en faglig veileder til hjelpeapparatet* [Youth Selling or Trading Sex]. Oslo, Norway: Akademika AS.

O'Connell, R., Price, J., & Barrow, C. (2004). *Cyber Stalking, Abusive Cyber Sex and Online Grooming: A Programme of Education for Teenagers*. University of Central Lancashire.

O'Donnell, A. B., Lutfey, K. E., Marceau, L. D., & McKinlay, J. B. (2007). Using focus groups to improve the validity of cross-national survey research: A study of physician decision making. *Qualitative Health Research*, *7*, 971–981. doi:10.1177/1049732307305257

Oblinger, D., & Oblinger, J. (2005). *Educating the Net Generation*. Boulder, CO: Educause.

Oddone, E., Genuis, M. L., & Violato, C. (2001). A meta-analysis of the published research on the effects of child sexual abuse. *The Journal of Psychology*, *135*, 17–36. doi:10.1080/00223980109603677

Ofsted (2010). The safe use of new technologies. Report no. 090231. *The Office for Standards in Education, Children's Services and Skills*. Manchester, UK: Ofsted.

Ogden, J. (1992). *Fat Chance! The Myth of Dieting Explained*. London: Routledge.

Oinas, E. (2001). *Making Sense of the Teenage Body – Sociological Perspectives on Girls, Changing Bodies, and Knowledge*. Åbo, Finland: Abo Akademi University Press.

Ojanen, K. (2005). Tallitytöt. Harrastus tyttöyksien tekemisenä. In Saarikoski, H. (Ed.), *Leikkikentiltä. Lastenperinteen tutkimuksia 2000-luvulta* (pp. 202–138). Helsinki, Finland: Suomalaisen Kirjallisuuden Seura.

Ojanen, K. (2008). Tyttötutkimuksen tytöt: keskusteluja moninaisuudesta ja tyttöjen vallasta. *Elore, 15*, 1/08.

Ojo, D. O., & Fasubaa, O. B. (2005). Adolescent Sexuality and family Life Education in South Western Nigeria: Responses from focus Group Discussion. *Journal of the Social Sciences*, *10*(2), 111–118.

Olin-Scheller, C., & Wikström, P. (2010). *Författande fans*. Lund, Sweden: Studentlitteratur.

Olin-Scheller, C. (2008). Trollkarl eller mugglare? Tolkningsgemenskaper i ett nytt medielandskap. *Didaktikens forum, 3*, 44-58.

Olin-Scheller, C., & Wikström, P. (2009). *Beyond the Boundaries of the Book - Young peoples' encounters with web basedfiction*. Karlstad: NordMedia09, the 19th Nordic Conference for Media and Communication Research.

Olupona, J. K. (2008). Sacred Ambiguity: Global African Spirituality, Religious Tradition, Social Capital and Self-Reliance. In Babawale, T., & Alao, A. (Eds.), *Global African Spirituality Social Capital and Self-Reliance in Africa* (pp. xvii–xxxii). Lagos, Nigeria: Malthouse Press Limited.

Olweus, D. (1978). *Aggression in the schools. Bullies and whipping boys*. Washington, DC: Hemisphere.

Olweus, D. (1993). *Bullying at school. What we know and what we can do*. Oxford, UK: Blackwell.

O'Reilly, T. (2005). *What Is Web 2.0: Design patterns and business models for the next generation of software*. Retrieved Januari 6, 2010, from http://oreilly.com/pub/a/web2/archive/what-is-web-20.html?page=1.

Orenstein, P. (1994). *Schoolgirls: young women, self-esteem, and the confidence gap* (1st ed.). New York: Doubleday.

Orlowski, A. (2003). Most bloggers 'are teenage girls' – survey. *The Register*. Retrieved 23/04/08 from http://www.theregister.co.uk/2003/05/30/most_bloggers_are_teenage_girls/

Orobio de Castro, B., Veerman, J. W., Koops, W., Bosch, J. D., & Monshouwer, H. J. (2002). Hostile attribution of intent and aggressive behavior: A meta-analysis. *Child Development, 73*, 916–934. doi:10.1111/1467-8624.00447

OSTWGOnline Safety and Technology Working Group. (2010). *Youth Safety on a Living Internet*. Washington, DC: National Telecommunications and Information Administration.

O'Sullivan, L. F. (2006). The Sexual Lives of Early Adolescent Girls. *Journal of Sex Research*, *43*(1), 6–2.

Oswell, D. (1999). The Dark Side of Cyberspace: Internet Content Regulation and Child Protection. *Convergence*, *5*(4), 42–62.

Owen, L., & Ennis, C. (2005). The ethic of care in teaching: An overview of supportive literature. *Quest*, *57*(4), 392–425.

Owuamanam, D. O. (1982). Sexual Activity of School-going Adolescents in Nigeria. *Adolescence*, *17*(65), 81–87.

Oyelaran-Oyeyinka, B., & Adeya, C. N. (2004). Internet Access in Africa: Empirical Evidence from Kenya and Nigeria. *Telematics and Informatics*, *21*(1), 67–81. doi:10.1016/S0736-5853(03)00023-6

Padva, G. (2005). Israel filmmaking. In Gerstner, D. (Ed.), *Routledge international encyclopedia of queer culture: Gay, lesbian, bisexual and transsexual contemporary cultures* (pp. 312–313). New York, London: Routledge.

Paikoff, R. L., & Brooks-Gunn, J. (1991). Do parent-child relationships change during puberty? *Psychological Bulletin*, *110*(1), 47–66. doi:10.1037/0033-2909.110.1.47

Palfrey, J., & Gasser, U. (2008). *Born digital: Understanding the first generation of digital natives. New-York.* Basic books.

Palm, G. (2000). *Den svenska högtidsboken.* Stockholm, Sweden: En bok för alla.

Palmer, P. J. (1998). *The courage to teach: Exploring the inner landscape of a teacher's Life*. New York: Jossey-Bass.

Palys, T. (2003). *Research Decisions: Quantitative and Qualitative Perspectives* (3rd ed.). New York: Thomson Nelson.

P*an European Game Information. (n.d.).* Retrieved 15.12.2009, from http://www.pegi.info/de

Panorama, (1998 - 2001). *The Wonderland Club.* Available at http://news.bbc.co.uk/1/hi/programmes/panoram/archive/1166945.stm

Park, M. R., & Floyd, K. (1996). Making friends in cyberspace. *Journal of Computer-Mediated Communication, 1*(4), 80–97.

Parker, R. (2004). Introduction to Sexuality and Social Change: Toward an Integration of Sexuality Research, Advocacy, and Social Policy in the Twenty-First Century. *Sexuality Research & Social Policy: Journal of NSRC, 1*(1), 7–14. doi:10.1525/srsp.2004.1.1.7

Patchin, J. W., & Hinduja, S. (2006). Bullies move beyond the school yard: A preliminary look at cyber bullying. *Youth Violence and Juvenile Justice, 4*(2), 148–169. doi:10.1177/1541204006286288

Patton, M. Q. (1990). *Qualitative Evaluation and Research Methods*. London: Sage.

Patton, M. Q. (1999). Enhancing the quality and credibility of qualitative analysis. (Part II). *Health Services Research, 34*(5), 1189e1208.

Paul, P. (2005). *Pornified: how the culture of pornography is changing our lives, our relationships, and our families*. New York: Times.

Pearson, J., Nelson, P., Titsworth, S., & Harter, L. (2003). *Human Communication*. Boston: McGraw Hill.

Perlman, N. B., Ericson, K. I., Esses, V. M., & Isaacs, B. J. (1994). The developmentally handicapped witness: Competency as a function of question format. *Law and Human Behavior, 18*, 171–187. doi:10.1007/BF01499014

Peter, J., & Valkenburg, P. M. (2006). Adolescents' Exposure to Sexually Explicit Material on the Internet'. *Communication Research, 33*(2), 178–204. doi:10.1177/0093650205285369

Philippsohn, S. (2001). Trends in Cybercrime - an overview of current financial crimes on the Internet. *Computers & Security, 20*(1), 53–69. doi:10.1016/S0167-4048(01)01021-5

Pickett, A. D., & Thomas, C. (2006). Turn OFF That Phone. *The American School Board Journal*, (April): 40–45.

Pilkington, N. W., & D'Augelli, A. R. (1995). Victimization of lesbian, gay and bisexual youth in community settings. *Journal of Community Psychology, 23*(1), 34–56. doi:10.1002/1520-6629(199501)23:1<34::AID-JCOP2290230105>3.0.CO;2-N

Pipher, M. B. (1994). *Reviving Ophelia: saving the selves of adolescent girls*. New York: Putnam.

Pizmony-Levy, O., Kama, A., Shilo, G., & Lavee, S. (2008). Do my teachers care I'm gay?: Israeli lesbigay school students' experiences at their schools. *Journal of Gay and Lesbian Youth, 5*(2), 33–61.

Plichta, P. (2010). Intellectualy disabled students – perpetrators and victims of electronic aggression. In Żółkowska, T., & Konopska, L. (Eds.), *Disability - the contextuality of its meaning* (p. 680). Szczecin, Poland: Wyd. Print Group.

Plog, F. (1980). *Anthropology: Decisions, Adaptation, and Evolution.* New York: Alfred A. Knopf.

Polman, H., Orobio de Castro, B., & van Aken, M. A. G. (2008). Experimental study of the differential effects of playing versus watching violent video games on children's aggressive behavior. *Aggressive Behavior, 34*(3), 256–264. doi:10.1002/ab.20245

Pratten, D. (2008). 'The Thief Eats His Shame': Practice and Power in Nigerian Vigilantism. *Africa*, *78*(1), 64–83. doi:10.3366/E0001972008000053

Prensky, M. (2001a). Digital Natives, Digital Immigrants Part 1. *Horizon*, *9*(5), 1–6. doi:10.1108/10748120110424816

Prensky, M. (2001b). Digital Natives, Digital Immigrants Part 2: Do They Really Think Differently? *Horizon*, *9*(6), 1–6. doi:10.1108/10748120110424843

Prezza, M., Guiseppina, M., & Dinelli, S. (2004). Loneliness and New Technologies in a Group of Roman Adolescents. *Computers in Human Behavior*, *20*, 691–709. doi:10.1016/j.chb.2003.10.008

Price, M., & Verhulst, S. (2005). *Self-regulation and the Internet*. Hague, the Netherlands: Kluwer Law International.

Price, J. M., & Dodge, K. A. (1989). Reactive and proactive aggression in childhood: Relations to peer status and social context dimensions. *Journal of Abnormal Child Psychology*, *17*, 455–471. doi:10.1007/BF00915038

Prinstein, M. J., & Cillessen, A. H. N. (2003). Forms and functions of adolescent peer aggression associated with high levels of peer status. *Merrill-Palmer Quarterly*, *49*, 310–342. doi:10.1353/mpq.2003.0015

Probyn, E. (2000). *Carnal Appetites: FoodSexIdentities*. London, New York: Routledge.

Prout, A. (2008). Culture-nature and the construction of childhood. In Drotner, K., & Livingstone, S. M. (Eds.), *The international handbook of children, media and culture*. London: Sage.

Prout, A., & James, A. (1997). A new paradigm for the sociology of childhood? Provenance, promise and problems. In James, A., & Prout, A. (Eds.), *Constructing and reconstructing childhood* (pp. 7–17). London: Falmer press.

Pugh, S. (2005). *The Democratic Genre: Fan fiction in a Literary Context*. Glawgow, UK: Seren.

Pyżalski, J. (2009). Agresja elektroniczna dzieci i młodzieży – różne wymiary zjawiska, *Dziecko krzywdzone. Teoria. Badania. Praktyka. „. Cyberprzemoc*, *1*(26), 12–26.

Pyżalski, J. (2009). Agresja elektroniczna wobec „pokrzywdzonych" – Inny jako ofiara agresji. In Chrzanowska, I., Jachimczak, B., & Podgórska-Jachnik, D. (Eds.), *Miejsce Innego we współczesnych naukach o wychowaniu. Trudy dorastania, trudy dorosłości* (pp. 54–59). Łódź: WSP/ Edukacyjna Grupa Projektowa.

Quan-Haase, A., & Collins, J. L. (2008). I'm there, but I might not want to talk you. *Information Communication and Society*, *11*(4), 526–543. doi:10.1080/13691180801999043

Quayle, E., & Taylor, M. (2001). Child Seduction and Self-representation on the internet. *Cyberpsychology & Behavior*, *4*(5), 597–610. doi:10.1089/109493101753235197

Quayle, E., Loof, L., & Palmer, T. (2008). *Child Pornography and Sexual Exploitation of Children Online: A Contribution of ECPAT International to the World Congress III against Sexual Exploitation of Children and Adolescents.* Bangkok: ECPAT International.

Qvortrup, J. (2003). *Barndom i et sociologiskt generationsperspektiv [Childhood from a sociological generations perspective].* (Working paper no. 123-03, Centre for Cultural Research, University of Aarhus). http://www.hum.au.dk/ckulturf/pages/publications/jq/barndom.htm

Radin, P. (2006). "To me it's my life": Medical communication, trust, and activism cyberspace. *Social Science & Medicine*, *62*(3), 591–601. doi:10.1016/j.socscimed.2005.06.022

Radway, J. A. (1984). *Reading the romance: women, patriarchy, and popular literature*. Chapel Hill, NC: University of North Carolina Press.

Rafaeli, S., Ariel, Y., & Katsman, M. (2010). *Adolescents online: Patterns of Internet usage and online consumption.* Ministry of Industry, Trade and Labor: Research and Economy Administration. Retrieved March 10, 2010 from http://www.moit.gov.il/NR/rdonlyres/6F6300A3-336D-4A8D-8D44-9213065DF8F7/0/9784.pdf.

Rainie, L. (January 5, 2010). *Internet, broadband, and cell phone statistics*. Pew Internet. Available from http://www.pewinternet.org/~/media//Files/Reports/2010/PIP_December09_update.pdf

Rambaree, K. (2008). Internet-Mediated Dating/Romance of Mauritian Early Adolescents: A Grounded Theory Analysis. *International Journal of Emerging Technologies & Society*, *6*(1), 34–59.

Rambaree, K. (2009). 'Internet, Sexuality & Development: Putting Early Adolescents First'. *The International Journal of Environmental, Cultural. Economic & Social Sustainability*, *5*(2), 105–119.

Rambaree, K. (2005) `*The Ecology of Sexuality in a Mauritian Internet Chat Room (MICR): An Internet Mediated Research (IMR)*. Paper presented for the IFRD Conference in Mauritius, January 2005

Rambaree, K. (2007).*The Ecology of Internet & Early Adolescent Sexuality in a Technology-driven Mauritian Society.* PhD Thesis submitted to The University of Manchester, UK

Rambaree, K. (2010) "Children and the Janus-faced Internet: Social Policy Implications for Mauritius as a Developing Country Case Study'. In I, Berson & M. Berson (Eds) *High-Tech Tots: Childhood in a Digital World.* Greenwich, CT: Information Age Publishing

Rapley, T. (2007). *Doing Conversation, Discourse and Document Analysis. The SAGE Qualitative Research Kit.* London: Sage Publications.

Raskauskas, J., & Stoltz, A. D. (2007). Involvement in traditional and electronic bullying among adolescents. *Developmental Psychology*, *43*(3), 564–575. doi:10.1037/0012-1649.43.3.564

Rasmussen, D., & Hansen, H. R. (2005). *Chat, chikane og mobning blandt børn og unge. Afsluttende rapport om skoleelevers brug af chat i deres interaktive sociale liv.* Konsulentgruppen AMOK. Retrieved from http://www.livsmodlab.dk/txt/chat_chikane_og_mobning_blandt_boern_og_unge.pdf

Rattleff, P., & Tønnesen, P. H. (2007). *Børn og unges brug af internettet i fritiden.* København, Denmark: The Danish School of Education - University of Aarhus, The Danish Media Council for Children and Young People. Retrieved from http://andk.medieraadet.dk/upload/brugafinternettetrapport.pdf

Rauner, D. (2000). *They still pick me up when I fall: The role of caring in youth development and community life*. New York: Columbia University Press.

Raynes-Goldie, K. (2010). Aliases, creeping, and wall cleaning: Understanding privacy in the age of Facebook. *First Monday*, *15*, 1–4.

Reaves, J. (2001). *Anorexia Goes High Tech.* Time Magazine 31 July 2001. Retreived October 13, 2009, from http://www.time.com.

Reid-Walsh, J., & Mitchell, C. (2004). Girls' web sites: a virtual "room of one's own"? In Harris, A. (Ed.), *All about the girl: power, culture, and identity* (pp. 173–182). New York: Routledge.

Renold, E. (2005). *Girls, boys and junior sexualities. Exploring children's gender and sexual relations in the primary school.* London: Routledge Falmer.

Resch, R. P. (1992). *Althusser and the renewal of Marxist social theory.* Berkeley, CA: University of California Press.

Reschly, D. (2009). Documenting the Developmental Origins of Mild Mental Retardation. *Applied Neuropsychology*, *16*(2), 124–134. .doi:10.1080/09084280902864469

Rheingold, H. (2003). *Smart Mobs: The Next Social Revolution New edition*. New York: Perseus Books.

Rheingold, H. (2000). *The virtual community: Homesteading on the electronic frontier*. Cambridge, MA: MIT Press.

Ribadu, N. (2004). *Obstacles to the effective prosecution of corrupt practices and financial crime cases in Nigeria.* Paper Presented at the Summit on Corrupt Practices and Financial Crimes in Nigeria, Kaduna, Nigeria.

Rice, F. P. (2001). *Human development*. Upper Saddle River, NJ: Prentice Hall.

Rich, A. (1993). Compulsory heterosexuality and the lesbian existence. In Abelove, H., Barale, M. A., & Halperin, D. M. (Eds.), *The lesbian and gay studies reader* (pp. 227–254). New York, London: Routledge. (Original work published 1986)

Rich, E., Harjunen, H., & Evans, J. (2006) Normal gone bad' – Health Discourses, Schools and the Female Body. In Peter Twohig & Vera Kalitzkus (Eds) *Bordering Biomedicine Interdisciplinary Perspectives on Health, Illness and Disease*. New York: Rodopi.

Rideout, V. (2001). *Generation Rx.com: How young people use the Internet for health information.* Available at http://www.kff.org/entmedia/upload/Toplines.pdf Accessed on 30th December 2009.

Rideout, V., Roberts, D. F., & Foehr, U. G. (2005). *Generation M: Media in the lives of 8-18 year-olds*. Washington DC: The Henry J. Kaiser Family Foundation.

Rigby, K. (2005). What children tell us about bullying in schools. *Children Australia Journal of Guidance and Counselling, 15*(2), 195–208. doi:10.1375/ajgc.15.2.195

Ritzer, G. (2000). *Classical Sociological Theory* (3rd ed.). Boston: McGraw Hill.

Riva, G. (2005). The Psychology of Ambient Intelligence: Activity, Situation and Presence. In Riva, G., Vatalaro, F., Davide, F., & Alcañiz, M. (Eds.), *Ambient Intelligence* (pp. 17–33). Amsterdam: ISO Press.

Rival, L., Slater, D., & Miller, D. (1998). Sex & Sociality: Comparative Ethnographies of Sexual Objectification. *Theory, Culture & Society, 15*(3-4), 295–321. doi:10.1177/0263276498015003015

Roberts, T. L. (1998). *Are newsgroups virtual communities*? Proceedings of the Annual ACM SIGCHI Conference on Human Factors in Computing Systems (CHI 1998). Pp. 360-367. New York: ACM Press.

Robinson, J. P., Neustadtl, A., & Kestnbaum, M. (2001). An Online Data Web Site for Internet Research: Some Features and an Example. *The American Behavioral Scientist, 45*(3), 565–568. doi:10.1177/00027640121957222

Robinson, L. (2007). The cyberself: the self-ing project goes online, symbolic interaction in the digital age. *New Media & Society, 9*(1), 93–110..doi:10.1177/1461444807072216

Rochlin, M. (1994). Sexual orientation of the therapist and therapeutic effectiveness with gay clients. In Gonsiorek, J. (Ed.), *A guide to psychotherapy with gay and lesbian clients* (pp. 21–29). New York: Harrington Park Press.

Rogers, E. M. (2003). *Diffusion of innovations* (5th ed.). New York: Free Press.

Ropelato, J. (2007). *Internet pornography statistics.* Available from http://Internet-filter-review.toptenreviews.com/Internet-pornography-statistics.html Accessed 23/12/09.

Rosen, L., Cheever, N., & Carrier, L. (2008). The association of parenting style and child age with parental limit setting and adolescent MySpace behavior. *Journal of Applied Developmental Psychology, 29*(6), 459–471. doi:10.1016/j.appdev.2008.07.005

Ross, M. W., & Kauth, M. R. (2002). Men who have sex with men and the internet: Emerging clinical issues and their management. In Al Cooper (Ed.) *Sex and the internet; A guidebook for clinicians,* pp 47-69. New York: Brunner-Routledge.

Ross, MW, Rosser, BRS, & Stanton, J (2004). Beliefs about cybersex & Internet-mediated sex of Latino men who have Internet sex with men: relationships with sexual practices in cybersex & in real life. *AIDS CARE*, 16 (8), 1002_1011

Runeborg, A. (2004). *Sexuality – a super force: Young People, Sexuality & Rights in the era of HIV/AIDS.* Stockholm: Sida, Department for Democracy and Social Development, Health Division.

Russell, D. (1998). *Dangerous Relationships: Pornography, misogyny, and rape*. Thousand Oaks, CA: Sage.

Rust, P. C. (2003). Finding sexual identity and community: Therapeutic implications and cultural assumptions in scientific models of coming out. In Garnets, L. D., & Kimmel, D. C. (Eds.), *Psychological perspectives on lesbian, gay and bisexual experiences* (pp. 227–269). New York: Columbia University Press.

Rustad, M. (2001). Private Enforcement of Cybercrime on the Electronic Frontier. *Southern California Interdisciplinary Law Journal, 11*(1), 63–116.

Saarikoski, H. (2001). *Mistä on huonot tytöt tehty?* Helsinki, Finland: Tammi.

Saban, K. A., McGivern, E., & Saykiewicz, J. N. (2002). A critical look at the impact of cyber crime on consumer internet behavior. *Journal of Marketing Theory and Practice, 10*(2), 29–37.

Sabina, C., Wolak, J., & Finkelhor, D. (2008). The nature and dynamics of Internet pornography exposure for youth. *Cyberpsychology & Behavior, 11*(6), 691–693. doi:10.1089/cpb.2007.0179

Safer Internet Programme. (2009). *Safer Internet Programme: Homepage - Europa - Information Society.* Retrieved January 15, 2010, from http://ec.europa.eu/information_society/activities/sip/index_en.htm

Sampson, R. J., & Wilson, W. J. (2005). Toward a theory of race, crime and urban inequality. In Gabbidon, L. S., & Greene, H. T. (Eds.), *Race, Crime and Justice: A Reader* (pp. 37–54). New York: Routledge.

Samuelsson, M. (2008). Att förhålla sig till institutionalisering: Utmanande för såväl lärare som elever. *LOCUS, tidskrift för forskning om barn och ungdomar, 3-4(4),* 83-99.

Sandemose, A. (1968). *En flykting korsar sitt spår.* Stockholm, Sweden: Forum.

Sandvoss, C. (2005). *Fans*. London: Polity Press.

Santos, R. L., Rocha, B. P., Rezende, C. G., & Loureiro, A. A. (2007). Characterizing the YouTube video-sharing community, (Technical report). Retrieved from http://security1.win.tue.nl/~bpontes/pdf/yt.pdf.

Sarbaugh-Thompson, J. S., & Feldman, M. S. (1998). Electronic mail and organizational communication: Does saying "hi" really matter? *Organization Science, 9*(6), 685–698. doi:10.1287/orsc.9.6.685

Saunders R. (2005). Happy slapping: transatlantic contagion or home-grown, mass-mediated nihilism, *The London Consortium,* 1-11.

Savin-Williams, R. C. (2005). *The new gay teenager*. Boston: Harvard University Press.

Schalock, R., & Luckasson, R. (2004). American Association on Mental Retardation's Definition, Classification, and System of Supports and its relation to international trends and issues in the field of intellectual disabilities. *Journal of Policy and Practice in Intellectual Disabilities, 1*(3-4), 136–146. .doi:10.1111/j.1741-1130.2004.04028.x

Schensul, S., Oodit, G., Schensul, J., Seebuluk, S. U., Bhowan, J., & Aukhojee, P. (1994). *Young Women, Work, & AIDS-Related Risk Behavior in Mauritius*. Washington, DC: International Center for Research on Women.

Schiller, E.-M., Strohmeier, D., & Spiel, C. (2009). Risiko Video- und Computerspiele? Eine Studie über Video- und Computerspielnutzung und Aggression bei 12- und 16-jährigen Jugendlichen. *Schweizerische Zeitschrift für Bildungswissenschaften, 31*(1), 75–98.

Schloss, P., Smith, M., & Kiehl, W. (1986). Rec Club: A community centered approach to recreational development for adults with mild to moderate retardation. [Retrieved from PsycINFO database.]. *Education and Training of the Mentally Retarded, 21*(4), 282–288.

Schneider, R. (1997). *The explicit body in performance*. New York: Routledge. doi:10.4324/9780203421079

Schodt, F. L. (1996). *Dreamland Japan: writings on modern manga*. Berkeley, CA: Stone Bridge Press.

Schoenberg, N. E., & Ravdal, H. (2000). Using vignettes in awareness and attitudinal research. *International Journal of Social Research Methodology, 3*(1), 63–74. doi:10.1080/136455700294932

Schofield Clark, L. (2005). The constant contact generation: exploring teen friendship networks online. In Mazzarella, S. R. (Ed.), *Girl wide web* (pp. 203–221). New York: Peter Lang.

Schopflin, G. (2001). *The construction of Identity: Learning-Theories*. Communication.

Schrøder, K. (2000). Pionérdagene er forbi! Hvor går receptionsforskningen hen? *MedieKultur*, Medieforskning til tiden: Rapport fra Statens Humanistiske Forskningsråds konference om dansk film- og medieforskning, (Vol. 31).

Schulze, S. (2003). Views on the combination of quantitative and qualitative research approaches. *Progressio, 25(2)*, 8e20.

Schwartz, H. (1986). *Never Satisfied. A Cultural History of Diets, Fantasies and Fat*. New York: Free Press.

Scollon, R. (2001b). *Mediated discourse: The Nexus of Practice*. New York: Routledge.

Scollon, R., & Scollon, S. W. (2004). *Nexus Analysis: Discourse and the Emerging Internet*. New York: Routledge.

Scollon, R., & Scollon, S. W. (2007). Nexus analysis: Refocusing ethnography on action. *Journal of Sociolinguistics, 11*, 608–625. doi:10.1111/j.1467-9841.2007.00342.x

Scollon, R. (2005). The rhythmic integration of action and discourse: work, the body and the earth. In Norris, S., & Jones, R. (Eds.), *Discourse in action: introducing mediated discourse analysis*. London: Routledge.

Scollon, R. (2001a). Action and Text: Towards an integrated understanding of the place of text in social (inter) action, mediated discourse analysis and the problem of social action. In Wodak, R., & Meyer, M. (Eds.), *Methods of Critical Discourse Analysis* (pp. 139–183). London: Sage Publications.

Scollon, R. (2007). *Geographies of Discourse*. Draft available at www.aptalaska.net/~ron/ron/downloads/Geographies%20of%20Discourse%20download%20draft.doc

Searle, J. (1983). *Intentionality: An essay in the philosophy of mind*. Cambridge, UK: Cambridge University Press.

Sedgwick, E. K. (1985). *Between men: English literature and male homosocial desire*. New York: Columbia University Press.

Selg, H. (2003). *Internet i den svenska högskolan våren 2003 (Eng. Internet at Swedish universities spring 2003)*. Användarstudier inom SUNET. Retrieved from http://basun.sunet.se/anvandare.pdf.

Selg, H., & Findahl, O. (2006). *File sharing in peer-to-peer networks - actors, motives and effects*. Broadband technologies transforming busienss models and challenging regulatory framworks - lessons from the music industry, Deliverable 4. Retrieved from http://w1.nada.kth.se/media/Research/MusicLessons/Reports/MusicLessons-DL4.pdf.

Selg, H., & Findahl, O. (2008). *Nya användarmönster. Jämförande analys av två användarstudier*. InternetExplorers, Delrapport 6. Nationellt IT-användarcentrum NITA, Uppsala universitet. Retrieved from http://www.internetexplorers.se/.

Serfaty, V. (2004). Online Diaries: Towards a Structural Approach. *Journal of American Studies, 38*(3). doi:10.1017/S0021875804008746

Seymour, W. (2005). ICTs and disability: Exploring the human dimensions of technological engagement. [Retrieved from Health Source: Nursing/Academic Edition database.]. *Technology and Disability, 17*(4), 195–204.

Shannon, D. (2007). *Vuxnas sexuella kontakter med barn via Internet [Adults' sexual contacts with children online]*. Stockholm, Sweden: Brottsförebyggande rådet.

Shapira, N. A., Goldsmith, T. D., & Keck, P. E. (2000). Psychiatric features of individuals with problematic internet use. *Journal of Affective Disorders, 57*, 267–272. doi:10.1016/S0165-0327(99)00107-X

Shapiro, C., & Varian, H. R. (1999). *Information Rules: A Strategic Guide to the Network Economy*. Boston: Harvard Business School.

Sharf, B. F. (1997). Communicating breast cancer online: Support and empowerment on the Internet. *Women & Health, 26*(1), 65–84. doi:10.1300/J013v26n01_05

Shariff, S. (2006). Cyber-Dilemmas: Balancing Free Expression and Learning in a Virtual School Environment. *International Journal of Learning, 12*(4), 269–278.

Shariff, S. (2008). *Cyber-bullying: Issues and solutions for the school, the classroom and the home*. New York: Routledge.

Shariff, S., & Johnny, L. (2007). Cyber-Libel and Cyber-bullying: Can Schools Protect Student Reputations and Free Expression in Virtual Environments? *Education Law Journal*, *16*(3), 307.

Shariff, S., & Gouin, R. (2005). *Cyber-Dilemmas: Gendered Hierarchies, Free Expression and Cyber-Safety in Schools*.paper presented at Oxford Internet Institute, Oxford University, U.K. International Conference on Cyber-Safety. Paper available at: www.oii.ox.ac.uk/cybersafety

Sharples, M., Graber, R., Harrison, C., & Logan, K. (2009). E-safety and Web 2.0 for children aged 11-16. *Journal of Computer Assisted Learning*, *25*(1), 70–84. doi:10.1111/j.1365-2729.2008.00304.x

Sharples, M., Graber, R., Harrison, C & Logan K. (2008) E-safety and Web 2.0. *Web 2.0 technologies for learning at Key Stages 3 and 4*. Becta report.

Sheets, R. H., & Gay, G. (1996). Student Perceptions of Disciplinary Conflict in Ethnically Diverse Classrooms. *NASSP Bulletin*, *80*(580), 84–95. doi:10.1177/019263659608058011

Shepherd, R., & Edelmann, R. (2005). Reasons for internet use and social anxiety. *Personality and Individual Differences*, *39*(5), 949–958. .doi:10.1016/j.paid.2005.04.001

Sherry, J. (2007). Violent video games and aggression: Why can't we find links? In Preiss, R., Gayle, B., Burrell, N., Allen, M., & Bryant, J. (Eds.), *Mass media effects research: Advances through meta-analysis*. Mahwah, NJ: Erlbaum.

Shillington, K. (1991). *Jugnauth: The Prime Minister of Mauritius*. London: McMillan.

Shilo, G. (2007). *Pink life: Gay, lesbian, bisexual and transgender youth*. TA, Israel: Resling. (Hebrew).

Shuler, C. (2009). *Pockets of potential: Using Mobile Technologies to Promote Children's Learning*. New York: The Joan Ganz Cooney Center at Sesame Workshop.

Sigsgaard, E. (1992). Forholdet mellem menneskesyn, paradigme, forforståelse og data i humanistisk forskning. *Nordisk Psykologi*, *44*(4), 9–19.

Silverblatt, A. (2004). Media as Social Institution. *The American Behavioral Scientist*, *48*(1), 35–41. doi:10.1177/0002764204267249

Silverstone, R. (Ed.). (2005). *Media, Technology, and Everyday Life in Europe: From Information to Communication. Aldershot*. Hants, England: Ashgate.

Silverstone, R. (2005). Introduction. In R. Silverstone (Ed.), *Media, Technology, and Everyday Life in Europe: From Information to Communication*. (pp. 1-18). Aldershot, Hants, England: Ashgate.

Simpson, G. (1963). *Emile Durkheim: Selections from His Work*. New York: Thomas Y. Crowell Co.

Singh, S., Bankole, A., & Woog, V. (2005). Evaluating the need for sex education in developing countries: sexual behaviour, knowledge of preventing sexually transmitted infections/HIV and unplanned pregnancy. *Sex Education*, *5*(4), 307–331. doi:10.1080/14681810500278089

Skrzydlewski, W. (2005). Kampanie społeczne w mediach-doświadczenia polskie. In Leppert, R., Melosik, Z., & Wojtasik, B. (Eds.), *Młodzież wobec niegościnnej przyszłości*. Wrocław, Poland: Wydawnictwo Naukowe DSW Edukacji TWP.

Slevin, J. (2000). *The internet and society*. Cambridge, UK: Polity Press.

Slonje, R., & Smith, P. (2008). Cyberbullying: Another main type of bullying? *Scandinavian Journal of Psychology*, *49*(2), 147–154. doi:10.1111/j.1467-9450.2007.00611.x

Smahel, D., & Subrahmanyam, K. (2007). Any Girls Want to Chat Press 911: Partner Selection in Monitored and Unmonitored Teen Chat Rooms. *Cyberpsychology & Behavior*, *10*(3), 346–353. doi:10.1089/cpb.2006.9945

Smith, C. (2009). Pleasure & Distance: Exploring Sexual Cultures in the Classroom. *Sexualities*, *12*(5), 568–585. doi:10.1177/1363460709340368

Smith, D. S. (2000). *Moral Geographies: Ethics in a World of Difference*. Edinburgh, UK: Edinburgh University Press.

Smith, J. (2007). ''Ye've got to 'ave balls to play this game sir!": Boys, peers and fears: The negative influence of school-based "cultural accomplices" in constricting hegemonic masculinities. *Gender and Education, 19*(2), 179–198. doi:10.1080/09540250601165995

Smith, P., Mahdavi, J., Carvalho, M., Fisher, S., Russell, S., & Tippett, N. (2008). Cyberbullying: its nature and impact on secondary school pupils. *Journal of Child Psychology and Psychiatry, and Allied Disciplines, 49*(4), 376–385. doi:10.1111/j.1469-7610.2007.01846.x

Smith, S. L., Lachlan, K., & Tamborini, R. (2003). Popular video games: Quantifying the presentation of violence and its context. *Journal of Broadcasting & Electronic Media, 47*, 58–76. doi:10.1207/s15506878jobem4701_4

Smith, A. D. (2004). Cybercriminal impacts on online business and consumer confidence. *Online Information Review, 28*(3), 224–234. doi:10.1108/14684520410543670

Snyder, H. N. (2000). *Sexual assault of young children as reported to law enforcement: victims, incident, and offender characteristics. Report NCJ 182990*. Washington, DC: US Department of Justice, National Center for Juvenile Justice, Bureau of Statistics.

Social Science Computer Review, 26(3), 379–388. doi:10.1177/0894439307311611

Solheim, J. (2001). *Den öppna kroppen. Om könssymbolik i modern kultur*. Göteborg, Sweden: Daidalos.

Sorapure, M. (2003). Screening Moments, Scrolling Lives: Diary Writing on the Web. *Biography, 26*(1). doi:10.1353/bio.2003.0034

Soyinka, F. (1979). Sexual Behaviour among University Students' in Nigeria. *Archives of Sexual Behavior, 8*(1), 15–26. doi:10.1007/BF01541209

Spalding, N. J., & Phillips, T. (2007). Exploring the Use of Vignettes: From Validity to Trustworthiness. *Qualitative Health Research, 17*(7), 954–962. doi:10.1177/1049732307306187

Spitzberg, B., & Hoobler, G. (2002). Cyberstalking and the technologies of interpersonal terrorism. *New Media & Society, 4*(1), 71. doi:10.1177/14614440222226271

Sprecher, S., & Metts, S. (1999). Romantic Beliefs: Their influence on relationships and patterns of change over time. *Journal of Social and Personal Relationships, 16*, 834–851. doi:10.1177/0265407599166009

Springhall, J. (1998). *Youth, popular culture and moral panics: penny gaffs to gangsta rap, 1830-1997*. Basingstoke, UK: Macmillan.

Sproull, L., & Kiesler, S. (1991). *Connections*. Cambridge, MA: MIT Press.

Stahl, C., & Fritz, N. (2002). Internet safety: adolescents' self-report. *The Journal of Adolescent Health, 31*(1), 7–10. doi:10.1016/S1054-139X(02)00369-5

Stallybrass, P., & White, A. (1986). *The politics and poetics of transgression*. London: Methuen.

Stanford Encyclopaedia. *Intentionality* (2003). Retrieved January 12, 2010 from http://plato.stanford.edu/entries/intentionality/.

Stanley, L. (1999). The Cultural Bias of Sex Surveys. In Nye, R. A. (Ed.), *Sexuality*. Oxford, UK: Oxford University Press.

Stassen-Berger, K. (2007). Update on bullying at school: science forgotten? *Developmental Review, 27*, 90–126. doi:10.1016/j.dr.2006.08.002

Stern, S. (1999). Adolescent girls' expression on web pages: spirited, sombre and self-conscious sites. *Convergence (London), 5*(4), 22–41. doi:10.1177/135485659900500403

Stern, S. (2002). Virtually speaking: girls' self-disclosure on the WWW. *Women's. Studies in Communications, 25*(2), 223–252.

Stern, S. (2008). Producing Sites, Exploring Identities: Youth Online Authorship. *Youth, Identity, and Digital Media*. In Buckingham, D. (Ed.), *The John D. and Catherine T. MacArthur Foundation Series on Digital Media and Learning* (pp. 95–118). Cambridge, MA: The MIT Press.

Stevens, E. D., & Payne, B. K. (1999). Applying deterrence theory in the context of corporate wrongdoing: Limitations on punitive damages. *Journal of Criminal Justice*, *27*(3), 195–207. doi:10.1016/S0047-2352(98)00060-9

Stevenson, M. R. (2002). 'Conceptualizing Diversity in Sexuality Research' in M.W. Wiederman & B.E. Whitley, Jr. (Eds) *H& book for Conducting Research on Human Sexuality*, (pp. 455-479). Mahwah, N.J: Lawrence Erlbaum Associates

Strandberg, L. (2000) *När mörkret kom – en bok om anorexia*. Falun: Författarhuset.

Strasburger, V. C. (2009). Why do adolescent health professionals ignore the impact of the media? *The Journal of Adolescent Health*, *44*, 203–205. doi:10.1016/j.jadohealth.2008.12.019

Subrahmanyam, K., Greenfield, P. M., & Tynes, B. (2004). Constructing sexuality & identity in an online teen chat room. *Applied Developmental Psychology*, *25*, 651–666. doi:10.1016/j.appdev.2004.09.007

Subrahmanyam, K., Reich, M. S., Waechter, N., & Espinoza, G. (2008). Online and offline social networks: Use of social networking sites by emerging adults. *Journal of Applied Developmental Psychology*, *19*, 420–433. doi:10.1016/j.appdev.2008.07.003

Subrahmanyam, K., & Greenfield, P. (2008). Online communication and adolescent relationships. *The Future of Children*, *18*(1), 119–146. doi:10.1353/foc.0.0006

Subrahmanyam, K., Kraut, R. E., Greenfield, P. M., & Gross, E. F. (2000). The impact of home computer use on children's activities and development. *The Future of Children*, *10*, 123–144. doi:10.2307/1602692

Subrahmanyam, K., Smahel, D., & Greenfield, P. M. (2006). Connecting developmental processes to the Internet: Identity presentation and sexual exploration in online teen chatrooms. *Developmental Psychology*, *42*, 1–12. doi:10.1037/0012-1649.42.3.395

Suler, J. (2005)... *International Journal of Applied Psychoanalytic Studies*, *2*(2), 184–188. doi:10.1002/aps.42

Sun, P., Unger, J., Palmer, P., Gallaher, P., Chou, P., & Baezconde-Garbanati, L. (2005). Internet accessibility and usage among urban adolescents in southern California: Implications for web-based health research. *Cyberpsychology & Behavior*, *8*(5), 441–453. doi:10.1089/cpb.2005.8.441

Sundby, J. (2006). Young people's sexual and reproductive health rights. *Best Practice & Research. Clinical Obstetrics & Gynaecology*, *20*(3), 355–368. doi:10.1016/j.bpobgyn.2005.12.004

Surf, S. (2007). Retrieved from http://www.safesurf.com/time.htm

Suzuki, L. K., & Calzo, J. P. (2004). The search for advice in cyberspace: An examination of online teen bulletin boards about health and sexuality. *Applied Developmental Psychology*, *25*, 685–698. doi:10.1016/j.appdev.2004.09.002

Sveningsson, M., Lövheim, M., & Bergquist, M. (2003). *Att fånga nätet – kvalitativa metoder för Internetforskning*. Lund, Sweden: Studentlitteratur.

Sveninsson, M. (Ed.). (2007). *Cyberfeminism in Northern lights: digital media and gender in a Nordic context*. Newcastle-upon-Tyne: Cambridge Scholars.

Swidler, A. (1986). Culture in Action: Symbols and Strategies. *American Sociological Review*, *51*(2), 273–286. doi:10.2307/2095521

Szpunar, M. (2006). Społecznośći wirtualne – realne kontakty w wirtualnym świecie. In Haber, L., & Niezgoda, M. (Eds.), *Społeczeństwo informacyjne. Aspekty funkcjonalne i dysfunkcjonalne* (pp. 158–167). Kraków, Poland: Wydawnictwo UJ.

Takayoshi, P., Huot, E., & Huot, M. (1999). No boys allowed: the World Wide Web as a clubhouse for girls. *Computers and Consumption*, *16*, 89–106. doi:10.1016/S8755-4615(99)80007-3

Tapscott, D. (1998). *Growing up Digital: The Rise of the Net Generation*. London: McGraw-Hill.

Tapscott, D. (1998). The Rise of the Net Generation. In *Meridian, January, 1998*. Growing Up Digital.

Tapscott, D. (1999). Educating the Net Generation. *Educational Leadership*, (Feb): 1999.

Tashakkori, A., & Teddlie, C. (2003). *Handbook of mixed methods in the social and behavioral research*. Thousand Oaks, CA: Sage Publications.

Taylor, M., Holland, G., & Quayle, E. (2001). Typology of paedophile picture collections. *The Police Journal*, *74*(2), 97–107.

Taylor, M., & Quayle, E. (2003). *Child Pornography: An Internet Crime*. Hove, UK: Brunner-Routledge.

Teh, Y. K. (2009). The best police force in the world will not bring down a high crime rate in a materialistic society. *International Journal of Police Science and Management*, *11*(1), 1–7. doi:10.1350/ijps.2009.11.1.104

The Association for Progressive Communications (APC). (2006). *Warning over African internet cable*. Retrieved from http://news.bbc.co.uk/2/hi/africa/4787422.stm.

The Ghanaian Journal. (2009). *Top Pastors in Ghana Patronise Sakawa Boys*. Retrieved May 29, 2010 from http://www.theghanaianjournal.com/2009/05/04/top-pastors-in-ghana-patronize-sakawa-boys/.

Thelwall, M. (2008). Social networks, gender, and friending: an analysis of MySpace member profiles. *Journal of the American Society for Information Science and Technology*, *59*(8), 1321–1330. doi:10.1002/asi.20835

Thiel, S. M. (2005). "IM me": identity construction and gender negotiation in the world of adolescent girls and instant messaging. In Mazzarella, S. R. (Ed.), *Girls wide web* (pp. 179–201). New York: Peter Lang.

Thio, A. (1989). *Sociology an Introduction* (2nd ed.). New York: Harper & Row Publishers.

Thomas, E. K., & Carpenter, B. H. (Eds.). (2001). *Mass Media in 2025: industries Organizations, People and Nations Westport*. Santa Barbara, CA: Greenwood Press.

Thomas, A. (2007). *Youth Online: Identity and Literacy in the Digital Age*. New York: Peter Lang Publishing Inc.

Thompson, K. M., & Haninger, K. (2001). Violence in E-rated video games. *Journal of the American Medical Association*, *286*, 591–598. doi:10.1001/jama.286.5.591

Thompson, K. M., Tepichin, K., & Haninger, K. (2006). Content and ratings of mature-rated video games. *Archives of Pediatrics & Adolescent Medicine*, *160*, 402–410. doi:10.1001/archpedi.160.4.402

Thompson, M. (2001). Coming out inside. In Berzon, B. (Ed.), *Positively gay: New approaches to gay and lesbian life* (pp. 32–38). Berkley, CA: Celestial Arts.

Thornburgh, D., & Lin, H. S. (Eds.). (2002). *Youth, pornography, and the Internet*. Washington, DC: National Academic Press.

Thulin, K., & Östergren, J. (1997). *X-märkt: flickornas guide till verkligheten*. Stockholm, Sweden: Rabén & Sjögren.

Tierney, S. (2006). The Dangers and Draw of Online Communication: Pro-Anorexia Websites and their Implications for Users, Practitioners, and Researchers. *Eating Disorders*, *14*(3), 181–190. doi:10.1080/10640260600638865

Tincknell, E., Chambers, D., Van Loon, J., & Hudson, N. (2003). Begging for it: "new femininities," social agency, and moral discourse in contemporary teenage and men's magazines. *Feminist Media Studies*, *3*(1), 47–63. doi:10.1080/14680770303796

Tippet, N., Thompson, F., & Smith, P. K. (2009). Research on Cyberbullying: Key findings and practical suggestions. *Education.com*. http://www.education.com/reference/article/cyberbullying-research/

Toffler, A. (1980). *The Third Wave*. New York: Collins.

Tolman, D., & Szalacha, L. (1999). Dimensions of Desire: Bridging Qualitative & Quantitative Methods in a Study of Female Adolescent Sexuality. *Psychology of Women Quarterly*, *3*, 7–39. doi:10.1111/j.1471-6402.1999.tb00338.x

Tolman, D. L., & Diamond, L. M. (2001). Desegrating sexuality research: Cultural and biological perspectives on gender and desire. *Annual Review of Sex Research*, *12*, 33–74.

Torres, S. (2009). Vignette methodology and culture-relevance: lessons learned from a study on Iranian's understandings of successful aging. *Journal of Cross-Cultural Gerontology*, *24*(1), 93–114. doi:10.1007/s10823-009-9095-9

Tracy, K. (1995). Action-implicative discourse analysis. *Journal of Language and Social Psychology*, *14*(1-2), 195–215. doi:10.1177/0261927X95141011

Traeen, B., Nilsen, T. S., & Stigum, H. (2006). Use of pornography in traditional media and on the internet in Norway'. *Journal of Sex Research*, *43*(3), 245–255. doi:10.1080/00224490609552323

Traeen, B., Spitznogle, K., & Beverfjord, A. (2004). Attitudes and use of pornography in the Norwegian population 2002. *Journal of Sex Research*, *41*(2), 193–200. doi:10.1080/00224490409552227

Trousdale, G. (2010). *An Introduction to English Sociolinguistics*. Edinburgh, UK: Edinburgh University Press.

Tsolidis, G. (1986). *Educating Voula: prepared by Georgina Tsolidis for Ministerial Advisory Committee on Multicultural and Migrant Education*. Melbourne, Victoria: Ministry of Education, Victoria.

Tsoulis-Reay, A. (2009). OMG I'M ONLINE...AGAIN! MySpace, MSN and the everyday mediation of girls. *Screen Education* (53), 48-55.

Tucker, K. (1998). *Anthony Giddens and Modern Social Theory*. London: Sage Publications.

Turkle, S. (1997). *Life on the screen: identity in the age of the Internet*. New York: Touchstone.

Turkle, S. (1999). Identity in the Age of the Internet. In Mackay, H., & O'Sullivan, T. (Eds.), *The Media Reader: Continuity & Transformation* (pp. 287–305). London: Sage.

Tynes, B. (2006). Internet Safety Gone Wild? Sacrificing the Educational and Psychosocial Benefits of Online Social Environments. *Journal of Adolescent Research*, *22*(6), 575–584. doi:10.1177/0743558407303979

Tynes, B. M., Giang, M. T., Williams, D. R., & Thopson, G. N. (2008). Online racial discrimination and psychological adjustment among adolescents. *The Journal of Adolescent Health*, *43*, 565–569. doi:10.1016/j.jadohealth.2008.08.021

Uche, L. U. (1996). *North-South Information Culture: Trends in Global Communications and Research Paradigms*. Lagos, Nigeria: Longman Plc.

Uduoakah, N., & Iwokwagh, N. S. (2008). Communication and HIV/AIDS Prevention among Adolescents in Benue State. *The Nigerian Journal of Communications*, *6*(1 & 2), 44–58.

UNESCO. (2004). *The Plurality of Literacy and its Implications for Policies and Programmes*. Paris: United Nations Educational Scientific and Cultural Organization.

UNICEF. (2008). *Convention on the Rights of the Child: An introduction*. Available at http://www.unicef.org/crc/index_30160 accessed on 5th February 2009.

United Nations. (1989). *Convention on the Rights of the Child, CRC*. New York: United Nations.

Uribe, V., & Harbeck, K. M. (1992). Addressing the needs of lesbian, gay and bisexual youth: The origins of Project 10 and school-based intervention. In Harbeck, K. M. (Ed.), *Coming out of the classroom closet: Gay and lesbian students, teachers and curricula* (pp. 9–29). New York: Haworth.

Ussher, J. M., & Mooney-Somers, J. (2000). Negotiating Desire and Sexual Subjectivity: Narratives of Young Lesbian Avengers. *Sexualities*, *3*(2), 183–200. doi:10.1177/136346000003002005

Valcke, M., Schellens, T., Van Keer, H., & Gerarts, M. (2007). Primary school children's safe and unsafe use of the Internet at home and at school: An exploratory study. *Computers in Human Behavior*, *23*(6), 2838–2850. doi:10.1016/j.chb.2006.05.008

van Dijk, T. A. (1990). Discourse analysis in the 1990's'. *Text*, *10*(1-2), 133–156. doi:10.1515/text.1.1990.10.1-2.133

Van Dijk, T. A. (1991). *Racism and the press*. London: Routledge.

VanOostrum, N., & Horvath, P. (1997). The effects of hostile attribution on adolescents' aggressive responses to social situations. *Canadian Journal of School Psychology*, *13*, 48–59. doi:10.1177/082957359701300105

Verghis, S. (2003). Girls stalk. *The Sydney Morning Herald,* Wednesday, January 8, p. 19.

Vincke, J., & van-Heering, K. (2004). Summer holiday camps for gay and lesbian young adults: An evaluation of their impact on social support and mental well-being. *Journal of Homosexuality*, *47*(2), 33–46. doi:10.1300/J082v47n02_02

Vitaro, F., Brendgen, M., & Barker, E. D. (2006). Subtypes of aggressive behaviors: A developmental perspective. *International Journal of Behavioral Development*, *30*(1), 12–19. doi:10.1177/0165025406059968

Vitaro, F., & Brendgen, M. (2005). Proactive and reactive aggression: A developmental perspective. In Tremblay, R. E., Hartup, W. W., & Archer, J. (Eds.), *Developmental origins of aggression*. New York: The Guilford Press.

Von Salisch, M., Kristen, A., & Oppl, C. (2007). *Computerspiele mit und ohne Gewalt. Auswahl und Wirkung bei Kindern*. Stuttgart: Kohlhammer.

Vonderau, P., & Snickars, P. (2009). *The YouTube Reader*. National Library of Sweden.

Wallenius, M., Punamäki, R.-L., & Rimpelä, A. (2007). Digital game playing and direct and indirect aggression in early adolescence: The roles of age, social intelligence, and parent-child communication. *Journal of Youth and Adolescence*, *36*, 325–336. doi:10.1007/s10964-006-9151-5

Walrave, M., & Wannes, H. (2009). Skutki cyberbullyingu – oskarżenie czy obrona technologii? *Dziecko krzywdzone. Teoria. Badania. Praktyka. „. Cyberprzemoc*, *1*(26), 27–46.

Walrave, M., Heirman, W. (2009). Skutki cyberbullyingu – oskarżenie czy obrona technologii?, *Dziecko krzywdzone. Teoria, badania, praktyka, 1* (26), 78-89.

Walsh, W. A., & Wolak, J. (2005). Nonforcible Internet-related sex crimes with adolescent victims: Prosecution issues and outcomes. *Child Maltreatment*, *10*, 260–271. doi:10.1177/1077559505276505

Walstrom, M. K. (2000). "You know, Who's the Thinnest?": Combating surveillance and creating safety in coping with eating disorders online. *CyberPsychology & Behaviour*, *3*, 761–783. doi:10.1089/10949310050191755

Walther, J. B. (2002). Research ethics in Internet-enabled research: Human subjects issues and methodological myopia. *Ethics and Information Technology*, *4*, 205–216. doi:10.1023/A:1021368426115

Walther, J. B., & Burgoon, J. K. (1992). Relational communication in Omputermediated interaction. *Human Communication Research*, *19*, 50–88. doi:10.1111/j.1468-2958.1992.tb00295.x

Wang, J., Iannotti, R., & Nansel, T. (2009). School bullying among adolescents in the United States: physical, verbal, relational, and cyber. *The Journal Of Adolescent Health: Official Publication Of The Society For Adolescent Medicine*, *45*(4), 368–375.

Ward, K. (2007). 'I Love You to the Bones': Constructing the Anorexic Body in 'Pro-Ana' Message Boards. *Sociological Research Online*, *12*(2). doi:10.5153/sro.1220

Ward, L. M. (1995). Talking about sex: common themes about sexuality in the prime-time television programs children and adolescents view most. *Journal of Youth and Adolescence*, *24*, 595–615. doi:10.1007/BF01537058

Ward, L. M. (2003). Understanding the role of entertainment media in the sexual socialization of American youth: A review of empirical research. *Developmental Review*, *23*, 347–388. doi:10.1016/S0273-2297(03)00013-3

Ward, T., & Hudson, S. M. (2000). Sexual offenders' implicit planning: a conceptual model. *Sexual Abuse*, *12*, 189–202. doi:10.1177/107906320001200303

Waskul, D., & Douglas, M. (1996). Considering the electronic participant: Some polemical observations on the ethics of online research. *The Information Society*, *12*(2), 129–139. doi:10.1080/713856142

Waters, M. (Director)(2004). *Mean Girls*.. Los Angeles: Paramount Pictures.

Watson, J. (1999). *Postmodern nursing and beyond*. Edinburgh, UK: Churchill Livingstone.

Webster, F. (1995). *Theories of the Information Society*. London: Routledge.

Weinstein, E., & Rosen, E. (1991). The development of adolescent sexual intimacy: Implications for counseling. *Adolescence*, *26*, 331–340.

Weiss, P., Bialik, P., & Kizony, R. (2003). Virtual reality provides leisure time opportunities for young adults with physical and intellectual disabilities. *Cyberpsychology & Behavior: The Impact Of The Internet, Multimedia And Virtual Reality On Behavior* [Retrieved from MEDLINE database.]. *Society*, *6*(3), 335–342.

Wellenius, B., & Townsend, D. (2005). Telecommunications and Economic Development. In Majumdar, S. K., Vogelsang, I., & Cave, M. E. (Eds.), *Handbook of Telecommunications Economics Technology Evolution and the Internet* (*Vol. 2*, pp. 557–621). London: Emerald Group Publishing.

Wellman, B., Haase, A. Q., Witte, J., & Hampton, K. (2001). Does the Internet Increase, Decrease, or Supplement Social Capital?: Social Networks, Participation, and Community Commitment. *The American Behavioral Scientist*, *45*(3), 436–455. doi:10.1177/00027640121957286

Wellman, B. (2001). Physical place and cyber place: The rise of personalized networking. *International Journal of Urban and Regional Research*, *25*(2), 227–252. doi:10.1111/1468-2427.00309

Wellman, B. (1988). Structural analysis: from method and metaphor to theory and substance. I B. Wellman & S. D. Berkowitz (Eds.), *Social Structures: A Network Approach*, Structural analysis in the social sciences. Cambridge, UK: Cambridge Univ. Press.

Wesch, M. (2008). An anthropological introduction to YouTube – Presented to the Library of Congress, USA. Retrieved August 28, 2009 from http://www.youtube.com/watch?v=TPAO-lZ4_hU.

Westlake, E. J. (2008). Friend me if you Facebook: generation Y and performative surveillance. *The Drama Review*, *52*(4), 21–40. doi:10.1162/dram.2008.52.4.21

Whelehan, I. (2000). *Overloaded: popular culture and the future of feminism*. London: Women's Press.

Whitney, I. (1994). Bullying and children with special needs. In Smith, P. K., & Sharp, S. (Eds.), *School bullying: Insights and perspectives* (pp. 213–240). London: Routledge.

Whitty, M. T., & Carr, A. N. (2003). Cyberspace as Potential Space: Considering the Web as a Playground to Cyber-Flirt. *Human Relations*, *56*(7), 861–891. doi:10.1177/00187267030567005

Whorton, J., et al. (1994). A Comparison of Leisure and Recreational Activities for Adults with and without Mental Retardation. In D. Montgomery (Ed.), *Rural Partnerships: Working Together. Proceedings of the Annual National Conference of the American Council on Rural Special Education (ACRES)* (14th, Austin, Texas, March 23-26, 1994); see RC 019 557. Retrieved from ERIC database.

Wigley, K., & Clark, B. (2000) *Kids.net London National Poll*. http://www.nop.co.uk

Wikipedia*: The free encyclopedia*. (2004, July 22). FL: Wikimedia Foundation, Inc. Retrieved August 10, 2004, from http://www.wikipedia.org

Willard, N. (2007). *Cyberbullying and cyberthreats responding to the challenge of online social aggression, threats, and distress*. Champaign, IL: Research Press.

Willard, N. (2006). Flame Retardant. *School Library Journal*, *52*(4), 55–56.

Willard, N. (2007). The authority and responsibility of school officials in responding to cyberbullying. *The Journal of Adolescent Health*, *41*(6), 64–65. doi:10.1016/j.jadohealth.2007.08.013

Williams, K., & Guerra, N. (2007). Prevalence and Predictors of Internet Bullying. *The Journal of Adolescent Health*, *41*(6), S14–S21. doi:10.1016/j.jadohealth.2007.08.018

Williams, C. (1995). *Invisible victims: Crime and abuse against people with learning difficulties*. London: Jessica Kingsley.

Wimmer, R. D., & Dominick, J. R. (2003). *Mass media research: An introduction*. Belmont, CA: Wadsworth.

Winnenden/Germany. German school gun man 'kills 15'. Retrieved 11.3.2009, from http://news.bbc.co.uk/2/hi/europe/7936817.stm

Winship, J. (2000). Women outdoors: advertising, controversy and disputing feminism in the 1990's. *International Journal of Cultural Studies*, *3*(1), 27–55. doi:10.1177/136787790000300103

Winterowd, C., Harrist, S., Thomason, N., Worth, S., & Carlozzi, B. (2005). The Relationship of Spiritual Beliefs and Involvement with the Experience of Anger and Stress in College Students. *Journal of College Student Development*, *46*(5), 515–529. doi:10.1353/csd.2005.0057

Winther Jorgensen, M., & Phillips, L. (2000). *Diskursanalys som teori och metod*. Lund, Sweden: Studentlitteratur.

Wishart, J., Oades, C., & Morris, M. (2007). Using online role play to teach internet safety awareness. *Computers & Education*, *48*(3), 460–473. doi:10.1016/j.compedu.2005.03.003

Witkowski, L. (2007). *Edukacja i humanistyka. Nowe (Kon) teksty dla nowoczesnych nauczycieli*. Warszawa: IBE.

Witting, M. (1992). *The straight mind and other essays*. Boston: Beacon Press.

Wojtasik, Ł. (2009). Przemoc rówieśnicza a media elektroniczne. *Dziecko krzywdzone. Teoria, badania, praktyka, 1* (26), 12-27.

Wolak, J., Finkelhor, D., & Mitchell, K. (2006). *Trends in arrests of "online predators"*. Durham, NC: Crimes Against Children Research Center.

Wolak, J., Kimberly, J. M., & Finkelhor, D. (2007). Does online harassment Constitute bullying? An Exploration of Online Harassment by known peers and online-only contacts. *Journal of Adolescent Health, 41,* S51-S58. Vandebosch, H., & Van Cleemput, K. (2008). Defining cyberbullying: A qualitative research into the perceptions of youngsters. *Cyberpsychology & Behavior*, *11*, 499–503.

Wolak, J., Finkelhor, D., Mitchell, K. J., & Ybarra, M. L. (2008). Online "Predators" and their victims: myths, realities and implications for prevention and treatment. *The American Psychologist*, *63*, 111–128. doi:10.1037/0003-066X.63.2.111

Wolak, J., Mitchell, K., & Finkelhor, D. (2003b). Escaping or connecting? Characteristics of youth who form close online relationships. *Journal of Adolescence*, *26*(1), 105–119. doi:10.1016/S0140-1971(02)00114-8

Wolak, J., Mitchell, K., & Finkelhor, D. (2007). Unwanted and wanted exposure to online pornography in a national sample of youth Internet users. *Pediatrics*, *119*(2), 247–257. doi:10.1542/peds.2006-1891

Wolak, J., Finkelhor, D., & Mitchell, K.J. (2004). Internet-initiated sex crimes against minors: implications for prevention based on findings from a national study. *Journal of Adolescent Health, 35(5),* 424.e11 - 424.e20.

Wolak, J., Mitchell, K., & Finkelhor, D. (2003). *Internet sex crimes against minors: The response of law enforcement*. Alexandria, VA: National Centre for Missing and Exploited Children, Publication No. 10-03-022.

Woolfolk, R. L. (1990). Intentional Explanation and Its Limits. *Psychological Inquiry*, *1*(3), 273–274. doi:10.1207/s15327965pli0103_24

Woolfolk Hoy, A., & Weinstein, C. S. (2006). Student and Teacher Perspectives on Classroom Management. In Evertson, C. M., & Weinstein, C. S. (Eds.), *Handbook of Classroom Management: Research, Practice and Contemporary Issues* (pp. 181–219). Mahwah, NJ: Lawrence Erlbaum Associates.

Worth, N. (2009). Understanding youth transition as 'Becoming': Identity, time and futurity. *Geoforum*, *40*, 1050–1060. doi:10.1016/j.geoforum.2009.07.007

Worthen, M. (2007). Education policy implications from the expert panel on electronic media and youth violence. *The Journal of Adolescent Health*, *41*(6), 61–62. doi:10.1016/j.jadohealth.2007.09.009

Wright, E. R., & Lawson, A. H. (2004). *Computer-mediated Communication and Student Learning in Large Introductory Sociology Courses.* Paper presented at the Annual Meeting of the American Sociological Association, Hilton San Francisco & Renaissance Parc 55 Hotel, San Francisco, CA. Retrieved from: http://www.allacademic.com/meta/p108968_index.html

Yalom, I. D. (1995). *The theory and practice of group psychotherapy*. New York: Basic Books.

Yan, Z. (2005). Age differences in children's understanding the complexity of the Internet. *Applied Developmental Psychology*, *26*, 385–396. doi:10.1016/j.appdev.2005.04.001

Yao, M., & Flanagin, A. (2006). A self-awareness approach to computer-mediated communication. *Computers in Human Behavior*, *22*, 518–544. doi:10.1016/j.chb.2004.10.008

Yates, A. (1978). *Sex without Shame: Encouraging the Child's Healthy Sexual Development*. New York: William Morrow & Company.

Ybarra, M. L., & Mitchell, K. J. (2005). Exposure to Internet Pornography among Children and Adolescents: A National Survey'. *Cyberpsychology & Behavior*, *8*(5), 473–486. doi:10.1089/cpb.2005.8.473

Ybarra, M., Diener-West, M., & Leaf, P. (2007). Examining the overlap in Internet harassment and school bullying: Implications for school intervention. *The Journal of Adolescent Health*, *41*(6), 42–50. doi:10.1016/j.jadohealth.2007.09.004

Ybarra, M., & Mitchell, K. (2004). Youth engaging in online harassment: associations with caregiver-child relationships, Internet use and personal characteristics. *Journal of Adolescence*, *27*(3), 319–336. doi:10.1016/j.adolescence.2004.03.007

Ybarra, M., & Mitchell, K. (2008). How Risky are Social Networking Sites? A comparison of places online where youth sexual solicitation and harassment occurs. *Pediatrics*, *121*(2), 350–357. doi:10.1542/peds.2007-0693

Ybarra, M. L., Espelage, D. L., & Mitchell, K. J. (2007). The co-occurrence of Internet harassment and unwanted sexual solicitation victimization and perpetration: associations with psychosocial indicators. *The Journal of Adolescent Health*, *41*(6), S31–S41. doi:10.1016/j.jadohealth.2007.09.010

Ybarra, M. L., Finkelhor, D., Mitchell, K. J., & Wolak, J. (2009). Associations between blocking, monitoring and filtering software on the home computer and youth-reported unwanted exposure to sexual material online. *Child Abuse & Neglect*, *33*(12), 857–869. doi:10.1016/j.chiabu.2008.09.015

YouTube(n.d.). Retrieved from http://www.youtube.com

Zhao, S., Grasmuck, S., & Martin, J. (2008). Identity construction on Facebook: Digital empowerment in anchored relationships. *Computers in Human Behavior*, *24*(5), 1816–1836. doi:10.1016/j.chb.2008.02.012

Zillmann, D., & Bryant, J. (1989). *Pornography: Research Advances and Policy Considerations*. Hillsdale, NJ: Erlbaum.

About the Contributors

Dunkels, Elza is a senior lecturer at the Department of Applied Educational Science at Umeå University, Sweden. Her PhD from 2007 deals with young people's own perceptions of online dangers. She is currently involved projects concerning online risk, adult's perceptions of online dangers and sexual exploitation of young people online.

Frånberg, Gun-Marie is Professor in Educational Work at the Department of Applied Educational Science, Umeå University, Sweden. Her research interests include social and cultural perspectives on contemporary educational work. There is a particular focus on social values and net cultures at the intersection of age, class, gender and ethnicity.

Hällgren, Camilla work as a senior lecturer at the department of Applied Educational Science at Umeå University in Sweden. She has a PhD in Educational Work, from the same university. Her research interest deals with the complexity of identity, young people, social values and online interactions.

***

Lena Adelsohn Liljeroth, Lena is the Swedish Minister for Culture, with responsibility for culture, media and sports. Minister Adelsohn Liljeroth has a degree in political science and sociology from Stockholm University and a university diploma from Stockholm School of Journalism. As a Minister she has a special interest in media and culture issues that concern young people.

Ayotunde, Titilayo is a Lecturer in the Department of Demography and Social Statistics, Obafemi Awolowo University, where he is presently completing his PhD program. His areas of research include migration and development, reproductive health, youth, and society. He has published in reputable local and international journals.

Brown, Karen, Ph.D. (Cand.) is an instructor in the School of Criminology at Simon Fraser University and the Projector Coordinator for the Centre for Education, Law and Society in the Faculty of Education at Simon Fraser University. Her primary research focuses on violence and threats against lawyers, law and legal citizenship in schools, cyber-bullying and cyber-crimes in schools, and studying the counterpoint to cyber-bullying – cyber-kindness, and reviewing the positive aspects of online exchange and interaction.

Cassidy, Wanda, Ph.D. is Associate Professor of Education and Director of the Centre for Education, Law and Society at Simon Fraser University in British Columbia, Canada, a centre which seeks to improve the legal literacy of youth through a program of research, teaching, program development and community-based initiatives. Her work focuses on the relationship between law/citizenship education and social justice, and the values that underpin a civil society, including the ethics of care, inclusion and social responsibility. Recently she completed a four-year SSHRC study examining the ethics of care in schools and is currently working on a project that examines students' and educators' understanding of human rights, identity, citizenship and environmental sustainability. She also researches cyber-bullying and cyber-kindness.

Daneback, Kristian is an Associate Professor of Social Work at the University of Gothenburg in Sweden. He is an expert on the topic of internet sexuality but his research interests also include how the internet is used by parents and how the internet can be used for qualitative and quantitative data collection. At the University of Gothenburg, Kristian teaches undergraduate and graduate courses on Philosophy of Science and Research Methods. He lectures regularly about his research at various departments in Gothenburg as well as at Swedish and international universities. Kristian's research has been published in several well known journals such as Archives of Sexual Behavior and CyberPsychology & Behavior. He is a member of the International Academy of Sex Research.

de Kaminski, Marcin is a junior researcher at Lund University, Sweden. His main focuses are within the field of sociology of law combined with Internet related issues, and net based interaction in young net cultures. He also lectures in digital media and net cultures at Uppsala University, Sweden.

Hendrick, S. Faye is a PhD candidate in English Linguistics at the Department of Language studies, and also affiliated with the Department of Culture and Media and HUMlab. Hendrick's thesis is entitled Negotiating the liminal space: How individual and social identities are performed in weblog communities of practice. In addition to Hendrick's PhD studies, she is involved in two, three-year research projects: YouTube as a performative arena and The role of the Internet as a surrogate social network in situations of domestic violence in Swedish context. Hendrick's research interests include how language is used to build and maintain online networks and communities, academic activism, and youth net culture. Hendrick's research methodology applied includes discourse analysis, sociolinguistic analysis, ethnography, and network analysis.

Jackson, Margaret, Ph.D. is Professor in the School of Criminology at Simon Fraser University, Co-Director of the SFU Institute for Studies in Criminal Justice Policy, and Director of FREDA, which undertakes research on violence against women and children. She was principal investigator for a Ministry of Justice study on child abuse. Other research areas of interest include bullying/cyber-bullying, problem-solving courts and sociocultural factors impacting marginalized girls and youth. These involvements resulted in her serving as a co-investigator/principal investigator for SSHRC, Status of Women, Heritage Canada, Metropolis-RIIM and NCPC grants, as well as currently for two CIHR grants.

Larsen, Malene is an Assistant Professor at Aalborg University (AAU), Department of Communication, where she is currently finishing her PhD on young people's use of social network sites. Her research focuses on the online experiences of young Danes between the age of 12 and 18 and aims at understanding how online social networking is integrated as part of everyday life and how everyday life is represented, mediated and acted out in the digital space. Her PhD project combines a wide range of data material such as online and multi-sided ethnography, open-ended questionnaires and scene and actions surveys. She is a widely used public speaker and commentator in Denmark on digital youth culture and online social networking and is working together with The Danish Media Council for Children and Young People in order to provide information about young people's use of the internet and social network sites.

Lindgren, Simon is a Professor of Sociology at Umeå University in Sweden. He works in the field of cultural sociology with issues relating to media, youth and popular culture. His current research deals with people's use of digital media, network cultures, and new emerging forms of online community. He has a broad interest in social and cultural theory, and uses a wide range of research methods; discourse analysis, ethnography, network analysis etc. At present, Lindgren heads a research project on online piracy, and another on youth culture and participation. Among other things, he also takes part in a project about YouTube as a performative arena, at HUMlab, Umeå University.

Löfberg, Cecilia has a Ph.D. in education and works as a researcher at the Department of Education, Stockholm University. Her academic work focuses on young people's informal learning, communication and socialization in on-line environments. Within this field she has a particular interest in the area of gender and sexuality. In 2008 she did her doctorate on young people's communication about gender, emotions and sexuality in on-line environments. Since 2009 Cecilia is studying young people's use of internet fora as arenas for support around issues that can feel otherwise challenging. The aim of the research project, with an ambition to support and guide young people in difficult matters related to gender and sexuality, is to recognize how web sites are shaped and used by the target groups.

Lögdlund, Ulrik lectures in behavioural sciences at Linköping University, Sweden. His main teaching interests are sociology and social-psychology. His research includes social and educational science, scrutinising the socio-materials and the actor-networks of local learning centres in Sweden. He is the author of Collaboration Revisited; Perspectives on Collaboration in the Context of Adult Education among other articles and reports in the field of adult education in Sweden.

Marciano, Avi is a graduate student at the Department of Communication, University of Haifa, Israel. He has been researching in the areas of gay and lesbian studies, queer theory, and cyberspace. His current project deals with online practices of transgender people with emphasis on the interaction between online and offline worlds.

McAloney, Kareena (BSc, PG Dip, PhD, C.Psychol) is a Research Fellow in longitudinal research methods on the Improving Children's Lives initiative, based in the School of Psychology at Queen's University Belfast. Kareena has been involved in longitudinal projects exploring adolescent risk-taking, substance use and deviancy, and worked for several years as a youth worker. Her research interests have focused on intergroup relations and well-being, the influence of group support on attitudes and behaviours, and adolescent risk-taking behaviours and deviancy.

Melvin, Agunbiade Ojo is a Lecturer in the Department of Sociology and Anthropology, Obafemi Awolowo University, where he is presently completing his PhD program with a specialization in medical sociology. Melvin teaches courses on sociological theory, social research methods, and sociology of health, healing, and Illness. His areas of research include sexual health, ageing, youth and society, gender and development. He has published in reputable local and international journals.

Ojo, Olugbenga David, a Certified Counselling Psychologist, took his Ph.D from the Obafemi Awolowo University, Ile-Ife, Nigeria. He is a member of several professional bodies in the fields of educational psychology, counselling and vocational psychology, including the Counselling Association of Nigeria, Social Science Research Council, and the Institute of Personnel Management. His work experience spanning over two decades covers both the Private and Public sectors in Nigeria. He is a senior academic at the National Open University of Nigeria and his research outputs are published in local and international journals. He has also made presentations at conferences, trainings and seminars both within and outside Nigeria on myriad of topics within the ancillary fields of vocational and counselling psychology, distance learning and quality assurance. He served as the Director of Examinations and Assessment of the National Open University of Nigeria between 2007 and 2009.

Olin-Scheller, Christina is a researcher and teacher at Karlstad University. Her main interest is young people's reading and writing in a new media landscape in general and has a special interest in how this landscape challenges traditional ways of regarding literacy. Her thesis, Mellan Dante och Big Brother. En studie om gymnasielevers textvärldar (2006), has been followed by other books and articles which turn to researchers, as well as teacher educators and teachers. Presently she is working with studies on fans, fan culture and fan fiction with focus on the relation between informal and formal learning settings. Also, she is often engaged as a lecturer for various groups of audiences.

Oyewole, Jaiyeola holds a B.A and M.A. degrees in Communication and Language Arts and M. Ed degree in Social Welfare, all from the University of Ibadan, Nigeria. Her research interest is on interpersonal communication, indigenous communication as well as health communication with special focus on women and children. She has published in reputable journals and contributed chapters to published books. She currently lectures at the Mass Communication Department of Bowen University, Iwo, Osun State of Nigeria.

Palmgren, Ann-Charlotte is a doctoral student in women's studies, Åbo Akademi University, Finland. She has a Master of Arts degree with a major in Ethnology. Her research interests include autoethnography, youth cultures, social media, the body and identity from a cultural perspective. Among recent publications are 'Bodily Lives in Virtual Worlds. Theoretical Approaches to the Self and Body in Relation to Blogs in a Swedish Context' in Ethnologia Scandinavica. vol. 39 (2009), and 'Posing my Identity. Today's Outfit in Swedish Blogs' in Observatorio (OBS*) vol 4, nr. 2 (2010).

Plichta, Piotr, Ph. D. Senior lecturer in The Pedagogy Academy in Lodz. M.A. in Special Education and Ph. D. in Education (thesis: "Burnout, meaning of life and value crisis in special educators"). Co-author of "Questionnaire of Occupational Burdens in Teaching (QOBT). Manual". Author and co-author of approximately 40 scientific publications. Research leader and team member in national grants. Team member in International grants (Acerish 2, Adults Mentoring, Daphne 3). Areas of scientific interests:

special education, professional stress and burnout in helping professions, ethics in helping, education and rehabilitation of intellectually disabled people.

Pyżalski, Jacek Ph.D. in Education, the vice chancellor of the Higher School of Pedagogy in Łódź and a researcher at the Nofer Institute of Occupational Medicine in Łódź – the National Centre Workplace Health Promotion, which serves as the national Contact Point of the European Centre for Workplace Health Promotion. The author of numerous publications in pedagogy and health promotion. Research interests associated with such issues as discipline in the classroom and coping by teachers/educators with aggression at school; occupational stress and professional burnout; difficult behaviour of students; health promotion; and cyberbullying (currently, represents Poland in the International Committee managing the COST (European Cooperation in the Field of Scientific and Technical Research) Action IS0801: Cyberbullying: coping with negative and enhancing positive uses of new technologies, in relationships in educational settings.

Rambaree, Komalsingh is currently a senior lecturer teaching International Social Work and Social Policy at the University of Gävle. For the last 7 years, he has been working as lecturer in Mauritius and in England. In addition, he has 15 years of working experience in the field of youth and sexuality as youth worker in Mauritius. His research interest includes Internet and Adolescent Sexuality, and he has published several articles within this area. Currently, he is also working on the setting up of an international group – 'Network on Internet Technologies and Adolescent Sexual Health (NITASH)'.

Ryberg, Thomas is an Associate Professor at Aalborg University (AAU), Department of Communication. Part of the research centre E-learning lab - Center for User Driven Innovation, Learning and Design. He holds a PhD from Aalborg University in Youth, Technology Enhanced learning and Problem Solving. He has participated in European and international research projects and networks. Primary research interests are within the field of Computer Supported Collaborative Learning (CSCL), ICT and learning for development and research on how new media and technologies transform our ways of thinking about and designing for learning. Has authored and co-authored several articles on ICT and Learning and Learning and New media.

Samuelsson, Marcus is a Senior Lecturer in Sloyd directed towards work in wood and metal and a post doctor in Educational Sciences at Linköping University, Sweden. His teaching and research are mainly directed towards classroom management and leadership, institutionalisation, net culture and youth culture, interpersonal relations between children and adults, late modernity as well as sloyd and arts and craft. His is recently engaged in a research project that includes several different perspectives on classroom management, such as teacher's proactive or reactive strategies. It also includes several different studies about institutionalisation and children or youth's use of mobile phones and Internet as tools for reporting about what happens in classroom and at school.

Schiller, Eva-Maria, psychologist, works at the Department of Economic Psychology, Educational Psychology and Evaluation at the Faculty of Psychology of the University of Vienna since 2007. Currently she holds a position as a PhD student in a research project which centers on the improvement of social and intercultural competencies in schools. Her main research interest centers on aggressive behavior in adolescents.

Schultes, Marie-Thérèse, psychologist, works at the Department of Economic Psychology, Educational Psychology and Evaluation at the Faculty of Psychology of the University of Vienna. Currently she holds a position as a scientific assistant in a research project which centers on the development of an online survey for the self-evaluation of violence in schools. Previously, she worked as an assistant for the national intervention evaluation study for the prevention of violence in schools (ViSC Program) and for the DAY Study (Development and Adaptation in Acculturating Youth) at the Faculty of Psychology of the University of Vienna. The focus of her research lies on the evaluation of educational programs, the impact of violent video- and computer games on youth aggression and psychological aspects of media education.

Selg, Håkan is senior research scientist at the Swedish IT-User Centre, Uppsala University. He has a long experience from research in the field of applied sciences, particularly in the analysis of the diffusion and use of new technologies. He has written extensively on the effects of digital technologies, beginning in the late 1970s at the Computers' and Electronics' Commission, a research unit at the Swedish Ministry of Industry. From the 1990s, he has been working as an independent analyst. After the millennium, his focus has been on Internet technologies, patterns of usage and impact on established organisations and structures.

Shields Dobson currently teaches subjects in youth culture and media culture. She also writes in both performative and academic genres. She has recently submitted her PhD thesis in Sociology at Monash University, where her research focuses on representations of femininity on social networking websites, and the implications of this for contemporary feminist politics and performance studies. Amy completed her Honours Degree in Performing Arts at Monash University in 2004, for which she researched and wrote her thesis on Internet cam girl communities.

Spiel, Christiane is Professor of Educational Psychology and Evaluation and department head at the Faculty of Psychology, University of Vienna. Actually, she is president of the Austrian Society for Psychology. She is and has been chair and member of various international advisory and editorial boards as e.g., chair of the ERIH expert panel of the European Science Foundation for psychology, president of the European Society for Developmental Psychology, and founding dean of the Faculty of Psychology at the University of Vienna. In several projects she is working together with the Austrian Federal Ministry for Education and the Austrian Federal Ministry for Science and Research. She has got several awards for research, university teaching, and university management and has published more than 190 original papers. Her research topics are on the border between developmental psychology, educational psychology and evaluation. Specific research topics are: Bullying und victimization, integration in multicultural school classes, lifelong learning, change measurement, evaluation research and quality management in the educational system.

Strohmeier, Dagmar received her Master´s degree in Psychology in 2001 at the University of Graz, Austria and her PhD in Psychology in 2006 at the University of Vienna, Austria. Dr. Strohmeier currently holds a position as an assistant professor at the Faculty of Psychology, University of Vienna. Dr. Strohmeier´s main research interests are peer relations in schools.

Wikström, Patrik (PhD) is research fellow and research manager at the Media Management and Transformation Centre at Jönköping International Business School. His primary research area is the innovative and adaptive behaviour of media organizations. Within this area he has done research on business models in the music industry; magazine publishers' use of social media; collaborative production of online fiction; and on the competitive behaviour of small and mid-sized TV producers.

Wilson, Joanne (BSc) is a Research Assistant in the Institute of Child Care Research at Queen's University Belfast. Joanne has worked on a number of research projects examining adolescent risk involvement, including the influence of peer networks on alcohol consumption and evaluations of sexual health service provision to teens.

Index

D

E

F

J

K

L

M

N

O

T

U